MAGILL'S ENCYCLOPEDIA OF SCIENCE

PLANT LIFE

MAGILL'S ENCYCLOPEDIA OF SCIENCE

PLANT LIFE

Volume 3

**Microbial Nutrition and Metabolism–
Sustainable Agriculture**

Editor
Bryan D. Ness, Ph.D.
Pacific Union College, Department of Biology

Project Editor
Christina J. Moose

SALEM PRESS, INC.
Pasadena, California
Hackensack, New Jersey

Some of the updated and revised essays in this work originally appeared in *Magill's Survey of Science: Life Science* (1991), *Magill's Survey of Science: Life Science, Supplement* (1998), *Natural Resources* (1998), *Encyclopedia of Genetics* (1999), *Encyclopedia of Environmental Issues* (2000), *World Geography* (2001), and *Earth Science* (2001).

∞ The paper used in these volumes conforms to the American National Standard for Permanence of Paper for Printed Library Materials, Z39.48-1992 (R1997).

Library of Congress Cataloging-in-Publication Data

Magill's encyclopedia of science : plant life / edited by Bryan D. Ness.

p. cm.
Includes bibliographical references (p.).
ISBN 1-58765-084-3 (set : alk. paper) — ISBN 1-58765-085-1 (vol. 1 : alk. paper) —
ISBN 1-58765-086-X (vol. 2 : alk. paper) — ISBN 1-58765-087-8 (vol. 3 : alk. paper) —
ISBN 1-58765-088-6 (vol. 4 : alk. paper)
1. Botany—Encyclopedias. I. Ness, Bryan D.
QK7 .M34 2002
580'.3—dc21

2002013319

First Printing

PRINTED IN THE UNITED STATES OF AMERICA

TABLE OF CONTENTS

LIST OF ILLUSTRATIONS, CHARTS, AND TABLES

ALPHABETICAL LIST OF CONTENTS

Volume 1

Volume 2

Volume 3

Volume 4

MAGILL'S ENCYCLOPEDIA OF SCIENCE

PLANT LIFE

MICROBIAL NUTRITION AND METABOLISM

Categories: Algae; bacteria; fungi; microorganisms; nutrients and nutrition; *Protista*

The diverse metabolic activities of microorganisms make them a critical component of all the earth's ecosystems and a source of many useful products for human industry.

Microorganisms—bacteria, fungi, algae, and protists—are found in every environment on the earth that supports life. Microorganisms have been found in hot springs where temperatures exceed 80 degrees Celsius as well as in rocks of Antarctic deserts. To ensure survival in a variety of habitats, microorganisms have developed a fascinating variety of strategies for survival. The study of microbial ecology involves consideration of the mechanisms employed by microorganisms to obtain nutrients and energy from their environment.

Nutritional Modes

To maintain life processes and grow, all cellular organisms require both a source of carbon (the principal element in all organic molecules) and a source of energy to perform the work necessary to transform carbon into all the molecular components of cytoplasm. Among plants and animals, two main nutritional modes have evolved to meet these requirements. All plants are *photoautotrophs*, fixing carbon from inorganic carbon and obtaining energy from light. All animals are *chemoheterotrophs*, meeting their carbon needs by taking preformed organic molecules from the environment and extracting energy from chemical transformation of the same organic molecules.

Both of these nutritional modes, photoautotrophy and chemoheterotrophy, are found among microorganisms; for example, all algae are photoautotrophs, while all fungi are chemoheterotrophs. In addition, certain specialized bacteria exhibit a mode of nutrition, chemoautotrophy, found in no higher organisms. Like photoautotrophs, chemoautotrophs are able to use carbon dioxide for all of their carbon requirements; however, they do not use light as an energy source. Instead, chemoautotrophic bacteria capture energy from inorganic chemical reactions, such as the *oxidation* of ammonia. Chemoautotrophic bacteria are highly specialized and can be found in unusual environments. The most spectacular display of chemoautotrophic energy metabolism is exhibited at the hydrothermal vents found in certain locations on the ocean floor. There, where sunlight cannot penetrate, chemoautotrophic bacteria serve as the producers for a rich and diverse ecosystem.

An appreciation of the metabolic diversity displayed by microorganisms enhances understanding of the ways in which matter and energy are transformed in the biosphere. Consideration of microbial contributions to the flow of carbon, nitrogen, and other elements is critical to defining the balance of ecosystems and the effects of changes in environmental chemistry and species composition. Microorganisms are, by definition, unseen, and many people become aware of them only in their negative manifestations as agents of disease and spoilage. In fact, however, the diverse metabolic activities of microorganisms make them a critical component of all the earth's ecosystems and a source of many useful products for human industry.

Cellulose Digestion

Even among chemoheterotrophs, microorganisms possess metabolic capabilities unknown in higher organisms. These include the ability of some bacteria and fungi to digest *cellulose*, a linear polymer of glucose that is the principal molecular constituent of paper. Sixty percent of the dry mass of green plants is in the form of cellulose, although no animal that eats the plants is directly able to obtain carbon or energy from cellulose. Microorganisms that digest cellulose do so by secreting *exoenzymes*, proteins that cause cellulose to be broken into simpler molecular units that are absorbed by the microorganism. Cellulose-digesting microorganisms are found in most terrestrial ecosystems and in the digestive tracts of animals, such as cattle and termites, that depend on cellulose-rich plant material as a nutrient source. By breaking down cellulose and other complex organic polymers, microorganisms make a significant contribution to the cycling of carbon in ecosystems.

Nitrogen Fixation

Digestion of complex organic polymers is only one of the ways in which microorganisms contribute to the cycling of elements in the environments they inhabit. Microorganisms also perform chemical transformations involving nitrogen, which is found in all cellular proteins and nucleic acids. Plants incorporate nitrogen from the soil in the form of nitrate or ammonium ions, and animals obtain nitrogen from the same organic compounds they use as carbon and energy sources.

When dead plant and animal tissue is decomposed by chemoheterotrophic microorganisms, the nitrogen is released as ammonia. A group of chemoautotrophic bacteria, the nitrifying bacteria, obtain their metabolic energy from the conversion of ammonia to nitrate; in this way, the *nitrifiers* convert the nitrogen released during decomposition to a form readily used by plants, thus contributing to soil fertility. A second group of bacteria converts nitrate to atmospheric nitrogen gas, which cannot be used by plants; these bacteria are called *denitrifiers* because (in contrast to the nitrifiers) their metabolic activities cause a net loss of nitrogen from the soil.

Nitrogen lost from the environment by denitrification is replaced by ammonia released during decomposition and by the metabolic activity of *nitrogen-fixing bacteria*, so called because they "fix" nitrogen gas from the atmosphere in the form of ammonia. Nitrogen fixation requires a great quantity of energy, and nitrogen-fixing bacteria are often

AP/Wide World Photos

Composting and other decomposition processes, including sewage treatment, are consequences of the metabolic activities of mixed populations of microorganisms.

found in symbiotic association with plants, especially legumes. The bacteria provide nitrogen in a usable form to the plant, while the plant provides carbon and energy in the form of organic compounds to the chemoheterotrophic nitrogen-fixing bacteria. The presence of nitrogen-fixing bacteria is often indicated by the formation of characteristic nodules on the roots of plants involved in the associations. Free-living nitrogen fixers are also known, and these may play a significant role in the nitrogen balance of aquatic ecosystems.

Respiration and Fermentation

Chemoheterotrophic microorganisms are found both in *aerobic* environments, where oxygen is available, and in *anaerobic* environments, where oxygen is lacking. The availability of molecular oxygen may determine the type of energy metabolism employed by a microorganism. Where oxygen is available, many microorganisms obtain energy by *respiration*. In respiration, electrons removed from organic nutrient sources are transferred through a complex sequence of reactions to molecular oxygen, forming water and carbon dioxide. In the process, energy is made available to the organisms. In the absence of oxygen, some microorganisms are able to carry out a form of anaerobic respiration using nitrate or sulfate in place of oxygen. Denitrification is an example of anaerobic respiration.

Other anaerobic microorganisms employ *fermentation*. In fermentation, electrons removed from organic nutrient sources are transferred to organic molecules, forming fermentation products, such as alcohols and organic acids, which may be used as nutrient sources by other chemoheterotrophs. A number of bacteria, the *facultative anaerobes*, are capable of performing either aerobic respiration or fermentation, depending on the availability of oxygen. These bacteria are able to achieve optimum growth in environments, such as soils, where the availability of oxygen may vary over time.

Effects and Uses

The contributions of microorganisms to the chemical transformations which characterize an ecosystem are many. Along with higher plants, photoautotrophic and chemoautotrophic microorganisms capture inorganic carbon dioxide and, using energy from sunlight or chemical reactions, synthesize organic molecules, which are used by animals and by chemoheterotrophic microorganisms as sources of carbon and energy. Through the processes of respiration and fermentation, chemoheterotrophs return inorganic carbon dioxide to the environment.

Much of this recycling of carbon from organic molecules to carbon dioxide depends on the activities of microbial decomposers, which are able to break down organic polymers, such as cellulose. Nitrogen, released from organic molecules by chemoheterotrophs in the form of ammonia, may be made available to chemoheterotrophs in the form of nitrate by nitrifying bacteria. Nitrate which is lost from an ecosystem through the activities of denitrifiers may be returned by nitrogen-fixing microorganisms.

Although the nature of the microbial world has been known only since about the turn of the twentieth century, the metabolic activities of microorganisms have been exploited throughout human history. The manufacture of alcoholic beverages, cheeses, vinegars, and linen depends on the metabolic activities of microorganisms. Farmers employed practices designed to optimize the availability of nitrogen to plants for centuries before the role of microorganisms in nitrogen cycling was understood. Composting and other decomposition processes, including sewage treatment, are consequences of the metabolic activities of mixed populations of microorganisms.

The number of organic compounds used as nutrients by one or another chemoheterotrophic microorganism is extraordinary. Some soil bacteria have been shown to use more than one hundred different organic molecules as their sources of carbon and energy. By using selective enrichment techniques, it has been possible to isolate microorganisms capable of degrading pesticides, complex petroleum by-products, and other toxic chemicals previously assumed to be resistant to natural decomposition processes. Through application of appropriate engineering technologies, these microorganisms may play a part in solutions to toxic waste disposal issues.

Kenneth A. Pidcock

See also: Algae; Anaerobes and heterotrophs; *Archaea*; Bacteria; Biofertilizers; Biopesticides; Biotechnology; Carbon cycle; Chemotaxis; Fungi; Legumes; Nitrogen cycle; Nitrogen fixation; Nutrients; Phosphorus cycle; Photosynthesis; *Protista*; Respiration.

Sources for Further Study

Caldwell, Daniel R. *Microbial Physiology and Metabolism*. Dubuque, Iowa: Wm. C. Brown, 1995. For undergraduate and beginning graduate students. Coverage includes subcellular structures of microbes, physiological implications of nutrition catabolic metabolism, small molecules, genetics, and regulation.

Creager, Joan G., Jacquelyn G. Black, and Vee E. Davison. *Microbiology: Principles and Applications*. Englewood Cliffs, N.J.: Prentice Hall, 1990. An accessible and beautifully illustrated textbook of microbiology. Chapters 26, "Environmental Microbiology," and 27, "Applied Microbiology," summarize the importance of microorganisms in ecosystems and their uses in industry. Includes references.

Gottschalk, Gerhard. *Bacterial Metabolism*. 2d ed. New York: Springer-Verlag, 1986. Written both as an advanced college textbook and as a microbiologist's desk reference, this monograph provides clear, concise outlines of all the major metabolic strategies employed by bacteria. The first place to look for the specifics of microbial metabolism. Includes references.

Kornberg, Arthur. *For the Love of Enzymes: The Odyssey of a Biochemist*. Cambridge, Mass.: Harvard University Press, 1991. Kornberg, a leading researcher in the field of biochemistry, provides a lucid introduction to the field in the form of an autobiographical account of his own work. Discusses both the workings of cells and the workings of science.

Margulis, Lynn, and Dorian Sagan. *Microcosmos: Four Billion Years of Evolution from Our Microbial Ancestors*. New York: Summit Books, 1997. An entertaining account of the evolution of life on earth that focuses on the roles played by microorganisms in developing all the basic metabolic pathways and on the continuing significance of microorganisms to the biosphere. One of the few books describing the history of life which gives microorganisms the attention they deserve.

White, David. *The Physiology and Biochemistry of Prokaryotes*. 2d ed. New York: Oxford University Press, 2000. Covers microbial and bacterial metabolism and physiology of *Archaea* and prokaryotes. Includes bibliographical references, index.

MICROBODIES

Categories: Anatomy; cellular biology; transport mechanisms

Microbodies, found in cells, are spherical, membrane-bound organelles that play a part in photorespiration and the conversion of fats into sucrose.

Peroxisomes and *glyoxysomes* are the two major types of microbodies in plant cells. Their *vesicles* ("packages") vary in size from 0.3 to 1.5 micrometers in diameter and are self-replicating. New microbodies are formed by incorporation of required proteins and lipids from the cytoplasm and subsequent splitting when they reach a certain size. Although structurally similar, their roles, and thus their contents, are different.

Functions of Peroxisomes

Peroxisomes are present in leaves, and their oxidative enzymes are involved in the breakdown of hydrogen peroxide and, more important, in photorespiration. Photorespiration occurs when carbon dioxide levels in the leaves drop and oxygen levels increase, a typical phenomenon on hot, sunny days when a plant is experiencing some level of water stress. Under these conditions, the enzyme (Rubisco) that normally catalyzes the attachment of carbon dioxide to ribulose bisphosphate (RuBP) begins to have a higher affinity for oxygen than for carbon dioxide. When oxygen is used instead of carbon dioxide, RuBP is split into two molecules, phosphoglycolate and 3-phosphoglycerate (PGA). PGA can be used in another part of the Calvin cycle, but phosphoglycolate must be extensively processed to be useful.

Phosphoglycolate is hydrolyzed and converted to glycolate in the chloroplast. Glycolate is then transported out of the chloroplast and into nearby peroxisomes. Peroxisomal oxidase converts glycolate to glyoxylate, and hydrogen peroxide is produced as a by-product. Because hydrogen peroxide is toxic, it is quickly converted by a catalase to water and oxygen. Glyoxylate goes through several more steps which involve reactions in the mitochondria and then back again in the peroxisomes. Eventually, glycerate is formed in the peroxisomes. Glycerate is then transported out of the peroxisomes and into a chloroplast, where it is converted to PGA, which can reenter the Calvin cycle.

Functions of Glyoxysomes

Glyoxysomes are found in the cells of fat-rich seeds. Fats are synthesized and stored as *oil bodies*, sometimes called *spherosomes*. Spherosomes are surrounded by a single layer of lipids instead of a lipid bilayer and are therefore not organelles in the strict sense. Glyoxysomes are responsible for converting fats and fatty acids into sucrose. The fat used by glyoxysomes comes from spherosomes.

Glyoxysomes are considered to be a type of peroxisome. In some plants, small glyoxysomes are found in the cotyledons of developing seeds. During germination and seedling development, they mature into fully functional glyoxysomes. They function until the fats are completely digested into sucrose. The energy from sucrose is required to drive early seedling development before photosynthesis begins. Large fat molecules are difficult to transport into the plant embryo; they must be converted to the more mobile sucrose molecules.

Breakdown of fats is a collaborative effort between enzyme-containing glyoxysomes and fat-containing spherosomes. Direct contact between spherosomes and glyoxysomes must occur. Fats from spherosomes leak out close to the membrane of glyoxysomes. Most of the activity of lipase enzymes does not take place in spherosomes but rather in or near the membrane of glyoxysomes. Lipases in glyoxysomes hydrolyze the ester bonds of fats and release the three fatty acids and one glycerol from each fat molecule. Glycerol is converted, at the cost of adenosine triphosphate (ATP), to glycerol phosphate, which is then oxidized by nicotinamide adenine dinucleotide (NAD^+) to dihydroxyacetone phosphate, most of which is converted to glucose.

Domingo M. Jariel

See also: ATP and other energetic molecules; Calvin cycle; Cytosol; Lipids; Membrane structure; Oil bodies; Peroxisomes; Vacuoles; Vesicle-mediated transport.

Sources for Further Study

Gunning, Brian E. S., and Martin W. Steer. *Plant Cell Biology: Structure and Function*. Sudbury, Mass.: Jones and Bartlett, 2000. Text integrates microscopy with plant cell and molecular biology. Includes more than four hundred micrographs and four pages of full-color plates.

Hay, R. K., and Alastair H. Fitter. *Environmental Physiology of Plants*. 3d ed. New York: Academic Press, 2001. Provides a synthesis of modern ecological and physiological thinking. Examines molecular approaches which can be harnessed to solve problems in physiology. Illustrations include color plates.

Hopkins, William G. G. *Introduction to Plant Physiology*. 2d ed. New York: John Wiley & Sons, 1998. Uses interactions between the plant and the environment as a foundation for developing plant physiology principles. Discusses the role of plants on specific ecosystems and global ecology and provides information on the cell, chemical background, plant growth regulators and biochemistry. Each chapter is illustrated with relevant graphs, figures, or photographs.

MICROSCOPY

Categories: Cellular biology; methods and techniques

In biology and botany, light microscopy is used for the observation of large numbers of cells or for the location of single cells. Scanning electron microscopy is used in the examination of the surface profiles of cells. Transmission electron microscopy is used to probe the interiors of cells to elucidate their structures. The scanning tunneling microscope can be used to examine still smaller structures. Each of the techniques described has been used extensively in the biological sciences.

Light Microscopes

The first microscopes used a beam of light to form an image and were probably invented in the Netherlands, where the devices were used in the manufacture of spectacles and cloth. Dutch microscopist Antoni van Leeuwenhoek (1632-1723) improved upon the cloth merchants' microscopes and used his version to study small objects from nature, such as single-celled organisms and red blood cells. English scientist Robert Hooke (1635-1703), using his own simple microscope, discovered "cells" in a slice of cork.

The human eye by itself is able to resolve images of about 100 micrometers (0.1 millimeter). This means that two objects (such as lines or dots) less than 100 micrometers apart will appear to blur into one object. The highest resolution available in light microscopes will improve upon the human eye five hundred times, allowing it to distinguish objects that are 0.2 micrometer (200 nanometers) apart. Resolution by a light microscope is limited because the shortest wavelength of visible light itself is about 0.4 micrometer (400 nanometers). This limits the effective magnification of a light microscope to about 2,000 times. At this magnification most bacteria are readily visible, as are a variety of organelles within plant cells, such as vacuoles, nuclei (including chromosomes during cell division), chloroplasts, and mitochondria. Smaller structures, such as ribosomes and microtubules (as well as other components of the cytoskeleton), are not visible using a light microscope.

Electron Microscopes

The *electron microscope* was invented in 1931 by Ernst Ruska, who won the Nobel Prize in Physics for this effort in 1986. Since the time of its invention, several different types of electron microscope have been developed. Electron microscopy uses a beam of electrons instead of a beam of light to form an image. Light is a form of *electromagnetic radiation*, a process that transfers energy as a wave without transferring matter. An electromagnetic wave can be visualized in terms of a wave traveling on the surface of a pool of water: The undulating pattern of peaks and troughs constitutes the wave. In the case of electromagnetic radiation, the undulations that constitute the wave occur in electric and magnetic fields that are perpendicular to each other.

Waves and Wavelengths

It is usual to consider light as a wave, or a ray, and to consider electrons as particles. According to quantum physics, however, waves and particles are two aspects of the same phenomenon. Light can be considered as a stream of particles, and a stream of electrons can be considered as a wave. This behavior is usually referred to as wave-particle duality.

One property of a wave is its *wavelength*. The wavelength of a wave is the distance between adjacent peaks (or adjacent troughs) in the waveform. Wavelength is important because when a wave interacts with matter, any structures that are smaller than one-quarter of the wavelength are "invisible" to the wave. In approximate terms, the smallest object that a wave can be used to image is equal in size to the wavelength of the wave used to form the image. The wavelength of visible light is between approximately 400 nanometers and 700 nanometers (a nanometer is one one-billionth of a meter).

A stream of electrons has a smaller wavelength than a beam of light, and electrons can therefore be

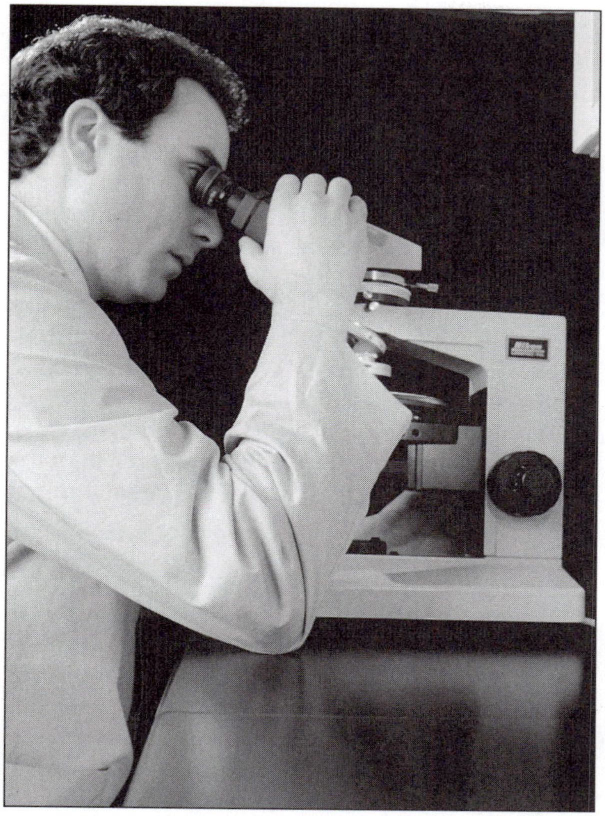

The human eye by itself is able to resolve images of about 100 micrometers. This means that two objects (such as dots) less than 100 micrometers apart will appear to blur into one object. The highest resolution available in a light microscope (shown here) will improve upon the human eye five hundred times, allowing it to distinguish objects that are 0.2 micrometer (200 nanometers) apart. The resolving power of a scanning tunneling microscope is sufficient to determine the positions of individual atoms in the surface layer of a material.

used to form images of small objects—such as cells. If distinct images of separate objects are formed, the images are said to be resolved. Because the wavelength of an electron is much smaller than the wavelength of light, the electron microscope can resolve images of objects that are a million times smaller than the objects seen in traditional optical microscopes. The resolving power of a scanning tunneling microscope (STM) is sufficient to render it capable of determining the positions of individual atoms in the surface layer of a material.

There are several different types of electron microscope. The basic types include the transmission electron microscope, the scanning electron microscope, and the scanning tunneling microscope. The specimen in an electron microscope is usually observed in a vacuum in order to prevent scattering of the electrons by air molecules; the need for a vacuum presents the greatest difficulty in the application of electron microscopy to biological systems.

Transmission Electron Microscopes

The simplest electron microscope is the *transmission electron microscope* (TEM). Electrons are produced by an electron gun and are accelerated by a potential difference (voltage). The electrons from the electron gun pass through a condenser lens and are then used to illuminate the specimen. The electrons which pass through the specimen are then allowed to pass through an electron lens objective. The objective magnifies the image, and then a second electron lens, which plays the role of the eyepiece in the standard microscope, is used to focus the image for observation. The "lenses" used in electron microscopes are not lenses in the usual sense; instead, they are electric and magnetic fields, and they are accordingly referred to as electrostatic or magnetic lenses.

The image can then be formed on a photographic plate or observed on a fluorescent screen, or the electrons can be collected by a charge-sensitive device to produce an image on a cathode-ray tube. Higher magnification can be achieved by using more lenses.

The sample thickness will affect the resolving ability of the TEM. Usually, at least in the case of biological samples, the sample should be no more than ten times thicker than the structures that are to be analyzed. The resolving power of the TEM is such that it can observe structures that are slightly larger than atoms, but since the development of other systems, it has become less used, even though its resolution often exceeds that of the scanning electron microscope. Typical resolutions are in the subnanometer range, with magnifications of up to 500,000 times.

The interpretation of the electron micrographs produced by a TEM is sometimes difficult. The major source of difficulty is that the image is produced by transmitted radiation. The eye is accustomed to interpreting images that are produced by reflection. In the absence of a sample, a TEM beam would saturate a film plate used to record an image. The sample prevents some of the electrons from reaching the film plate; the image produced by a TEM is somewhat similar to a negative produced in a normal camera. Much of the difficulty can be removed by photographing the micrograph and converting it to a positive image—there are, however, some residual interpretation difficulties caused by shadows.

High-Voltage and Scanning Electron Microscopes

The *high-voltage electron microscope* (HVEM) is a variant of the TEM. The conventional TEM works best on particles less than 0.5 nanometer thick; electrons of higher speed can be produced by increasing the voltage used to accelerate them, and thicker samples can then be analyzed. The wavelength of the electrons decreases as their speed (hence, their kinetic energy) increases. These short-wavelength electrons are less likely to collide with atoms as they pass through the specimen, and they are therefore able to render a sharper image of a thicker sample.

The *scanning electron micrscope* (SEM) works on a different principle. Electrons are again produced and accelerated by an electron gun, but in an SEM the beam is focused by electron lenses and used to scan a sample. The scanning will result in two different electron beams being emitted by the sample, a primary beam of backscattered electrons produced by reflection and a secondary beam of electrons emitted by the atoms of the sample. By scanning the entire sample and collecting the primary and secondary electrons, the operator can produce an image of a sample on a cathode-ray tube.

Scanning Tunneling Electron Microscopes

The *scanning tunneling electron microscope* (STM) works on a completely different principle. It was developed by Gerd Binnig and Heinrich Rohrer at an International Business Machines (IBM) research laboratory in the 1980's. Binnig and Rohrer shared the 1986 Nobel Prize in Physics with Ernst Ruska for their work. The STM uses *quantum tunneling*. Quantum tunneling is the penetration of a barrier by a particle that, when analyzed by classical physics, does not have enough energy to pass through the barrier. Quantum mechanics predicts that there is a finite probability of a particle passing through a barrier, even if it lacks the energy to pass over it. The number of particles passing through the barrier will vary with the barrier thickness and the particle energy.

This principle is used in the STM by allowing electrons to tunnel across a vacuum, from a stylus to a sample. The quantum tunneling of electrons sets up a "tunneling current" that increases as the vacuum gap between the stylus and the sample decreases. The variation in tunneling current can be used to map the surface of the sample as the stylus moves across its surface. The STM is capable of detecting structures 0.1 nanometer in size in the direction parallel to the motion of the scanning stylus. This means that atoms can be easily detected. The performance in the vertical direction is even more impressive: The STM can detect irregularities on the order of 0.01 nanometer. Thus, the STM, with its higher resolving power than the SEM or TEM, has little difficulty in the imaging of atoms.

Preparing Samples

The principal accommodations that must be made in the examination of biological samples using electron microscopy occur in the preparation of the sample. A commonly used method is the construction of replicas, made by the vacuum deposition of thin layers of carbon, metals, or alloys on the surface of the sample. These films provide a replica of the surface, which can be scanned. Another useful technique, which is the standard technique of sample preparation used in optical microscopy, is the sectioning of samples and their impregnation with stains. The stains that are useful in electron microscopy are usually chemical compounds of heavy metals. These are effective stains because they strongly scatter electrons.

Many methods of sample preparation have been developed to enable electron microscopy to be more widely used on biological samples. *Freeze-fracture* and *freeze-etch* are methods of sample preparation that have been widely used. Freeze-fracture involves the freezing and splitting of a water-containing sample. Freeze-etch is a second step, in which ice is allowed to sublime (vaporize without forming a liquid) before the sample is analyzed in an electron microscope. Both techniques are used to examine the internal structure of materials without subjecting them to chemical changes. Freeze-fracture and freeze-etch allow the interior layers of water-containing samples to be investigated without the straining and chemical preparations that are needed to render internal structures visible in a standard light microscope. In particular, these techniques allow the observation of cell walls and cell membranes in a state which is as close as possible to the living state.

Uses

While light microscopy remains important in the identification, location, and observation of cells both singly and in groups, the electron microscope has revolutionized scientists' understanding of the microscopic world in the biological, medical, and physical sciences. Although it is not possible to observe living materials using electron microscopy, freeze-fracture and freeze-etch have allowed the observation of biological materials in an almost natural state. *Correlative microscopy*, an increasingly popular method, involves the examination of a sample using both light microscopy and electron microscopy; this method allows the acquisition of a variety of views of the same structure, and it removes the ambiguities that may result from views produced by a single microscope.

The transmission electron microscope (TEM) was the first electron microscope to be developed, and it is the most common type. The TEM has more in common with light microscopes than it does with any other electron microscope. It also shares the chief disadvantage of the optical microscope—that is, it gives little impression of the vertical scale of the specimen under observation. This means that the structures present in the specimen are imaged, but the subtleties of the surface of the specimen are lost. Furthermore, the TEM imposes severe constraints on the type of specimens which can be analyzed. The sample must be thin enough to permit the beam of electrons to pass through it, and it must be resilient enough to resist being damaged by the imaging electrons.

Most biological materials are too thick to be observed under the TEM, so it is necessary to prepare *ultra-thin sections* of samples prior to their analysis. Both plant and animal samples have been examined using the TEM, and their analysis has led to the discovery of a variety of internal structures. The internal structure of the mitochondria, which had previously been discovered with light microscopy, has been probed. The TEM has also been used to examine the interiors of the nuclei of cells. The examination of the cell nuclei has enabled the investigation of chromosome organization and gene structure. This examination of the microstructure of cells has contributed to the development of molecular biology and genetic engineering.

The TEM has also been used to examine bacteria and viruses. The TEM detected the presence of nucleic acids in bacteria and produced the first images of viruses; most viruses are too small to be resolved in light microscopes. In plants, the TEM has been used in the study of chloroplasts and the walls of cells. The observation of chloroplasts led to the discovery of the internal membranes called *thylakoids*, which absorb light for the process of photosynthesis. The discovery of the thylakoids has enhanced the understanding of photosynthesis.

The scanning electron microscope (SEM) has been widely used in the biological sciences. Physical laws impose no constraints on the size of the sample to be examined—they are, instead, imposed by the available sample chambers. The SEM is usually used in the magnification range of 10 to 100,000 times. When compared with the light microscope, the main advantage of the SEM is that it is able to produce three-dimensional images. These images are possible because the entire sample can be observed in focus at the same time, and the sample can be observed from a variety of angles.

The SEM has allowed images to be formed of algae, bacteria, spores, molds, and fungi. These images have enabled the structure and function of these samples to be determined. The xylem and phloem cells that transport water through the stems of plants have been examined in the SEM, thus allowing the water transportation process to be better understood. The SEM also has applications in exploring the pathology of cells. Many structures formed by living cells are better understood because sample preparation techniques such as freeze-fracture and freeze-etch have allowed the examination of lifelike samples under the SEM.

Stephen R. Addison, updated by Bryan Ness

See also: Cell theory; Fluorescent staining of cytoskeletal elements.

Sources for Further Study

Bozzola, John J., and Lonnie D. Russell. *Electron Microscopy: Principles and Techniques for Biologists*. 2d ed. Boston: Jones and Bartlett, 1999. Covers the theory of scanning and transmission electron microscopes, specimen preparation for both, digital imaging and image analysis, laboratory safety, and interpretation of images and provides an atlas of ultrastructure.

Harris, Robin, ed. *Electron Microscopy in Biology: A Practical Approach.* New York: IRL Press, 1991. Part of the Practical Approach series; includes bibliographical references and index.

Hey, Tony, and Patrick Walters. *The Quantum Universe.* New York: Cambridge University Press, 1988. This work, written for the layperson, is devoted to quantum physics. All physical processes necessary for a good understanding of electron microscopy are covered, and there is a section devoted to microscopy. The main ideas are developed pictorially rather than through the use of mathematical equations.

Koehler, James K. *Advanced Techniques in Biological Electron Microscopy.* 3 vols. New York: Springer-Verlag, 1973-1986. As the title suggests, this is an advanced work, intended for the reader who wishes to learn to use the electron microscope. An invaluable reference for those who wish to know how samples are prepared for use in an electron microscope.

Shipman, James T., Jerry D. Wilson, and Arron W. Todd. *An Introduction to Physical Science.* 6th ed. Boston: Houghton Mifflin, 2000. For readers who lack an understanding of wave-particle duality and quantum physics. It is an elementary college-level text requiring no mathematical prerequisites. Recommended for those who wish to develop a thorough understanding of the advantages of electron microscopy.

Watt, Ian M. *The Principles and Practice of Electron Microscopy.* 2d ed. New York: Cambridge University Press, 1997. An easily understood survey of the entire field of electron microscopy. This book is a good resource for those who need more information about the field. Many references for further study are provided.

MITOCHONDRIA

Categories: Anatomy; cellular biology

An organelle of eukaryotic cells, a mitochondrion is bounded by a double membrane. It is the major source of adenosine triphosphate (ATP), which is derived from the breakdown of organic molecules and contains the enzymes used in the Krebs cycle and the electron transport system.

With the exception of a few metabolically inert types, such as the red blood cells of many higher animals, eukaryotic cells of animals, plants, fungi, and protozoa contain mitochondria. Most cells contain several hundred. The efficiency of mitochondria in *adenosine triphosphate* (ATP) production provides the energy source that powers all the varied activities of eukaryotic cells. For these reasons, mitochondria have been aptly termed the "powerhouses" of the cell.

Structure

In most cells, mitochondria appear as spherical, elongated bodies about 0.5 micrometer in diameter and 1 to 2 micrometers in length. Their size is roughly the same as that of bacterial cells. In some cells mitochondria may measure several micrometers in diameter, with lengths up to 10 micrometers. The name "mitochondrion" comes from Greek words meaning "thread" (*mitos*) and "granule" (*chondros*). Mitochondria appear in either elongated, threadlike forms or more spherical, granular shapes. Under the light microscope, mitochondria can be seen to grow, branch, divide, and fuse together.

Mitochondria are surrounded by two separate membranes. The outer membrane is a continuous, unbroken membrane that completely covers the surface of the organelle. It is smooth in appearance. The inner membrane is also continuous and unbroken. It is convoluted, with infoldings called *cristae* that greatly increase its surface area.

Structure of a Mitochondrion

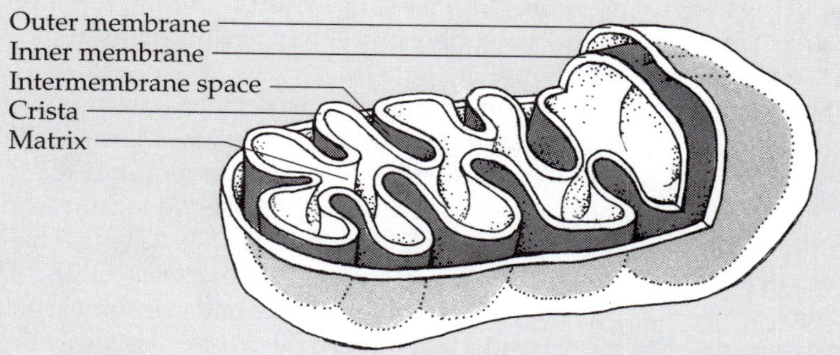

Mitochondria are special cellular organelles where respiration—the release of energy from organic molecules as ATP—occurs. Mitochondria are surrounded by two separate membranes. The outer membrane is a continuous, unbroken membrane that completely covers the surface of the organelle. The inner membrane is also continuous and unbroken but is convoluted, with infoldings called cristae that greatly increase the inner membrane's surface area.

The outer and inner membranes separate the mitochondrial interior into two distinct compartments. The *intermembrane space* is located between the outer and inner membranes. The innermost compartment, enclosed by both membranes, is the *mitochondrial matrix*. Each membrane and compartment carries out specific functions important to the production of ATP.

The outer membrane is more permeable than the inner membrane and allows molecules up to the size of small proteins to pass freely from the surrounding cytoplasm into the intermembrane space. Larger proteins are prevented from escaping into the surrounding cytoplasm, and certain cytoplasmic proteins, such as potentially destructive enzymes, are prevented from entering.

Electron Transport

The innermost compartment, the matrix, contains a battery of enzymes that catalyze the *oxidation* of fuel substances of many types, including simple sugars, fats, amino acids, and other organic acids. The primary goal of oxidation is the removal of high-energy electrons and the use of them to perform chemical work.

High-energy electrons are used in the membrane that immediately surrounds the matrix, the inner membrane. This membrane contains a group of proteins that work as electron-driven protons (hydrogen ions or H^+) pumps. The proteins, called *electron transport carriers*, accept electrons at a higher energy level and release them at a lower level. The energy released by the electrons causes the shape of the carrier proteins to change, which allows them to transport protons across the inner membrane. The various electron transport carriers accept and release electrons at different energy levels, allowing them to act in a series called the *electron transport chain*. The electrons released by one carrier have sufficient energy to power the pumping activity of the next one in line. When operating at peak efficiency, a mitochondrion of average size conducts about 100,000 electrons through the electron transport chain per second. After passing through several carriers, most of the energy of the electrons has been tapped off.

As a consequence of proton pumping, protons become depleted in the matrix and become more concentrated in the intermembrane space. This creates an *electrochemical gradient* between the intermembrane space and the matrix. It is called an electrochemical gradient because there is a concentration gradient across the inner membrane and because protons have an electrical charge, so there is also an *electrical potential* (a charge difference) across the membrane. Electrochemical gradients represent a form of stored energy capable of doing work and in this case is used to drive the synthesis of ATP.

At the very end of the electron transport chain, the electrons have so little energy that the last carrier molecule donates these low-energy electrons to oxygen that is already in the matrix. The oxygen eukaryotes need to survive thus has its primary biological role as the final acceptor of spent electrons released by the electron transport chain. When an oxygen molecule receives four electrons, it then picks up two protons from the matrix, and water (H_2O) is produced. The protons used in this process cause an additional reduction in proton concentration in the matrix, increasing the electrochemical gradient across the inner membrane.

ATP Synthesis

A protein complex embedded in the inner membrane uses the proton gradient to make ATP. This complex, called *ATP synthase*, is a molecular "machine" that acts as a hydrogen-driven ATP synthesizer. It takes ADP (adenosine diphosphate) from the matrix and combines it with inorganic phosphate (P_i) to make ATP.

The proteins of the inner mitochondrial membrane thus work in two coordinated groups. One, the electron transport chain, uses the energy of electrons removed in oxidative reactions to pump protons hydrogens from the matrix to the intermembrane space. The second group, the ATP synthases, uses the proton gradient created by the electron transport chain as an energy source to make ATP from ADP and P_i. ADP comes from the matrix itself and from other regions of the cell in which ATP is used to facilitate cellular activities such as growth and movement. Much of the ATP produced in this way is then transported out of the mitochondria for use in other parts of the cell.

mRNA and DNA

When examined with an electron microscope, a number of structures are visible in mitochondrial matrix, including granules of various sizes, fibrils, and crystals. Among the granules are *ribosomes*. These structures, like their counterparts in the cytoplasm, are capable of protein assembly, using directions encoded in *messenger ribonucleic acid* (mRNA) molecules as a guide. The mitochondrial ribosomes are more closely related in structure and function to prokaryotic ribosomes than to ribosomes in the cytoplasm.

Also in the mitochondrial matrix are molecules of *deoxyribonucleic acid* (DNA). *Mitochondrial DNA* (mtDNA) stores the information required for synthesis of some of the proteins needed for mitochondrial functions. Unlike nuclear DNA, which is linear, mtDNA is circular, like bacterial DNA. The presence of bacterialike DNA and ribosomes inside mitochondria has given rise to the *endosymbiotic theory*, which proposes that mitochondria may have evolved from bacteria that invaded the cytoplasm of other prokaryotic cells and established a symbiotic relationship. Over long periods of time, these bacteria are believed to have lost their ability to live independently and gradually became transformed into mitochondria. The evolutionary advantage to the host provided by the bacterial invaders may have been greater efficiency in ATP production.

Stephen L. Wolfe, updated by Bryan Ness

See also: ATP and other energetic molecules; Chloroplast DNA; Chloroplasts and other plastids; Cytoplasm; DNA in plants; DNA replication; Extranuclear inheritance; Krebs cycle; Membrane structure; Mitochondrial DNA; Oxidative phosphorylation; RNA.

Sources for Further Study

Alberts, Bruce, Dennis Bray, Julian Lewis, Martin Raff, Keith Roberts, and James D. Watson. *Molecular Biology of the Cell*. 4th ed. New York: Garland, 2002. The chapter "Energy Conversion: Mitochondria and Chloroplasts" describes the structure of mitochondria, oxidative reactions taking place inside them, and the mechanisms synthesizing ATP. Discusses the DNA of mitochondria and its hereditary functions. The text is clearly written at the college level and is illustrated by numerous diagrams and photographs. Includes an extensive bibliography at the end of the chapter.

Darnell, James, Harvey Lodish, and David Baltimore. *Molecular Cell Biology*. 3d ed. New York: Scientific American, 1995. An excellent textbook written at the college level. The chapter "Energy Conversion: The Formation of ATP in Chloroplasts, Mitochondria, and Bacteria" describes all aspects of ATP production in these organelles and in bacteria. Many diagrams and photographs are included.

Tzagoloff, Alexander. *Mitochondria*. New York: Plenum Press, 1982. This book presents a survey of experimental information in all phases of mitochondrial structure and function and the history of mitochondrial research. Although written at a technical level for a scientific audience, much of the information in this book is accessible to a lay reader. The book is copiously illustrated with photographs of mitochondria, diagrams of mitochondrial reactions, and many informative tables.

MITOCHONDRIAL DNA

Categories: Cellular biology; genetics

Plant cells have three sets of DNA to code for proteins: one set in the chromosomes of the nucleus, another in the chloroplasts, and a third genome in mitochondria. The mitochondrial genomes of higher plants are larger than those of animals and form a complex series of linear and circular molecules of different sizes.

Mitochondria play an essential role in the generation of energy in eukaryotic cells. Mitochondria are the organelles that are the main "chemical factories" of the cell where cellular aerobic respiration—using the Krebs (citric acid) cycle and respiratory electron transport to produce NADH (nicotinamide adenine dinucleotide) and ATP (adenosine triphosphate)—occurs. In the light microscope, mitochondria look like short rods or thin filaments about 0.5 to 2 microns long. A mitochondrion is made up of a smooth outer membrane and an inner membrane that is folded into tubular shapes called cristae. Many aerobic respiration reactions are catalyzed by enzymes that are bound to mitochondrial membranes. Other reactions occur in the space between the inner and outer mitochondrial membranes. Cells may contain several hundred mitochondria. Cells that are dividing and cells that are metabolically active need larger amounts of ATP and usually have large numbers of mitochondria.

Size and Structure

All eukaryotic cells except some primitive protozoans contain mitochondria. All mitochondria contain their own DNA (genomes). There are typically between twenty and one hundred copies of the mitochondrial genome per mitochondrion. The mitochondria of multicellular animals contain genomes of 14 to 20 kilobases (kb), present as single circles. The mitochondrial DNA of some organisms, such as some protozoa, algae, and fungi, is organized in linear molecules with ends of chromosomes (telomeres) much like nuclear chromosomes.

In contrast, the mitochondrial DNA of higher plants is larger and more complex—from 200 to 2,500 kb—and is present in many different molecules. The size and organization of the mitochondrial genome vary widely from one plant species to another. Electron micrographs of mitochondrial DNA show linear and circular DNAs of a variety of sizes and complex, branched molecules that are larger than the size of the genome.

Cloning the mitochondrial DNA and comparing the sequences of the clones show that the entire complexity of a plant mitochondrial genome can be represented as a "master circle." Also, it has been learned that sequences are repeated on the master chromosome. The repeated sequences differ for different plant species. A series of recombination events between these identical repeated sequences results in a series of rearrangements of mitochondrial DNA and forms the complex, multiple molecules of varying sizes that are the physical structure of the plant mitochondrial genome.

Adding to the complexity of mitochondrial DNAs in higher plants is the fact that some plants, such as corn, contain extrachromosomal mitochondrial nucleic acids. Plasmid-like DNAs (circular double-stranded molecules) and double-stranded and single-stranded RNAs have been found in some corn strains.

Genes Encoded by Mitochondrial DNA

In addition to containing their own genomes, mitochondria contain enzymes for DNA replication and transcription, and ribosomes and transfer RNAs for protein synthesis. (Transfer RNA, or tRNA, carries the building blocks of proteins, called amino acids, to the ribosome, where they are assembled according to the instructions found in messenger RNA.) The ribosomes of mitochondria are different from those of chloroplasts and the cytoplasm, using a slightly different genetic code (a sequence of three bases that codes for a particular amino acid). Mitochondrial genomes code for all of the ribosomal RNAs found in mitochondria and for most of the tRNAs. Mitochondria make only a small number of proteins that are needed for electron transport and ATP production. The other proteins needed in mitochondria are coded by nuclear DNA, translated in the cytoplasm of the cell, then transported into the mitochondria. Plant mitochondria do not encode a full set of tRNAs, and some are imported from the cytoplasm.

Even though the mitochondrial genome of higher plants is much larger than that of animals, the plant mitochondrial genome codes for only a few more genes. The mitochondrial genome of *Arabidopsis* has been sequenced and contains thirty-two protein-coding genes, twenty-two tRNA genes, and three ribosomal RNA genes.

Exchange of DNA

Mitochondrial DNA from plants also differs from that of animals in that mitochondrial DNAs contain segments of DNA that originally were in nuclear and chloroplast DNAs. There appear to have been exchanges of DNAs between all three of the higher plant genomes. There is evidence that mitochondrial genes have been transferred to the nucleus and some mitochondrial tRNAs appear to be of chloroplast origin. Changes in nuclear genes have been shown to lead to changes in the copy number of the different mitochondrial DNA configurations.

RNA Editing

Mitochondria and chloroplasts contain the biochemical machinery to alter the sequence of the final messenger RNA (mRNA) product in a process called *RNA editing*. The most common editing is changing a cytosine to a uracil (two of the bases found on the "rungs" of DNA molecules and which are responsible for determining the nucleotide sequences that form the genetic code).

Inheritance of Mitochondrial DNA

Given the complex branched network of plant mitochondrial DNA, it is difficult to see how the inheritance of a complete genome is ensured. It is still not clearly understood how this complex network of DNAs is passed to daughter cells in a way that assures that all of the genetic information is maintained.

Susan J. Karcher

See also: Chloroplast DNA; Chloroplasts and other plastids; Chromosomes; Cytoplasm; DNA in plants; DNA replication; Extranuclear Inheritance; Gene regulation; Genetic code; Genetics: post-Mendelian; Mitochondria; Nucleus; Oxidative phosphorylation; RNA.

Sources for Further Study

Kung, Shain-Dow, and Shang-Fa Yang, eds. *Discoveries in Plant Biology*. Vol. 2. River Edge, N.J.: World Scientific, 1998. Contains a chapter on the discovery of plant mitochondrial genomes. For advanced students. Illustrations, references.

Lea, Peter J., and Richard C. Leegood, eds. *Plant Biochemistry and Molecular Biology*. West Sussex, England: John Wiley, 1993. A good overview that contains a chapter on genome organization, protein synthesis, and processing in plants. Illustrations, references, index.

Mauseth, James D. *Botany: An Introduction to Plant Biology*. 2d ed. Boston: Jones and Bartlett, 1998. This introductory textbook includes a well-explained summary of mitochondrial DNA. Illustrations, problems sets, glossary, index.

Moore, Randy, W. Dennis Clark, Kingsley R. Stern, and Darrell Vodopich. *Botany*. Dubuque, Iowa: Wm. C. Brown, 1995. Another introductory textbook with a basic overview. Illustrations, questions, suggested readings, glossary, index.

Reece, Jane B., and Neil A. Campbell. *Biology*. 6th ed. Menlo Park, Calif.: Benjamin Cummings, 2002. Introductory textbook. Illustrations, problems sets, glossary, index.

MITOSIS AND MEIOSIS

Categories: Cellular biology; reproduction

Mitosis is the process of cell division in multicellular eukaryotic organisms. Meiosis is the process of cell division that produces haploid gametes in sexually reproducing eukaryotic organisms.

Organisms must be able to grow and reproduce. Prokaryotes, such as bacteria, duplicate deoxyribonucleic acid (DNA) and divide by splitting in two, a process called *binary fission*. Cells of eukaryotes, including those of animals, plants, fungi, and protists, divide by one of two methods: mitosis or meiosis. Mitosis produces two cells, called *daughter cells*, with the same number of chromosomes as the parent cell, and is used to produce new somatic (body) cells in multicellular eukaryotes or new individuals in single-celled eukaryotes. In sexually reproducing organisms, cells that produce gametes (eggs or sperm) divide by meiosis, producing four cells, each with half the number of chromosomes possessed by the parent cell.

Chromosome Replication

All eukaryotic organisms are composed of cells containing *chromosomes* in the nucleus. Chromosomes are made of DNA and proteins. Most cells have two complete sets of chromosomes, which occur in pairs. The two chromosomes that make up a pair are homologous, and contain all the same *loci* (genes controlling the production of a specific type of product). These chromosome pairs are usually referred to as *homologous pairs*. An individual chromosome from a homologous pair is sometimes called a *homolog*. For example, typical lily cells contain twelve pairs of homologous chromosomes, for a total of twenty-four chromosomes. Cells that have two homologous chromosomes of each type are called *diploid*. Some cells, such as eggs and sperm, contain half the normal number of chromosomes (only one of each homolog) and are called *haploid*. Lily egg and sperm cells each contain twelve chromosomes.

DNA must replicate before mitosis or meiosis can occur. If daughter cells are to receive a full set of genetic information, a duplicate copy of DNA must be available. Before DNA replication occurs, each chromosome consists of a single long strand of DNA called a *chromatid*. After DNA replication, each chromosome consists of two chromatids, called sister chromatids. The original chromatid acts as a template for making the second chromatid; the two are therefore identical. Sister chromatids are attached at a special region of the chromosome called the *centromere*. When mitosis or meiosis starts, each chromosome in the cell consists of two sister chromatids.

Mitosis and meiosis produce daughter cells with different characteristics. When a diploid cell undergoes mitosis, two identical diploid daughter cells are produced. When a diploid cell undergoes meiosis, four unique haploid daughter cells are produced. It is important for gametes to be haploid, so that when an egg and sperm fuse, the diploid condition of the mature organism is restored.

Cellular Life Cycles

Mitosis and meiosis occur in the nuclear region of the cell, where all the cell's chromosomes are found. *Nuclear control mechanisms* begin cell division at the appropriate time. Some cells rarely divide by mitosis in adult organisms, while other cells divide constantly, replacing old cells with new. Meiosis occurs in the nuclei of cells that produce gametes. These specialized cells occur in reproductive organs, such as flower parts in higher plants.

Cells, like organisms, are governed by life cycles. The life cycle of a cell is called the *cell cycle*. Cells spend most of their time in *interphase*. Interphase is divided into three stages: *first gap* (G_1), *synthesis* (S), and *second gap* (G_2). During G_1, the cell performs its normal functions and often grows in size. During the S stage, DNA replicates in preparation for cell division. During the G_2 stage, the cell makes materials needed to produce the mitotic apparatus and for division of the cytoplasmic components of the cell.

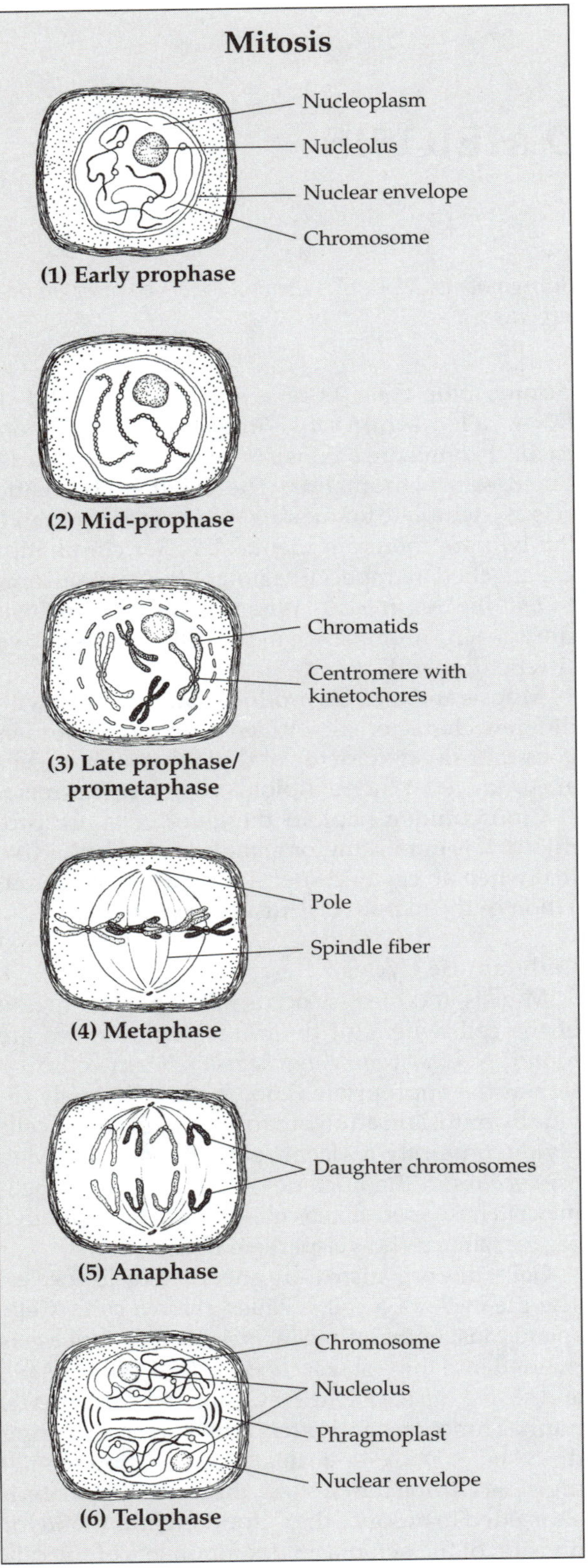

Kimberly L. Dawson Kurnizki

At the end of interphase, the cell is ready to divide. Although each chromosome now consists of two sister chromatids, this is not apparent when viewed through a microscope. This is because all the chromosomes are in a highly relaxed state and simply appear as a diffuse material called *chromatin*.

Mitosis

Mitosis consists of five stages: *prophase*, *prometaphase*, *metaphase*, *anaphase*, and *telophase*. Although certain events identify each stage, mitosis is a continuous process, and each stage gradually passes into the next. Identification of the precise state is therefore difficult at times.

During prophase, the chromatin becomes more tightly coiled and condenses into chromosomes that are clearly visible under a microscope, the *nucleolus* disappears, and the *spindle apparatus* begins to form in the cytoplasm. In prometaphase, the nuclear envelope breaks down, and the spindle apparatus is now able to invade the nuclear region. Some of the spindle fibers attach themselves to a region near the centromere of each chromosome called the *kinetochore*. The spindle apparatus is the most obvious structure of the mitotic apparatus. The nuclear region of the cell has opposite poles, like the North and South Poles of the earth. Spindle fibers reach from pole to pole, penetrating the entire nuclear region.

During metaphase, the cell's chromosomes align in a region called the *metaphase plate*, with the sister chromatids oriented toward opposite poles. The metaphase plate traverses the cell, much like the equator passes through the center of the earth. Sister chromatids separate during anaphase. The sister chromatids of each chromosome split apart, and the spindle fibers pull each sister chromatid (now a separate chromosome) from each pair toward opposite poles, much as a rope-tow pulls a skier up a mountain. Telophase begins as sister chromatids reach opposite poles. Once the chromatids have reached opposite poles, the spindle apparatus falls apart, and the nuclear membrane re-forms. Mitosis is complete.

Meiosis

Meiosis is a more complex process than mitosis and is divided into two major stages: *meiosis I* and *meiosis II*. As in mitosis, interphase precedes meiosis. Meiosis I consists of prophase I, metaphase I, anaphase I, and telophase I. Meiosis II consists of prophase II, metaphase II, anaphase II, and telophase II. In some cells, an interphase II occurs between meiosis I and meiosis II, but no DNA replication occurs.

During prophase I, the chromosomes condense, the nuclear envelop falls apart, and the spindle apparatus begins to form. Homologous chromosomes come together to form *tetrads* (a tetrad consists of four chromatids, two sister chromatids for each chromosome). The arms of the sister chromatids of one homolog touch the arms of sister chromatids of the other homolog, the contact points being called *chiasmata*. Each chiasma represents a place where the arms have the same loci, so-called homologous regions. During this intimate contact, the chromosomes undergo *crossover*, in which the chromosomes break at the chiasmata and swap homologous pieces. This process results in *recombination* (the shuffling of linked *alleles*, the different forms of genes, into new combinations), which results in increased variability in the offspring and the appearance of character combinations not present in either parent.

Tetrads align on the metaphase plate during metaphase I, and one spindle fiber attaches to the kinetochore of each chromosome. In anaphase I, instead of the sister chromatids separating, they remain attached at their centromeres, and the homologous chromosomes separate, each homolog from a tetrad moving toward opposite poles. Telophase I begins as the homologs reach opposite poles, and similar to telophase of mitosis, the spindle apparatus falls apart, and a nuclear envelope re-forms around each of the two haploid nuclei. Because the number of chromosomes in each of the telophase I nucleus is half the number in the parent nucleus, meiosis I is sometimes called the reduction division.

Meiosis II is essentially the same as mitosis, dividing the two haploid nuclei formed in meiosis I. Prophase II, metaphase II, anaphase II, and telo-

Meiosis: Selected Phases

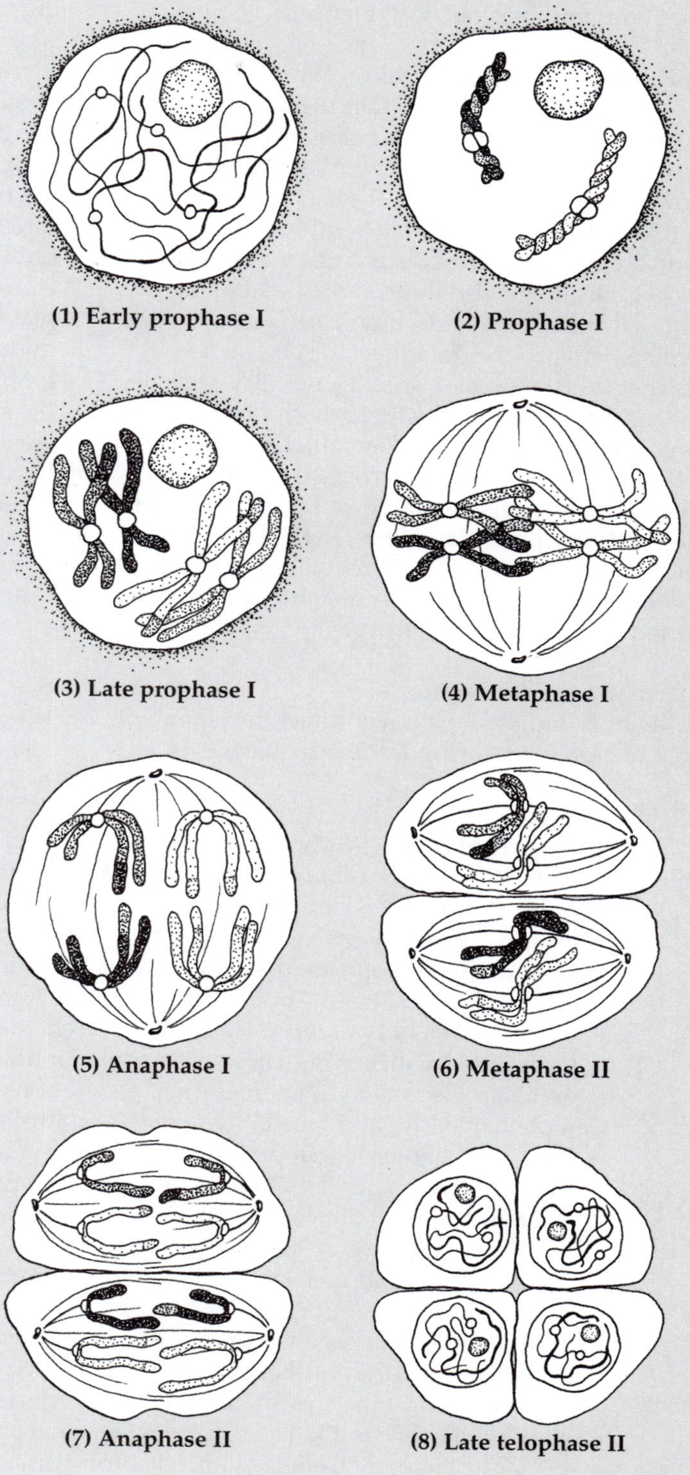

Kimberly L. Dawson Kurnitzki

phase II are essentially identical to the stages of mitosis. Meiosis II begins with two haploid cells and ends with four haploid daughter cells.

Nuclear Division and Cytokinesis

Mitosis and meiosis result in the division of the nucleus. Nuclear division is nearly always coordinated with division of the cytoplasm. Cleaving of the cytoplasm to form new cells is called *cytokinesis*. Cytokinesis begins toward the middle or end of nuclear division and involves not just the division of the cytoplasm but also the organelles. In plants, after nuclear division ends, a new cell wall must be formed between the daughter nuclei. The new cell wall begins when vesicles filled with cell wall material congregate where the metaphase plate was located, producing a structure called the *cell plate*. When the cell plate is fully formed, cytokinesis is complete. Following cytokinesis, the cell returns to interphase. Mitotic daughter cells enlarge, reproduce organelles, and resume regular activities. Following meiosis, gametes may be modified or transported in the reproductive system.

Alternation of Generations

Meiotic daughter cells continue development only if they fuse during fertilization. Mitosis and meiosis alternate during the life cycles of sexually reproducing organisms. The life-cycle stage following mitosis is diploid, and the stage following meiosis is haploid. This process is called *alternation of generations*. In plants, the diploid state is referred to as the *sporophyte generation*, and the haploid stage as the *gametophyte generation*. In nonvascular plants, the gametophyte generation dominates the life cycle. In other words, the plants normally seen on the forest floor are made of haploid cells. The sporophytes, which have diploid cells, are small and attached to the body of the gametophyte. In vascular plants, sporophytes are the large, multicellular individuals (such as trees and ferns), whereas gametophytes are very small and either are embedded in the sporophyte or are free-living, as are ferns. The genetic variation introduced by sexual reproduction has a significant impact on the ability of species to survive and adapt to the environment. Alternation of generations allows sexual reproduction to occur, without changing the chromosome number characterizing the species.

Joyce A. Corban and Randy Moore

See also: Cell cycle; Chromosomes; DNA replication; Gene regulation; Genetic equilibrium: linkage; Genetics: Mendelian; Genetics: mutations; Genetics: post-Mendelian; RNA.

Sources for Further Study

Alberts, Bruce, Dennis Bray, Julian Lewis, Martin Raff, Keith Roberts, and James D. Watson. *Molecular Biology of the Cell*. 4th ed. New York: Garland, 2002. The chapter "How Cells Are Studied" gives extensive information regarding study methods in cell biology. Light and electron microscopy are discussed as well as staining techniques and tissue culture.

Audesirk, Teresa, Gerald Audesirk, and Bruce E. Myers. *Biology: Life on Earth*. 6th ed. Upper Saddle River, N.J.: Prentice Hall, 2001. The chapter "Cellular Reproduction and the Life Cycles of Organisms" is a brief overview of mitosis, meiosis, and the cell cycle. Includes excellent discussion of alternation of generations.

Campbell, Neil A., and Jane B. Reece. *Biology*. 6th ed. San Francisco: Benjamin Cummings, 2002. The chapter "Reproduction of Cells" provides extensive information regarding mitosis and the cell cycle. The phases of mitosis, the mitotic spindle, cytokinesis, control mechanisms, and abnormal cell division are discussed in detail. The chapter "Meiosis and Sexual Life Cycles" addresses the stages of meiosis, sexual life cycles, and a comparison of mitosis and meiosis. This text is intended for use in introductory biology and is very readable and informative.

John, Bernard. *Meiosis*. New York: Cambridge University Press, 1990. Review and discussion of meiosis, the antithesis of fertilization. Discusses the scheduling, mechanisms, biochemistry, and the genetic control of the events in meiosis.

Keeton, William T., and James L. Gould. *Biological Science*. 5th ed. New York: W. W. Norton, 1993. The chapter "Cellular Reproduction" discusses in detail the stages of mitosis and meiosis. Excellent diagrams allow visualization of cell division.

MITOSPORIC FUNGI

Categories: Fungi; microorganisms; taxonomic groups

Mitosporic fungi—also known as Deuteromycota, Deuteromycotina, *fungi imperfecti, and deuteromycetes—are fungi that are unable to produce sexual spores and are therefore placed in a separate phylum.*

The term "mitosporic" is a combination of the words "mitosis" and "sporic." Mitosis is the process of asexual cell division, which results in the daughter cells having the same genetic makeup as the mother cell. "Sporic" is used to denote the creation of spores. Therefore, the fungi in *Deuteromycota* do not produce sexual spores. The groupings in this phylum are artificial.

Conidium

The asexual spore produced by mitosporic fungi is the *conidium*. Conidia are produced from cells called condiogenic cells without the combination of nuclei; therefore the conidia are a product of mitosis. The conidial phase is a repetitive one. Thousands of conidia can be produced under adequate environmental conditions. These conidia will then continue the repetitive cycle.

Parasexuality

The genetic content of most fungi is dikaryotic. Each fungal cell has two individual and genetically distinct haploid nuclei. This state is originally formed by the combination of haploid mycelia. During *parasexuality*, the formation of diploid nuclei occurs, and then these revert back to the haploid state. Parasexuality permits the exchange of genetic information between the two haploid nuclei. The resultant haploid nuclei may be genetically modified. This is of great importance with plant pathogens, as parasexuality may help reduce genetic resistance of the host. Parasexuality is not considered to be a meiotic state, as no resultant sexual structure (such as an ascus or basidium) is formed.

Teleomorphs vs. Anamorphs

The *teleomorph* is the form of a fungus that occurs when sexual reproduction has occurred. These forms are found in the *Ascomycota* and the *Basidiomycota*. The *anamorph* is the form of a fungus that occurs when no sexual recombination has occurred. Anamorphs are important, as they are often a repetitive phase of the fungus and are able to produce spores without the additional requirement of sexual recombination.

When a specific teleomorph has been identified by scientists, the anamorphic phase is referred to as a subset of the teleomorph. For example, the anamorph fungus that produces blight of rice is known as *Pyricularia oryzae*. The teleomorph of this fungus is know as *Magnaporthe grisea*. The proper name for the anamorph is then the *Pyricularia anamorph* (or state) of *Magnaporthe grisea*.

Classification

Classification of the fungi in this group is based on the presence of conidia, kind of conidia (shape, color, size), and whether or not the conidia are produced in fungal structures called *conidiomata*. Conidiomata may have the generalized shape of a flask made of fungal tissue (a pycnidium); a pin cushion (sporodochium); or a mass of conidiophores located under either the epidermis or cuticle of a plant host (aecervulus).

There are three classes in this grouping. The *hyphomycetes* contain the fungi that produce conidia and conidiophores on hyphae or groups of hyphae. The *agonomycetes* do not produce conidia. The *coelomycetes* contain the fungi that produce conidia in distinct conidiomata.

Classification is dependent on the fungus meeting the following requirements. First, there must be the absence or presumed absence of the teleomorph. Second, there may be the absence or presumed absence of any meiotic or mitotic reproductive structure (as in agonomycetes). Finally, there may be the production of conidia from mitosis. There must not be any sexual structures.

Challenges

One of the challenges of the classification of mitosporic fungi is that these are form-genera of fungi that include both fungi that do not have teleomorphic states and fungi that do. For example, the teleomorph genus *Cochliobolus* has anamorphs in both the anamorphic form-genera *Bipolaris* and *Curvularia*. At the same time, there are several form-species in both anamorphic genera *Bipolaris* and *Curvularia* that have no teleomorph stage.

The reason the teleomorph can be found in some species and not in others is not known. Some isolates of anamorph species are distinct mating types and can be crossed with the other distinct mating type. When distinct mating types occur, it is possible that over time the mating types could be separated or that one mating type could be destroyed. Over time, without having the pressure to sexually reproduce, it is possible that the ability to sexually reproduce may have declined. Also, it is possible that the proper environmental conditions for sexual reproduction may not be known. Whether these fungi are best covered in the form-phylum mitosporic fungi or the form-phylum *Deuteromycota* is a question that will continue to be discussed. Because only fungi that form sexual stages are recognized as legitimate species, the distinct species and genera of the anamorphic fungi will have to be covered in this artificial phylum.

J. J. Muchovej

See also: Ascomycetes; Basidiomycetes; Basidiosporic fungi; Deuteromycetes; Fungi.

Sources for Further Study

Alexopoulos, Constantine J., and Charles E. Mims. *Introductory Mycology.* New York: John Wiley, 1995. The text gives detail into the biology, anatomy, and physiology of fungi. Includes illustrations, bibliographic references, and index.

Manseth, James D. Botany: *An Introduction to Plant Biology.* 2d ed. Sudbury, Mass.: Jones and Bartlett, 1998. General botany text with section on fungi, their importance, and their biology. Includes illustrations, bibliographic references, and index.

Raven, Peter H., Roy F. Evert, and Susan E. Eichhorn. *Biology of Plants*. 6th ed. New York: W. H. Freeman/Worth, 1999. Introductory college-level text. Chapter 15, "Fungi," covers their importance in the environment and their biology. Includes photographs, drawings, appendixes, and index.

MODEL ORGANISMS

Category: Methods and techniques

Practitioners in various areas of experimental biology recognize particular organisms as standard objects of study within their specialties. These model organisms are species that are used to exemplify a given category of organisms. For example, the laboratory rat is a common model used to study mammals, and Arabidopsis thaliana *is commonly used as a model to study plants.*

The characteristics of well-chosen model organisms, such as their genetic makeup or their development, make them suited to be the subjects of biological research. They are often developed specifically for laboratory study of pure-breeding strains that can be relied upon to provide a consistent medium for experimentation or examination. Biologists planning an experiment might specify a particular breed of rat, for example, whose traits are well defined and that are familiar to their peers. When a given model organism has become the standard in a particular field, it can become a kind of common currency that facilitates exchange among scientists.

Under these circumstances, a great deal of knowledge about individual model organisms may be accumulated rapidly. The crucial assumption, based on the theory of evolution, that underlies the use of specific organisms as models is that species sharing a common ancestor will have fundamental similarities of physiology and biochemistry. Among ubiquitous model organisms are the bacterium *Escherichia coli*, the yeast *Saccharomyces cerevisiae*, the roundworm *Caenorhabditis elegans*, the fruit fly *Drosophila melanogaster*, and the plant *Arabidopsis thaliana*.

Common Features

Above all else, model organisms must be practical to observe and use in experiments. They must be easy to breed or propagate and resilient enough to withstand manipulation. For knowledge gained in the study of model organisms to be applicable on a larger scale, the organisms must be representative of the taxonomic group in question. Clearly, the applicability of studies performed on a particular model organism varies, depending on the nature of the inquiry. For example, the yeast *S. cerevisiae* is broadly representative of the fungi as a whole, but its study may also provide insights into specific molecular processes common to all eukaryotes, including humans. Model organisms are often chosen because they are among the simplest examples of the group being studied. They may have a particularly small genome, a short life cycle, or even a small size that makes them convenient organisms with which to work. They may also lend themselves very well to the study of specific features. For example, fruit flies are commonly used in the study of genetics because they have a small genome from which it is easy to induce and detect mutations.

Well-chosen model organisms, those that possess some or all of the aforementioned characteristics, have been valuable tools for scientific research. Given these characteristics, it is easy to identify two types of scientific inquiry that are well served by the use of model organisms. There are studies in which a category of organisms is investigated by studying one of its simplest members and those in which a particular feature or biological process is illuminated by examining an organism in which it is especially accessible. Thus, the mouse is frequently used as a model for all mammals, and the green alga *Chlamydomonas* is used to study photosynthesis.

If a specific organism becomes the consensus model for a given category, the situation lends itself well to a speedy advancement of knowledge. The fact that many clusters of researchers choose to focus on the same model promotes collaboration and the more rapid accumulation of a body of knowledge about the organism, enhancing the likelihood of broader insights or theoretical advances. Having a research subject organism in common facilitates communication among researchers and leads to the formation of standard terminology. The widespread study of a single organism promotes the development and propagation of effective techniques for its use and allows for the introduction of standard experimental practices. Many observers have argued that the very success of science as a collaborative activity relies on scientists having some consensus about the tools and objects of their research and the terminology with which they describe it.

Although there are many advantages of a model organism becoming widespread in a particular field, there are some limitations to what can be achieved by the study of model organisms. There must always be a question of the applicability to other species of knowledge gained from the study of a model organism. A poor choice of an organism for a model can hinder the production of scientific knowledge just as much as research on a valid model can be beneficial. There is also the risk that focusing a discipline on one or a few models may inhibit understanding of diversity. As the botanist Dina Mandoli has pointed out, "flowering plants have an estimated 300,000 species . . . no one plant, not even *Arabidopsis thaliana*, can encompass this enormous diversity at the whole plant, physiologic, chemical, genetic, or molecular level." It is important, therefore, that research be carried out on enough model organisms to produce adequate breadth of knowledge. To that end, there are dozens of model plants in use representing a cross-section of the kingdom, of which *Arabidopsis* has been the most widely and successfully employed.

Arabidopsis

Arabidopsis is a genus of the mustard family that is closely related to food plants such as canola, cabbage, cauliflower, broccoli, radish, and turnip. Furthermore, although *Arabidopsis* is not used in agriculture, it is assumed that its study can lead to better knowledge of crop plants such as corn and soybeans because of evolutionary similarities among the genomes of all angiosperms. In the 1980's *Arabidopsis thaliana* (thale cress) became the

Elliot Meyerowitz/Courtesy of the National Science Foundation

Among plants, the most ubiquitous model organism is the plant Arabidopsis thaliana, *thale cress.*

primary model organism used in botany. Many characteristics lend it to such use, including its small size—plants are a few inches tall when mature—and its short life cycle of less than six weeks. The short life cycle allows researchers to see the effects of experimentation across successive generations in a relatively short span of time. *A. thaliana* also has a small genome and the least amount of DNA (deoxyribonucleic acid) per haploid cell of any known flowering plant. The haploid complement of *A. thaliana* is 117 million base pairs, compared to 1.6 billion in tobacco. As a result, it is comparatively easy to trace effects of experimentation to specific genes. It is valuable in the laboratory because of its prolific seed production and the availability of numerous mutations. It may be efficiently transformed with the bacterium *Agrobacterium tumefaciens*, which is used as a vector for the introduction of foreign DNA to the plant genome.

Arabidopsis was publicly recognized for its potential as a model organism in the 1960's. In 1985 it was first promoted as a model for molecular genetic research, and the first molecular map of one of the five *Arabidopsis* chromosomes was published in 1988. In 1990 the *Arabidopsis* Genome Project was begun (in part because of support by the codiscoverer of DNA James Watson). Thanks to a multinational effort, by the year 2000 the *Arabidopsis* gene sequence was fully decoded. The sequencing project was itself acclaimed as a model, because the researchers strove to be systematic and comprehensive in their investigation of the genome. It is now known that the *Arabidopsis* has slightly more than twenty-five thousand genes, making it comparable in its genetic complexity to a fruit fly.

Prior to the widespread use of *Arabidopsis*, many prominent scientists claimed that progress in botanical research was hindered by the study of too many organisms at once. Since *Arabidopsis* became a principal subject of research, botanical knowledge has advanced markedly. Researchers concentrating on *Arabidopsis* have helped to unify the studies of classical and molecular genetics, plant development, plant physiology, and plant pathology. These advances have led to a more fundamental understanding of many processes of plant growth and development at a molecular level.

Some specific areas in which *Arabidopsis* research has produced important advances are light perception, floral induction, flower development, and response to pathogenic and environmental stresses. For example, the functions of individual phytochromes, which are photoreceptors involved in many aspects of plant growth and development, were elucidated in *Arabidopsis*. Likewise, the first hormone receptor isolated in plants, that for ethylene, was discovered as a result of using *Arabidopsis* mutants. The next goal for the community of *Arabidopsis* scientists is to assign functions to all of the plant's genes by the year 2010.

Chlorella and Chlamydomonas

Chlorella pyrenoidosa and *Chlamydomonas reinhardtii* are unicellular green algae that have been used extensively as model organisms. They have many features in common with other model organisms, including short and simple life cycles and easily isolated mutants. Although not strictly members of the plant kingdom, they have been important tools for botanically related research because they are photosynthetic eukaryotic organisms. They therefore offer less complex subjects through which to study many processes that are central to plant life. There are no unicellular members of the plant kingdom, so study of many important botanical processes may be more easily undertaken on *Chlorella* or *Chlamydomonas* than on any plant.

In the mid-twentieth century Melvin Calvin used *Chlorella* in his Nobel Prize-winning research, which elucidated the cycle involved in photosynthetic carbon fixation which now bears his name, the Calvin cycle. This is a perfect example of model organisms' value in research. It is often easier to work out a mechanism in a simple organism and see whether it operates the same way in complex organisms (the understanding of which may be the ultimate purpose of the research) than to attempt the investigation on a complex organism in the first place. Once the Calvin cycle had been explained in *Chlorella*, it was shown to be ubiquitous in the chloroplasts of higher plants.

Chlamydomonas is the green alga most commonly used as a model organism in contemporary research. A *Chlamydomonas* genome project is under way with efforts taking place in the United States and Japan. Among the topics of research in which *Chlamydomonas* is the model organism of choice, one of the most compelling is that of chloroplast biogenesis and inheritance. *Chlamydomonas* is often referred to as the "green yeast," and like the yeast *S. cerevisiae*, it is an important eukaryotic model system. For studying certain aspects of cell biology to which yeast is not applicable, *Chlamydomonas* is chosen in preference. Such areas include cell motility caused by flagella, phototaxis (phototaxy), photosynthesis, and the study of centrioles, basal bodies, and chloroplasts.

Other Prominent Model Organisms

For areas other than those just mentioned, yeast is the most commonly used simple eukaryotic model organism. In 1996 *S. cerevisiae* became the first eukaryote to have its genome fully sequenced; its size is approximately one-tenth of that of *Arabidopsis*, or twelve million base pairs. Around the same time, the genome project was completed for the preeminent prokaryotic model organism, *E. coli*. This bacterium has become crucial not only as a focus of experiment but also as a biotechnological workhorse. Genes can be cloned by their insertion into *E. coli*, and gene products can therefore be mass-produced in large-scale fermentations of the bacteria.

Alistair Sponsel

See also: Calvin cycle; Cell cycle; Chromatin; Chromosomes; Eukaryotic cells; Green algae; History of plant science; Mitochondrial DNA; Plant biotechnology; Thigmomorphogenesis; Yeasts.

Sources for Further Study

Bowman, John, ed. *Arabidopsis: An Atlas of Morphology and Development*. New York: Springer Verlag, 1994. Documents the morphology and development of *Arabidopsis thaliana*, covering its embryogenesis, vegetative growth, root growth, reproductive structures, and host-pathogen interactions. Extensively illustrated.

Creager, Angela N. H. *The Life of a Virus: Tobacco Mosaic Virus as an Experimental Model, 1930-1965*. Chicago: Chicago University Press, 2002. This is a detailed history of the wide-ranging experimental use of the plant virus TMV as a model organism. Creager's account demonstrates how the use of such a model system allows specific biological knowledge generated within a single research laboratory to "travel" and become generally applicable.

Kohler, Robert E. *Lords of the Fly*. Chicago: Chicago University Press, 1994. A historical treatment of the fruit fly as a model organism in T. H. Morgan's "*Drosophila* school" at Columbia University. Kohler contends that model organisms such as the fly essentially become laboratory instruments.

Meyerowitz, Elliot M., ed. *Arabidopsis*. Cold Spring Harbor, N.Y.: Cold Spring Harbor Laboratory, 1994. Monograph provides a comprehensive account of what is known about the plant, presented in the context of general plant biology.

MOLECULAR SYSTEMATICS

Categories: Cellular biology; classification and systematics; disciplines; evolution

Molecular systematics is the discipline of classifying organisms based on variations in protein and DNA in order to make fine taxonomic categorizations not solely dependent on morphology.

Taxonomy, sometimes called *systematics*, is the study of categorizing organisms into logically related groupings. Historically, the way to perform taxonomy was to examine physical characteristics of organisms and classify species according to the most commonly held traits. Unfortunately, this method of systematizing plants and animals assumed that because they have common physical traits, they have common ancestry. A gross form of this miscategorization might take place, for example, if one suggested that since both mushrooms and ivy can grow on the sides of trees, they are closely related. The two species certainly have common physical traits but only vaguely resemble each other.

It is such a realization that motivated systematists to begin using molecular differences to compare species and populations. Molecular systematics uses variations in *protein* and *deoxyribonucleic acid* (DNA) molecules to determine how similar, or dissimilar, sets of organisms are. These molecular differences provide a much more accurate taxonomic picture.

Systematics and Evolution

The real power of molecular systematics is that it allows the examination of how species have changed over evolutionary time, as well as of the relationships between species that have no common physical characteristics. Molecular changes can be used to explore *phylogenetics* (how populations are related evolutionarily and genetically). It has been suggested that the amount of change that takes place in DNA over time can act as a *molecular clock*, gauging how much evolutionary time has passed. The clock is set by first examining geological and historical records to determine how long two species have been physically separated. By examination of the number of molecular changes that have occurred between those species over that known time, a time frame of change can be established. Genes are thought to evolve and mutate at a constant, predictable rate, giving rise to this evolutionary clock hypothesis.

There are three major domains of life: prokaryotes (modern bacteria), *Archaebacteria* (descendants of ancient bacteria), and eukaryotes (cellular organisms with nuclei and organelles). All these organisms share a common ancestry of hundreds of millions of years. All species over time are connected to one another through a web of interlacing DNA as they reproduce, separate to become new species, and reproduce again. All organisms carry their ancestors' genetic information with them as a bundle in each cell, and the more closely related organisms are to one another, the more similar the contents of that bundle will stay over time. Humans share common genes, unchanged over millennia, with all other organisms—from the bacterium *Escherichia coli* to barley to gophers. The more important the job of a gene, the less it changes over time; this concept is called *conservation*. Conservation is the force that keeps a biological or genetic link between every species on earth.

Protein-Level Analysis

Proteins were the earliest biomolecules used to study phylogenetics. Initially, protein differences could be studied only at the grossest levels. It was found that populations of organisms could be distinguished based on possessing different alleles (genetic sites) that made proteins possessing the same function but with different chemical structures. These enzymes were called *isozymes*. Isozymes can be separated and compared for size by employing a technique called gel electrophoresis. Gel electrophoresis uses a slab of gelatin-like medium and an electric field to separate molecules on the basis of size and electric charge. The genetic similarity of two different species can be determined based on common molecular weight of the isozymes.

Proteins are composed of strings of the twenty amino acids common to all life on earth. It is possible to ascertain the amino acid sequence of a protein. If the amino acid sequence of the same protein is ascertained among several different species, that sequence should be more similar between closely related species than more distantly related species. These differences allow taxonomists to gauge similarity of populations.

Antibodies are biomolecules that are able to recognize and bind very specifically to other molecules. Biologists employ antibodies that specifically recognize molecules at the surface of cells to test relationships between species. Antibodies that recognize cell-surface molecules on one species should recognize those same molecules in closely related species, but not from distantly related species, allowing a researcher to gauge similarity between species.

DNA-Level Analysis

The most common method used to establish taxonomic relationships is to compare DNA sequences between species. DNA is the double-stranded, polymeric molecule that encodes the proteins that direct the inner workings of all cells. The DNA molecule is structure like a ladder, with rungs formed by pairings of of four molecules, the bases guanine (G), adenine (A), thymine (T), and and cytosine (C). These bases, arranged in unique order, are read by special enzymes and encode messages that are translated into proteins. Sequences encoding for the same protein can change between species. In taxonomy, DNA sequences are obtained from several populations of organisms. Analysis of these sequences allows one to obtain a picture of how different populations have changed over time. This *DNA sequencing* may be used to compare many different types of DNA: regions that encode for genes, do not encode for genes, reside in chloroplast DNA, or reside in mitochondrial DNA.

Another common method of DNA phylogenetic analysis is called *restriction mapping*. In this method, DNA from different species is subjected to enzymatic treatment from proteins called *endonucleases*. These endonucleases have the ability to cleave DNA into fragments. Where the enzymes cleave the DNA is determined by the DNA sequence itself. The size and pattern of the fragments created by this treatment should be more similar in related species than in unrelated species.

A fairly new method of DNA analysis examines repetitive DNAs, called *microsatellite sequences*, that are found in all eukaryotic organisms. Microsatellite sequences are short arrangements of bases, such as GATC, repeated over and over. The number of repeats at a particular genetic location is usually more similar in related species than in unrelated ones. The differences in these repeated sequences are called "simple sequence polymorphisms" and are detected by a special enzymatic reaction called the *polymerase chain reaction*. Once detected, the fragments are separated and compared for size by means of gel electrophoresis.

James J. Campanella

See also: Cladistics; DNA in plants; Electrophoresis; Evolution of plants; Genetics: mutations; Nucleic acids; Proteins and amino acids; Systematics and taxonomy; Systematics: overview.

Sources for Further Study

Avise, John. *Molecular Markers, Natural History, and Evolution*. New York: Chapman and Hall, 1994. Examines relationships between genetic changes, taxonomy, and evolution.

Freeman, Scott, and Jon Herron. *Evolutionary Analysis*. New York: Prentice Hall, 2000. Covers the connections between evolution and taxonomy.

Nei, Masatoshi, and Suhir Kumar. *Molecular Evolution and Phylogenetics*. New York: Oxford University Press, 2000. Covers the mathematics of phylogenetic analysis.

Page, Roderic, and Edward Holmes. *Molecular Evolution: A Phylogenetic Approach*. Malden, Mass.: Blackwell Science, 1998. Concentrates on molecular evolution in populations and how species arise.

MONOCOTS VS. DICOTS

Categories: Angiosperms; *Plantae*; reproduction and life cycles

Within the angiosperms (flowering plants), two classes have been traditionally recognized by botanists: monocots and dicots. The terms connote differences between these groups' seed embryos. The recently introduced class, eudicots, literally meaning "true dicots," is increasingly used in place of the term "dicots."

Of the large number of plant species which currently inhabit the earth, most (about one-quarter million) are *angiosperms*, plants that reproduce by means of flowers. The famous naturalist Charles Darwin called angiosperms "an abominable mystery." Even today, much remains to be learned about the ancestry of angiosperms and when and where angiosperms first appeared. It is now believed that the first extant angiosperm is a single species of the genus *Amborella*. This plant, found on an island in the South Pacific, is a shrub with cream-colored flowers.

All angiosperms are assigned to the phylum *Anthophyta*. Within this phylum, considerable diversity occurs. However, two large lineages have traditionally been recognized: class *Monocotyledones* (monocots) and class *Dicotyledones* (dicots).

The monocots include grasses, cattails, irises, lilies, orchids, and palm trees. The dicots include the vast majority of seed plants: herbs, vines, shrubs, and most trees (cone-bearing trees are not angiosperms).

The terms "monocot" and "dicot" reflect the number of *cotyledons*, one or two, respectively, possessed by seeds of the plants. A cotyledon is the central portion of a seed embryo to which the epicotyl (immature shoot) and radicle (immature root) are attached. However, one need not examine the seed of a particular flowering plant in order to assign it to the correct class. Fortunately, each angiosperm possesses a "syndrome" of features, any one of which may be used for the purpose of classification.

Vegetative Parts

Monocots have leaves with parallel veins: The large, easily visible veins are parallel and extend the length of the (usually) linear leaves. In contrast, leaves of dicots are net-veined: The veins branch repeatedly as they extend into the various portions of the leaf. Vascular bundles (clusters of conducting cells) within stems of monocots are scattered, in contrast to those of dicots, which are arranged into a cylinder. These are best seen in a cross-section of the root. Furthermore, stems of monocots rarely produce wood, whereas those of many dicots (trees

Characteristics of Monocots vs. Dicots

Feature	*Monocots*	*Dicots*
Embryos	One cotyledon	Two cotyledons
Pollen	One pore	Three pores
Flowers	Parts in threes	Parts in fours or fives
Leaf veins	Parallel	Netted or reticulated
Vascular bundles	Scattered, complex	Ordered in rings
Woody growth	In 10% of species	In 50% of species
Root system	Adventitious	Primary+adventitious
Species	~65,000	~165,000
Examples	Bromeliads Grasses Irises Lilies Orchids Palms	Asters Cacti Herbs Shrubs Roses Trees

and shrubs) increase their diameter yearly, as wood accumulates within them.

Reproductive Parts

Flowers of monocots typically have their parts arranged in threes and are said to be "3-merous." An example would be a lily flower, which has these parts (designated from the outside toward the center of the flower): three sepals, three petals, six stamens, and one pistil. Flowers of dicots have parts arranged in fours or fives. Also, pollen grains produced by flowers of monocots each have one pore in the outer covering, as compared to three pores in the case of dicot pollen grains. Thus, it is apparent that nearly any part of a flowering plant can be used to place it into the correct subgroup of angiosperms. However, it is the flowers and leaves that are most easily considered (not requiring microscopic examination).

PHOTO AGORA

Flowers of monocots typically have their parts arranged in threes. An example would be a lily flower, which has these parts (designated from the outside toward the center of the flower): three sepals, three petals, six stamens, and one pistil.

New Interpretations

Like all areas of science, botany is subject to modification and reinterpretations as a result of the accumulation of new information. One group of plants traditionally considered to be dicots, the magnoliids, have long been problematic. They have some features of dicots, but their floral parts are free (unattached to one another) and arranged spirally. Also, they produce pollen grains, each of which has only a single pore. These features are characteristic of monocots. These traits have thus caused those who study angiosperms to place them into a new, third category, the class *magnoliids*. This most primitive group of angiosperms is thought to be ancestral to both of the other classes. The southern magnolia, *Magnolia grandiflora*, with its huge, white flowers, is a good example of this class. The name *eudicots* is applied to the still-large group that includes the remainder of the dicots.

Thomas E. Hemmerly

See also: Angiosperm evolution; Eudicots; Evolution of plants; Flower structure; Flower types; Leaf anatomy; *Monocotyledones*; Paleobotany; Reproduction in plants; Seeds; Stems; Wood.

Sources for Further Study

Hemmerly, Thomas E. *Appalachian Wildflowers*. Athens: University of Georgia Press, 2000. Representative angiosperms of this mountain range and adjacent areas are presented in full color. Uses the traditional two-class system.

Heywood, V. H. *Flowering Plants of the World*. New York: Mayflower Books, 1978. Well-illustrated overview of world angiosperms. Uses the traditional two-class system.

Raven, Peter H, Ray F. Evert, and Susan E. Eichhorn. *Biology of Plants*. 6th ed. New York: W. H. Freeman/Worth, 1999. The rationale for the three-class system is presented in this classic college-level botany textbook.

Strickberger, Monroe W. *Evolution*. 3d ed. Boston: Jones and Bartlett, 2000. Presents an overview of plant evolution along with that of animals.

MONOCOTYLEDONES

Categories: Angiosperms; *Plantae*; taxonomic groups

The Monocotyledones, *or monocots, are a large and very distinctive class of angiosperms or flowering plants, phylum* Anthophyta, *consisting of some 133 families, 3,000 genera, and 65,000 species.* Monocotyledones *form one of the two major subdivisions of angiosperms, the other being the* Eudicotyledones *(eudicots), with about 165,000 species.*

Typical monocots have a single *cotyledon* (seedling leaf), stems with scattered vascular bundles, root systems composed entirely of adventitious roots (arising directly from stem tissues), leaves with parallel venation and sheathing bases, and flower parts in threes. Monocots lack the ability to produce secondary growth (wood). In most monocots, stems remain at or below ground level and take the form of rhizomes (horizontal stems), bulbs (short, vertical stems covered with modified, fleshy leaves), corms (short, wide stems), or tubers (wood), which produce new adventitious roots continually or seasonally.

The growing tips (*apical meristems*) or buds of these plants remain below ground, except when they rise to produce flowers, and thus are protected against seasonal cold, drought, fire, and grazing animals. The growing tips are also surrounded and protected by the sheathing bases of the leaves. Monocot leaves are typically long and strap-shaped with numerous parallel veins that connect individually to the stem. Leaves grow primarily from their bases, increasing in length and pushing the earlier formed tips upward but not increasing much in width.

Most monocot flower parts occur in threes, such as three sepals, three petals, six stamens, and three carpels. Some members of the largely aquatic subclass *Alismatidae* have other patterns and show the primitive condition of *apocarpy* (carpels remain free from one another). In most other monocots, carpels are completely fused together into a three-chambered pistil. (Exceptions occur also in some palms). Monocots thus appear to have evolved from primitive anthophytes and have had a long, separate history.

Monocot Families Common in North America

Common Name	*Scientific Name*
Arum family	*Araceae*
Cattail family	*Typhaceae*
Daffodil family	*Amaryllidaceae*
Grass family	*Poaceae*
Iris family	*Iridaceae*
Lily family	*Liliaceae*
Orchid family	*Orchidaceae*
Sedge family	*Cyperaceae*

Note: For a full list of angiosperm families, see the tables that accompany the essay "Angiosperms," in volume 1.
Source: U.S. Department of Agriculture, National Plant Data Center, *The PLANTS Database*, Version 3.1, http://plants.usda.gov. National Plant Data Center, Baton Rouge, LA 70874-4490 USA.

Ecology and Variation

Because of their predominantly underground stem structures and basally regenerating leaves, monocots predominate in open habitats with strong seasonal contrasts or unpredictable droughts, such as grasslands. Grasses and similar monocots are particularly well adapted to survive drought, fire, and overgrazing, the three primary threats in grasslands. Bulbs, corms, and tubers are other forms of monocot stems that allow for underground survival, both in grasslands and in other regions that experience long winters or dry seasons.

The numerous monocots that inhabit marshes and aquatic habitats often have elongate petioles that grow from the base to lift their expanded leaf blades above the water level (such as *Sagittaria* and *Aponogeton*). The spectacular Egyptian papyrus plants have highly specialized upright stems that consist of a single *internode* (stem segment between nodes that elongates from the base), lifting a tuft of leaves and reproductive branches above fluctuating water levels. Other submerged aquatic monocots may remain rooted, producing conventional strap-shaped leaves (such as sea grasses) or may drift rootless, with short, whorled leaves produced along the stem.

Many monocots have adapted to live as *epiphytes* (plants that live on top of other plants), particularly on the upper branches of tropical rain-forest trees, a habitat that is often quite dry. The *orchid*, *bromeliad*, and *aroid* (*Philodendron, Anthurium*) families in particular have evolved largely in this habitat. Epiphytic orchids have succulent leaves or roots that store water for use during the dry periods between rains. A bromeliad grows as a rosette of tightly overlapping leaves radiating from the central growing point, forming a water-holding cup, or tank, in the center. Aroids can be found climbing up tree trunks or perched on upper branches and are unusual both for their flower structure and their leaves. Flowers are tiny and crowded on the clublike spadix, which in turn is enveloped by a large, often colored, leaflike structure called a *spathe*, while the leaves are broad, often dissected or divided into separate leaflets, and net-veined—quite unlike the typical monocot leaf.

Lack of the ability to form conventional tree trunks has led to some other novel growth forms in monocots. One of these is the *pseudostem* (false stem). In banana plants, and to a less obvious extent in gingers, heliconias, and other giant tropical herbs, an apparent stem forms from the elongate, tubular, concentric leaf sheaths, and gets taller as each new leaf grows up through the center.

Monocot Trees

Some monocots have become *arborescent* (treelike) by growing upward and developing thick, fibrous trunks without conventional secondary growth. In most arborescent monocots (palms, screw pines), the thickening growth of the stems occurs at the growing tip, in what is known as the *primary thickening meristem*. In these specialized meristems, the dividing cells expand laterally more than vertically. Stems reach their full thickness in this initial growth at the tip of the plant and expand very little after that.

In members of the *Dracena* family, however, a new way of thickening stems has evolved. Extensions from the apical meristem remain active along the sides of the stem and produce new parenchyma tissue and whole vascular bundles (not layers of wood). This allows the plants continually to increase the thickness of the stem and the amount of vascular tissue.

Branching is limited or absent in arborescent monocots, and leaves therefore tend to be large. Screw pines and dracenas have very long, narrow leaves, while palm leaves are broad and dissected into folded linear leaflets in either a *pinnate* (featherlike) or a *palmate* (fanlike) arrangement. Palm leaves are the largest in the world, reaching more than 60 feet (18 meters) in length in some species, and are highly fibrous. Development of these compound leaves is complex, involving multiple meristematic areas.

Frederick B. Essig

See also: Angiosperms; Eudicots; Monocots vs. dicots.

Sources for Further Study

Dahlgren, R. M. T. *The Families of the Monocotyledons: Structure, Evolution, and Taxonomy*. New York: Springer-Verlag, 1985. This comprehensive, technical reference includes an excellent introductory background to the biology and evolution of the monocots.

Jones, David L. *Palms Throughout the World*. Washington, D.C.: Smithsonian Institution Press, 1995. This comprehensive and beautifully illustrated reference on a large and distinctive family of monocots is an example of the many books that can be found on specific groups of plants.

Raven, Peter H., Ray F. Evert, and Susan E. Eichhorn. *Biology of Plants*. 6th ed. New York: W. H. Freeman/Worth, 1999. The various features of monocots are discussed and illustrated, in comparison with other groups of plants, throughout this widely respected textbook.

Takhtajan, A. *Diversity and Classification of Flowering Plants*. New York: Columbia University Press, 1997. This is one of the most respected references on the flowering plants. Contains a good discussion of the features of monocots.

MONOCULTURE

Categories: Agriculture; economic botany and plant uses; environmental issues

Monoculture is the agricultural practice of repetitively planting a single plant species rather than growing a variety of types of plants.

There has been considerable debate regarding the advantages and disadvantages of monoculture agriculture. This type of plant production is a system in which a single plant species, typically one producing grain (such as corn, wheat, or rice), forage (such as alfalfa or clover), or fiber (such as cotton), is grown in the same field on a repetitive basis, to the exclusion of all other species. In its most extreme version, a single variety of a plant species is grown, and all plants are virtually identical to one another. Monoculture can be contrasted with other agricultural production practices, such as multiple

PhotoDisc

In the most extreme version of monoculture agriculture, a single variety of a plant species is grown, and all plants are virtually identical to one another. Monoculture can also apply to perennial produce systems, such as coffee plantations like this one.

cropping (in which sequential monoculture crops are grown in the same year) or intercropping (in which two or more different crops are grown at the same time and place). Monoculture can also apply to perennial produce systems, such as fruiting trees, citrus crops, tea, coffee, and rubber trees.

Advantages

Monocultures are unnatural ecological occurrences. They are maintained through the use of resources such as labor, energy, and capital (fertilizers, chemicals, and so on). Left to itself, a monoculture crop will quickly revert to being a mixed plant community. However, monoculture agriculture has several advantages that caused its widespread adoption from the moment agriculture began. Monocultures allow agriculturalists to focus their energy on producing a single crop best adapted to a particular environment or to a particular market. For example, a premium is paid for white corn, used in making snack foods.

Monoculture is an appropriate agricultural strategy to optimize crop yield per unit of land when either temperature or water conditions limit the growing season. Monoculture agriculture lends itself to mechanization, an important consideration when labor is expensive relative to energy costs. Consequently, monoculture agriculture in the United States has developed in concert with the resources required to support it—markets, credit, chemicals, seed, and machinery—and with the social conditions that have caused the United States to change from a rural to a largely urban and suburban population.

Disadvantages

The disadvantages of monoculture are numerous. There are apparent limits to the increase in crop yields brought about by new hybrid seeds, *fertilizers*, and *pesticides*. Yield increases in monoculture agriculture have diminished since the 1980's. There is an economy of scale at which farm size becomes too small to permit effective mechanization or for which insufficient markets exist for reliance on a single crop. Focus on production of a single crop may lead to unbalanced diets and nutritional deficiencies in agricultural communities where no external supplies of produce are available.

More important, monoculture crops are biologically unstable, and considerable effort must be made to keep other plants and pests out. When every plant under cultivation is the same, these systems are inherently susceptible to natural events (storms, drought, and wind damage) and to *biological invasions* by insects and plant pathogens. The classic example of overreliance on monoculture was the Irish Potato Famine (1845-1850). The famine was instigated by natural climatic conditions that allowed the plant pathogen *Phytophthora infestans* (potato late blight) to destroy successive potato crops in a population too impoverished to afford other food staples that were available.

Mark S. Coyne

See also: Agriculture: modern problems; Green Revolution; High-yield crops; Hybridization; Multiple-use approach; Nutrition in agriculture; Pesticides; Slash-and-burn agriculture; Sustainable agriculture.

Sources for Further Study

Coughenour, C. Milton, and Shankariah Chamala. *Conservation Tillage and Cropping Innovation: Constructing the New Culture of Agriculture*. Ames: Iowa State University Press, 2000. Explores causes, conditions, and significance of the changes in tillage and cropping practices in the United States and Australia.

Odemerho, F. O., and A. Avwunudiogba. "The Effects of Changing Cassava Management Practices on Soil Loss: A Nigerian Example." *Geographical Journal* 159, no. 1 (March, 1993): 63-69. Study compares erosion between monoculture and polyculture systems. Assesses planting practices in reducing erosion under monoculture.

Prihar, S. S., P. R. Gajri, D. K. Benbi, and V. K. Arora. *Intensive Cropping: Efficient Use of Water, Nutrients, and Tillage*. New York: Food Products Press, 2000. Chapters address intensive cropping and related concerns, including efficient input use and sustainability of high-production systems.

MOSSES

Categories: Nonvascular plants; paleobotany; *Plantae*; taxonomic groups

Members of the ten thousand species of the class Bryopsida *or* Musci *in the order* Bryophyta, *mosses are usually tiny, fragile, nonflowering, spore-bearing plants. They are related to hornworts and liverworts.*

Mosses evolved from green algae before the existence of reptiles and flying insects. Fossils from the Devonian period, about 410 million to 353 million years ago, contain mosses. The first known mosses lived 286 million to 245 million years ago, during the Permian period. By the Tertiary period, approximately 66.4 million to 1.6 million years ago, at least one hundred moss species existed. Fossilized mosses, including *Muscites*, *Palaeohypnum*, and *Protosphagnum*, resemble modern moss species structurally, indicating that mosses have not evolved quickly.

Found globally, mosses prefer damp, shady, sparsely vegetated environments and often grow

Gary Bass/Photo Agora

Grass grows sparsely amid this moss.

Classification of Mosses

Class *Andreaeopsida* (granite mosses)
Families:
Andreaeaceae
Andreaeobryaceae

Class *Sphagnopsida* (peat mosses)
Families:
Sphagnaceae

Class *Bryopsida* (true mosses)
Families:
Archidiaceae
Aulacomniaceae
Bartramiaceae
Bryaceae
Catoscopiaceae
Hypopterygiaceae
Meesiaceae
Mniaceae
Pseudoditrichaceae
Racopilaceae
Rhizogoniaceae
Timmiaceae
Buxbaumiaceae
Bruchiaceae
Bryoxiphiaceae
Dicranaceae
Ditrichaceae
Leucobryaceae

Class *Bryopsida* (continued)
Fissidentaceae
Disceliaceae
Ephemeraceae
Funariaceae
Gigaspermaceae
Oedipodiaceae
Splachnaceae
Splachnobryaceae
Grimmiaceae
Ptychomitriaceae
Scouleriaceae
Callicostaceae
Daltoniaceae
Hookeriaceae
Amblystegiaceae
Brachytheciaceae
Entodontaceae
Fabroniaceae
Helodiaceae
Hylocomiaceae
Hypnaceae
Leskeaceae
Myriniaceae
Plagiotheciaceae
Pleuroziopsidaceae
Pterigynandraceae

Class *Bryopsida* (continued)
Rhytidiaceae
Sematophyllaceae
Stereophyllaceae
Thamnobryaceae
Theliaceae
Thuidiaceae
Fontinalaceae
Anomodontaceae
Climaciaceae
Cryphaeaceae
Hedwigiaceae
Leptodontaceae
Leucodontaceae
Meteoriaceae
Neckeraceae
Pterobryaceae
Erpodiaceae
Orthotrichaceae
Rhachitheciaceae
Polytrichaceae
Calymperaceae
Encalyptaceae
Pottiaceae
Schistostegaceae
Seligeriaceae
Tetraphidaceae

Source: U.S. Department of Agriculture, National Plant Data Center, *The PLANTS Database*, Version 3.1, http://plants.usda.gov. National Plant Data Center, Baton Rouge, LA 70874-4490 USA.

on land adjacent to bodies of fresh water and in woods, sometimes appearing carpetlike. Some mosses prefer rotting wood, while others thrive in soil saturated with water or bare soil. Although hardy, mosses are sensitive to pollutants. Mosses need external sources of moisture to transport nutrients. They can adapt to environments, exhibiting variations. Although many mosses live in temperate zones, they also have been found living in tundra. Mosses can grow on rocks, mountains, and brick and cement structures. They are frequently found on tombstones and other memorials.

Mosses are categorized into three groups, the true mosses (*Bryopsida* or *Musci*), the peat mosses (*Sphagnopsida*), and the granite mosses (*Andreaopsida*). The greatest number of moss species are in the subclass *Bryidae*. Some subclasses contain only a few species.

Most mosses are short. Cape pygmy moss (*Ephemerum capensi*), the size of a pencil dot, is the tiniest moss that can be seen without a microscope. The largest mosses are 1 meter (40 inches) long. Water mosses in the order *Bryales* live in streams and ponds, sometimes adhering to tree roots or rocks. Brook moss grows shoots ranging from 30 to 100 centimeters (12 to 40 inches) long. Common mosses include hair-cap, sheet, top, slit, granite, broom, and wind-blown mosses. Luminous mosses, whose curved cells concentrate light on chloroplasts for photosynthesis, grow in dark spaces.

Mosses consist of green stems and leaves that are usually one cell thick and sometimes have midribs. Leaf cell shape and pattern are used to identify mosses, and differences can be detected microscopically. Instead of roots, each moss stem has *rhizoids* which secure the plant to the ground. Lacking a vascular system to conduct fluid, mosses need a watery environment to distribute male gametes.

Life Cycle

Mosses have two reproductive forms, referred to as the alternation of generations. During the gametophytic, or sexual, generation, a gametophyte creates a plant body known as the thallus. Two specialized organs are involved in sexual reproduction. The antheridium is a sac that produces male gametes, each with two flagella, and the archegonium is the bottle-shaped female sexual organ. Short cell filaments, called the paraphyses, are associated with the antheridium. Usually both sexual organs are located on moss leaves.

Sperm fertilizes the egg to create a maroon-colored stalk called a seta, which has a sporangium, or spore capsule, in the sporophytic generation. The sporangium consists of an urn, peristome, annulus, operculum, and calyptra. Capsule components vary by species. This sporangium relies on the gametophyte to deliver necessary nourishment and moisture. When the sporophyte dries, spores are scattered like dust and germinate to become threadlike *protonema*, an immature moss form, which form new gametophytes.

During wet conditions, the capsule can withhold spores by closing from one to two rows of peristome teeth. Because spores are sensitive to moisture, temperature, and soil chemistry and have limited supplies of energy for protonema to grow, odds are that only one spore per million forms a mature moss plant. Mosses also reproduce asexually by growing from fragments of stems and leaves that germinate into new plants.

Uses

Ecologically, mosses serve several purposes. They help secure soil by forming patches of dense mats of ground cover and absorb water to prevent erosion. Mosses also provide nutrients that enrich soils for future plant growth. Mosses are popular terrarium plants. They are often illegally harvested from public lands, including national forests.

The word "moss" in plant names does not always indicate a moss species. The club moss is, in fact, an evergreen herb from the family *Lycopodiaceae*, and its relatives, the spike mosses, are primitive vascular plants in the family *Selaginellaceae*. Often, people in the Northern Hemisphere use moss to tell direction because it grows on the north-facing sides of trees and structures, where sunlight does not shine directly. Green algae that grow on trees' northern sides are referred to as moss but usually are not. Other types of algae, such as pond moss, are incorrectly described as mosses. Spanish moss is one of the best-known plants called a moss, but it actually is a lichen and air plant belonging to the pineapple family, *Bromeliaceae*.

Some moss species in the genus *Sphagnum*, of the order *Sphagnales*, partially decompose into carbonized tissues in acidic, watery areas such as bogs to become peat, which is significant economically as a fuel and as a commercial gardening product. Empty moss cell spaces absorb water, and bogs become drier. Peat mosses float in mats, prevent fungi and bacteria from multiplying, and create an environment that can preserve carcasses. Prior to the development of modern medical techniques, sphagnum moss was used to dress surgical sites because of its antiseptic qualities.

Elizabeth D. Schafer

See also: Algae; Bioluminescence; *Bromeliaceae*; Bryophytes; Hornworts; Liverworts; Lycophytes; Peat.

Sources for Further Study

Conard, Henry S. *How to Know the Mosses and Liverworts*. 2d ed., edited by Paul L. Redfearn, Jr. Dubuque, Iowa: Wm. C. Brown, 1979. A classic guide to identifying North American moss species. Includes illustrations, index.

Crosby, Marshall R., and Robert E. Magill. *A Dictionary of Mosses*. St. Louis: Missouri Botanical Garden, 1981. Brief list of moss genera, with taxonomic information for classification of specimens. Includes index and corrections and additions to supplement two previous printings.

_______. *Index of Mosses, 1990-1992*. St. Louis: Missouri Botanical Garden, 1994. Complements authors' previous works by updating moss nomenclature and providing citations to significant moss research publications. Includes bibliography.

Crum, Howard A., and Lewis E. Anderson. *Mosses of Eastern North America*. 2 vols. New York: Columbia University Press, 1981. Comprehensive guide to mosses indigenous to

Canada and the eastern United States. Includes illustrations, bibliographies, and index.
Schenk, George. *Moss Gardening: Including Lichens, Liverworts, and Other Miniatures*. Portland, Oreg.: Timber Press, 1997. Explains how to use mosses and similar plants to landscape decoratively. Includes color illustrations, bibliographical references, and index.

MULTIPLE-USE APPROACH

Categories: Agriculture; economic botany and plant uses; environmental issues; forests and forestry

The multiple-use approach is a concept of resource use in which land supports several concurrent managed uses rather than single uses over time and space.

The multiple-use approach is a management practice that is teamed with the concept of sustained yield. Multiple use began as a working policy, generally associated with forestry, and was enacted as law in 1960. As a concept of land-use management, it has most often been applied to the use of forestlands. Historically, multiple use has been linked with another concept, that of *sustained yield*.

Historical Background

The history of the intertwined multiple-use and sustained-yield approaches to land management in the United States dates from the late 1800's. Prior to that time, forestlands were used for timber production, rangeland for grazing, and parklands for recreation. Little attention was given to the interrelated aspects of land use. By the late 1800's, however, some resource managers began to see land as a resource to be managed in a more complex, integrated fashion which would lead to multiple use. This awakening grew out of the need for conservation and sustained yield, especially in the forest sector of the resource economy.

Sustained Yield

Since the earliest European settlement of North America, forest resources had been seen both as a nearly inexhaustible source of timber and as an impediment to be cleared to make way for agriculture. This policy of removal led to serious concern by the late 1800's about the future of American forests. By 1891 power had been granted to U.S. president Benjamin Harrison to set aside protected forest areas. Both he and President Grover Cleveland took actions to establish forest reserves. To direct the management of these reserves, Gifford Pinchot was appointed chief forester. Pinchot was trained in European methods of forestry and managed resources, as noted by Stewart Udall in *The Quiet Crisis* (1963), "on a sustained yield basis." The sustained-yield basis for forest management was thus established. Essentially, the sustained-yield philosophy restricts the harvesting of trees to no more than the ultimate timber growth during the same period.

Multiple Use

Properly managed, forestlands can meet needs for timber on an ongoing, renewable basis. However, land in forest cover is more than a source of timber. *Watersheds* in such areas can be protected from excessive runoff and sedimentation through appropriate management. Forest areas are also *wildlife habitat* and potential areas of outdoor recreation. The combination of forest management for renewable resource production and complex, interrelated land uses provided the basis for the development of multiple use-sustained yield as a long-term forest management strategy.

Multiple Use Joins Sustained Yield

The merging of these two concepts took shape over a period of many years, beginning in the early twentieth century. The establishment of national forests by Presidents Harrison and Cleveland provided a base for their expansion under President Theodore Roosevelt in the early 1900's. With the active management of Pinchot and the enthusiastic support of Roosevelt, the national forests began to be managed on a long-term, multiple-use, sustained-

AP/Wide World Photos

When properly managed, forestlands such as the Tongass National Forest in Alaska are renewable sources of timber. However, land in forest cover is more than a source of trees. The combination of forest management for renewable resource production and complex, interrelated land uses provided the basis for the development of multiple use-sustained yield as a long-term forest management strategy.

yield basis. The desirability of this approach eventually led to its formalization by law: On June 12, 1960, Congress passed the Multiple Use-Sustained Yield Act. To some, this act was the legal embodiment of practices already in force. However, the act provides a clear statement of congressional policy and relates it to the original act of 1897 that had established the national forests.

The 1960 act specifies that "the national forests are established and shall be administered for outdoor recreation, range, timber, watershed, and wildlife and fish purposes." Section 2 of the act states that the "Secretary of Agriculture is authorized and directed to develop and administer the renewable resources of the national forests for multiple use and sustained yield of the several products and services obtained therefrom." The act gives no specifics, providing a great deal of freedom in choosing ways to meet its provisions. It also refrains from providing guidelines for management. In practice, the achievement of a high level of land management under the act has called for advocating a *conservation* ethic, soliciting citizen participation, providing technical and financial assistance to public and private forest owners, developing international exchanges on these management principles, and extending management knowledge.

Jerry E. Green

See also: Agriculture: modern problems; Forest management; Forests; Grazing and overgrazing; High-yield crops; Monoculture; Rangeland; Sustainable agriculture; Sustainable forestry; Timber industry.

Sources for Further Study

Cutter, Susan, Hilary Renwick, and William Renwick. *Exploitation, Conservation, Preservation: A Geographical Perspective on Natural Resource Use*. 3d ed. New York: J. Wiley & Sons, 1999. Integrates physical, economic, social, and political considerations of the major natural resource issues facing the world.

Hewett, Charles E., and Thomas E. Hamilton, eds. *Forests in Demand: Conflicts and Solutions*. Boston: Auburn House, 1982. A look at the United States' evolving philosophy governing renewable resource use.

Lovett, Francis. *National Parks: Rights and the Common Good*. Lanham, Md.: Rowman & Littlefield, 1998. Explores issues regarding the proper balance between individual and subgroup rights and the needs and value of the community.

Sedjo, Roger A., ed. *A Vision for the U.S. Forest Service: Goals for Its Next Century*. Washington, D.C.: Resources for the Future, 2000. At the centenary of the Forest Service, ten papers cover such topics as rethinking scientific management, state trust lands, and conservation.

Udall, Stewart. *The Quiet Crisis*. 1963. Reprint. Salt Lake City, Utah: Gibbs Smith, 1998. Good historical material.

MUSHROOMS

Categories: Economic botany and plant uses; food; fungi; poisonous, toxic, and invasive plants

Members of the kingdom Fungi *and division* Mycota, *mushrooms are members of the class* Basidiomycetes. *This class includes both the hymenomycetes, which reproduce spores using a layer structure called the hymenium, and the gasteromycetes, which include the stinkhorns and puffballs.*

Fossil evidence suggests that mushrooms existed ninety million years ago. The order *Agaricales* represents approximately four thousand species in sixteen families of commonly found mushrooms. The order *Polyporus* consists of fungi that grow on tree limbs or stumps. Although they prefer moist environments, mushrooms are sometimes found in deserts, beaches, and occasionally snowdrifts.

Characteristics

The mushrooms familiar to most people are *hymenomycetes*. They are either edible or poisonous structures lacking chlorophyll. That is, like other fungi, they do not make their own food through photosynthesis but instead are heterotrophs, which feed off other organisms. They are classified as thousands of species, distributed globally. Most mushrooms consist of a fleshy fruiting body called the *sporophore* and a cylindrical *stalk*.

Most mushrooms with the familiar umbrella-shaped sporophore, called the *cap* or *pileus*, belong to the family *Agaricaceae*. The pileus has narrow sheets called *gills* that contain the spores. The pileus sits atop the stalk, known as the *stipe*. Pilei and stipes vary in size.

Sporophores grow from a mat of thin strands called the *mycelium*, or spawn, which is located below the soil surface. Each mycelium grows new sporophores during the annual fruiting season. Mycelial life spans range from several months to centuries, depending on nutrient and moisture sources and suitable temperature. A honey mushroom mycelium in Michigan once spread across 40 acres over fifteen hundred years.

Mushrooms grow quickly, often maturing within several days. A *membrane* joins the cap's edges to the stem. As mushrooms mature, the membrane breaks, revealing the cap's gills. When the agaric mushroom ripens, its color changes from white to pink, then brown. Some mushrooms' colors change when they are exposed to air or water. Mushrooms are *geotropic*, keeping their caps upright and gills vertical and turning the pileus if placed on their sides after being picked.

Species

Wild mushrooms grow in fields, forests, lawns, and gardens. The waxy caps are among the most colorful mushrooms. In spring, morels emerge, often thriving on burned land. Their gills resemble honeycombs. The genus *Gyromitra* includes the false morels, which can be toxic.

The family *Boletaceae* includes mushrooms that contain layers of spores in tubes on the pileus's undersurface. The *Hydnum* mushrooms appear to have teeth, and the *Clavarias* look like coral. Occasionally, sporophores grow in circular patterns popularly called fairy rings.

The *Coprinus comatus*, or shaggy mane, is another spring mushroom which appears through the fall in open areas, with individuals or groups of this species often reemerging in the same place annually. This mushroom has tall, brown caps that appear shaggy because of soft scales on its surface. Unlike most mushrooms, members of *Coprinus* have caps that do not expand at maturity, which causes the compacted gills to liquefy into a black fluid, resulting in the collective name for the one hundred species of this genus, "inky caps."

The species *Agaricaceae campestris* profusely grows in rural areas, especially during the summer. The orange or yellow funnel-capped *Cantherellus cibarius* thrives in European forests, tastes nutty, and has been a preferred edible mushroom since the days of the Roman Empire. By autumn, large mushrooms representing the genus *Boletus* ripen in wooded areas. Veins on caps of *Boletus* species change color from white to yellow, indicating their readiness for picking.

Several mushroom species live on decaying wood hosts. Rotten timber is the habitat for *Polyporus sulfurreus*, which form as orange-yellow shelves that can be meters wide and weigh many kilograms. Pores beneath the layers create spores.

Poisonous Mushrooms

Between sixty and seventy mushroom species are poisonous. If consumed, poisonous mushroom species can cause intense physical reactions, rang-

Kevin and Bethany Shank/Photo Agora

The mushroom genus Amanita *includes the most dangerous mushrooms, including* Amanita muscarine. *Only the* Amanita caesarea *is edible. Shown is* Amanita flavonconia, *or the yellow patches mushroom.*

ing from sickness to death. The genus *Amanita* includes the most dangerous mushrooms, including *Amanita muscarine*. Only the *Amanita caesarea* is edible. The fly mushroom, *Amanita muscaria*, is a large, colorful mushroom that is deadly to insects. The white *Amanita phalloides*, known as the death cap or death angel, are the most toxic mushrooms and are widely distributed. Other hazardous mushrooms are *Boletus satana* (Satan's mushroom) and *Clitocybe illudens* (jack-o'-lantern), which glows in the dark and is physically similar to the benign *Clitocybe gigantea*.

Alkaloid mycotoxins in mushrooms attack nerve, muscle, blood, and organ cells. Victims suffer digestive symptoms within eight to twelve hours after ingesting mushrooms, then slip into a coma and die several days later. Patients undergo gastrointestinal tract purging and antidote treatment in an effort to counter poisoning.

Uses

Humans have cultivated mushrooms for thousands of years. Some mushrooms lack flavor or are too bitter or woody to consume. Others, such as *Ithyphallus impudicus*, have a foul odor. The *Pleurotus ostreatus* tastes like oysters. Although mushrooms are not especially nutritious because they are 90 percent water, many have pleasing tastes and textures and are low in fat.

Annually, about ten million North Americans hunt common mushrooms, especially morels. The mushroom industry relies on mushrooms picked and sold at wholesale prices totaling millions of dollars. Brokers ship the mushrooms to distributors or retail markets.

Commercial growers produce a safe source of edible mushrooms. These mushrooms are cultivated in specially designed structures, caves, or cellars in which the darkness, humidity, and temperature are regulated. Beds of soil-covered straw and manure are planted with mycelia. The *Agaricus bisporus*, a hybrid developed by researchers, is the most common commercial species. Exotic wild mushrooms, including portobello and shiitake, are cultivated artificially. Liquid nitrogen is used to store spawns and enhances their endurance and quality. Scientists are genetically designing improved mushroom strains.

Some mushrooms are hallucinogenic, and most governments restrict their use as a narcotic, although mushrooms are smuggled into countries that have bans. Various cultures incorporate mushrooms into spiritual rituals, believing the mushrooms have magical powers. Researchers have determined that some mushrooms, specifically the *Gleophyllum odoratum* and *Clitocybe gibba* species, can be therapeutic if thrombin inhibitors are extracted to manufacture anticoagulant pharmaceuticals.

Elizabeth D. Schafer

See also: Ascomycetes; Basidiosporic fungi; *Basidiomycetes*; Bioluminescence; Culturally significant plants; Fungi; Medicinal plants; Plant biotechnology; Poisonous and noxious plants; Tropisms.

Sources for Further Study

Benjamin, Denis R. *Mushrooms: Poisons and Panaceas*. New York: W. H. Freeman, 1995. Comments on mushrooms' medicinal and toxic properties. Includes illustrations, bibliographical references, and index.

Bessette, Alan E., William C. Roody, and Arleen R. Bessette. *North American Boletes: A Color Guide to the Fleshy Pored Mushrooms*. Syracuse, N.Y.: Syracuse University Press, 2000. Focuses on identification and uses of a specific mushroom family. Includes illustrations, bibliographical references, and indexes.

Fisher, David W., and Alan E. Bessette. *Edible Wild Mushrooms of North America*. Austin: University of Texas Press, 1992. Guide with recipes for consuming mushrooms. Includes illustrations, bibliographical references, and index.

Phillips, Roger. *Mushrooms of North America*. Boston: Little, Brown, 1991. Provides details for identification of common and rare edible and poisonous mushrooms. Includes illustrations, bibliographical references, and index.

Schaechter, Elio. *In the Company of Mushrooms: A Biologist's Tale*. Cambridge, Mass.: Harvard University Press, 1997. Narrates author's experiences, supplemented with information about mushroom science, history, and myths. Includes illustrations, bibliographical references, tables, and index.

MYCORRHIZAE

Categories: Ecosystems; fungi

"Mycorrhiza" (from the Greek mukes, *meaning "fungus," and* rhiza, *meaning "root") is a term describing a mutualistic, symbiotic relationship between plant roots or other underground organs and fungi. Mycorrhizae are among the most abundant symbioses in the world.*

Mycorrhizal associations have been described in virtually all economically important plant groups. Investigators in Europe detected fungal associations in most European species of flowering plants, all gymnosperms, ferns, and some bryophytes, especially the liverworts. Similar patterns are predicted in other ecosystems. Continuing studies of ecosystems, from boreal forests to temperate grasslands to tropical rain forests and agroecosystems, also suggest that most plant groups are intimately linked to one or more species of fungus.

It is theorized that most of the plants in stable habitats where competition for resources is common probably have some form of mycorrhizal association. Species from all of the major taxonomic groups of fungi, including the *Ascomycotina*, *Basidiomycotina*, *Deuteromycotina*, and *Zygomycotina*, have been found as partners with plants in mycorrhizae.

Considering the prevalence of mycorrhizae in the world today, botanists theorize that mycorrhizae probably arose early in the development of land plants. Some suggest that mycorrhizae may have been an important factor in the colonization of land.

The fungal partner (or *mycobiont*) in a mycorrhizal relationship benefits by gaining a source of carbon. Often these mycobionts are poor competitors in the soil environment. Some mycobionts have apparently coevolved to the point that they can no longer live independently of a plant host.

The plant partner in the mycorrhizal relationship benefits from improved nutrient absorption. This may occur in different ways; for example, the mycobiont may directly transfer nutrients to the root. Infected roots experience more branching, thus increasing the volume of soil that the plant can penetrate and exploit. Evidence also suggests that mycorrhizal roots may live longer than roots without these associations. Comparison of the growth of plants without mycorrhizae to those with fungal partners suggests that mycorrhizae enhance overall plant growth.

Types of Mycorrhizae

Mycorrhizae may be classified into two broad groups: *endomycorrhizae* and *ectomycorrhizae*. Endomycorrhizae enter the cells of the root cortex. Ectomycorrhizae colonize plant roots but do not invade root cortex cells.

The most common form of endomycorrhizae are the *vesicular-arbuscular mycorrhizae*. The fungi involved are zygomycetes. These mycorrhizae have internal structures called *arbuscules*, which are highly branched, thin-walled tubules inside the root cortex cells near the vascular cylinder. It is estimated that 80 percent of all plant species may have vesicular-arbuscular mycorrhizae. This type of mycorrhiza is especially important in tropical trees.

There are several other subtypes of endomycorrhizae. *Ericoid mycorrhizae*, found in the family *Ericaceae* and closely related families, supply the host plants with nitrogen. These are usually restricted to nutrient-poor, highly acidic conditions, such as heath lands. *Arbutoid mycorrhizae*, found in members of the *Arbutoideae* and related families, share some similarities with ectomycorrhizae in that they form more developed structures called the sheath and Hartig net (described below).

Monotropoid mycorrhizae, found in the plant family *Monotropaceae*, are associated with plants that lack chlorophyll. The host plant is completely dependent on the mycobiont, which also has connections to the roots of a nearby tree. Thus the host, such as *Monotropa*, indirectly parasitizes another plant by using the mycobiont as an intermediate. *Orchidaceous mycorrhizae* are essential for orchid seed germination.

Ectomycorrhizae are common in forest trees and shrubs in the temperate and subarctic zones. Well-developed fungal sheaths characterize these mycorrhizae, along with special structures called Hartig nets. Basidiomycetes are the usual mycobionts and often form mushrooms or truffles. Ectomycorrhizae help protect the host plant from diseases by forming a physical fungal barrier to infection.

Anatomy and Development

Individual filaments of a fungal body are called *hyphae*. The entire fungal body is called a *mycelium*. Root infection may occur from fungal spores that germinate in the soil or from fungal hyphae growing from the body of a nearby mycorrhiza. When infection occurs, hyphae are drawn toward certain chemical secretions from a plant root.

In ectomycorrhizae, root hairs do not develop in roots after infection occurs. Infected roots have a *fungal sheath*, or *mantle*, that ranges from 20 to 40 micrometers thick. Fungal hyphae penetrate the root by entering between epidermal cells. These hyphae push cells of the outer root cortex apart and continue to grow outside of the cells. This association of hyphal cells and root cortex cells is called a *Hartig net*. In ectomycorrhizae, the mycobionts never invade plant cells, nor do they penetrate the endodermis or enter the vascular cylinder. The root tip may be ensheathed by fungi, but the apical meristem is never invaded. Main roots experience fewer anatomical changes than lateral roots after infection. Lateral roots become thickened, may show the development of characteristic pigments, and grow very slowly. Infected roots also show different branching patterns than those of uninfected roots.

Endomycorrhizae are highly variable in structure. Many endomycorrhizae do not have sheaths or Hartig nets. In all endomycorrhizae, hyphae penetrate into root cortex cells, while portions of the mycelium remain in contact with the soil. The hyphae that remain in the soil are important in fungal reproduction and produce large numbers of haploid spores. Fungi do not invade root meristems, vascular cylinders, or chloroplast-containing cells in the plant.

Some of the host cells contain fungal extensions called *vesicles* that are filled with lipids. Vesicles are specialized structures that are often thick-walled and may serve as storage sites or possibly in reproduction. Vesicles are also produced on the hyphae that grow in the soil. Near the vascular cylinder, the hyphae branch dichotomously and form large numbers of thin-walled tubules called *arbuscules* that invade host cells. The arbuscules cause the host membranes to fold inward, creating a plant-fungus interface that has a very large surface area. The arbuscules last for about fourteen days before they break down on their own or are digested by the host cell. Host cells whose fungal arbuscules have broken down may be reinvaded by other hyphae.

Darrell L. Ray

See also: Ascomycetes; *Basidiomycetes;* Coevolution; Deuteromycetes; Fungi; Legumes; Nitrogen fixation; Roots; Zygomycetes.

Sources for Further Study

Deacon, J. W. *Modern Mycology*. 3d ed. Malden, Mass.: Blackwell Science, 1997. An introductory text in mycology. Includes photos, diagrams.

Harley, J. L., and S. E. Smith. *Mycorrhizal Symbiosis*. New York: Academic Press, 1983. An authoritative text on the formation and biology of mycorrhizae.

Raven, Peter H., Ray F. Evert, and Susan E. Eichhorn. *Biology of Plants*. 6th ed. New York: W. H. Freeman/Worth, 1999. A standard text for introductory botany courses at the university level. Discusses the significance of major types of mycorrhizae. Includes photos.

NASTIC MOVEMENTS

Categories: Animal-plant interactions; movement; physiology

Plants, unlike animals, are sedentary organisms, but they are capable of some limited movements. These include nastic movements, which enable plants to adapt rapidly to changes in their environment by changing orientation.

Nastic movements and tropisms, or growth movements, are two important, but different, kinds of movements in plants. In nastic movements, the direction of movement is determined by the anatomy of the plant rather than by the position of the origin of the stimulus. In tropisms, the movement is in a direction either toward or away from the origin of the stimulus. In addition, the orientational changes that occur in nastic movements are temporary; they are reversible and repeatable. The tropisms, in contrast, are generally irreversible.

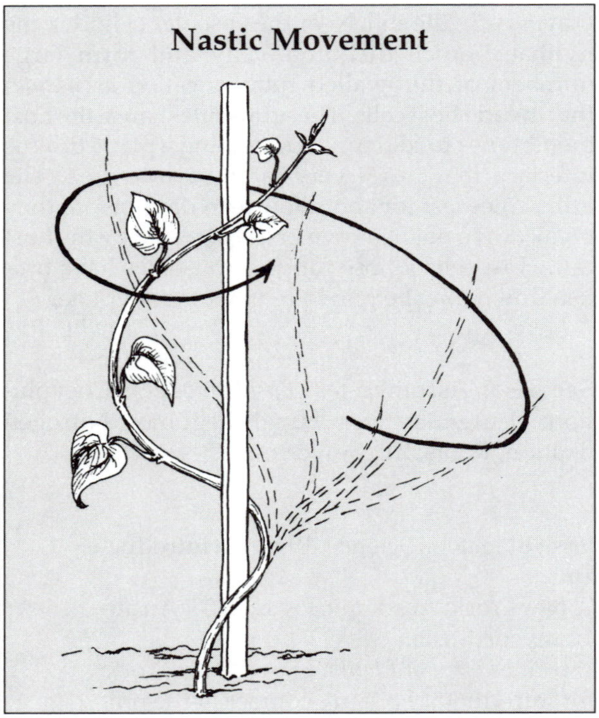

Kimberly L. Dawson Kurnizki

Nastic movements are controlled by hormones that result in cell growth across stem tips at changing positions. When a twining plant grows, such alterations in growth push the stem tip in different directions. When the stem touches an obstacle such as a stake or other support structure, it will spiral around that structure. Such movement, driven by hormones, is not the same as a tropism, which is a response to external stimuli.

Nastic movements are common in certain plant families, especially among the legumes (family *Fabaceae*, formerly *Leguminosae*). In legumes having leaves that are composed of many leaflets—that is, compound—both the leaflets and the leaves may exhibit the movements. The most widely occurring nastic movements are probably day-and-night movements, known as *nyctinastic movements*. Another important kind is *thigmonastic movements*, triggered by touch or other mechanical stimuli.

Nyctinastic Movements

The leaves of many plants respond to the daily alternation between light and darkness by moving up and down. In these nyctinastic, or sleep, movements, the leaves extend horizontally (open) to intercept sunlight during the day and fold together vertically (close) at night. Leguminous plants exhibiting nyctinastic movements include the sensitive plant (*Mimosa pudica*) and the silk tree (*Albizzia julibrissin*). The movements also occur in some species of oxalis (family *Oxalidaceae*). In the prayer plant, maranta species (family *Marantaceae*), the leaves, which are simple, fold at night into a vertical configuration that suggests praying hands.

Thigmonastic Movements

Mechanical disturbances that may trigger thigmonastic movements include touch, shaking, or electrical or thermal stimulation. The stimulus is transmitted from touch-sensitive cells to responding cells located elsewhere in the plant. Many of the plants that display thigmonastic movements are ones that also exhibit nyctinastic ones, such as some members of the *Fabaceae* and *Oxalidaceae*.

The sensitive plant exhibits pronounced thigmonastic movements. If even a single leaflet is touched, it and the other leaflets of the leaf fold upward in pairs until their surfaces touch. The signal moves down the leaf stalk, or petiole, which droops, and then proceeds to the rest of the shoot. If the plant is allowed to rest, the leaves return to their original orientation in fifteen to twenty minutes. The adaptive significance of thigmonastic movements to most plants is not well understood. Some evidence suggests that the movements may scare off leaf-eating insects.

Many plants require a stronger stimulus than does the sensitive plant in order for a response to be generated. A notable exception is Venus's flytrap (*Dionaea muscipula*, in the family *Droseraceae*). The leaves of this carnivorous plant respond in a rapid, highly specialized way, and the purpose is predatory, not defensive. An insect alighting on a leaf, which has two lobes, stimulates sensitive "trigger" hairs on the leaf epidermis, or surface layer of cells. Within about a half-second, the two lobes of the leaf snap shut. Enzymes digest the insect in one to several days, and the empty trap then reopens.

Physiological Mechanisms

The "motor" that drives most nastic movements is a controlled change in turgor pressure, which is the pressure exerted on a cell wall due to movement of water into the cell. In the sensitive plant and many other thigmonastically responsive plants, these turgor changes occur in certain large, thin-walled cells that function as motor cells, located at the bases of the leaf blades or petioles, and leaflets, if present. The motor cells surround a central strand of vascular, or conducting, tissue, forming a jointlike thickening called a pulvinus.

Movement occurs when there are differential turgor changes in the thin-walled cells on opposite sides of a pulvinus, resulting in differential contraction and expansion of these cells. The turgor changes are triggered when potassium ions (K^+) and other ions move into or out of the cells, and water follows by osmosis (as in the opening and closing of stomata in the leaf epidermis). Cells on one side of the pulvinus function as extensors, and cells on the opposite side function as flexors. Leaves or leaflets open when extensor cells accumulate K+ and then swell with water and flexor cells lose K+ and then shrink from loss of water. Conversely, the leaf or leaflet closes when the extensor cells shrink and the flexor cells swell.

The mechanisms causing these ion fluxes vary. In nyctinastic movements, the fluxes and the pulvinal turgor changes that they trigger are rhythmic and are regulated by interactions between light and the plant's innate biological clock. Phytochrome, a plant pigment involved in many timing processes, including flowering, plays a role in this regulation. In thigmonastic movements, the touching of a leaf or leaflet generates an electrical signal called an action potential. This signal typically moves along the stalk of the leaflet and leaf. It is then translated into a chemical signal that causes the ion fluxes and pulvinal turgor changes.

In Venus's flytrap, touching of the hairs on a leaf surface generates an action potential. There are no pulvini to respond, however. The underlying biochemical mechanisms of trap closure are not well understood. They may involve turgor changes in a layer of photosynthetic cells immediately beneath the leaf's upper epidermis.

Jane F. Hill

See also: Carnivorous plants; Circadian rhythms; Leaf anatomy; Osmosis, simple diffusion, and facilitated diffusion; Photoperiodism; Thigmomorphogenesis; Tropisms; Water and solute movement in plants.

Sources for Further Study

Attenborough, David. *The Private Life of Plants*. Princeton, N.J.: Princeton University Press, 1995. Written for the layperson. Discusses, with photos, thigmonasty and prey capture by Venus's flytrap. Includes index.

Moore, Randy, et al. *Botany*. Dubuque, Iowa: Wm. C. Brown, 1995. A comprehensive, introductory college textbook. Includes a chapter on plant responses to environmental stimuli, with a section on nastic movements. Includes references, color photos, diagrams, glossary, tables, and index.

Raven, Peter H., Ray F. Evert, and Susan E. Eichhorn. *Biology of Plants*. 6th ed. New York: W. H. Freeman/Worth, 1999. A comprehensive, introductory college-level textbook.

Contains a chapter on external factors and plant growth, with a section on nastic movements. Includes color photos, diagrams, index, and bibliography.
Salisbury, Frank B., and Cleon W. Ross. *Plant Physiology*. 4th ed. Belmont, Calif.: Wadsworth, 1992. College-level textbook, with a chapter on plant movements. Includes black-and-white illustrations, references, and indexes.

NITROGEN CYCLE

Categories: Biogeochemical cycles; cellular biology; ecology; nutrients and nutrition

The nitrogen cycle outlines the movement of the element nitrogen from one chemical state to another as it makes its way through a series of complex physical and biological interactions.

Nitrogen (N) is one of the most dynamic elements in the earth's biosphere; it undergoes transformations that constantly convert it between organic, inorganic, gaseous, and mineral forms. Nitrogen is an essential element in all living things, where it is a crucial component of organic molecules such as proteins and nucleic acids. Consequently, nitrogen is in high demand in biological systems.

Unfortunately, most nitrogen is not readily available to plants and animals. Although the biosphere contains 300,000 terrograms (a terrogram is a billion kilograms) of nitrogen, that amount is one hundred times less nitrogen than is in the hydrosphere (23 million terrograms) and ten thousand times less nitrogen than is in the atmosphere (about 4 billion terrograms). Atmospheric nitrogen is almost all in the form of nitrogen gas (N_2), which composes 78 percent of the atmosphere by volume. The greatest reservoir of nitrogen on the earth is the lithosphere (164 billion terrograms). Here the nitrogen is bound up in rocks, minerals, and deep ocean sediments.

Even though living things exist in a "sea" of nitrogen gas, it does them little good. The bond between the nitrogen atoms is so strong that nitrogen gas is relatively inert. For living things to use nitrogen gas, it must first be converted to an organic or inorganic form. The nitrogen cycle is the collection of processes, most of them driven by microbial activity, that converts nitrogen gas into these usable forms and later returns nitrogen gas back to the atmosphere. It is considered a cycle because every nitrogen atom can ultimately be converted by each

AP/Wide World Photos

Nitrogen-fixing microbes can grow in association with higher organisms such as legumes, including these soybeans, in a process called symbiotic nitrogen fixation. This process is one reason legumes are highly valued as a natural resource. Because they are able to form symbiotic associations with nitrogen-fixing bacteria, legumes can produce seeds and leaves with more nitrogen than those of other plants. When they die, they return much of that nitrogen to the soil, enriching it for future growth.

process, though that conversion may take a long time. It is estimated, for example, that the average nitrogen molecule spends 625 years in the biosphere before returning to the atmosphere to complete the cycle.

Nitrogen Fixation

The first step in the nitrogen cycle is *nitrogen fixation*. Nitrogen fixation is the conversion, by bacteria, of nitrogen gas into ammonium (NH_4^+) and then organic nitrogen (proteins, nucleic acids, and other nitrogen-containing compounds). It is estimated that biological nitrogen fixation adds about 160 billion kilograms of nitrogen to the biosphere each year. This represents about half of the nitrogen taken up by plants and animals. The microorganisms that carry out nitrogen fixation are highly specialized. Each one carries a special enzyme complex, called *nitrogenase*, that allows it to carry out fixation at temperatures and pressures capable of permitting life—something industrial nitrogen fixation does not allow.

Nitrogen-fixing microbes, may be either free-living or growing in association with higher organisms such as legumes (in which case the process is called *symbiotic nitrogen fixation*). Symbiotic nitrogen fixation is a very important process and is one reason legumes are so highly valued as a natural resource. Because they are able to form these symbiotic associations with nitrogen-fixing bacteria, legumes can produce seeds and leaves with more nitrogen than other plants. When they die, they return much of that nitrogen to soil, enriching it for future growth.

Mineralization and Nitrification

When plants and animals die they undergo a process called *mineralization* (also called *ammonification*). In this stage of the nitrogen cycle, the organic nitrogen in decomposing tissue is converted back into ammonium. Some of the ammonium is taken up by plants as they grow. This process is called *assimilation* or *uptake*. Some of the ammonium is taken up by microbes in the soil. In this case the nitrogen is not available for plant growth. If this happens, it is said that the nitrogen is immobilized. Some nitrogen is also incorporated into the clay minerals of soil. In this case it is said that the nitrogen is fixed—it is not immediately available for plant and microbial growth, but it may become available at a later date.

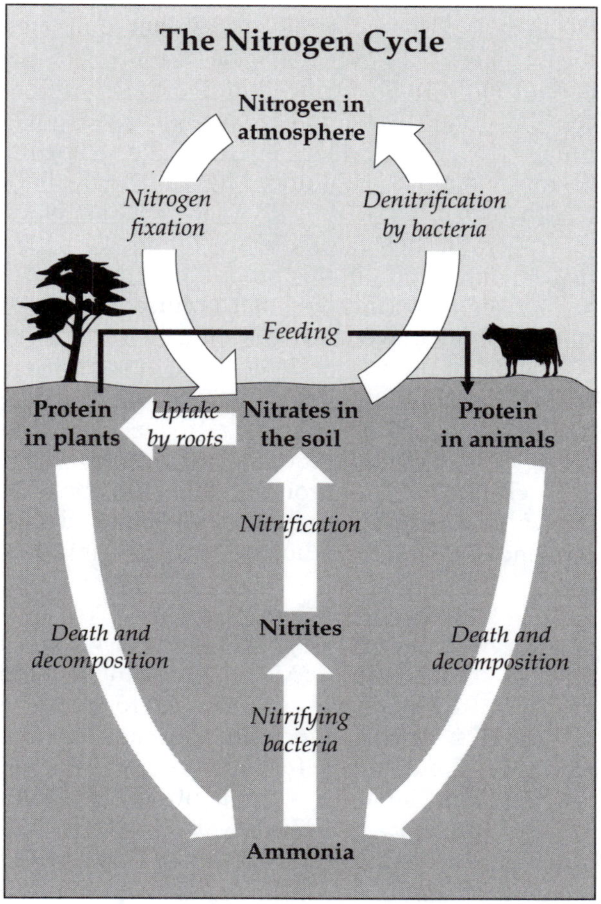

Ammonium has another potential fate, and this step in the nitrogen cycle is *nitrification*. In nitrification the ammonium in soil is oxidized by bacteria (and some fungi) to nitrate (NO_3^-) in a two-step process. First, ammonium is oxidized to nitrite. Next, nitrite is rapidly oxidized to nitrate. Nitrification requires oxygen, so it occurs only in well-aerated environments. The nitrate that forms during nitrification can also be taken up by plants and microbes. However, unlike ammonium, which is a cation (positively charged ion) and readily adsorbed by soil, nitrate is an anion (negatively charged ion) and readily leaches or runs off of soil. Hence, nitrate is a serious water contaminant in areas where excessive fertilization or manure application occurs.

Denitrification

Obviously some process is responsible for returning nitrogen to the atmosphere; otherwise organic and inorganic nitrogen forms would accumulate in the environment. The process that completes the nitrogen cycle and replenishes the nitrogen gas is *denitrification*. Denitrification is a bacterial process that occurs in anaerobic (oxygen-limited) environments such as waterlogged soil or sediment. Nitrate and nitrite are reduced by denitrifying bacteria, which can use these nitrogen oxides in place of oxygen for their metabolism. Wetlands are particularly important in this process because at least half of the denitrification that occurs in the biosphere occurs in wetlands.

The major product of denitrification is nitrogen gas, which returns to the atmosphere and approximately balances the amount of nitrogen gas that is biologically fixed each year. In some cases, however, an intermediate gas, nitrous oxide (N_2O), accumulates. Nitrous oxide has serious environmental consequences. Like carbon dioxide, it absorbs infrared radiation, so it contributes to global warming. More important, when nitrous oxide rises to the stratosphere, it contributes to the catalytic destruction of the ozone layer. Besides the potential for fertilizer nitrogen to contribute to nitrate contamination of groundwater, there is the concern that some of it can be denitrified and contribute to ozone destruction.

The nitrogen cycle is a global cycle involving land, sea, and air. It circulates nitrogen through various forms that contribute to life on earth. When the cycle is disturbed—as when an area is deforested and nitrogen uptake into trees is stopped, or when excessive fertilization is used—nitrogen can become an environmental problem.

Mark S. Coyne

See also: Nitrogen fixation; Nutrient cycling; Nutrients.

Sources for Further Study

Anderson, Stanley H., Ronald Beiswenger, and P. Walton Purdom. *Environmental Science*. 4th ed. New York: Maxwell Macmillan International, 1993. Probably the most readable introductory-level text, intended for a college course in environmental science. Discusses ecosystems and includes sections on the nitrogen cycle, ammonification, and nitrogen fixation. Suitable for the average reader; black-and-white illustrations, a glossary of terms, and an index. A good introduction to the subject.

Burt, T. P., A. L. Heathwaite, and S. T. Trudgill, eds. *Nitrate: Processes, Patterns, and Management*. Chichester, N.Y.: John Wiley, 1993. Explores nitrate transport patterns and considers various strategies for nitrate pollution, including legislative, land use, and water treatment policies.

Clark, F. E., and T. Rosswall, eds. *Terrestrial Nitrogen Cycles: Processes, Ecosystem Strategies, and Management Impacts*. Stockholm: Swedish Natural Science Research Council, 1981. A collection of technical papers that is very readable with clear and concise abstracts to head each article. Extensive coverage of the terrestrial nitrogen cycle with excellent references and index for further investigation into this complex topic. Suitable for the college-level reader or serious layperson who wishes to read about how the nitrogen cycle is being studied by modern environmental scientists.

Goodwin T. W., and R. M. S. Smeille, eds. *Nitrogen Metabolism in Plants*. London: Biochemical Society, 1973. A collection of papers discussing the role of nitrogen in plant metabolism, with extensive coverage of biochemical reactions within the cell. The text is technical and suited to a college-level audience. Well indexed to locate specific areas of interest. A good volume for those interested in the biological processes involved in the nitrogen cycle.

Krebs, Charles J. *Ecology: The Experimental Analysis of Distribution and Abundance*. 5th ed. San Francisco: Benjamin Cummings, 2001. Extensive coverage of the biogeochemical cycles in chapter dealing with nutrient cycling. Referenced, with an index including many index entries on nitrogen; the nutrient cycle is discussed throughout the book. This is an excellent college-level text with some mathematical models of nutrient cycles and algae growth. Numerous illustrations, figures, graphs, and diagrams.

Lasserre, P., and J. M. Martin, eds. *Biogeochemical Processes at the Land-Sea Boundary*. New York: Elsevier, 1986. A collection of very accessible technical papers dealing with nutrient cycling interactions at the continental-ocean boundary. Contains excellent discussions of ecosystem and nutrient-cycle modeling. Suitable for the advanced student or college-level reader with an interest in land-sea interactions. Unfortunately, the book contains no index, but it is organized to facilitate finding topics of interest.

Mlot, Christine. "Tallying Nitrogen's Increasing Impact." *Science News* 151, no. 7 (February 15, 1997): 100. Discusses research findings that the rate at which nitrogen is made available to the land has risen to the point that humans, through the use of commercial fertilizer, global cultivation of legumes, and clearing of wildlands, have become the major force in the nitrogen cycle.

Tamm, Carl Olof. *Nitrogen in Terrestrial Ecosystems: Questions of Productivity, Vegetational Changes, and Ecosystem Stability.* New York: Springer-Verlag, 1991. Local and global changes in nitrogen are examined, with emphasis on changes that are a result of shifts in land use and automobile and industrial pollution, found to be damaging to plant life.

NITROGEN FIXATION

Categories: Cellular biology; nutrients and nutrition; physiology

Nitrogen fixation is the process whereby elemental nitrogen from the atmosphere is converted to ammonium, an ionic form of nitrogen available to higher plants.

Nitrogen, the fourth most abundant element in most organisms, can account for as much as 4 percent of a plant's dry weight. The majority of this nitrogen is present as a constituent of protein structure, but it is also a component of numerous other biological compounds, such as the chlorophylls and nucleic acids. Thus, for normal plant growth and development, nitrogen must be maintained at fairly high levels in the soil.

The earth's atmosphere is about 79 percent nitrogen, and the vast majority of atmospheric nitrogen exists in the elemental state. Unfortunately, the elemental form of nitrogen is of no direct value to higher plants; they must acquire their nitrogen in the form of either ammonium or nitrate. These two forms of nitrogen can be supplied to the soil as fertilizer by humans or by nature, as the product of microbial action.

There are three microbial processes that render nitrogen into forms usable by higher plants. These are *ammonification*, *nitrification*, and *nitrogen fixation*. Ammonification is the process whereby various forms of organic nitrogen, such as is present in the proteins in plant and animal residues and animal wastes (manures), are converted to ammonium. Nitrification is the process by which ammonium is converted to nitrate. Both these processes are carried out by populations of free-living soil microorganisms.

In the nitrogen fixation process, atmospheric nitrogen is converted to ammonium. While some of the nitrogen fixers are free-living microbes, bacteria that live symbiotically within the roots of a number of plant species are also responsible for much of this conversion in terrestrial ecosystems.

Nitrogen-Fixing Bacteria

Nitrogen-fixing bacteria have been shown to coexist with a variety of lower plants, including lichens, liverworts, mosses, and ferns. Among the more advanced seed-bearing plants, nitrogen fixers have been found to be associated with some tropical grasses and a number of shrubs and trees, such as the alders. Agriculturally, the legumes are the most important group of plants coexisting with nitrogen-fixing bacteria. Some fifteen hundred species of legumes, including peas, beans, clover, and alfalfa, have been shown to live symbiotically with nitrogen-fixing bacteria called *Rhizobium*. A different species of *Rhizobium* infects each different species of legume.

The bacteria penetrate the root by entering the filamentous projections of epidermal cells called *root hairs*. The epidermal cells respond to the inva-

Nitrogen-fixing bacteria can coexist with a variety of plants. Some fifteen hundred species of legumes, such as these beans shown planted together with corn, can live symbiotically with nitrogen-fixing bacteria called Rhizobium. *These bacteria are responsible for the biochemical conversion of elemental nitrogen to ammonium, which is used by the legumes and by other plants that are growing nearby.*

sion by enclosing the bacterium in a threadlike structure referred to as an *infection thread*. The infection thread begins to grow and branch, and, as it does so, the Rhizobia reproduce numerous times inside the thread. After penetrating several layers of cells, the infection thread eventually reaches the *root cortex*, where it ruptures and releases the encased bacteria.

The release of the bacteria induces the secretion of plant hormones that stimulate specialized root cortical cells to divide several times. As these cells divide, the Rhizobia are encapsulated, and a nodule is formed. Within the *cytosol* of the nodule cells, the bacteria become nonmotile, increase in size, and accumulate in groups of four to eight *bacteroids*. These bacteroids are responsible for the biochemical conversion of elemental nitrogen to ammonium.

Chemistry of Nitrogen Fixation

Chemically, the fixation of nitrogen requires that six electrons and eight hydrogen ions be transferred to the atmospheric nitrogen molecule. This reaction is an energy-requiring process; therefore, *adenosine triphosphate* (ATP), the cell's form of stored energy, must be available for the reaction to take place. The electrons, hydrogen ions, and ATP are supplied by the cellular respiration process that takes place in the root cells.

The electrons and hydrogen ions are transferred to the atmospheric nitrogen atom by an enzyme called *nitrogenase*. This enzyme consists of two subunits. One subunit takes the electrons and hydrogen ions from the respiratory products and transfers them to the other subunit. The ATP binds with part II of the nitrogenase, and the hydrolysis of the ATP releases the energy stored in the molecule.

This energy drives the reaction and makes it even easier to pass the electrons and hydrogen ions on to the second subunit. In the last step, the nitrogenase transfers the electrons and hydrogen ions to the nitrogen atom. This final transfer results in the production of ammonium. The ammonium moves out of the bacteroids into the cytosol, where it is converted to an organic form of nitrogen that can be transported throughout the plant.

Rates of Fixation

Not all species fix nitrogen at the same rate. A number of factors can account for these differences. Some plants form nodules much more abundantly than others. Because of their more extensive nodule formation, these plants will fix more nitrogen than those that produce fewer nodules.

The nitrogenase of all rhizobial species has a tendency to transfer electrons to hydrogen ions rather than to nitrogen. As a result, hydrogen gas, which escapes into the atmosphere, is produced. This represents a loss of electrons that could have been used to produce ammonium. Some Rhizobia species, however, have a second enzyme, called hydrogenase, which uses the hydrogen gas to produce water. ATP is produced as a byproduct of this process. Consequently, these rhizobial species are more efficient because less energy is wasted.

In addition, the fixation rate and the amount of nitrogen fixed will vary with age or the stage of plant development. In most cases, fixation rates are highest when the fruits and seeds are being produced. The seeds of many plants, and especially legumes, are high in protein. Hence, nitrogen fixation and transport out of the nodules must be higher at the time the seeds are developing. In fact, more than 85 percent of the total nitrogen fixation in legumes occurs at such times.

Plant-Bacteria Mutualism

The nitrogen-fixing bacteria exist in a symbiotic relationship with their plant hosts. The bacteria supply the plant with much-needed nitrogen, while the plant supplies the bacteria with carbohydrates and other nutrients. Some of the energy, derived from the plant-supplied carbohydrates, is used in nitrogen fixation, but there is ample left over to supply the bacteria with all the energy necessary for their survival.

Providing Nitrogen

Plant production throughout the world is limited more by the shortage of nitrogen than by any other nutrient. The root zone that encompasses the upper 15-centimeter layer of soil contains from 100 to 6,000 kilograms of total nitrogen per hectare. This includes all forms of nitrogen, many of which are not available to plants. The total nitrogen content is determined by a number of factors, such as the minerals making up the soil, the kinds of vegetation, and the extent to which these factors are affected over time by climate, topography, and the presence of people.

Ammonium and nitrate are the only forms readily available to plants. These two molecules, in addition to those such as organic nitrogen compounds that can easily be converted to the available forms, are the only ones of ecological or agricultural importance. Unfortunately, these forms of nitrogen are continually being removed from the soil. Crop removal, *leaching* (the removal of minerals as water percolates through the soil), *denitrification* (the process by which anaerobic microbes convert nitrates to gaseous nitrogen-containing compounds that escape the soil), and erosion account for a total loss of approximately 125 kilograms per hectare annually. While some of the lost nitrogen can be replaced by available forms falling to the earth in rain, that amount is much too low to be of value in plant growth. Microbial fixation and the application of fertilizers are the only sources that supply sufficient nitrogen for plant growth.

Research Applications

The nonsymbiotic nitrogen fixers are of extreme ecological importance, especially in nonagricultural soils. Forest, desert, and prairie ecosystems are dependent on nitrogen fixation by free-living species to replace the annual nitrogen loss. Without it, growth of a number of plant species would suffer, and food chains in these ecosystems would soon be disrupted. A number of studies have investigated the advantages of incorporating free-living nitrogen fixers into nonlegume crop production, but clear benefits are uncertain.

On the other hand, knowledge of symbiotic nitrogen fixation have resulted in definite improvements in the production of legume crops. When *Rhizobium* is included with seeds as they are planted, increased yields have been observed in nearly every case. There is considerable interest in enhancing the efficiency of the nitrogen-fixing process by increasing nodulation in the roots of some species or by incorporating the hydrogenase system into species that do not have it. An increase in biological nitrogen fixation could enhance the nitrogen content of the soil and decrease the dependency on commercial nitrogen fertilizers.

The application of nitrogen fertilizers to nonleguminous crops was once one of the best investments a farmer could make. Commercial fertilizers, however, have become very expensive because of increased energy costs. In addition, a number of environmental problems have developed from the accumulation of nitrates from fertilizers in rivers, ponds, and lakes. Consequently, there is a renewed interest in symbiotic nitrogen fixation. For years, before the extensive use of nitrogen fertilizers, farmers planted legume crops alternately with other crops. The legume crops were plowed under to supply nitrogen to the soil. There has been a resurgence in this technique. In fact, there is considerable interest in growing plants containing symbiotic nitrogen fixers in the same fields with plants lacking symbiotic nitrogen fixers, to improve the natural nitrogen balance in certain environments. There are a number of studies directed at incorporating nitrogen fixation into nonlegume crop plants; this research is difficult, however, because the genetics of the process is so complex.

D. R. Gossett

See also: ATP and other energetic molecules; Eutrophication; Fertilizers; Lichens; Mycorrhizae; Nitrogen cycle; Nutrients.

Sources for Further Study

Legocki, Andrezej, Hermann Bothe, and Alfred Pühler, eds. *Biological Fixation of Nitrogen for Ecology and Sustainable Agriculture*. New York: Springer, 1997. From the proceedings of the NATO Advanced Research Workshop held in Poznań, Poland, in 1996. Includes bibliographical references.

Salisbury, Frank B., and Cleon W. Ross. *Plant Physiology*. 4th ed. Belmont, Calif.: Wadsworth, 1992. An intermediate college-level textbook for science students. Chapter 13, "Assimilation of Nitrogen and Sulfur," gives an in-depth view of the role of nitrogen fixation in plant physiology. An excellent discussion of the biochemistry of the fixation reaction is provided. Includes detailed bibliography.

Stacey, Gary, Robert H. Burris, and Harold J. Evans, eds. *Biological Nitrogen Fixation*. New York: Chapman and Hall, 1992. Leading researchers contribute reviews on all aspects of nitrogen fixation, including molecular biology and genetics, physiology, and ecology.

Tortora, Gerard J., Berdell R. Funke, and Christine L. Case. *Microbiology: An Introduction*. 7th ed. San Francisco: Benjamin/Cummings, 2001. An introductory college textbook for science students. The chapter "Soil and Water Microbiology" provides a good general discussion of the process from the microbiologist's point of view. A very readable, easy-to-understand text with sufficient illustrations to explain the process. Includes list of suggested readings; glossary.

NONRANDOM MATING

Categories: Genetics; reproduction and life cycles

Nonrandom mating occurs in plants when there is inbreeding or assortative mating. The consequence of nonrandom mating is that population genotype frequencies do not exist in Hardy-Weinberg equilibrium. Although nonrandom mating does not directly drive evolutionary change—meaning allele frequencies are not altered—it can have a profound effect on genotype frequencies and hence can indirectly affect evolution.

There are three main ways for nonrandom mating to occur. First, *positive assortative mating* results when individuals and their mates share one or more phenotypic characteristics with themselves. *Negative assortative mating* occurs when individuals and their mates are dissimilar phenotypically.

Inbreeding is a third form of nonrandom mating that occurs when individuals mate with relatives more often than would be expected by chance. For both inbreeding and assortative mating, genes combine in such a way that offspring genotypes differ from those that are predicted by the most basic population genetic model, described by the *Hardy-Weinberg theorem*. One assumption of the Hardy-Weinberg theorem, which predicts unchanging equilibrium values for genotype and allele frequencies, is that individuals mate at random. Unlike other violations of this model (such as natural selection, genetic drift, mutation), nonrandom mating affects genotype but not allele frequencies.

Inbreeding

Inbreeding is very common in many plant species for two main reasons. First, seed dispersal tends to follow a leptokurtic distribution, such that most seed falls near the parent plant. This results in near neighbors that are closely related and increases the probability that short-distance pollen movement will result in mating among relatives. In small populations with a limited number of potential mates, such matings between relatives are also common. Second, most flowering plants are hermaphroditic or monoecious. Thus, individual plants produce both male and female gametes and are capable of self-fertilization, the most extreme form of inbreeding.

The degree to which inbreeding occurs in a population depends upon the probability that an individual will mate with a relative or with itself. A plant's mating system is characterized by the degree to which *self-fertilization* occurs and can range from complete outcrossing to complete self-fertilization, or selfing. While certain plant families tend to be characterized by a particular mating system (such as the inability to self-fertilize in the passionflower family, *Passifloraceae*), others exhibit great diversity in the levels of inbreeding among species (the grasses, *Poaceae*, and the legumes, *Fabaceae*).

For species with a mixed mating system, and which therefore engage in both selfing and outcrossing, the degree to which individual offspring are inbred is highly variable. Flowers with multiple ovules within an ovary can produce fruits with both selfed and outcrossed seeds. Some plants, such as violets (*Viola*) and jewelweed (*Impatiens*), produce morphologically distinct flowers for selfing and outcrossing.

Consequences of Inbreeding

Inbreeding has a larger evolutionary impact than assortative mating because it can affect all genes in the population. It can have negative consequences for plant survival and reproduction (fitness) because it tends to increase homozygosity and decrease heterozygosity. In response to these negative effects, collectively known as *inbreeding depression*, plants have evolved numerous adaptations to reduce levels of inbreeding. Although evidence for inbreeding depression in plants has been found, many species of plants are almost completely self-fertilizing and do not appear to suffer fitness consequences.

Under certain conditions, inbreeding may be advantageous. For example, rare plants or plants with rare pollinators may have few opportunities for outcrossing, and thus, self-fertilization provides a level of reproductive assurance. Many weedy plant species that tend to colonize disturbed sites are, in fact, capable of self-fertilization. Common crop weeds such as velvetleaf (*Abutilon theophrasti*) and shepherd's purse (*Capsella bursa-pastoris*) predominantly self-fertilize. In these species, inbreeding may provide benefits that outweigh any associated costs. It has also been suggested that inbreeding species are better able to adapt to local environmental conditions because fewer maladapted genes from other populations would enter through outcrossing.

Reducing Inbreeding Depression

Inbreeding depression (which occurs when alleles that decrease fitness drift to fixation, causing a decrease in average fitness within a population) is reduced when plants are genetically or morphologically unable to self-fertilize. Genetic self-incompatibility, which is thought to occur in more than forty different plant families (for example, *Brassicaceae*, *Solanaceae*, and *Asclepiadaceae*) prevents mating between individuals that share certain genes that are involved in the interaction between pollen grains and the stigma or style. Morphological adaptations that reduce self-fertilization include those that separate anther and stigma maturation in time (protandry and protogyny) or space (heterostyly).

The individual hermaphroditic flowers of protandrous plants, such as phlox, shed their pollen prior to the time when the stigma on the same flower is receptive. Protogyny, which is less common than protandry, occurs when stigma receptivity occurs first (as in *Plantago lanceolata*). Dioecious plant species, such as date palms (*Phoenix*) and marijuana (*Cannabis sativa*), avoid selfing by having unisexual male and female flowers on separate individuals. In some hermaphroditic species, selfing is avoided when flower morphology favors crosses between certain flower phenotypes, as in the case of heterostyly.

Assortative Mating

Assortative mating generally affects only those traits important for reproduction. Many primrose (*Primula*) species are distylous, having two types of flowers. Flowers with the pin morphology have a tall style and relatively short stamens, while flowers with thrum morphology have a short style and long stamens. Insect-mediated pollen transfer results in matings between pin and thrum but not between two pin or two thrum plants. The result is nonrandom negative assortative mating. Unlike inbreeding, negative assortative mating tends to increase the level of heterozygosity in a population, at least for those traits that are involved in mate choice (such as relative style length).

Positive assortative mating, like inbreeding, results in increased homozygosity and decreased heterozygosity. Positive assortative mating for flowering time, for example, is common in many plant populations because individuals that flower early in the season will tend to mate with other early-flowering individuals.

Cindy Bennington

See also: Angiosperm life cycle; Flower types; Genetics: mutations; Genetics: post-Mendelian; Hardy-Weinberg theorem; Hybridization; Pollination; Population genetics; Reproduction in plants; Species and speciation.

Sources for Further Study

Attenborough, D. *The Private Life of Plants*. Princeton, N.J.: Princeton University Press, 1995. Describes self-fertilization and plant adaptations that promote outcrossing. Includes color photographs.

Raven, Peter H., Ray F. Evert, and Susan E. Eichhorn. *Biology of Plants*. 6th ed. New York: W. H. Freeman/Worth, 1999. This textbook describes effects of inbreeding and mechanisms plants employ to promote outbreeding.

Willson, M. F. *Plant Reproductive Ecology*. New York: John Wiley and Sons, 1983. Coverage of the evolution of sex and mating systems in plants. Topics include inbreeding, heterostyly, dioecy, and self-incompatibility.

NORTH AMERICAN AGRICULTURE

Categories: Agriculture; economic botany and plant uses; food; world regions

North American crops include grains, legumes, fruits, vegetables, and plants for clothing and other nonfood uses. A discussion of modern agriculture in North America cannot be complete without some attention to agribusiness, the system of businesses associated with agricultural production in an industrialized society.

In North America, agriculture generally has become mechanized and heavily dependent upon an integrated system of supporting *agribusinesses*, although traditional practices continue in Mexico. In the United States and Canada, most farmers and ranchers depend heavily upon technology, although groups such as the Amish have rejected automation and continue to use animal power for traction. Most farmers practice *monoculture*, relying upon a single crop for their primary income, and have expanded to very large acreages in order to take advantage of economies of scale. Such farms are referred to in terms of the primary crop, for example, a dairy farm, a cattle ranch, or a wheat farm. Some small farms are run by part-time farmers who also have other occupations.

Selected Agricultural Products of North America

Within the United States, there were 2,192,000 farms, cultivating a total of 954,000,000 acres, in the 1990's. These farms produced net returns of $44.1 billion. Although farmers represented less than 2 percent of the U.S. population, they successfully fed the country at a high standard of living, produced grain and other products for export, and still maintained a surplus carryover of as much as 2 percent of the total grown.

Regional Crops and Cultivation

Modern farming techniques in the United States and Canada require specialization in a single *cash crop*. Such specialized farms tend to cluster by region, where the climate and soil quality are appropriate to a given crop. The supporting agribusinesses—such as suppliers of implements, chemical *fertilizers* and *pesticides*, and grain elevators—tend to specialize in products and activities that support the primary crops of their given area.

Wheat, the most important cereal grain in Western diets, grows in the broad, open lands of the Great Plains, in Kansas, Nebraska, North and South Dakota, and the Canadian provinces of Alberta and Saskatchewan.

In the southern part of this region, the primary crop is winter wheat, which is planted in the fall, is dormant during the winter, completes its growth in spring, and is harvested in midsummer. In many of these areas, a farmer then can plant a crop of soybeans, a practice known as *double-cropping*. The soybeans often can be harvested in time to plant the following year's wheat crop in the fall. Farther to the north, where the weather is too harsh for wheat to survive the winter, farmers plant spring wheat, which completes its entire growth during the spring and summer and is harvested in the fall. Wheat is used to make bread, pasta, and many breakfast cereals and is an ingredient in numerous other products.

Corn, which originally was domesticated by American Indians, is the best producer per acre. It requires a longer growing season than wheat, so areas where it can be grown economically are limited. The Midwestern states—Iowa, Illinois, Indiana, Ohio—are the principal areas for cultivation of corn and frequently are referred to as the Corn Belt states. Much corn is used as livestock feed, although a considerable amount is processed into human foods as well, often in the form of cornstarch and corn-syrup sweeteners.

Rice requires flooded fields for successful cultivation, so it can be grown only in areas such as Louisiana, where large amounts of water are readily available. Because labor costs are the primary limiter in U.S. agriculture, American rice growers use highly mechanized, single-field growing techniques rather than the labor-intensive transplantation technique used in Asian countries. Laser levels and computerized controls tied into the Global Positioning System enable farmers to prepare smooth fields with a slight slope for efficient flooding and drainage. Because the ground is usually wet during tilling and harvesting, the machinery typically used in growing rice is fitted with tracks instead of wheels to reduce soil compaction.

Rye, oats, and barley are other major grain crops, although none form the backbone of an area's economy to the extent that wheat, corn, and rice do. Oats, once a staple feed grain for horses, now is used mainly for breakfast cereals, while most barley is malted for brewing beer. Rye typically is used in the production of specialty breads.

Legumes, such as soybeans and alfalfa, form the next major group of crops produced in North America. In addition to being an important source of protein in human and livestock diets, legumes are important in maintaining soil fertility. Nodules on their roots contain bacteria that help to transform nitrogen in the soil into compounds that plants can use. Because of this, soybeans have also become a regular *rotation crop* with corn in much of the U.S. Midwest.

Both corn and soybeans can be grown with the same machinery and sold to the same markets, although harvesting corn requires a specialized cornheader that pulls down the stalks and breaks loose the cob on which the corn kernels grow, rather than the generalized grain platform used with soybeans and small grains. Soybeans for human consumption generally are heavily processed and become filler in other foods, although there is a market for tofu (bean curd) and other soybean products.

Other crops include edible oil seeds, such as sunflower seeds and safflower seeds, which are generally grown as rotation crops with corn or wheat. Sugarcane is grown in Louisiana and other areas on the coast of the Gulf of Mexico that have the necessary subtropical climate. Many varieties of fruits and vegetables are grown in California's irrigated valleys; Florida grows much of the United States' juice oranges. Other citrus crops are grown in Alabama, Mississippi, and Texas, where these warmth-loving trees will not be damaged by frost. Fruits such as apples and pears, which require a cold period to break dormancy and set fruit, are grown in northern states such as Michigan and Washington and the eastern provinces of Canada.

In the United States and Canada, most farmers and ranchers depend heavily upon technology and mechanized equipment, such as this combine.

Fibrous Plants

In addition to food plants, the production of fibers for textiles is an important part of American agriculture, although such artificial fibers as nylon and polyester have taken a share of the market. Cotton and flax also are important sources of natural fibers. Cotton requires a long growing season and relatively high rainfall levels; therefore, it generally is grown in an area in the southern United States often referred to as the Cotton Belt. Flax has a shorter growing season and is often planted in rotation with such small grains as wheat and oats. Flax stems are used to produce linen, and edible oils and meal are obtained from its seeds.

Trees have become a cultivated species, although their long growth cycle has limited the ability of humans to create particular varieties. During the twentieth century, concerns about the environmental damage done by the *clear-cutting* of virgin forests for lumber and paper encouraged many companies to reseed the cut areas with tree species that could be harvested thirty or forty years later. Another form of tree farming, although on a much smaller scale, is the production of small evergreens for Christmas trees.

The Business of Farming

Because of the intense specialization of modern mechanized agriculture, farming has become a business interlocked with a number of supporting businesses. Farm management—the control of capital outlay, production costs, and income—has become as vital to a farmer's economic survival as skill in growing the crops themselves. Such organizations as Farm Business/Farm Management help farmers develop the skills needed to farm more productively and economically.

Farmers also have had to become actively involved in the marketing of their crops to ensure an adequate income. In many areas of the United States and Canada, farmers have banded together in cooperatives to gain economic leverage in buying supplies and selling their products. Some of these cooperatives have taken on some of the preliminary and intermediate steps in transforming the raw farm products into consumer goods, thus increasing the prices farmers receive from buyers.

Modern farmers also are concerned with the management of the resources that support agriculture. In earlier generations, it often was assumed that natural resources were unlimited and could be used and abused without consequence. The result of this ignorance was ecological destruction such as the Dust Bowl of the 1930's, in which the *topsoil* over large areas of U.S. Plains States dried up and blew away in the wind, rendering the land unfarmable. To prevent more disasters and the economic dislocation they produced, various soil conservation measures were introduced through government programs that gave farmers financial incentives to change their practices. The use of contour plowing and terracing on steeply sloping hillsides helped to slow the movement of water that could carry away soil, thus preventing the formation of gulleys. Reduced tillage techniques allowed more plant residue to remain on the surface of the soil, protecting it from the ravages of both wind and water.

Leigh Husband Kimmel

See also: African agriculture; Agriculture: modern problems; Asian agriculture; Australian agriculture; Central American agriculture; Corn; Erosion and erosion control; European agriculture; Grazing and overgrazing; Green Revolution; High-yield crops; Legumes; Monoculture; North American flora; Rice; South American agriculture; Wheat.

Sources for Further Study

Bonnano, Alessandro. *From Columbus to ConAgra: The Globalization of Agriculture and Food*. Lawrence: University Press of Kansas, 1994. Discusses the history of the food industry, agricultural businesses, and international trade.

Danborn, David B. *Born in the Country: A History of Rural America*. Baltimore: Johns Hopkins University Press, 1995. Integrates social, political, and economic history, focusing on farm people rather than residents of small towns. Treats agriculture, rural culture and society, and agrarian policy as an inseparable web.

Fussell, Betty Harper. *Story of Corn*. New York: Alfred A. Knopf, 1992. An account of the history of corn in the New World. Blends history and myth, science and art, anecdote and image, personal narrative and epic to tell the story of this amazing crop and the people who for centuries have planted, eaten, worshiped, processed, and profited from it.

Goreham, Gary, et al., eds. *Encyclopedia of Rural America*. Santa Barbara, Calif.: ABC-Clio, 1997. Covers a broad range of topics, such as agriculture, economics, the environment, health, and political and social science.

Hassanien, Neva. *Changing the Way America Farms: Knowledge and Community in the Sustainable Agriculture Movement*. Omaha: University of Nebraska Press, 1999. Describes two organizations, the Ocooch Grazers Network and the Wisconsin Women's Sustainable Farming Network, to trace how some farmers are developing their own collective local knowledge as a basis for moving toward an ecologically, economically, and socially sound agricultural system.

NORTH AMERICAN FLORA

Category: World regions

The world's major biomes are all represented in the diverse vegetation of North America, from Arctic tundra in the north to deserts in the Southeast and the grasslands, wetlands, and various forest biomes between.

Forest is the native vegetation of almost half of mainland Canada and the United States. Before European settlement, forestlands dominated the eastern, and much of the northern, part of the continent. Grasses covered a large part of the continental interior. Desert vegetation is native in the Southwest, *tundra* in the far north. Over much of the continent, however, human activity has virtually eliminated native vegetation.

Coniferous Forests of the West

Along the Pacific coast from Alaska to Northern California, evergreen *coniferous forest* grows luxuriantly, watered by moisture-laden winds blowing from the ocean. This lowland forest includes some of the largest and longest-lived trees in the world. North of California, characteristic trees include Sitka spruce, western hemlock, and western red cedar. Douglas fir, one of the major timber species in

North America, is also common. The northwest coastal coniferous forest is sometimes called temperate rain forest because, in its lushness, it resembles the tropical rain forests. Many of the trees of the coastal forest have been cut for timber.

In California, the dominant coastal conifer species is the coast redwood. The tallest tree in the world, coast redwood reaches 330 feet (100 meters) and can live two thousand years. Coniferous forest also grows along the Cascade Mountains and the Sierra Nevada. Trees of the Cascades include mountain hemlock and subalpine fir at high elevations and western hemlock, western red cedar, and firs somewhat lower. Sierra Nevada forests include pines, mountain hemlock, and red fir at high elevations; and red and white firs, pines, and Douglas fir somewhat lower. Ponderosa pine is dominant at low mountain elevations in both of these Pacific ranges.

The giant sequoia, long thought to be the largest living organism on earth, grows in scattered groves in the Sierra Nevada. (The largest organism actually may be a very old tree root-rot fungus that covers 1,500 acres in Washington State.) Although shorter than the coast redwood, the giant sequoia is larger in trunk diameter and bulk. It can reach 260 feet (80 meters) tall and 30 feet (10 meters) in circumference.

Coniferous forest also dominates the Rocky Mountains and some mountainous areas of Mexico. In the Rockies, Englemann spruce and subalpine fir grow at high elevations, and Douglas fir, lodgepole pine, and white fir somewhat lower. Ponderosa pine grows throughout the Rockies at low elevations and is a dominant tree in western North America.

Boreal Coniferous Forest

Just south of the Arctic tundra in North America is a broad belt of *boreal, coniferous, evergreen forest*. It is often called *taiga*, the Russian name for similar coniferous forests growing in northern Eurasia. However, in large areas of northeastern Siberia, the dominant tree is larch, which is deciduous, whereas the North American taiga is mostly evergreen. Taiga is the most extensive coniferous forest in North America, covering nearly 30 percent of the land area north of Mexico. It grows across Alaska and Canada and southward into the northern Great Lakes states and New England. White spruce and balsam fir dominate much of the Canadian taiga.

Eastern Deciduous Forest

A forest of mainly broad-leaved, deciduous trees is the native vegetation of much of eastern North America. Narrow fingers of this forest, growing along rivers, penetrate westward into the interior grasslands. Early settlers from Europe cut most of the eastern forest, but second-growth forest now covers considerable areas. The plants are closely related to plant species of the temperate deciduous forests in Europe and Asia. In contrast, the plants of other biomes in North America are generally not closely related to the plants that occur in the same biomes elsewhere in the world, although they look similar.

In the eastern deciduous forest, maple and oak are widespread—maples especially in the north, oaks in the south. There are major subdivisions within the forest. These include oak and hickory forests in Illinois, Missouri, Arkansas, eastern Texas, and also in the east—Pennsylvania, Virginia, and West Virginia—where oak and chestnut forest formerly predominated; beech and maple forest in Michigan, Indiana, and Ohio; and maple and basswood forest in Wisconsin and Minnesota. The forest in parts of Michigan, Wisconsin, Minnesota, and New England contains not only deciduous trees but also evergreen conifers, including pines and hemlock. Vast native pine stands in the Great Lakes states have been cut for lumber.

Plant diseases have changed the composition of the eastern forest. American chestnut was once an important tree but has now nearly disappeared as a result of an introduced fungal disease. Dutch elm disease is similarly devastating American elms.

Other Forests

The southeastern United States, excluding the peninsula of Florida, once supported open stands of pine and also mixed evergreen and deciduous forest. The *mixed forest* included a variety of pines, evergreen oaks, and deciduous trees. In much of Florida, the native vegetation is a mixture of deciduous and evergreen trees that are subtropical rather than temperate. In many parts of the Southeast, people have replaced the native vegetation with fast-growing species of pines for timber production. In Mexico, tropical rain forests are prominent on the west coast, in the south and east, and in Yucatán. On the south coasts of Mexico and Florida, swamps of mangrove trees are common.

Central Grasslands

The central plains of North America, a wide swath from the Texas coast north to Saskatchewan, Canada, were once a vast *grassland*, the prairie. The climate there is too dry to support trees, except along rivers. From west to east, there is a transition from the more desertlike short-grass prairie (the Great Plains), through the mixed-grass prairie, to the moister, richer, tall-grass prairie. This change is related to an increase in rainfall from west to east. Grasses shorter than 1.5 feet (0.5 meter) dominate the short-grass prairie. In the tall-grass prairie, some grasses grow to more than 10 feet (3 meters). Colorful wildflowers brighten the prairie landscape.

Grassland soil is the most fertile in North America. Instead of wild prairie grasses, this land now supports agriculture and the domesticated grasses corn and wheat. The tall-grass prairie, which had the best soil in all the grasslands, has been almost entirely converted to growing corn. Much of the grassland that escaped the plow is now grazed by cattle, which has disturbed the land and aided the spread of *invasive*, nonnative plants.

Other outlying grasslands occur in western North America. Between the eastern deciduous forest and the prairie is *savanna*, a grassland with scattered deciduous trees, mainly oaks. Savanna also occurs over much of eastern Mexico and southern Florida.

Scrub and Desert

In the semiarid and arid West, the natural vegetation is grass and shrubs. Over a large part of California, this takes the form of a fire-adapted *scrub community* called *chaparral* or *Mediterranean scrub*, in which evergreen, often spiny shrubs form dense thickets. The climate, with rainy, mild winters and hot, dry summers, is like that around the Mediterranean Sea, where a similar kind of vegetation, called *maquis*, has evolved. However, chaparral and maquis vegetation are not closely related genetically. Humans have greatly altered the chaparral through *overgrazing* of livestock and other disturbances.

The North American deserts, which are located between the Rocky Mountains and the Sierra Nevada, cover less than 5 percent of the continent. Shrubs are the predominant vegetation, although there are many species of annuals. Desert plants,

PhotoDisc

Wildflowers grow along a trail at the Grand Canyon in Arizona.

commonly cacti and other *succulents*, are sparsely distributed.

In Southern California, Arizona, New Mexico, West Texas, and northwestern Mexico, there are three distinct deserts. The Sonoran Desert stretches from Southern California to western Arizona and south into Mexico. A characteristic plant of the Sonoran is the giant saguaro cactus. To the east of the Sonoran, in West Texas and New Mexico, is the Chihuahuan Desert, where a common plant is the agave, or century plant. North of the Sonoran Desert, in southeastern California, southern Nevada, and northwestern Arizona, is the Mojave Desert, where the Joshua tree, a tree-like lily, is a well-known plant. It can reach 50 feet (15 meters) in height. Creosote bush is common in all three deserts. To the north, the Mojave Desert grades into the Great Basin Desert, which is a cold desert, large and bleak. The dominant plant of the Great Basin Desert is big sagebrush. Plant diversity there is lower than in the hot American deserts.

Deserts dominated by grasses rather than shrubs once occurred at high elevations near the Sonoran and Chihuahuan Deserts. Much of this area has been overtaken by desert scrub, including creosote bush, mesquite, and tarbush. Cattle grazing may have been a factor in this change. Desert is very fragile; even one pass with a heavy vehicle causes lasting damage.

Tundra

Tundra vegetation grows to the northern limits of plant growth, above the Arctic Circle, in Canada. The flora consists of only about six hundred plant species. In contrast, tropical regions that are smaller in area support tens of thousands of plant species. *Arctic tundra* is dominated by grasses, sedges, mosses, and lichens. Some shrubby plants also grow there. Most tundra plants are perennials. During the short Arctic growing season, many of these plants produce brightly colored flowers. Like desert, tundra is exceptionally fragile, and it takes many years for disturbed tundra to recover. Tundra also occurs southward, on mountaintops, from southern Alaska into the Rocky Mountains, the Cascades, and the Sierra Nevada. This *alpine tundra* grows at elevations too high for mountain coniferous forest.

Coastal Vegetation

Along the coasts of the Atlantic Ocean and the Gulf of Mexico, the soil is saturated with water and is very salty. Tides regularly inundate low-lying vegetation. The plants in these *salt marsh* areas consist mainly of grasses and rushes. Marshes are a vital breeding ground and nursery for fish and shellfish. They play an important role in absorbing and purifying water from the land. Coastal marshes are being lost to development at a rapid rate.

Jane F. Hill

See also: Arctic tundra; Biological invasions; Biomes: types; Forests; Deserts; Grasslands; Grazing and overgrazing; Mediterranean scrub; North American agriculture; Savannas and deciduous tropical forests; Taiga; Tundra and high-altitude biomes.

Sources for Further Study

Braun, E. Lucy. *Deciduous Forests of Eastern North America*. New York: Macmillan, 1950. Reprint. New York: Free Press, 1985. Botanist Braun's devotion to land preservation was one of the pivotal influences in the then-developing field of ecology. This book describes in detail the trees and shrubs in the deciduous forests of Kentucky, Tennessee, Ohio, Virginia, West Virginia, and Pennsylvania.

Foster, David R. *New England Forests Through Time: Insights from the Harvard Dioramas*. Cambridge, Mass.: Harvard University Press, 2000. Today, in many ways, the New England region is more natural than at any time since the American Revolution. Historical and environmental lessons in this book are told through world-renowned dioramas in Harvard University's Fisher Museum.

MacMahon, James A. *Deserts*. The Audubon Society Nature Guides. New York: Random House, 1998. Desert ecology, with life histories of characteristic flora and fauna, range maps, and illustrations.

Randall, John M., and Janet Marinelli, eds. *Invasive Plants: Weeds of the Global Garden*. Brooklyn, N.Y.: Brooklyn Botanic Garden, 1996. Refreshingly simple explanation of environmental impacts and how to control the problem of invasive species. With excellent illustrations.

NUCLEAR ENVELOPE

Categories: Anatomy; cellular biology; physiology; transport mechanisms

The nuclear envelope is the outer covering of the nucleus in plant and other eukaryotic cells that acts as a barrier separating the nuclear contents from the surrounding cytoplasm.

The nuclear envelope is a double membrane system, consisting of two concentric membranes. The membranes are separated by a fluid-filled space called the perinuclear cisterna that measures about 20 to 40 nanometers. Like other plant cell membranes, the nuclear envelope consists of two bilayers, both made of phospholipids, in which numerous proteins are embedded. Attachment sites for protein filaments are stitched on the innermost surface of the nuclear envelope. These protein filaments anchor the molecules of deoxyribonucleic acid (DNA) to the envelope and help to keep them organized. The network of filaments that enmesh the nuclear envelope provides stability.

Large numbers of ribosomes are located on the outer surface of the envelope. The outermost membrane is continuous with the organelle called *rough endoplasmic reticulum* (RER) in the *cytoplasm* of the plant cell, which also has ribosomes attached to it. The space between the outer and inner membranes is also continuous with the rough endoplasmic reticulum space and can fill with newly synthesized proteins, just as the RER does. When the nuclear envelope breaks down during cell division, its fragments are similar to portions of the endoplasmic reticulum. This can be commonly observed at the root and shoot apices of a plant, where active mitosis (regular cell division for growth) occurs. Both bilayers of the envelope are fused at intervals to form many nuclear pores, which consist of protein complexes in clusters. The two membranes enclose a flattened sac and are connected at the nuclear pore sites.

Functions

The nuclear envelope surrounds the fluid portion of the nucleus, called the *nucleoplasm*, in all plant cells. The ribosomes on the outer surface of the envelope serve as sites for protein synthesis in addition to ribosomes located in the cytoplasm. The envelope is selectively permeable and therefore regulates the passage of materials and energy between the nucleoplasm and the cytoplasm. The envelope allows certain cell activities to be localized within the nucleus or outside in the cytoplasm. It also permits many different activities to go on simultaneously within and outside of the nucleus.

Like other membranes of the plant cell, the nuclear envelope membranes serve as important work surfaces for many chemical reactions in plants that are carried out by enzymes bound to the membranes. These functions are essential in the plant for the transport and use of minerals and water from the roots to cells of the stem and leaves. They are also important for the movement and use of carbohydrates, proteins, lipids, nucleic acids, and other chemical compounds in plant cells after photosynthesis, protein synthesis, and other biochemical activities have occurred.

Nuclear Pores

Nuclear pores, of about 100 nanometers in diameter, perforate the nuclear envelope. These pores look like wheels with eight spokes when observed from the top. Each contains eight subunits over the region where the inner and outer membranes join. They form a ring of subunits that are 15 to 20 nanometers in diameter. At the lip of each pore, the inner and outer membranes of the nuclear envelope are fused.

Each nuclear pore serves as a water-filled channel, and the arrangement allows transport in and out of the nucleus to occur in several ways. The nuclear pores allow the passage of materials to the cytoplasm from the interior of the nucleus, and vice versa, but the process is highly selective, permitting only specific molecules to pass through these openings. An intricate protein structure called the pore complex lines each pore and regulates the entry and exit of certain large macromolecules and particles. Large molecules, including the subunits of ribosomes, cross the bilayers at the pores in highly controlled ways. Ions and small, water-soluble molecules cross the nuclear envelope at the pores.

Studies show that the pore can actually dilate more when it gets the appropriate signal. Studies have also shown that the signal is in the peptide sequences of the molecules. These signals are recognition sequences rich in the amino acids lysine, arginine, and proline. Nuclear pores in plants therefore exert control over the movement of materials. This is, for example, demonstrated in the fact that if a nucleus is extracted from a cell and placed into water, it swells; this can happen only if the pores prevent material from oozing out as the nucleus absorbs water.

Nuclear Lamina

The *nuclear lamina* is a layer of specific proteins, called lamins, attached to inside membrane of the nuclear envelope. The layer consists of thin filaments (intermediate filaments) that are 30 to 40 nanometers thick. Each filament is a polymer of lamin. There are two types of lamin: A-type lamins are inside, next to the nucleoplasm, and B-type lamins are near the inner part of the nuclear membrane. The nuclear lamina surrounds the nucleus, except at the nuclear pores. The lamina serves as a skeletal framework for the nucleus. It may be involved in the functional organization of the nucleus and may also play an important role in the breakdown and reassembly of the nuclear envelope during mitosis.

Samuel V. A. Kisseadoo

See also: Cell cycle; Cell wall; Cytoskeleton; Cytosol; Endoplasmic reticulum; Eukaryotic cells; Liquid transport systems; Membrane structure; Mitosis and meiosis; Nucleus; Plasma membranes; Water and solute movement in plants.

Sources for Further Study

Campbell, Neil A., and Jane B. Reece. *Biology*. 6th ed. New York: Benjamin Cummings, 2002. Basic textbook has clear definition and description of the nuclear envelope and its components. Includes illustrations, pictures, glossary, bibliography, and index.

Fawcett, Don W. *A Textbook of Histology*. 12th ed. New York: Chapman & Hall, 1994. Good account of structure and function of the pores of the nuclear envelope. Includes illustrations and pictures.

Mauseth, James D. *Botany*. Boston: Jones and Bartlett, 1998. Throws light on the functions of the nuclear envelope in plants. Includes photographs, illustrations, glossary, and index.

Solomon, Eldra P., Linda Berg, and Diana Martin. *Biology*. 6th ed. Pacific Grove, Calif.: Brooks/Cole, 2002. Focuses on functions of the membranes and gives general structure of the nuclear envelope. Includes review questions, glossary, bibliography, and index.

Starr, Cecie. *Biology*. 4th ed. Pacific Grove, Calif.: Brooks/Cole, 2002. Focuses on structure of the nuclear envelope and transport across the membranes and pores. Includes good photographs, illustrations, micrographs, glossary, bibliography, and index.

NUCLEIC ACIDS

Categories: Cellular biology; genetics; reproduction and life cycles

Nucleic acids are the genetic material of cells, including DNA and the various types of RNA.

Nucleic acids were discovered in the mid-nineteenth century, but their role as genetic material was not substantiated until the mid-twentieth century. When chromosomes were discovered at the beginning of the twentieth century, they were quickly identified as the genetic material of the cell. Chromosomes were found to be composed of nucleic acids and proteins. Through the experiments of Fred Griffith on transformation in pneumonia bacteria and the work of Alfred Hershey and Martha Chase on bacteriophages, by 1952 most biologists recognized *deoxyribonucleic acid* (DNA) as containing the genes. James Watson and Francis Crick provided the capstone to science's initial understanding of nucleic acids when they determined the double helix structure of DNA in 1953.

Heredity is the process by which the physical traits of an organism are passed on to its offspring. At the molecular level, DNA contains the information necessary for the transmission of genetic characteristics from one generation to the next, as well as the information required for the new organisms to grow and to live. DNA is the chemical basis of heredity and provides the synthesis of new proteins, such as enzymes.

DNA and RNA

There are two types of nucleic acids within cells, the single-stranded *ribonucleic acid* (RNA) and the double-stranded DNA. Each kind has specific roles. DNA was isolated in 1869 by German chemist Friedrich Miescher. The substance that Miescher found was white, sugary, and slightly acidic, and it contained phosphorus. Because it occurred only within the nuclei of cells, he called it "nuclein." The name was later changed to deoxyribonucleic acid, to distinguish it from ribonucleic acid, which is also found in cells.

In eukaryotic cells, DNA is present in the *chromosomes* of the nucleus and within the *mitochondria* and *chloroplasts*. Bacteria, yeasts, and molds, in addition to the chromosomes, contain circular strands of DNA, called *plasmids*, within the cytoplasm of their cells. Plasmids are relatively small, circular strands of DNA that exist independently of the chromosome. Plasmids typically have only twenty-five or thirty genes, which are not essential to the host cell but often confer antibiotic resistance, the ability to pass DNA to other bacterial cells, and other useful functions. Some plasmids are only found as single copies, whereas others occur as many copies. The mitochondria and chloroplasts of eukaryotic cells are self-replicating and contain a tiny circular chromosome (DNA) resembling a plasmid of a bacterium.

Viruses (minute parasites that infect specific hosts) contain only one type of nucleic acid—either DNA or RNA, never both, and the DNA or RNA can be single- or double-stranded. The DNA of some viruses can integrate into the DNA of the host cell. In this state, the viral DNA replicates as the host DNA replicates. The genetic apparatus of a virus, whether RNA or DNA, is much the same as that of bacteria but is far less complex. Even large viruses (such as the pox virus) have only a few hundred genes. Smaller viruses (such as the polio virus) have considerably fewer.

Chemical Structure

Both DNA and RNA are long-chained polymers made up of nucleotides. The nucleotides, in turn, are made up of a nitrogenous base, a ribose or deoxyribose sugar, and a phosphate. All the bases of DNA and RNA are heterocyclic amines. Two, adenine and guanine, are called *purines*; the other three, cytosine, thymine, and uracil, are called *pyrimidines*. The two purines and one of the pyrimidines, cytosine, occur in both RNA and DNA. Uracil is found only in RNA, while thymine occurs only in DNA.

The purines are nine-membered heterocyclic rings with nitrogen occurring in place of carbon at several positions. Adenine and guanine differ in the *functional groups* attached to them. The pyrimidines are six-membered heterocyclic rings, with nitrogen in place of two of the carbon atoms. Like the purines, the three pyrimidines also differ in the specific functional groups attached to them.

The ribose sugars are made up of a five-membered heterocyclic ring containing one oxygen atom between carbons one and four. A fifth carbon (number five) is not part of the ring and is bonded to carbon number four. Along with hydrogen atoms attached to each carbon, there is one hydroxyl group (OH) attached to each of the four heterocyclic carbon atoms in the ribose sugar of RNA. (The fifth carbon has a phosphate group attached to it.) The sugar of DNA is called *D-deoxyribose* because a hydroxyl group is missing from the second carbon, having been replaced by a hydrogen atom—thus the name deoxyribonucleic acid.

The sugar-base combination is called a *nucleoside*. The purines are linked to carbon one of the sugars with the nitrogen at position one. The nucleoside of guanine and ribose is guanosine; it is adenosine for adenine and D-ribose. The pyrimidines of RNA, when attached to ribose, are uridine and cytosine. In DNA the nucleoside names are deoxyadenine, deoxyguanosine, deoxythymidine, and deoxycytidine.

Nucleotides are phosphate esters of nucleosides. In these molecules, a phosphate group (phosphoric acid) is attached to carbon five (called the 5′ carbon) of the sugar (ribose or deoxyribose) in the nucleoside. Nucleotides are named by combining the name of their nucleoside with a word describing the numbers of phosphates attached to it. Guanosine monophosphate, for example, is the name of the phosphate ester of guanosine, which is often abbreviated as GMP.

Individual nucleotides also occur in cells. These free nucleotides usually exist as diphosphates or triphosphates. Examples of these are adenosine diphosphate (ADP) and adenosine triphosphate (ATP). ATP is the universal energy source for the anabolic processes of all cells, including the formation of the DNA and RNA.

DNA Structure and Function

DNA can be an extremely long molecule that is tightly wound within the nuclei of eukaryotic cells and within the cytoplasm of prokaryotic cells. Nuclear DNA is linear, whereas prokaryotic DNA is circular. (If the DNA in a human cell could be stretched out, it would measure roughly 2 meters, or 6 feet long; bacterial DNA would be about 1.5 millimeters long, or just over 0.5 inch.) Using a typical lily as a point of reference, DNA is packaged into twenty-four chromosomes, twelve of which are contributed by the pollen and twelve by the egg. Every cell derived from the fertilized egg (zygote) will have exactly the same amount of DNA containing exactly the same genetic information. Within the cytoplasm, several mitochondria (the sites of respiration) and chloroplasts (the sites of photosynthesis) are found, both of which contain their own DNA, which is circular and resembles prokaryotic DNA in many respects.

DNA is a double-stranded spiral; its shape is called the *double helix*. Structurally, it may be compared to a ladder, with the rails or sides of the ladder consisting of alternating deoxyribose sugar and phosphate molecules connected by phosphodiester bonds between the 5′ carbon of one sugar and the 3′ carbon of the other. The rungs of the ladder consist of purine (adenine and guanine, often abbreviated as A and G, respectively) and pyrimidine (cytosine and thymine, often abbreviated as C and T, respectively) building blocks from the opposite strands, held together by hydrogen bonds. The building blocks pair with each other consistently in what are called complementary pairs: Adenine always pairs with thymine with two hydrogen bonds, and cytosine always pairs with guanine with three hydrogen bonds. Consequently, the attraction between cytosine and guanine is stronger than that between adenine and thymine.

Because of this arrangement, the sequence of the purine and pyrimidine building blocks on one strand is complemented by the sequence of building blocks on the other strand. The specificity of the base pairing between the two strands allows strands to fit neatly together only when such pairing exists. Each DNA strand has a 5′ end with a hydroxyl group attached to the 3′ carbon of a deoxyribose sugar. When connected, the two strands are actually in an opposite orientation and are referred to as being *antiparallel*. This is best observed by looking at one end of the double-stranded molecule. One strand terminates with a 5′ phosphate group and the other with a 3′ hydroxyl group.

The specific nucleotide composition in a species is essentially constant but can vary considerably among organisms. Regardless, the amounts of adenine and thymine are always the same, as are the amounts of guanine and cytosine because of the required complementary pairing. Due to the greater strength of G-C bonds, organisms with a high GC content have DNA that must be heated to a higher temperature to *denature*, or separate, the strands. Some bacteria that live in hot springs have an especially high GC content.

The instructions contained within the DNA molecules occur in segments called *genes*. Most genes instruct the cell about what kind of *polypeptide* (molecule composed of amino acids used to make functional proteins) to manufacture. These polypeptides lead to the formation of enzymes and other proteins necessary for survival of the cell. Other genes are important in coding for the production of antibodies, RNA, and hormones.

RNA Types

The DNA acts as a template to make three kinds of RNA: *messenger RNA* (mRNA), *transfer RNA* (tRNA), and *ribosomal RNA* (rRNA). Each kind of RNA has a specific function. RNA is not found in chromosomes and is located elsewhere in the nucleus and in the cytoplasm.

The largest and most abundant RNA is rRNA. Between 60 and 80 percent of the total RNA in cells is rRNA, and it has a molecular weight of several million atomic mass units. The rRNA combines with proteins to form *ribosomes*, which are the sites for the synthesis of new protein molecules. About 60 percent of the ribosome is rRNA, and the rest is protein. Although single-stranded, rRNA molecules fold into specific functional shapes that involve the pairing of portions of the molecule to form double-stranded regions. The precise shape of rRNAs is important for their function, and some of them actually have catalytic properties, just as enzymes do. RNAs of this type are sometimes called ribozymes.

Molecules of mRNA carry the genetic information from DNA to the ribosomes. The process of converting the DNA code of a gene into an mRNA is called *transcription*. When attached to the ribosomes, mRNAs direct protein synthesis in a process called *translation*. The size of the mRNA molecule depends upon the size of the protein molecule to be made. In prokaryotes (such as bacterial cells), as well as in mitochondria and chloroplasts, mRNAs are ready to take part in translation even while transcription of the remainder of the mRNA is taking place. In eukaryotes (cells of most other forms of life), the mRNAs transcribed from nuclear DNA are initially much larger than they are later, when they participate in translation. These mRNAs must be processed to removed large pieces of noncoding RNA, called *introns*, and to modify both ends of the mRNA in specific ways. After introns are removed, the remaining codon regions, called *exons*, are spliced together by *splicesomes* (a complex system composed of proteins and small RNAs). Once all the modifications are complete, the mRNA is ready to be translated. A small number of mRNAs are translated in the nucleus, but most are transported to the cytoplasm first.

The smallest of the three main kinds of RNA is tRNA. Each of the tRNA molecules consists of about one hundred nucleotides in a single chain that loops back upon itself in three places, forming double-stranded regions that result in a structure that, when viewed in two dimensions, could be compared to a cross or clover leaf. The function of tRNA is to bring amino acids to the ribosomes to be used in the formation of new proteins. Each of the twenty amino acids found in proteins has at least one particular tRNA molecule to carry it to the site of protein synthesis.

The cloverleaf shape of the tRNA molecule is maintained by hydrogen bonds between base pairs. The other parts of the molecule that do not have hydrogen-bonded base pairs exist as loops. Two parts of every tRNA molecule have significant biological functions. The first is the place where the specific amino acid to be transferred is attached. This is located at the longest free end of the three-looped structure, often called the stem, where it is specifically attached to an adenine of an adenine monophosphate nucleotide. The second important site is the loop at the opposite end of the molecule from the stem. This loop contains a specific three-base sequence that represents a code for the amino acid that is being transferred by the tRNA. This three-base sequence is called an anticodon and plays an important role in helping place the amino acid in the correct position in the protein molecule under construction.

Jon P. Shoemaker, updated by Bryan Ness

See also: Chloroplast DNA; Chromosomes; DNA in plants; DNA replication; Extranuclear inheritance; Gene regulation; Genetic code; Genetic equilibrium: linkage; Mitochondrial DNA; Proteins and amino acids; RNA.

Sources for Further Study

Bettelheim, Frederick A., William H. Brown, and Jerry March. *Introduction to General, Organic, and Biochemistry*. 6th ed. New York: Saunders College Publishing, 2001. Excellent illustrations of the nucleic acids, along with a good historical background.

Blackburn, G. Michael, and Michael J. Gait, eds. *Nucleic Acids in Chemistry and Biology*. 2d ed. New York: Oxford University Press, 1996. Discusses the impact of sophistication of nucleic acids synthesis on the biotechnology industry.

Diamond, R., et al., eds. *Molecular Structures in Biology*. New York: Oxford University Press, 1993. Authoritative surveys of techniques used in structure determination, modeling studies, and specific groups of biomacromolecules.

Hecht, Sidney M., ed. *Bioorganic Chemistry: Nucleic Acids*. New York: Oxford University

Press, 1996. Part of the Topics in Bioorganic and Biological Chemistry series; includes methods of preparation, mapping, analysis, binding, and catalysis of nucleic acids.

Prescott, Lansing M., John P. Harley, and Donald A. Klein. *Microbiology*. 5th ed. Boston: McGraw-Hill, 2002. A textbook that contains excellent illustrations and good discussions of nucleic acids.

Seager, Spencer L., and Michael R. Slabaugh. *Chemistry for Today: General, Organic, and Biochemistry*. 4th ed. Pacific Grove, Calif.: Brooks/Cole, 2000. A good presentation of the historical background leading to the discovery of DNA structure; good coverage of nucleic acids.

Watson, James D. *Molecular Biology of the Gene*. 4th ed. Menlo Park, Calif.: Benjamin/Cummings, 1987. This is a concise, well-written, and well-illustrated popular text.

NUCLEOLUS

Categories: Anatomy; cellular biology; physiology

The most prominent organelle of the eukaryotic cell is the nucleus. Within the nucleus, substructures called nucleoli are visible by light microscopy, each nucleolus as a small spot.

Generally, nucleoli are most prominent in non-dividing cells, and they are not visible in dividing cells in which the nuclear membrane has disassembled. The nucleolus is composed of genes that encode the *ribonucleic acid* (RNA) molecules found in ribosomes (*ribosomal RNA*, or rRNA), the "working" copies or transcripts of these genes (the rRNA itself), and proteins. Nucleoli are the sites of ribosome synthesis in eukaryotic cells.

The *ribosome* is the site of protein synthesis in the cell. All proteins made by cells are made on ribosomes, so cells need an abundant supply of ribosomes to support the synthesis of all the cellular proteins. Mature ribosomes are found in the cytoplasm of eukaryotic cells, either attached to the membrane of the endoplasmic reticulum or "free" within the cytoplasm. Although ribosomes function outside the nucleus in the cytoplasm, they are synthesized in the nucleus at the nucleolus.

Nucleolar Structure

Nucleoli are clearly visible with light microscopy. They disassemble and reassemble with the cell cycle and exhibit different staining properties than the rest of the nucleus. They range in diameter from 1 micrometer in small yeast cells to 10 micrometers in larger cells, such as the root cells of pea plants and wheat.

When observed using electron microscopy, nucleoli appear to have roughly three distinct regions: the *fibrillar center* (FC), the *dense fibrillar center* (DFC), and the *granular component* (GC). In some plant cells, the nucleolus also has a centrally located clear region, sometimes called the *nuclear vacuole*. Although the precise molecular structures within nucleoli have not yet been thoroughly described, the regions of the nucleolus appear to correspond roughly to content and function, rRNA transcription, processing, and assembly.

rRNA Genes

A typical cell contains around one million ribosomes. To produce this number of ribosomes, a cell needs to be able to mass-produce the ribosomal components. To provide enough ribosomes for daughter cells produced from cell divisions, a cell must synthesize new ribosomes at the rate of several hundred per second. Additionally, these ribosomes must be exported from the nucleus into the cytoplasm, where proteins are synthesized.

In order to mass-produce the RNA components of ribosomes, the three rRNA genes that are transcribed (copied) in the nucleolus are present in multiple copies. The genes for the 5.8S, 18S, and 28S rRNAs (rRNAs are named for their sedimentation rate, "S," a measure of their size) are arranged in tandem repeats on the chromosomes of the cells. Typically, each repeat contains the 18S, the 5.8S, and the 28S genes, in that order, preceded by a region called a *nontranscribed spacer* (NTS) and separated by internal nontranscribed spacers. The NTS's are typically as small as two thousand to three thousand base pairs of deoxyribonucleic acid (DNA) in yeast and plants. These spacer regions may be involved in regulating gene expression in these regions of DNA or in attaching these areas of the chromosomes to protein structures inside the nucleolus.

Among different organisms, the number of copies of rRNA genes varies greatly. In fact, the number of copies of rRNA genes can even vary within different cells of a single organism. The human genome (haploid set) contains approximately one hundred copies of these genes, but many organisms, especially plants, have several thousand copies of the rRNA genes. By having hundreds or thousands of copies of the genes, the cells are able to mass-produce the rRNAs that these genes encode.

Nucleolar Organizing Regions

The genes that encode the rRNAs are found along metaphase chromosomes at constrictions called *nucleolar organizing regions*, or NORs. When the cell completes division and reenters interphase, NORs are the sites where the nucleoli will form. In plants, experiments to stain the specific rRNA genes reveal a complex organization of the genomic DNA. Large areas of inactive or untranscribed rRNA gene repeats appear at the periphery of the plant nucleoli, and several spots containing inactive rRNA genes also appear at the center of the nucleoli. However, many areas of active transcription of a single copy of the rRNA genes appear dispersed throughout the nucleolus. These areas of active rRNA gene transcription may correspond to the fibrillar center regions.

Ribosome Assembly and rRNA Processing

The three rRNA genes are initially copied as one large molecule of RNA, called the pre-rRNA. This large piece of RNA is processed to make the three smaller 18S, 5.8S, and 28s rRNAs that will become part of the ribosome. During processing the nontranscribed spacer sequences are removed. Each pre-rRNA molecule moves away from the genomic material as it is processed in several steps. Processing of the pre-rRNA molecule appears to require many accessory proteins and other RNA molecules, called small, nucleolar RNAs (snoRNAs).

The nucleolus contains many proteins that function in rRNA transcription and processing as well as transport of ribosome components into and out of the nucleus. One of these proteins, nucleolin, which is specific to the nucleolus, is found in the dense fibrillar center and appears to be involved in different stages of ribosome synthesis. Nucleolin travels between the nucleus and cytoplasm and therefore may also be involved in transport of ribosome components between the two cellular compartments.

For ribosome synthesis, ribosome proteins must be assembled with the three nucleolar rRNAs. The ribosomal proteins are synthesized, like other cellular proteins, on existing ribosomes in the cytoplasm. These proteins must be transported into the nucleus, where they associate with the pre-rRNA in the nucleolus even while it is being processed. A fourth rRNA, the 5S RNA, is transcribed in a separate part of the nucleus and transported to the nucleolus to be assembled with the other ribosomal components.

The ribosome consists of a small subunit and a large subunit. The small subunit, which contains the 18s rRNA and many proteins, is produced and exported from the nucleus separately from the large subunit, which contains the 28S, 5.8S, and 5S rRNAs. The two subunits continue to mature in the cytoplasm, and matured subunits recognize messenger RNA within the cytoplasm to assemble a functional protein-synthesizing ribosome.

Michele Arduengo

See also: DNA in plants; DNA replication; Nucleic acids; Nucleus; Plant cells: molecular level; Proteins and amino acids; RNA.

Sources for Further Study

Cooper, Geoffrey M. *The Cell: A Molecular Approach*. Sunderland, Mass.: Sinauer Associates, 1997. Standard college-level text contains an entire chapter on the nucleus, with considerable information on the nucleolus and ribosome synthesis.

Raven, Peter H., Ray F. Evert, and Susan E. Eichhorn. *Biology of Plants*. 6th ed. New York: W. H. Freeman/Worth, 1999. Introductory botany text provides a chapter introducing eukaryotic cells in general and plant cells specifically.

NUCLEOPLASM

Categories: Anatomy; cellular biology; physiology

The nucleoplasm is the protoplasm contained within the plant cell nucleus and can best be described as the fluid-filled matrix that is contained within the nuclear membrane.

The nucleoplasm is a distinct entity of the plant cell *nucleus*, bounded, protected, and separated from the cytoplasm by the double-membrane *nuclear envelope*. Connections between the *cytoplasm* and nucleoplasm are closely regulated by nuclear pores that penetrate the nuclear envelope at intervals. Interiorly, these *nuclear pores* connect to a complex of nucleoplasmic channels that lead into the pores and serve simultaneously to direct and regulate exchange of materials between the nucleoplasm and the cytoplasm.

The nucleoplasm consists of a viscous mix of water, in which various substances and structures are dissolved or carried, and an underlying intranuclear ultrastructure. Comparisons of the aqueous phase of nucleoplasm with that of cytoplasm using a technique called the Stokes-Einstein equation reveal that diffusion rates are 1.2 times slower through cytoplasm than nucleoplasm, indicating that nucleoplasm is a less viscous fluid than cytoplasm. Substances in the nucleoplasm include ions, enzymes, minerals, and some organic molecules and macromolecules. The nucleoplasm is especially rich in protein enzymes and protein constituents involved in the synthesis of deoxyribonucleic acid (DNA) and the various types of ribonucleic acid (RNA), the precursor molecules of RNA, and the nucleotides from which they are assembled. Some of these proteins direct initial transcription, while others function in the further modification of the RNA molecules for packaging and transport to the cytoplasm.

Prominent structures located within the interphase nucleoplasm (the resting cell or the nonreplicating cell) include organelles called *nucleoli* and the unwound DNA, called *chromatin*. The nucleoli resemble miniature nuclei and are the sites of synthesis of precursor RNA molecules and their assembly.

The other major components in nucleoplasm include the DNA chromosomes seen during mitosis. During cell interphase most of the DNA chromosomes exist as unwound chromatin that extend through the nucleoplasm. Two distinct types of chromatin are recognized. Diffuse, or uncondensed, chromatin is called *euchromatin* and exists as thin threads that extend throughout much of the nucleoplasm. Small sections of DNA that remain uncondensed during cell interphase are called heterochromatin. The chromatin can be further subdivided into gene-rich R bands concentrated within the nucleus and gene-poor G bands found in the peripheral regions of the nucleus.

The space between these chromatin elements is called the *interchromatin space*. Ultrastructure studies have revealed a surprisingly complex structure within this interchromatin space. It consists of several components, including a nuclear lamina, nuclear matrix, thousands of highly organized sites called foci, and nuclear speckles.

The *nuclear lamina* and *nuclear matrix* consist of protein microfilaments—fine tubular structures—which apparently provide structural support for the interior of the nucleus as well as a framework for further nuclear structure and function. The lamina consists of lamin proteins that help shape and maintain the inner membrane of the nuclear envelope. The nuclear matrix functions in pore formation and may also be continuous with the nuclear lamina in conferring an interior structure or shape to the nuclear envelope. The nuclear matrix also binds chromosomes, or parts of chromosomes, to specific sites on the inner nuclear membrane.

Foci and *nuclear speckles* are both involved in the production of the several types of RNA molecules manufactured by the nucleus. Foci are apparently sites that are functionally associated with gene-rich chromatin. Some scientists argue, in fact, that foci are actually functional domains on active chromatin. Whether active chromatin sites or separate entities, the foci contain protein enzymes and assembly components involved in the transcription and splicing of ribonucleoprotein (RNP) molecules associated with the production of RNA precursor or subassembly molecules. The production of a specific RNP molecule may involve the activity of hundreds or thousands of discrete foci. Once completed, the molecular products of the foci are transported to nuclear speckles for further modification.

Nuclear speckles appear to be clusters of RNP. They are highly organized sites that are separated from other nuclear compartments by nucleoplasm. Their role and importance is still uncertain, but they are probably nuclear structures directly involved in the final assembly and modification of the various types of RNA molecules from RNP molecules supplied by the foci.

Although much of the structure and role of the nucleoplasm remains to be investigated, it is increasingly clear that the nucleoplasm has a complex ultrastructure underlying and facilitating its myriad functions.

Dwight G. Smith

See also: Chromatin; Cytoplasm; Cytosol; Mitosis and meiosis; Nuclear envelope; Nucleic acids; Nucleolus; Nucleus; Plant cells: molecular level.

Sources for Further Study

Busch, Harris. *The Cell Nucleus.* Vol. 1. New York: Academic Press, 1974. This somewhat older volume provides a very detailed summary of the nuclear interior, including the role and structure of the nucleoplasm.

Lodish, Harvey, et al. *Molecular Cell Biology.* 4th ed. New York: W. H. Freeman, 2000. One of the definitive books on cell biology. All topics about cells are well and thoroughly covered. This provides a good general introduction to the structure and functioning of all nucleus components in a highly readable style.

Smith, C. A., and E. J. Wood. *Cell Biology.* 2d ed. New York: Chapman and Hall, 1999. Advanced textbook of cell biology provides a basic but thorough coverage of the latest information about all aspects of cell biology. Well illustrated and well written for both undergraduate and graduate students.

NUCLEUS

Categories: Anatomy; cellular biology; physiology

A eukaryotic cell's nucleus directly or indirectly controls virtually all cellular physiological activities, including initiation, regulation, and termination of enzymatic events. It is also the repository of genetic information (the genome), housing and protecting the chromosomes and the genes that they carry. In all eukaryotic cells the nucleus is generally the largest and most centrally located cell structure, although in plant storage cells certain kinds of vacuoles may be larger and more conspicuous.

The nucleus, from the Greek word *nucle* (meaning "pit" or "kernel") is the command and control center of the cell. The six basic functions of the nucleus are, first, to protect and store genes, ultimately protecting the deoxyribonucleic acid (DNA) on which the genes are organized from the rest of the cell; second, to organize genes into chromosomes to facilitate their movement and distribution during cell division; third, to organize the uncoiling of DNA during the copying of genes for the production of thousands of proteins; fourth, to manufacture and transport regulatory molecules, mostly enzymes and other gene products, into the cytoplasm; fifth, to manufacture subunits of ribosomes; and sixth, to respond to hormones and other chemical signals received via the nuclear pores.

Components

Structurally, the nucleus consists of several distinct parts: a *nuclear envelope*, *nucleoplasm*, *chromatin*, and one or more suborganelles called *nucleoli*.

The nuclear envelope forms a protective barrier that isolates the nucleus from the cytoplasm of the cell. The envelope consists of two unit membranes (a double-unit membrane) which are structurally similar to other membranes of the cell. The outer membrane is closely associated with the cell's endoplasmic recticulum (ER) and may be continuous with it. Like the rough ER of the cytoplasm, the outer nuclear membrane has ribosomes embedded in it. Some scientists, in fact, suggest that the nuclear envelope is just a localized and specialized version of the ER. The inner nuclear membrane is lined with a fibrous layer, called the nuclear lamina, which provides strength and structure to the nucleus shape and may also serve as a binding site for some chromatin.

At intervals, the nuclear envelope is perforated by small pores which function as communication channels for the controlled exchange of materials between the nucleus and the cytoplasm. Collectively, the *nuclear pores* cover about 10 percent of the surface of the nucleus. Each nuclear pore is a complex consisting of a central pore that has been estimated at 30-100 nanometers in diameter. The selectively permeable nuclear pores function as entry and exit ways for a variety of water-soluble molecules, mostly nuclear products, such as ribosome subunits, messenger RNA (ribonucleic acid) molecules, and chromosomal proteins.

The protoplasm within the nucleus is called nucleoplasm. Like cytoplasm, it consists of a jellylike mix of substances and organelles but differs in having a higher concentration of nucleotides and other organic molecules that are used in the synthesis of DNA and RNA.

Major structures within the nucleoplasm include the DNA and usually one organelle—but sometimes several—called the nucleolus. Except during cell division, the molecules of DNA occur as a network of unwound fibers called chromatin. During cell division molecular strands of DNA coil and supercoil around histone proteins to condense and form the chromosomes. The number of chromosomes found within the nucleus are specific for each species of plant and animal. Humans, for example, have forty-six chromosomes, tobacco has forty-eight, corn has twenty, carrots have eighteen, and peas have fourteen chromosomes.

The nucleolus is the largest visible organelle within the nucleus. It is typically associated with specific regions of chromosomes, called nuclear organizer regions, which contain genes that direct synthesis of ribosomal subunits. The main products of nucleolus activity are the units of ribosomal RNA (rRNA). These subunits are eventually complexed with ribosomal proteins and transported from the nucleus into the cytoplasm by special carrier proteins. Other sites within the nucleus called *functional domains* control the synthesis of messenger (premRNA), transfer (tRNA) molecules. Once formed, these molecules are then complexed with proteins and transported as nucleoproteins to the cytoplasm.

Cell Division

Although seemingly both stable and durable, the nucleus disappears from normal view and is reformed during cell division in almost all plants except yeasts, which retain a clearly defined nucleus throughout the division process. In other eukaryotic plant cells the nucleus disappears early during the prophase of mitosis, when the nuclear envelope is enzymatically fragmented into small, nearly invisible vesicles. These are not reassembled until the final events of telophase, when they reform around the chromosomes and are controlled by the lamina of the daughter cells.

Dwight G. Smith

See also: Chromosomes; Chromatin; Cytoplasm; Cytosol; DNA in plants; DNA replication; Eukaryotic cells; Mitosis and meiosis; Nuclear envelope; Nucleolus; Nucleoplasm.

Sources for Further Study

Becker, W. M., L. J. Kleinsmith, and J. Hardin. *The World of the Cell*. 4th ed. San Francisco: Benjamin/Cummings, 2000. This very well-written college-level textbook describes the nucleus within the larger context of the structure and function of the cell.

Ross, M. H., L. J. Romrell, and G. I. Kaye. *Histology: A Text and Atlas*. 3d ed. Philadelphia: Williams and Wilkins, 1995. A detailed and well organized discussion of the structure of the cell nucleus and its role within the eukaryotic cell.

Strauss, Phyllis R., and Samuel H. Wilson. *The Eukaryotic Nucleus: Molecular Biochemistry and Macromolecular Assemblies*. Vol. 1. Caldwell, N.J.: Telford Press, 1990. This volume includes seventeen papers which provide an exhaustive and very technical description of the structure and function of the nucleus.

Strouboulis, J., and A. P. Wolffe. 1996. "Functional Compartmentalization of the Nucleus." *Journal of Cell Science* 109 (1991). This somewhat technical article presents the latest findings on the structure and function of the nucleus.

NUTRIENT CYCLING

Categories: Biogeochemical cycles; ecosystems; nutrients and nutrition

Within an ecosystem, nutrients move through biogeochemical cycles. Those cycles involve chemical exchanges of elements among the earth's atmosphere, water, living organisms, soil, and rocks.

All biogeochemical cycles have a common structure, sharing three basic components: inputs, internal cycling, and outputs.

Input of Nutrients

The input of nutrients to an ecosystem depends on the type of biogeochemical cycle. Nutrients with a gaseous cycle, such as carbon and nitrogen, enter an ecosystem from the atmosphere. For example, carbon enters ecosystems almost solely through photosynthesis, which converts carbon dioxide to organic carbon compounds. Nitrogen enters ecosystems through a few pathways, including lightning, nitrogen-fixing bacteria, and atmospheric deposition. In agricultural ecosystems, nitrogen fertilization provides a great amount of nitrogen influx, much larger than by any other influx pathways.

In contrast to carbon and nitrogen with input from the atmosphere, the input of nutrients such as calcium and phosphorus depends on the weathering of rocks and minerals. Soil characteristics and the process of soil formation have a major influence on processes involved in nutrient release to recycling pools. Supplementary soil nutrients come from airborne particles and aerosols, as wet or dry depositions. Such atmospheric deposition can supply more than half of the input of nutrients to some ecosystems.

The major sources of nutrients for aquatic ecosystems are inputs from the surrounding land. These inputs can take the forms of drainage water, detritus and sediment, and precipitation. Flowing aquatic systems are highly dependent on a steady input of detrital material from the watershed through which they flow.

Internal Cycling

Internal cycling of nutrients occurs when nutrients are transformed in ecosystems. Plants take up mineral (mostly inorganic) nutrients from soil through their roots and incorporate them into living tissues. Nutrients in the living tissues occur in various forms of organic compounds and perform different functions in terms of physiology and morphology. When these living tissues reach senescence, the nutrients are usually returned to the soil in the form of dead organic matter. However, nitrogen can be reabsorbed from senescent leaves and transferred to other living tissues. Various microbial decomposers transform the organic nutrients into mineral forms through a process called *mineralization*. The mineralized nutrients are once again available to the plants for uptake and incorporation into new tissues. This process is repeated, forming the internal cycle of nutrients. Within the internal cycles, the majority of nutrients are stored in organic forms, either in living tissues or dead organic matter, whereas mineral nutrients represent a small proportion of the total nutrient pools.

Output of Nutrients

The output of nutrients from an ecosystem represents a loss. Output can occur in various ways, depending on the nature of a specific biogeochemical cycle. Carbon is released from ecosystems to the atmosphere in the form of carbon dioxide via the process of *respiration* by plants, animals, and microorganisms. Nitrogen is lost to the atmosphere in gaseous forms of nitrogen, nitrous oxide, and ammonia, mostly as by-products of microbial activities in soil. Nitrogen is also lost through leaching from the soil and carried out of ecosystems by groundwater flow to streams. Leaching also results in export of carbon, phosphorus, and other nutrients out of ecosystems.

Output of nutrients from ecosystems can also occur through surface flow of water and soil erosion. However, loss of nutrients from one ecosystem may represent input to other ecosystems. Output of organic matter from terrestrial ecosystems constitutes the majority of nutrient input into stream ecosystems. Organic matter can also be transferred between ecosystems by herbivores. For example, moose feeding on aquatic plants can transport nutrients to adjacent terrestrial ecosystems and deposit them in the form of feces.

Considerable quantities of nutrients are lost permanently from ecosystems by harvesting, especially in farming and logging lands, when biomass is directly removed from ecosystems. Fire usually results in the loss of large amounts of nutrients. Fire kills vegetation and converts portions of biomass and organic soil matter to ash. Fire causes loss of nutrients through volatilization and airborne particulate. After fire, many nutrients become readily available, and nutrients in ash are subject to rapid mineralization. If not taken up by plants during vegetation recovery, nutrients are likely to be lost from ecosystems through leaching and erosion.

The Hubbard Brook Example

Nutrient cycling has been studied in several intact ecosystems. One of the most notable experiments was conducted in the Hubbard Brook experimental forest in New Hampshire. The experimental forest was established initially for forest hydrology research. Begun in the early 1960's, one of the longest-running studies of water and nutrient dynamics of forest ecosystems has been on the Hubbard Brook site. Both water and nutrient concentrations in precipitation inputs

PhotoDisc

Supplementary soil nutrients come from airborne particles and aerosols, as dry or wet depositions, such as this fog. Such atmospheric deposition can supply more than half of the input of nutrients to some ecosystems.

and stream outputs were regularly monitored, allowing estimations of nutrient balances over the watershed ecosystems.

One of the major findings from the Hubbard Brook study was that undisturbed forests exhibit regularity and predictability in their input-output balances for water and certain chemical elements. Nitrogen, however, shows a more complex, but still explicable, pattern of stream concentrations. Losses of nitrates from the control watershed are higher in the dormant season, when biological activity is low. Losses are near zero during the growing season, when biological demand for nitrogen by plants and microbes are high. Removal of vegetation in the Hubbard Brook forest had a marked effect on water and nutrient balances. Summer stream flow during the devegetation experiment was nearly four times higher than in the control watershed. The increase in stream flow, combined with increases in the concentration of nutrients within the stream, resulted in increases in loss rates of nitrate much higher than those of undisturbed areas. Similarly, loss of potassium used in large quantities by plants showed a great increase.

Nutrient Uptake and Competition

Ecosystem nutrient cycling is critical for plant growth and ecosystem productivity. Plant uptake of essential nutrient elements is related to nutrient availability, root absorption surface, rooting depth, and uptake kinetics of roots. A nutrient-rich site usually supports more plants of different species than a site with fewer available nutrients. Nutrient competition among plants is usually manifested through physiological, morphological, and ecological traits. Usually grasses and forbs can coexist in one grassland ecosystem, for example, through different rooting depth. To compete for less soluble nutrients such as phosphorus, plants usually extend their root surfaces using symbiotic relationships with mycorrhizae. Differential seasonality in nutrient uptake and rooting depth become more critical to compete limited nutrients.

Yiqi Luo

See also: Carbon cycle; Ecosystems: studies; Erosion and erosion control; Hydrologic cycle; Nitrogen cycle; Nutrients; Nutrition in agriculture; Phosphorus cycle; Root uptake systems; Soil.

Sources for Further Study

Aber, J. D., and J. M. Melillo. *Terrestrial Ecosystems*. 2d ed. San Diego: Academic Press, 2001. Advanced ecosystem ecology book contains a comprehensive discussion of nutrient cycling.

Likens, G. E., et al. *Biogeochemistry of a Forested Ecosystem*. New York: Springer-Verlag, 1977. The book describes studies of nutrient cycling in the Hubbard Brook watershed.

Schlesinger, W. H. *Biogeochemistry: An Analysis of Global Change*. 2d ed. San Diego: Academic Press, 1997. A comprehensive book on nutrient cycling.

NUTRIENTS

Categories: Cellular biology; nutrients and nutrition

Plant nutrients are the molecular compounds necessary to maintain plant life.

Plants, like animals and other forms of life, require a great diversity of compounds. Molecules of these compounds are used to build and maintain cells and to perform other necessary life processes, such as growth and reproduction, as well as respiration and photosynthesis.

Plants manufacture, by the process of photosynthesis, simple sugars. From these are formed polysaccharides (starches, cellulose) and, eventually, a variety of lipids (fats, oils, and waxes), hundreds of different protein molecules, and many other organic compounds. Normal functioning of plants re-

Plant Nutrients

Nutrient	*Functions*
Macronutrients	
Calcium	Formation of cell walls, regulator of membrane permeability, enzyme regulator
Carbon	Basic to all structures and functions
Hydrogen	Basic to all structures and functions
Magnesium	Chlorophyll component, enzyme activator
Nitrogen	Formation of amino acids, proteins, nucleotides, nucleic acids, chlorophyll, coenzymes
Oxygen	Basic to all structures and functions
Phosphorus	Formation of ADP and ATP, nucleic acids, coenzymes, phospholipids
Sulfur	Component of proteins, coenzyme A, thiamine, biotin
Micronutrients	
Boron	Calcium metabolism, nucleic acid synthesis, cell membranes, carbohydrate synthesis
Chlorine	Osmosis, photosynthesis
Copper	Redox enzymes
Iron	Chlorophyll synthesis, electron transport system, cellular respiration
Manganese	Enzyme activator, chloroplast membrane activity, photosynthesis
Molybdenum	Nitrogen fixation
Zinc	Enzyme activator, auxin synthesis
Some species need:	Aluminum, cobalt, silicon, sodium

quires that they absorb, usually by roots from soil, a variety of chemical elements. From these elements, which are combined with water, carbon dioxide, and other compounds, the more complex compounds are formed.

Plant Nutrition

Plant nutrition involves the uptake from the environment of the raw materials needed for performing a variety of biochemical processes, the conduction of these substances to all parts of the plant, and their use in growth and metabolic processes. Most of these materials are in the form of ions dissolved in soil water. The water occupies the spaces between solid soil particles and is absorbed through the roots and then conducted upward by means of *xylem* tissue located in roots, stems, and leaves.

Present in soil water, and therefore in a plant growing in that soil, are more than sixty chemical elements. Some of these, however, are present only incidentally and perform no known function. Others, known as *essential elements*, are used in some particular way by the plant. These elements have been the object of study by many plant physiologists.

Essential Elements

By the mid-1800's, chemistry had become sufficiently advanced that botanists could analyze the chemical content of plants. They discovered that in the ash remaining after a plant had been burned were a variety of minerals. In order to be considered an essential element, that element must be necessary for normal metabolic processes, growth, and reproduction; another element cannot replace it. Essential elements are classified as either *macronutrients* or *micronutrients*. Although both types of nutrients are required, macronutrients are needed in much larger amounts.

To recognize the difference between the two categories of minerals, one may consider the following. When freshly harvested plants are heated in an oven to remove the water, the dry matter remaining can be analyzed. Macronutrients are those required in concentrations of at least 1,000 milligrams per kilogram of dry plant matter. In contrast, micronutrients are required in much smaller or even trace amounts, generally less than 100 milligrams per kilogram of dry matter.

Among the macronutrients, nitrogen is a key element needed in large amounts to form proteins, nucleotides, chlorophyll, and coenzymes. Phosphorus is also required to build nucleic acids, but it is also necessary to form adenosine triphosphate (ATP) and adenosine diphosphate (ADP), energy-carrying compounds vital to all cells. Among the several functions of calcium are its roles as a component of cell walls and in changing the permeability of cell membranes. Magnesium is necessary to activate certain enzymes; also a single atom of magnesium occupies a central position in each molecule of chlorophyll. Sulfur is essential to synthesizing proteins, coenzyme A, thiamine, and biotin.

Among the micronutrients, iron functions in the electron transport system, playing a role in cellular respiration; also it is vital for chlorophyll synthesis. Chlorine helps to maintain an ionic balance by controlling osmosis. Manganese is involved in activating many enzymes. Boron is believed to be involved in carbohydrate synthesis; also it is required for nucleic acid synthesis. Zinc is required for the synthesis of the plant hormone auxin; it also is involved in enzyme activation. Copper is present in the active site of redox enzymes and electron carriers. Nickel plays a role in enzyme functioning in the metabolism of nitrogen. Molybdenum is required for nitrogen fixation, the process by which free nitrogen (N_2) in the air is converted into nitrates.

Identifying Essential Elements

To identify a particular mineral as essential, a classic *protocol* (series of steps) is followed. The method used today was developed by Julius von Sachs, an early German plant physiologist, in 1860: Two seedlings of the same kind of plant are grown in separate containers. One container contains what is considered to be a complete growth medium. The other container contains all but a single mineral. If the growth of the seedling in the latter container is abnormal, or if its normal flowering and seed production are not successful, the missing element is determined to be essential.

Just as animals suffer symptoms when certain vitamins or minerals are absent in their diets, so plants are affected when they suffer a deficit in a particular essential element. Such symptoms are used by farmers and gardeners to determine which types of fertilizer need to be added to the soil in order to correct the problem. Some of the more common mineral deficiencies include nitrogen, iron, magnesium, calcium, and phosphorus deficiencies. For example, because plants require large amounts of nitrogen for growth, this mineral is often not present in sufficient amounts. A typical symptom is a uniform yellowing of the older leaves, a condition called chlorosis. An iron deficiency is diagnosed when the youngest leaves turn yellow. In the case of a magnesium deficiency, the older leaves are yellow between the veins. A calcium deficiency causes the growing points to die back; young leaves are yellow and crinkly. Plants with a phosphorus deficiency turn dark green and have leaves with purple veins.

Alternative Nutrient Sources

Whereas most plants, both wild or cultivated, absorb required nutrients from soil water, some plants obtain nutrients by other means. Plants such as Venus's fly trap, the pitcher plant, and various other carnivorous plants supplement soil nutrients by trapping insects or other small animals. Once an animal is caught, its protein is digested, yielding nitrogen that is made available to the plant. Intricate traps and other means of attracting and capturing insects have evolved over long periods of time. The traps are commonly modified leaves. Carnivorous plants generally inhabit sunny habitats with soils that have low levels of nitrogen. Thus nitrogen from animal protein is necessary to supplement that obtained in the more usual way from the soil.

Because of the special role of the proteins in plant (and animal) cells, nitrogen, a key element required to form proteins, is of special concern. Many plant species, especially those of the pea or legume family (*Fabaceae*) have root nodules, or mycorrhizae, filled with bacteria that are able to convert free nitrogen into compounds of nitrogen called nitrates. This process, called *nitrogen fixation*, makes nitrates available to the plant (as well as to the bacteria) as a usable source of nitrogen. Also, the surrounding soil is enriched, making nitrogen available to other plants growing in the soil. Thus, nitrogen taken from the soil in large quantities, especially by cultivated crops, is replaced. The common agricultural practice of rotating crops, in which a leguminous crop such as alfalfa or clover is alternated with corn or wheat, which removes large amounts of nitrogen from the soil, takes advantage of the nitrogen that has been fixed by legumes.

Hydroponic cultures allow for the growing of plants without soil. The plants survive on solutions that contain water and essential minerals. Plants are typically grown in greenhouses with their roots bathed in circulating water containing the minerals. Provision must be made to keep the water aerated. Hydroponic plants can be grown year-round, but the added expense of hydroponic culture limits its use to fruits and vegetables made available to specialty markets.

Thomas E. Hemmerly

See also: Biofertilizers; Carbohydrates; Carnivorous plants; Fertilizers; Hydroponics; Lipids; Nitrogen fixation; Nutrition in agriculture; Photosynthesis; Proteins and amino acids; Root uptake systems; Soil; Water and solute movement in plants.

Sources for Further Study

Purves, W. K., et al. *Life: The Science of Biology*. 6th ed. New York: W. H. Freeman, 2001. Chapter 36 of this excellent college text deals with plant nutrients.

Raven, Peter H., Ray F. Evert, and Susan E. Eichhorn. *Biology of Plants*. 6th ed. New York: W. H. Freeman, 1999. This comprehensive and lavishly illustrated textbook of botany explores various aspects of the relationships of plants to soil and minerals.

Taiz, Lincoln, and Eduardo Zeiger. *Plant Physiology*. 3d ed. Sunderland, Mass.: Sinauer Associates, 2002. This standard textbook covers all aspects of plant physiology, including plant nutrients and their use.

NUTRITION IN AGRICULTURE

Categories: Agriculture; economic botany and plant uses; nutrients and nutrition

Extensive research has been conducted to investigate the nutritional requirements of different crops and ways to enhance soil fertility, which has greatly benefited agricultural production.

Even though people have known for more than two thousand years that adding mineral elements, such as plant ash or lime, to the soils can improve plant growth, the systematic study of plant nutrition is a relatively young science, considering humanity's long history of cultivating crops. About 250 years ago, farmers and gardeners started to ask the question, "What makes plants grow?" It was widely believed that soil humus, a brown or black organic substance resulting from the partial decay of plant and animal matter, provided plants with carbon for making sugar and starch, and substances such as saltpeter, lime, and phosphates simply helped the humus to be more useful. It was not until around 1840 that the German chemist Justus von Liebig (1803-1873) helped to compile and summarize the scattered information on the importance of mineral elements for plant growth, and plant nutrition began to be established as a scientific discipline. Since then, great progress has been made in the study of plant nutrition.

It is now known that, aside from carbon, hydrogen, and oxygen, which plants get from air and water, about a dozen other nutrients are needed for plant growth. They can be divided into three classes. The primary (or major) nutrients, including *nitrogen*, *phosphorus*, and *potassium*, are needed in larger quantities than are the secondary nutrients, *calcium*, *magnesium*, and *sulfur*. These in turn are required in greater quantity than the trace (or minor) nutrients iron, boron, manganese, copper, zinc, molybdenum, chloride, and nickel. These nutrients are contained in the minerals and organic matter in the soil. Many more elements are found in both soils and plants—for instance, aluminum, cobalt, fluorine, iodine, and sodium. They may not be needed by all plants and may be either beneficial or toxic to plant growth. Silicon, sodium, and cobalt are beneficial to some plants.

Soil and Plant Nutrients

Soil is the natural medium in which crops grow. The nutrient content of a soil and the availability of the nutrients to crops are important factors that determine a soil's productivity. Soil nutrients are mostly present in forms not immediately available to plants, such as being adsorbed onto or as constituents of soil mineral particles, or in organic matter. They become available through slow processes, such as biological decomposition of organic matter called *mineralization*, chemical reactions on soil minerals called *weathering*, and release from soil particles.

Soil nutrients in agricultural land may be gradually lost with the removal of harvested products, such as grain and straw, which take with them considerable amounts of all nutrients. In addition, nutrients may be lost from the soil through *leaching* and *erosion*. Therefore, it is common for field crops to suffer from nutrient deficiencies. On the other hand, certain soils may present the crops with problems of mineral toxicity, that is, excess amounts of particular elements. To maintain healthy growth of crops, it is necessary to correct these problems.

Soil testing, visual diagnosis of the plant, and plant tissue analysis can all be used to evaluate the fertility status of the soil and detect deficiencies and toxicities in soil nutrients. The results of such tests provide the basis for recommendations on the application of fertilizer and soil amendments.

Coping with Nutrient Deficiencies

Based on knowledge of plant nutrition and experience accumulated over a long period of time, nutrient management practices have been established to enhance soil fertility and overcome crop nutritional problems by balancing the use of mineral fertilizers combined with organic and biological sources of plant nutrients. In practice, different methods may have both advantages and disadvantages, depending on the particular set of local conditions.

Organic sources of plant nutrients include farmyard manure (animal waste products), green manure (plant products), and compost. They contain small amounts of nutrients and are often bulky in nature. When added to the soil, their main value is to provide organic matter that promotes microbial activity and improves soil structure, aeration, and water-holding capacity, enabling the soil to respond better to fertilizers and irrigation. The organic matter may also supply micronutrients and help to make the phosphate in the soil more available to crops. The disadvantages of applying organic manures are that they may be expensive, difficult to handle, or likely to release excess nutrients into the environment.

Nitrogen availability is one of the most limiting factors in crop productivity. Although nitrogen is abundantly available in the air, this form of it is not directly usable by the plants. Some crops, such as legumes, can form a beneficial relationship called *symbiosis* with certain bacteria capable of converting atmospheric nitrogen to ammonia, a process known as biological *nitrogen fixation*. The bacteria supply the plant with nitrogen (ammonia), while the plant provides the bacteria with organic compounds for use as energy. Legumes can be used as a source of nitrogen when planted with cereals. Fast-growing legumes can be grown early in the season and then ploughed under to provide nitrogen for the main crop. Recent research has identified the mechanisms controlling the expression of the nitrogen-fixation genes at the molecular level, and one of the goals of ongoing research is to explore various approaches to constructing a viable nitrogen-fixing system for use with nonlegumes.

Low soil phosphorus availability is another constraint on plant growth. Phosphorus is progressively lost from soil through weathering, and reactions with various soil constituents substantially reduce phosphorus available to plants. One research effort has been directed toward investigating various root characteristics that are useful in phosphorus uptake by plants. The findings may help farmers to breed for crops that are better able to grow in soils with low phosphorus.

Nutrient removal from cultivated land usually exceeds the natural rate of nutrient input. One remedy is to add appropriate and balanced fertilizers back to the soil. Mineral fertilizers are widely used to supply either single nutrients or multiple nutrients in combination. Foliar spray is effective for correcting deficiencies in micronutrients. Timing and dosage of applications is important, because the nutritional requirements of a crop vary with its stage of growth. Insufficient supply reduces crop yields, but applying too much or at a wrong time is not only wasteful but also potentially harmful to the environment.

Overcoming Nutrient Toxicities

Soil can become acidified as a result of its own physical properties, microbial activity, climate, vegetation, and the addition of acidifying fertilizers. As a result, important nutrients can be lost, and toxicities from aluminum and manganese may occur. Liming, the application to land of a material containing calcium, usually chalk or limestone, is often used as a standard measure in order to reduce problems of soil acidity.

Salts introduced in irrigation water, blown inland from oceans, or produced by weathering may accumulate in topsoil and cause toxicity to crops. This is normally controlled by applying more water than the crop can use, so that excess salts are leached downward below the root zone. It is important, however, that the irrigation water does not have high salt content and that drainage is not a problem. The addition of calcium salts, such as calcium sulfate, together with organic manures, is also effective in treating salt-affected soils.

One focus of research in the field of plant nutrition has been the study of the mechanisms plants employ in avoiding or tolerating toxic elements, such as aluminum, manganese, and salts, and accessing scarce nutrients, such as phosphorus, in the soil. Knowing how plants cope with various nutrient stresses will help the effort of breeding crops that are better able to withstand adverse soil nutrient conditions and achieve high yields by making use of their own genetic potential.

Zhong Ma

See also: Agriculture: modern problems; Agriculture: traditional; Agriculture: world food supplies; Agronomy; Biofertilizers; Composting; Eutrophication; Fertilizers; Green Revolution; High-yield crops; Hydroponics; Nitrogen cycle; Nitrogen fixation; Nutrients; Organic gardening and farming; Phosphorus cycle; Roots; Root uptake systems; Slash-and-burn agriculture; Soil management; Sustainable agriculture.

Sources for Further Study

Glass, A. D. M. *Plant Nutrition: An Introduction to Current Concepts*. Boston: Jones and Bartlett, 1989. This is an introductory textbook to plant mineral nutrition. Includes charts, illustrations, photographs, and index.

Lagreid M., O. C. Bockman, and O. Kaarstad. *Agriculture, Fertilizers, and the Environment*. Wallingford, England: CABI, 1999. Presents environmental and sustainability issues related to fertilizer use and its role in ensuring adequate food supplies. Includes illustrations, tables, bibliographical references, and index.

Ludwick, A. E., et al., eds. *Western Fertilizer Handbook*. 8th ed. Danville, Ill.: Interstate, 1995. An educational text for developing a better understanding of agronomic principles and practices, with convenient references on soil, water, plants, and fertilizer. Includes charts, tables, photographs, and index.

Marschner, H. *Mineral Nutrition of Higher Plants*. 2d ed. San Diego: Academic Press, 1995. Comprehensive discussion of plant nutritional physiology and plant-soil relationships. Includes illustrations, tables, bibliographical references, and index.

Prasad, R., and J. F. Power. *Soil Fertility Management for Sustainable Agriculture*. Boca Raton, Fla.: CRC Lewis, 1997. Presents the principles of controlling soil fertility in the various climates of the world, with examples of soil fertility management for sustainable agriculture. Includes illustrations, tables, bibliographical references, and index.

OIL BODIES

Categories: Anatomy; cellular biology; physiology

Oil bodies, also called oleosomes, are plant cell organelles that serve as storage structures for lipids, primarily triacylglycerols.

Oil bodies are found in many plant seeds and represent an efficient storage form of energy for a germinating embryo. Other, non-oilseed, plants rely primarily on starch as a storage form of energy in their seeds. On a weight basis, oils contain more the twice the amount of energy as starches, because they contain more carbon and hydrogen atoms and fewer oxygen atoms than starch. Oil thus represents a more compact storage form of energy. Some high oil content species include castor beans, canola, safflower, peanuts, sunflower, macadamia nuts, and hazelnuts. All of these contain about 50 percent or more oil on a dry-weight basis. Despite the ubiquitous presence of corn and soybean oil in the marketplace, these oils contain only 5 percent and 17 percent oil, respectively. Although oil bodies are found predominantly in seeds, they may also be present in fruits, such as olives and avocados.

Structure

Oil bodies are unique in structure. They are surrounded by single-layer phospholipid membrane, as opposed to the bilayer membrane found around most other plant organelles. The single-layer membrane is derived from the endoplasmic reticulum (ER), from which oil bodies are believed to originate. *Triacylglycerols* are synthesized in the ER and accumulate within the two layers of its phospholipid membrane. Eventually the two layers are separated as oil accumulates and pinches off to form the oil body. The hydrophobic (water-avoiding) tails of the resulting phospholipid monolayer are associated with the oily interior, and the polar heads face outward. In addition, proteins called *oleosins* are present in the single-layer membrane and are thought to prevent fusion with other oil bodies. Oleosomes appear to be consistent in size, about 1 micron. Thus, a cell with a large amount of oil will have more oil bodies present.

Function

When seeds imbibe water and germination is initiated, many metabolic activities are activated. In order for a germinating seed to utilize the energy in an oil body, the triacylglycerols must be converted to a form of sugar, typically sucrose, that can be used by the growing embryo. The sequence of enzymatic events that makes this possible involves the close coordination of three organelles: the oleosome, the mitochondrion, and the highly specialized *glyoxysome*. It is in the glyoxysome where the many carbon atoms present in the fatty acids of the triacylglycerols are removed, two at a time, in a process called oxidation. These reactions, together with those of the mitochondria, are often referred to as the *glyoxylate cycle*. The end products allow for *reverse glycolysis*, which produces sugars that can be metabolized by typical respiratory pathways in the growing seedling.

Thomas J. Montagno

See also: Cytoplasm; Endoplasmic reticulum; Glycolysis and fermentation; Lipids; Membrane structure; Mitochondria; Plasma membranes; Respiration.

Sources for Further Study

Dennis, David T., David B. Layzell, Daniel D. Lefebre, and David H. Turpin, eds. *Plant Metabolism*. 2d ed. Singapore: Longman, 1997. Advanced textbook with three chapters devoted to lipid formation and breakdown. Includes illustrations, bibliographic references, and index.

Salisbury, Frank B., and Cleon W. Ross. *Plant Physiology*. 4th ed. Belmont, Calif.: Wadsworth, 1992. Textbook with a chapter devoted to lipids and other natural products. Includes illustrations, photographs, bibliographic references, and index.

OLD-GROWTH FORESTS

Categories: Forests and forestry; ecosystems; environmental issues

Ancient ecosystems, old-growth forests consist of trees that have never been harvested. These forests are, in some cases, the only habitat for a number of plant and animal species.

The timber industry views the large, old trees as a renewable source of fine lumber, but environmentalists see them as part of an ancient and unique ecosystem that can never be replaced. In the 1970's scientists began studying the uncut forests of the Pacific Northwest and the plants and animals that inhabited them. In a U.S. Forest Service publication, *Ecological Characteristics of Old-Growth Douglas-Fir Forests* (1981), Forest Service biologist Jerry Franklin and his colleagues showed that these forests were not just tangles of dead and dying trees but rather a unique, thriving ecosystem made up of living and dead trees, mammals, insects, and even fungi.

Old-Growth Forest Ecosystem

The forest usually referred to as old growth occurs primarily on the western slope of the Cascade Mountains in southeast Alaska, southern British Columbia, Washington, Oregon, and Northern California. The weather there is wet and mild, ideal for the growth of trees such as Douglas fir, cedar, spruce, and hemlock. Studies have shown that there is more biomass, including living matter and dead trees, per acre in these forests than anywhere else on earth. Trees can be as tall as 300 feet (90 meters) with diameters of 10 feet (3 meters) or more and can live as long as one thousand years. The forest community grows and changes over time, not reaching *biological climax* until the forest primarily consists of hemlock trees, which are able to sprout in the shade of the sun-loving Douglas fir.

One of the most important components of the old-growth forest is the large number of standing dead trees, or *snags*, and fallen trees, or *logs*, on the forest floor and in the streams. The fallen trees rot very slowly, often taking more than two hundred years to decompose completely. During this time they are important for water storage, as wildlife habitat, and as "nurse logs" where new growth can begin. In fact, seedlings of some trees, such as western hemlock and Sitka spruce, have difficulty competing with the mosses on the forest floor and need to sprout on the fallen logs.

Another strand in the complex web of the forest consists of mycorrhizal fungi (mycorrhizae), which attach themselves to the roots of the trees and enhance their uptake of water and nutrients. The fruiting bodies of these fungi are eaten by small mammals such as voles, mice, and chipmunks, which then spread the spores of the fungi in their droppings. There are numerous species of plants and animal wildlife that appear to be dependent on this ecosystem to survive.

Protecting the Forest

By the 1970's most of the trees on timber industry-owned lands had been cut. Their replanted forests, known as second growth, would not be ready for harvest for several decades, so the industry became increasingly dependent on public lands for their raw materials. Logging of old growth in the national forests of western Oregon and Washington increased from 900 million board feet in 1946 to more than 5 billion board feet in 1986.

Environmentalists claimed that only 10 percent of the region's original forest remained. Determined to save what was left, they encouraged the use of the evocative term "ancient forest" to counteract the somewhat negative connotations of "old growth." Then they were given an effective tool in the northern spotted owl. This small bird was

AP/Wide World Photos

The ancient 5,500-acre Headwaters Forest in Humboldt County, California, is among the largest remaining coastal redwood forests in the world.

found to be dependent on old growth, and its listing under the federal Endangered Species Act in 1990 caused a decade of scientific, political, and legal conflict.

Under law, the U.S. Forest Service was required to protect enough of the owl's habitat to ensure its survival. An early government report identified 7.7 million acres of forest to be protected for the bird. Later, the U.S. Fish and Wildlife Service recommended 11 million acres. In 1991 U.S. District Court judge William Dwyer placed an injunction on all logging in spotted owl habitat until a comprehensive plan could be finalized. The timber industry responded with a prediction of tens of thousands of lost jobs and regional economic disaster. In 1993 President Bill Clinton convened the Forest Summit conference in Portland, Oregon, to work out a solution. The Clinton administration's plan, though approved by Judge Dwyer, satisfied neither the industry nor the environmentalists, and protests, lawsuits, and legislative battles continued.

As the twentieth century came to an end, timber harvest levels had been significantly reduced, the Northwest's economy had survived, and additional values for old-growth forests were found: habitat for endangered salmon and other fish, a source for medicinal plants, and a repository for benefits yet to be discovered. The decades-long controversy over the forests of the Northwest had a deep impact on environmental science as well as natural resource policy and encouraged new interest in other native forests around the world, from Brazil to Malaysia to Russia.

Joseph W. Hinton

See also: Conifers; Forests; Forest management; Logging and clear-cutting; Mycorrhizae; Sustainable forestry; Timber industry.

Sources for Further Study

Dietrich, William. *The Final Forest: The Battle for the Last Great Trees of the Pacific Northwest*. New York: Penguin Books, 1993. Studies the old-growth controversy with a balanced look at individuals affected by it, conservationists and loggers alike.

Durbin, Kathie. *Tree Huggers: Victory, Defeat, and Renewal in the Northwest Ancient Forest Campaign*. Seattle: Mountaineers, 1996. A detailed examination of the politics of the old-growth controversy, with special emphasis on events between 1987 and 1996.

Kelly, David, and Gary Braasch. *Secrets of the Old Growth Forest*. Salt Lake City, Utah: Gibbs-Smith, 1988. A coffee-table book with color photographs and extensive text that provides an excellent introduction to the topic.

Maser, Chris. *Forest Primeval: The Natural History of an Ancient Forest*. Corvallis: Oregon State University Press, 2001. A fascinating biological biography of a hypothetical thousand-year-old forest.

OOMYCETES

Categories: Molds; *Protista*; taxonomic groups; water-related life

Oomycetes are a diverse group of fungus-like eukaryotic microorganisms that form the phylum Oomycota. *First classified within the kingdom* Fungi, *oomycetes are now unambiguously recognized as distinct from fungi and more closely related to heterokonts, such as brown algae, phylum* Phaeophyta.

The more than five hundred species of oomycetes, commonly known as water molds, white rusts, or downy mildews, are essentially saprophytic but include pathogens of plants, insects, crustaceans, fish, vertebrate animals, and various microrganisms. Plant pathogenic oomycetes cause devastating diseases on several crop, ornamental, and native plants. Animal pathogenic oomycetes can cause severe losses in aquaculture and fisheries. Both have had a significant impact on human history.

Evolutionary History

Traditionally, and due essentially to their filamentous growth habit, oomycetes have been classified in the kingdom *Fungi*. However, modern molecular and biochemical analyses as well as morphological features suggest that oomycetes share little taxonomic affinity with filamentous fungi but are more closely related to brown algae (phylum *Phaeophyta*) in the kingdom *Protista*. This position is supported by molecular phylogenies based on ribosomal RNA (ribonucleic acid) sequences, compiled amino acid data for mitochondrial proteins, and protein encoding chromosomal genes. The oomycetes also display a number of biochemical and morphological characteristics that distinguish them from the fungi and confirm their affinity to brown algae and other heterokonts.

The cell walls of oomycetes are composed mainly of glucans and cellulose and, unlike fungal cell walls, contain little or no chitin. The zoospores display two flagella, with an ultrastructure similar to that of the flagella of the motile spores of heterokont algae. The oomycetes also contain the energy storage chemical mycolaminarin, a molecule that is also found in kelps and diatoms.

Taxonomic Classes

The subdivison of oomycetes into taxonomic classes remains under debate. Typically four classes of oomycetes are identified. These include *Saprolegniales*, *Leptomitales*, *Lagenidales*, and *Peronosporales*. Some authors elevated the plant pathogenic genera *Phytophthora* and *Pythium* to a separate class, named *Pythiales*. Recent molecular phylogenetic studies using ribosomal and mitochondrial sequences have started to unravel the evolutionary relationships between the different classes of oomycetes. The *Peronosporales* and

Pythiales, which together account for the majority of plant pathogenic genera, form an ancient monophyletic group, suggesting that acquisition of plant pathogenicity probably occurred early in the evolution of this lineage. Most of the saprophytic and animal pathogenic species are restricted to the other classes. *Aphanomyces*, a genus with strong affinity to the *Saprolegniales*, includes both animal and plant pathogenic species.

General Features

The oomycetes inhabit primarily aquatic and moist soil habitats. They are often very abundant and can be easily cultured from both freshwater and saltwater ecosystems, as well as from a variety of agricultural or natural soils. However, several species are mainly terrestrial, including obligate biotrophic pathogens of plants that depend on air currents to disperse their spores.

The basic somatic structure of a majority of oomycete species is an extending funguslike thread, the *hypha*, that grows into a branched network of filaments, the *mycelium*. Oomycetes are known as *coenocytic* organisms; that is, their mycelium lacks septa or crosswalls that divide the hypha, except to separate it from the reproductive organs.

Both asexual and sexual reproductive structures occur. The primary asexual reproductive organ is the sporangium that differentiates at the tip of a vegetative hypha to produce and release motile zoospores with two flagella. The zoospores can germinate directly or indirectly to produce a vegetative mycelium or can differentiate into secondary zoospores. Sexual reproduction involves the interaction of a male antheridia with a female oogonia through a fertilization tube that allows the male nuclei to migrate into the oogonium. Some oomycetes are self-fertile or homothallic, whereas others are self-sterile or heterothallic and require that strains with different mating types come into contact to achieve sexual reproduction. The sexual spores are the oospores, which can survive desiccation and starvation over long periods of time. Under favorable environmental conditions, the oospores germinate to form vegetative mycelium or to release zoospores. Oospores are also the structures that gave the oomycetes their name of egg fungi. Oomycetes are diploid in the dominant vegetative phase with meiosis occurring only during gametogenesis.

Economic Importance

Saprophytic oomycetes play an important role in the decomposition and recycling of decaying matter in aquatic and soil environments. In addition, both plant and animal pathogenic oomycetes can cause serious economic impact by destroying crop, ornamental, and native plants as well as fish and other aquatic organisms.

Typically, oomycete diseases are difficult to manage and require the use of specific chemicals (fungicides or oomycides). Sources of sustainable genetic resistance in plants to oomycetes are limited. In addition, most oomycetes, such as *Phytophthora*, exhibit tremendous ability to adapt to chemical and genetic resistance through the development of new resistant strains. For example, the appearance in the 1990's of *Phytophthora infestans* strains resistant to the chemical metalaxyl resulted in potato late blight epidemics in the United States that were severe and destructive. Most modern research focuses on innovative approaches for the management of oomycete diseases, including the use of plant breeding, genetic engineering, and genomic technologies.

Plant-Pathogenic Oomycetes

Plant-associated oomycetes may be facultatively or obligately pathogenic. Pathogenic oomycetes form specialized infection structures, also found in fungi, such as *appressoria* (penetration structures) and *haustoria* (feeding structures). Plant-pathogenic oomycetes include about sixty species of the genus *Phytophthora*, several genera of the biotrophic downy mildews, and more than one hundred species of the genus *Pythium*. *Phytophthora* species cause some of the most destructive plant diseases in the world and are arguably the most devastating pathogens of dicotyledonous plants.

The most notable plant-pathogenic oomycete is *Phytophthora infestans*, the Irish Potato Famine pathogen, which causes late blight, a disease of potato and tomato. Introduction of this pathogen to Europe in the mid-nineteenth century resulted in the potato blight famine and the death and displacement of millions of people. Today, *Phytophthora infestans* remains a prevalent pathogen causing multibillion dollar losses in potato production worldwide.

Other economically important *Phytophthora* diseases include root and stem rot caused by *Phytophthora sojae*, which hampers soybean production in several continents, and black pod of cocoa caused by *Phytophthora palmivora* and *Phytophthora megakarya*, a recurring threat to worldwide chocolate production. The introduction of exotic plant pathogenic oomycetes to natural ecosystems can also cause devastating effects. For example, *Phytophthora cinnamomi* has decimated native plants in Australia and South America. More recently, sudden oak death, a disease caused by a new species, *Phytophthora ramorum*, has emerged as a severe disease of oak trees along the Pacific Coast of the United States. Other notorious oomycete pathogens include the obligate biotrophs *Plasmopara viticola*, the agent of downy mildew of grapevine, as well as *Albugo* and *Peronospora* species, which cause white rust and downy mildews on several crops.

Animal-Pathogenic Oomycetes

Animal-pathogenic oomycetes are common. At least one oomycete species, *Pythium insidiosum*, is known to infect mammals, sometimes including humans, fatally. However, most economic impact on animals is caused by oomycetes that infect fish, fish eggs, and crustaceans. Examples include *Saprolegnia parasitica*, a ubiquitous pathogen of fish that is common in aquaria and can cause severe losses in aquaculture, particularly when fish density is too high or fish diet is unbalanced. Another important animal pathogenic oomycete is *Aphanomyces astaci*, the agent of crayfish plague that decimated European crayfish populations following its introduction from North America. Animal-pathogenic oomycetes can also have beneficial effects. At least one species of oomycetes, the insect pathogen *Lagenidium giganteum*, has been commercialized by a California company as a biocontrol agent for mosquitoes.

Sophien Kamoun

See also: Algae; Biopesticides; Brown algae; Cladistics; Diseases and disorders; Eukaryotic cells; Flagella and cilia; Fungi; Heterokonts; Molecular systematics; *Protista*; Resistance to plant diseases.

Sources for Further Study

Alexopoulos, C. J., C. W. Mims, and M. Blackwell. *Introductory Mycology*. 4th ed. New York: John Wiley & Sons, 1996. This textbook contains one chapter on oomycetes. Includes illustrations, bibliographical references, and index.

Erwin, D. C., and O. K. Ribeiro. *Phytophthora Diseases Worldwide*. St. Paul, Minn.: APS Press, 1996. This monumental treatise is the main reference work for *Phytophthora*. It includes a description of methods of study and chapters detailing each of the more than sixty species of *Phytophthora*. Includes illustrations, bibliographical references, and index.

Margulis, L., and K. V. Schwartz. *Five Kingdoms: An Illustrated Guide to the Phyla of Life on Earth*. New York: W. H. Freeman, 2000. This book includes one chapter on oomycetes that discusses the general features of their life cycles and habits. Includes illustrations, bibliographical references, and index.

ORCHIDS

Categories: Angiosperms; economic botany and plant uses; gardening; taxonomic groups

Orchids are the largest family of monocots among the angiosperms (flowering plants), with between twenty-five thousand and thirty thousand species, and new species are continually being described.

There are numerous natural and artificial hybrids of orchids (family *Orchidaceae*). The floral and vegetative variation in this family is enormous, and most orchid specialists consider the orchids to be undergoing a great deal of evolutionary activity as well. Orchids follow the general monocot pattern of having flower parts in three's. They differ from other monocots in several ways, however.

First, the stamens are fused together. Orchids are divided into two major groups based on their anthers and stamens: the *diandrous* orchids, which have two functional anthers and a sterile stamen, which in the lady's slipper orchids (such as *Cypripedium, Selenipedium, Paphiopedilum,* and *Phragmipedium*) becomes modified into a staminode, and the *monandrous* orchids (all other orchids), which have one functional anther.

Second, the stamens are fused with the pistil to form a structure called a *column*, which has at its tip a *rostellum*, a gland that separates the pollinia from the stigmatic surface. The fusion of the stamens and pistils into a column also occurs in the milkweed family (*Asclepiadaceae*) in the dicots.

Third, the pollen grains are formed together into packets called *pollinia*.

Fourth, flowers have bilateral symmetry or asymmetry, as in *Tipularia*.

Fifth, *resupination* is the most common condition for orchid flowers. In resupinate orchids, the pedicel and ovary undergo a 180 degree twisting process so that as the flower matures the lip is lowermost and the column uppermost. Although some orchids (such as *Calopogon, Polystachya, Malaxis,* and *Platanthera nivea*) are nonresupinate, most are resupinate.

Sixth, the third petal, or lip, is usually highly modified in size, shape, or color when compared to the other two petals.

Seventh, the ovary is inferior.

Eighth, orchids produce large numbers of tiny seeds that lack endosperm. Most orchids have bisexual flowers; however, some (*Catasetum, Cycnoches, Mormodes*) produce unisexual (male or female) flowers. Flowers may last only a day (*Sobralia*), or for more than a month (some *Paphiopedilium* and *Dendrobium*).

Richard Martin/Photo Agora

*This species of lady's slipper orchid (*Cypripedium calceolus*) is rare over much of its natural range because of collecting.*

Pollination

Pollination ranges from promiscuous to pseudocopulation. Promiscuous pollination occurs when almost any appropriate insect can transfer the pollinia successfully from one flower to another. Pseudopollination involves flowers which mimic the size, shape, smell, or movements of female wasps or bees (*Ophrys, Caladenia, Drakaea*). These flowers attract the male wasp or bee, which attempts to mate with the flower. Pollinia are thereby attached to the insect's body. Eventually the frustrated male leaves the flower and finds another flower, again trying to mate with it but succeeding only in depositing the pollinia from the first flower onto the second flower's stigmatic surface.

Mycorrhizae and Seed Germination

To germinate, orchid seeds, which lack endosperm, must form a symbiotic relationship with the mycelium of an appropriate fungus (mycorrhiza) which provides the seeds with nutrients and water.

Vegetative Plant Structure

Vegetative plants are herbs or, rarely, vines (*Vanilla*), perennials or, rarely, annuals (*Zeuxine strateumatica*). Vegetative parts are normally roots, stems, and leaves, which can undergo many different modifications. The roots of terrestrial orchids are similar to other terrestrial monocots and are nonphotosynthetic. Some are modified into food storage structures (tuberoids). Roots of epiphytic orchids have a specialized outer multilayered tissue called the velamen, which forms a spongy, whitish sheath around the root. The velamen allows absorption of water and dissolved mineral nutrients.

Stems are frequently modified into creeping rhizomes and often into water- and food-storage structures, such as corms, tubers, or pseudobulbs. These structures are usually aboveground and photosynthetic. Plants may be minute (some *Bulbophyllum* and *Platystele*) or gigantic (about 44 feet tall, such as *Sobralia altissima*) or may form large clusters of pseudobulbs weighing several hundred pounds (*Grammatophyllum*).

Growth Patterns, Distribution, and Habitats

There are two basic growth forms in orchids. In the *sympodial* form, successive new stem growth originates from the base of the preceding stem growth (as in *Cattleya* and *Cypripedium*). In the *monopodial* form, new growth comes from an apical meristem; these plants often become quite tall (such as *Vanda*).

Orchids occur worldwide, from the Arctic to the tropical rain forests but do not grow wild in Antarctica. Most tropical orchids are epiphytic (growing on other plants) or epilithic (growing on rock surfaces), and a relatively small number, including the temperate and Arctic orchids, are terrestrial. The genus *Rhizanthella* in Australia is subterranean.

Economic Uses

Uses include a complex starch (called salep) used as food in Turkish and other Middle-Eastern confections. Salep is made from the dried tuberoidal roots of *Dactylorhiza*, *Eulophia*, and *Orchis*. In the Americas, the Mayans and the Aztecs cultivated *Vanilla* vines for the fleshy fruit, or bean, which was fermented to make vanilla for use as food flavoring and perfume. Some orchids have been used as sources of fibers for weaving or basket making. Currently, many different orchids are important as cut flowers or as potted plants, including *Cattleya*, *Dendrobium*, *Cymbidium*, *Paphiopedilium*, and *Laelia*. Advances made during the 1990's in mericloning have greatly reduced the cost of producing high-quality plants.

Lawrence K. Magrath

See also: Angiosperm evolution; Animal-plant interactions; Garden plants: flowering; Monocots vs. dicots; *Monocotyledones*; Mycorrhizae.

Sources for Further Study

Brown, P. M. *Wild Orchids of the Northeastern United States: A Field Guide*. Ithaca, N.Y.: Comstock, 1997. Includes keys, distribution maps, line drawings, and color plates.

Flora North America Editorial Committee. *Orchidaceae*. Vol. 26 in *Flora of North America North of Mexico*. New York: Oxford University Press, 2001. This volume contains keys to genera and species, descriptions of genera and species, line drawings, and distribution maps.

Northen, R. T. *Miniature Orchids*. New York: Van Nostrand Reinhold, 1980. Covers miniature orchids, how to grow them, and a selection of recommended species. Includes black-and-white and color illustrations.

Richter, W. *Orchid Care: A Guide to Cultivation and Breeding*. New York: Van Nostrand Reinhold, 1972. Basic guide to cultivation, discussing light, humidity, temperature, watering, potting, and how to pollinate orchids. Includes good line drawings.

ORGANIC GARDENING AND FARMING

Categories: Agriculture; economic botany and plant uses; environmental issues; food; gardening

Organic growing strives to produce healthy soils and plants through practices that replenish and maintain soil fertility. Organic farmers avoid the use of synthetic and often toxic fertilizers and pesticides.

At the beginning of agricultural history, farmers believed that plants "ate the soil" in order to grow. During the nineteenth century, German chemist Justus von Liebig discovered that plants extracted nitrogen, phosphorus, and potash from the soil. His findings dramatically changed agriculture as farmers found they could grow crops in any type of soil, even sand and water solutions, if the right chemicals were added.

By the late twentieth century, diversified family farms in industrialized nations had given way to huge, specialized operations. Crop yields were raised with the use of chemicals, but farms and their soil and water were not being used efficiently. Over time, the organic quality of the soil was lost, even though the agricultural chemicals remained in the soil.

Chemicals were found to leach into the water supply. In 1988 the U.S. Environmental Protection Agency (EPA) found that groundwater in thirty-two states was contaminated with seventy-four different agricultural chemicals. Leaching causes once-friable and fertile soils to become nonproductive. Further, the use of chemical insecticides was having toxic effects on both the foods grown and the farmworkers encountering them. About forty-five thousand accidental pesticide poisonings occur in the United States each year; in 1987 the EPA ranked pesticides as the third leading environmental cancer risk. A National Academy of Science study estimated that twenty thousand people each year get cancer because of pesticides alone. Growing public awareness of the effects of the traditional agricultural reliance on synthetic chemicals is reflected in a growing demand for organically produced foods.

Tenets of Organic Farming

Organic farming represents the use of diversified farming practices that emphasize working with nature to create an ecologically sound and sustainable system of agriculture. Certified organic farmers are bound by practices that are free of synthetic chemicals and genetic engineering. Lands must be chemical-free for at least three years before products grown on them can be certified as organic. The organic certification further requires that both plants and livestock be raised without the use of chemicals, antibiotics, hormones, or synthetic feed additives.

Certified organic farmers must comply with both organic regulations in their state and the 1990 Organic Foods Production Act (OFPA). In the late 1990's the U.S. Department of Agriculture (USDA), through the National Organic Standards Board (NOSB), began working to develop standards and regulations that would ensure consistent national standards for organic products. There are six areas of standards that the NOSB is examining: crops; livestock and livestock products; processing, packaging and labeling; accreditation; international issues; and materials. There is interest in developing global standards for the importation and exportation of organic products.

Soil Fertility

Like all farmers, organic farmers must ensure soil fertility and control unwanted plants and pests. Organic farmers build organic materials in the soil by the addition of *green manure*, *compost*, or *animal manure*. Green manure is the term used for crops that are grown to introduce organic matter and nutrients into the soil. Such crops are raised expressly for the purpose of being plowed under rather than sold to the consumer. Green manure crops protect soil against erosion, cycle nutrients from lower levels of the soil into the upper layers, suppress weeds, and keep much-needed nutrients in the soil.

Legumes are an excellent green manure crop because they are efficient at extracting nitrogen from the air and transferring it into the soil, leaving a supply of nitrogen for the next crop. The legumes have nitrogen nodules on their roots; when the legumes are tilled under and decompose, they add more nitrogen to the soil. As plants decay, they also make insoluble plant nutrients such as carbon dioxide and organic acids available in the soil.

Much of the organic material derived from green manure comes in the form of decaying roots. Alfalfa, one type of legume, sends its roots several feet down into the soil. When alfalfa plants are turned, the entire root system decomposes into organic material. Thus they help improve water retention and soil quality at the same time. Some examples of legumes used for green manure are sweet clover, ladino clover, and trefoil. Grasses used for green manure include rye, redtop, and timothy grass.

Greater fertility can be achieved if green manure is grazed by animals and animal manure is deposited on the soil. When grass is grazed, a proportion of the roots die and rot to form *humus*. This humus is the stable organic material that acts as a catalyst for allowing plants to find nutrients.

Animal manure is also used as an organic fertilizer. Rich in nitrogen, the best animal manures come from animals with high-protein diets. For ex-

A farmer tends organically grown cucumber plants.

ample, beef cattle being fattened for market consume a higher level of protein than dairy cattle that are producing milk for market. The application of manure to fields improves the structure of the soil, raises organic nitrogen content, and stimulates the growth of soil bacteria and fungi necessary for healthy soils.

Compost is particularly useful to the small-scale farmer or gardener who does not raise livestock or have the acreage to grow a green manure crop to plow under. Composting is a natural soil-building process that began with the first plants that existed on earth. It continues to be employed as a natural process today. As leaves and dead plants, animals, and insects decompose into the soil, they form a rich, organic layer. Compost can be made by farmers and gardeners by using alternating layers of carbohydrate-rich plant cuttings and leaves, animal manure, and topsoil and allowing it to decay and form a rich, organic matter to be added to the soil.

Crop Rotation

Planting the same crop year after year on the same piece of ground results in depleted soil. Crops such as corn, tobacco, and cotton remove nutrients, especially nitrogen, from the soil. In order to keep the soil fertile, a legume should be planted the year after a cash crop in order to add nitrogen and achieve a balanced nutrient level. Planting a winter-cover crop such as rye grass will help protect land from erosion and will, when plowed under in the spring, provide a nutrient-rich soil for the growth of a cash crop. *Crop rotation* also improves the physical condition of the soil because different crops vary in root depth and respond to either deep or shallow soil preparation.

Organic Insect and Weed Control

Monoculture puts a large amount of the same crop in proximity to pests that are destructive to that particular crop. Insect offspring can multiply out of proportion when the same crops are grown in the same field year after year. Because insects are drawn to the same home area, they are less able to proliferate if the crop is changed to one they cannot eat. For this reason, organic farmers rely on crop rotation as one aspect of insect control. Rotating crops can also help control weeds. Some crops and cultivation methods inadvertently allow certain weeds to thrive. Crop rotation can incorporate a successor crop that eradicates those weeds. Some crops, such as potatoes and winter squash, work as "cleaning crops" because of the different styles of cultivation that are used on them.

Organic farmers and gardeners believe that plants within a balanced ecosystem are resistant to disease and pest infestation. Therefore, the whole premise of organic farming is working with nature to help grow healthy, unstressed plants. Rather than using chemicals, which often create resistant generations of pests and thus the need for newer and stronger chemicals, organic farmers have found natural ways to diminish pest problems.

Organic farmers strive to maintain and replenish soil fertility to produce healthy plants resistant to insects. They also try to select plant species that are naturally resistant to insects, weeds, and disease. Crop rotation, as already discussed, is one method of keeping infestation down. Rows of hedges, trees, or even plants that are not desirable to insects planted in and around the crop field can act as barriers to pests. They can also provide habitat for pests' natural enemies, including birds, beneficial insects, and snakes.

Beneficial insects can be ordered through the mail to help alleviate insect pests. The ladybug and the praying mantis are two insects that can help rid farms or gardens of aphids, mites, mealy bugs, and grasshoppers. Just one ladybug can consume fifty or more aphids per day. Because most people order ten thousand to twenty thousand ladybugs for their garden, and one ladybug can lay up to one thousand fertile eggs, the overall cost of *biopesticides* is much less than that of buying *insecticides*. Trichogramma, also available by mail order, is a small wasp that will destroy moth eggs, squash borers, cankerworms, cabbage loopers, and corn earworms.

Organic farmers and gardeners also rely on what is called an *insecticide crop*. Garlic planted near lettuce and peas will deter aphids. Geraniums or marigolds grown close to grapes, cantaloupes, corn, and cucumbers will deter Japanese and cucumber beetles. Herbs such as rosemary, sage, and thyme planted by cabbages will deter white butterfly pests. Potatoes will repel Mexican bean beetles if planted near beans, and tomatoes planted near asparagus will ward off asparagus beetles. Natural insecticides such as red pepper juice can be used for ant control, while a combination of garlic oil and lemon may be used against fleas, mosquito larvae, houseflies, and other insects.

Organic farming relies on the physical control of weeds, especially through the use of cutting (cultivation) or smothering (mulching and hilling). Cultivation is the shallow stirring of surface soil to cut off small developing weeds and prevent more from growing. Cultivating tools include tractors, wheel hoes, tillers, or, for small gardens, hand hoes.

Mulch is a soil cover that prevents weeds from getting the sunlight they need for growth. Mulching with biodegradable materials can help build soil fertility while controlling weeds. Plastic mulches can be used on organic farms as long as they are removed from the field at the end of the growing or harvest season.

Dion Stewart and Toby R. Stewart

See also: Agriculture: modern problems; Biopesticides; Composting; Herbicides; Hydroponics; Legumes; Monoculture; Nitrogen cycle; Nitrogen fixation; Nutrients; Nutrition in agriculture; Pesticides.

Sources for Further Study

Board of Agriculture of the National Research Council. *Alternative Agriculture*. Washington, D.C.: National Academy Press, 1989. Provides interesting information on organic farming.

Carson, Rachel. *Silent Spring*. 1962. Reprint. Boston: Houghton Mifflin, 1994. Classic book on the detriments of the agricultural use of pesticides and other chemicals.

Coleman, Eliot. *The New Organic Grower*. White River Junction, Vt.: Chelsea Green, 1995. An easy-to-read manual on organic concepts for home or market gardeners.
Community Alliance with Family Farmers. *National Organic Directory*. Davis, Calif.: Author, 2001. Provides information on organic farming and lists producers and products.
Engelken, Ralph, and Rita Engelken. *The Art of Natural Farming and Gardening*. Greeley, Iowa: Barrington Hall Press, 1985. An interesting story of two decades of organic farming in Iowa.
Hamilton, Geoff. *The Organic Garden Book*. New York: D. Kindersley, 1994. An excellent book for gardeners who wish to grow vegetables, flowers, and fruits organically.

OSMOSIS, SIMPLE DIFFUSION, AND FACILITATED DIFFUSION

Categories: Cellular biology; transport mechanisms

Osmosis, simple diffusion, and facilitated diffusion are the processes by which water and other substances—usually small molecules and ions—cross cell membranes.

Transport of materials across cellular membranes is essential to the functioning of plants and other living organisms. It is the movement of materials across these semipermeable barriers that provides the conditions necessary for life, not only for the plasma membrane separating a cell from its environment but also for membranes surrounding organelles within cells.

Unless cells are able to maintain a stable internal environment (homeostasis), growth, development, and metabolism are not possible. Thus, understanding how substances move across membranes and how membranes select which substances to admit and exclude leads to a better understanding of homeostasis and its maintenance.

Some substances require metabolic energy to cross membranes in a process called *active transport*. All other movement across membranes, however, is through either simple or facilitated diffusion. Osmosis is a special case of simple diffusion involving just the movement of water.

Cell Membranes

Cell membranes, typically about eight nanometers thick, serve as barriers between the cell's cytoplasm and the extracellular environment. The cell membrane's lipid constituents (primarily phospholipids) make it fundamentally insoluble in water and therefore impermeable to all but the smallest uncharged polar molecules and a variety of lipid soluble molecules. It is necessary, though, for some of these substances to cross this barrier.

The currently accepted structure of the cell membrane was proposed by Jonathan Singer and Garth Nicolson in 1972. According to their *fluid mosaic model*, a membrane consists of a double layer of phospholipids, called a *lipid bilayer*. Phospholipids are more or less linear molecules with a hydrophilic end (water-soluble or, literally, "water-loving"), often called the "head," and a hydrophobic end (water-insoluble or "water-fearing"), often called the "tail." Within the bilayer, the lipids naturally orient themselves with their hydrophilic heads facing toward the aqueous fluid on each side—the cytoplasm on one side, the extracellular fluid on the other—and their hydrophobic tails touching one another within the bilayer. Phospholipids on each side of the membrane are free to move around, a property called fluidity. Among thephospholipids float various proteins. They, too, are free to move unless constrained by anchors to cytoplasmic or extracellular structures or by limits on the fluidity of the lipid. Lipids and proteins that are exposed to the outside of the cell may also have oligosaccharides (short chains of sugar molecules) attached to them, making them glycolipids and glycoproteins. The carbohydrate portion recognizes and is recognized by specific binding substances on other cells or in the environment.

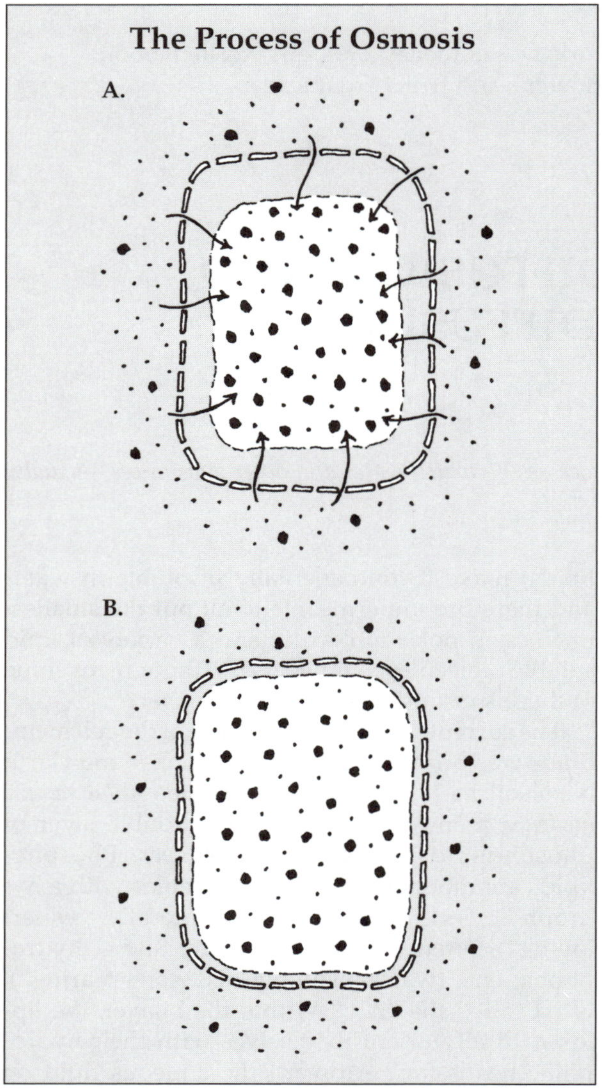

KIMBERLY L. DAWSON KURNIZKI

A. In osmosis, water (small dots) outside the plant cell moves from a region of greater concentration, outside, across the cell wall and the cell membrane to a region of less concentration initially occupied by more solutes (large dots) inside the cell.

B. The cell's uptake of water increases the volume of the cytoplasm and presses the cell membrane against the cell wall. The resulting turgor pressure or turgidity keeps the plant from wilting. The opposite occurs in dry conditions and causes the plant to wilt.

Concentration and Electrochemical Gradients

Substances crossing the membrane barrier by osmosis or diffusion will cross only from the side of the membrane with a higher concentration of the substance to the side with a lower concentration of the substance. If the substance is an ion (all of which are electrically charged), the concentration gradient of the ion and the difference in electrical charge across the membrane (called the *membrane potential*) together determine the direction of diffusion. The concentration gradient and the membrane potential together are referred to as an *electrochemical gradient*. Ions will only go down their electrochemical gradient.

If the concentration gradient and the membrane potential are lined up in the same direction, it is easy to determine which way the ion involved will travel. For example, if chloride ions (Cl^-), which have a negative charge, are in higher concentration outside the cell than inside the cell, they will be able to move into the cell by facilitated diffusion. This is because all cells have a negative membrane potential, which is to say, the electrical charge in the cytoplasm is more negative than the electrical charge on the outside. Thus, for this Cl^- example, both the concentration gradient and the membrane potential are aligned in the same direction. On the other hand, if the Cl^- concentration is higher inside the cell than outside, Cl^- could potentially travel in either direction. If the concentration gradient is very large and the membrane potential is only slightly negative, Cl^- would likely move out of the cell, whereas if the concentration gradient is small, and the membrane potential is very negative, Cl^- will likely move into the cell.

Routes Across the Membrane

Regardless of the direction substances will potentially be able to travel across the membrane, they will be able to do so by only one of three different processes: osmosis, simple diffusion, or facilitated diffusion. Water molecules, which are very small, can penetrate the lipid bilayer by passing between the phospholipids in a process called osmosis, although the exact mechanism is poorly understood. A few other hydrophilic molecules, such as methanol and ethanol, can cross a lipid bilayer. Apparently they, too, can penetrate the otherwise inhospitable hydrophobic environment of the membrane because of their small size and lack of charge. Hydrophobic substances, including dissolved gases such as oxygen, can also cross the membrane essentially unrestrained, and because they are lipid-soluble they simply "dissolve" into the membrane and through to the other side.

Hydrophilic substances, particularly charged molecules (ions), however, cannot cross a lipid bilayer by simple diffusion. They can cross membranes only with assistance from specialized *transport proteins*. These proteins penetrate both sides of the bilayer, with one part exposed to the cytoplasm and the opposite part to the extracellular fluid. Many transport proteins appear to work by a "shuttle" type of mechanism, binding to recognized substances on one side of the membrane and releasing them on the other side. Like many other processes involving proteins, the activity of transport proteins depends critically on the shape of the protein itself and on its ability to recognize a limited group of substances. Some membrane proteins form complexes of several proteins that form aqueous channels of the proper dimensions and charge distribution to serve as pores for specific molecules or ions.

Osmosis

Because living cells occupy an aqueous environment, water will always be present on both sides of a membrane. If one side has a higher concentration of *solutes* (dissolved material) in it than the other side, it will have a correspondingly lower concentration of water. Driven by random molecular movement, more water molecules will bump into one another and move out of the region of lower solute concentration than move into it, until eventually the water is uniformly distributed on both sides and the solute concentrations are the same. This net movement driven by the difference in solute concentration is called osmosis.

Osmosis is of fundamental importance in a variety of living systems. Organisms originally evolved in the ocean, so the cytoplasm of most cells has a solute concentration similar to seawater. Even terrestrial organisms have evolved mechanisms that maintain proper solute concentrations in tissue fluids. If the balance deviates from relatively narrow limits, there can be serious consequences.

For example, the freshwater green alga *Spirogyra* has a higher solute concentration than the surrounding water. Consequently, the net movement of water is from outside the algal cells to the cytoplasm. If this movement of water were to continue indefinitely, it would eventually cause such a buildup of pressure inside the algal cells that they would burst, except that algal cells (as well as all plant cells) are surrounded by a rigid cell wall. As the pressure builds, the cytoplasm swells in volume and presses the cell membrane against the cell wall, and the cell is said to be *turgid*. This same process occurs in terrestrial plants when roots absorb water from the soil, and this enables stem and leaf cells to maintain the turgidity required to keep the plant from wilting. In these examples, the cytoplasm is said to have a lower (more negative) water potential than the water outside. A solution with a lower water potential than another solution across a semipermeable membrane is said to be *hypertonic* (or hyperosmotic), while the other solution is said to be *hypotonic* (or hypo-osmotic). If the solute concentrations are the same on both sides of a semipermeable membrane, they are said to be *isotonic* (or iso-osmotic).

If the same green algae is placed in concentrated saltwater, the cytoplasm will now be hypotonic (and the saltwater will be hypertonic), and the net flow of water will be out of the algal cells. When this occurs in plants, the cytoplasm shrinks in volume, and the cell membrane pulls away from the cell wall in a process called plasmolysis. As a result, turgor pressure drops, and in the case of terrestrial plants in salty or dry soil, they begin to wilt. Some freshwater protozoa have specialized structures called contractile vacuoles that serve as pumps to maintain hypertonic conditions in the cytoplasm.

Diffusion

Both simple and facilitated diffusion involve the movement of a substance down a concentration or electrochemical gradient between the two compartments separated by a membrane The main difference between the two mechanisms is that substances moving by simple diffusion move through the lipid portion of the membrane, while substances using facilitated diffusion require specialized membrane proteins to allow their transport.

In simple diffusion, dissolved gases, such as oxygen and carbon dioxide, can cross membranes, as can lipid soluble substances. One type of important small molecule that crosses membranes by free diffusion are steroid hormones. These extremely potent regulators of cell metabolism readily diffuse through cellular membranes because their hydrophobic nature makes them dissolve readily in lipids. Within cells, they bind to specific receptors, which pass messages on to other proteins that control various biochemical events.

Facilitated Diffusion

All other materials that enter and leave cells down their chemical or electrochemical gradients do so by facilitated diffusion. Forming this pathway is the role of the particular membrane proteins, called carrier proteins, transport proteins, or *permeases*. Such proteins allow larger, polar molecules such as sugars and amino acids to be taken up by cells. They control the response of cells to certain growth factors and hormones, whose binding to the cell membrane causes channels for facilitated diffusion to open or close selectively.

Extremely specific recognition by the transport protein enables material to be transported. For this reason, membranes may readily transport one type of molecule but be completely impermeable to another molecule, even a closely related one. Membranes with selective permeability probably represent one of the most important innovations in living organisms. Without membranes, separate compartments could not be maintained to separate the various incompatible biochemical reactions of living systems. Without the semipermeability of biological membranes, the environment within the various compartments could not be continuously replenished and would be unable to respond to a continuously changing environment.

Mary Lee S. Ledbetter, updated by Bryan Ness

See also: Active transport; Cell wall; Cells and diffusion; Liquid transport systems; Membrane structure; Plasma membranes; Vesicle-mediated transport; Water and solute movement in plants.

Sources for Further Study

Beyenbach, Klaus W., ed. *Cell Volume Regulation*. New York: Karger, 1990. Discusses cellular control mechanisms, biological transport, cell membranes, cell physiology, osmotic pressure, and water-electrolyte balance. Includes bibliographic references and index.

Campbell, Neil A. *Biology*. 6th ed. San Francisco: Benjamin/Cummings, 2001. This introductory textbook is an excellent source of information about membrane structure and function. It is distinguished by extraordinary clarity of explanation, beautiful illustrations, including both line drawings and micrographs, and an engaging, lively style that captures readers' interest.

Karp, Gerald. *Cell and Molecular Biology: Concepts and Experiments*. 3d ed. New York: J. Wiley, 2002. Oriented toward the physiological aspects of cell biology, this textbook has a rigorous yet accessible discussion of the thermodynamics of the electrochemical gradient. It will be most useful to those not deterred by mathematical expressions describing physical events.

Kleinsmith, Lewis J., and Valerie Kish. *Principles of Cell and Molecular Biology*. New York: HarperCollins, 1995. Describes the experimental basis for the current understanding of cell structure and function. Discusses osmosis, simple diffusion, and facilitated diffusion. Though some of the discussion is rather advanced, the illustrations of mechanisms of facilitated diffusion are particularly helpful.

Petty, Howard R. *Molecular Biology of Membranes: Structure and Function*. New York: Plenum, 1993. Introduces the molecular biology of cell membranes. Familiarity with biochemistry is assumed; knowledge of elementary physics would be helpful.

Vance, Dennis, E., and Jean E. Vance, eds. *Biochemistry of Lipids, Lipoproteins, and Membranes*. New York: Elsevier, 1991. Advanced textbook covers the major areas in the fields of lipid, lipoprotein, and membrane biochemistry as well as molecular biology. Provides a clear summary of these research areas for working scientists.

OXIDATIVE PHOSPHORYLATION

Categories: Cellular biology; transport mechanisms

Oxidative phosphorylation is the sequence of reactions in mitochondria that convert energy from food into cellular energy by synthesizing ATP, the primary energy currency of cells. To drive the second step in oxidative phosphorylation, electrons must be passed to one of the electron carrier molecules of the electron transport system.

The ability to convert the energy from food molecules into cellular energy efficiently is crucial to cell survival. The central conversion system is oxidative phosphorylation, a sequence of reactions that take place in mitochondria (a type of organelle found in plant cells). These reactions take high-energy electrons and use them to make *adenosine diphosphate* (ADP) and inorganic phosphate (P_i). The name "oxidative phosphorylation" derives from the fact that organic molecules are oxidized to provide the electrons that are used as an energy source to facilitate the phosphorylation of ADP.

Oxidative phosphorylation may be divided into four general steps:

- obtaining high-energy electrons
- transferring energy from the electrons into cellular energy, accomplished via the electron transport system
- using the cellular energy from the electron transport system to establish a proton (H^+) gradient across the inner mitochondrial membrane
- using the stored energy of the proton gradient to drive the synthesis of *adenosine triphosphate* (ATP)

Finding Electrons: Oxidation

The main source of the electrons for the first step of oxidative phosphorylation is the chemical oxidation of organic molecules. Sources can include *glycolysis, fatty acid oxidation*, and the *Krebs cycle* (citric acid cycle). Chemical oxidation results in a loss of electrons by the oxidized molecule; when the electrons are removed from a molecule they are picked up by an electron acceptor, or *electron carrier*. Electron carriers are molecules that can transport electrons between molecules in the cell, much as a package delivery service will carry a box between two addresses. The most common electron carriers are *nicotinamide adenine dinucleotide* (NAD^+) and *flavin adenine dinucleotide* (FAD). Each one of these molecules can carry two electrons at a time. Once NAD^+ or FAD molecules accept electrons, they are said to be chemically reduced and are denoted as NADH or $FADH_2$.

Electron Transport

To drive the second step in oxidation phosphorylation, the electrons carried by NADH and $FADH_2$ must be passed to one of the electron carrier molecules of the electron transport system. The carriers that receive the electrons then pass them to other carriers in the system. Every time one carrier gives a pair of electrons to another, a small amount of energy is released. This energy is used by the *mitochondrion* to pump H^+ across the inner mitochondrial membrane. The electron pair continues to be transferred through a series of electron carriers until any extra energy they carry has been used for pumping H^+. The last carrier transfers what are now low-energy electrons to oxygen, which is the final electron acceptor in the series. When the electrons are accepted by the oxygen, it combines with two H^+ to form a molecule of water.

The components of the electron transport system are embedded in the inner mitochondrial membrane, and they are arranged in four large complexes. Each complex contains a component responsible for picking up the electrons and a protein portion that delivers the electrons to the next carrier in the chain. Each complex also contains additional proteins—in some cases as many as twenty—that attach the complex to the inner mitochondrial membrane.

In addition to these four complexes, there are two smaller electron carriers that can transport electrons between the larger complexes. One of these, *cytochrome c*, is a small protein. The other electron carrier is called *ubiquinone*, or coenzyme Q. Every time an electron pair is delivered to a new carrier, it loses a certain amount of energy, and each of the carriers can only accept electrons that have a particular amount of energy. Therefore, an electron pair moves through the electron transport system in an exact order, going only to carriers that are able to accommodate electrons of a precise energy level.

The order in which electrons move through all the components of the electron transport system has been determined by Britton Chance and several other investigators. An electron pair entering the system would proceed as follows. (The four complexes are identified by Roman numerals I-IV.)

NADH → complex I → ubiquinone → complex III → cytochrome c → complex IV → oxygen

Complex II accepts electrons directly from $FADH_2$, found in the matrix of the mitochondrion, then passes them to ubiquinone.

Establishing the Proton Gradient

The energy harvested during electron transport is used in the third step of the process to create an H^+ gradient. Three of the complexes (I, III, and IV) contain an additional protein component that is able to use the harvested energy to move protons from the matrix across the inner membrane, into the space between the inner and outer membranes, which is called the intermembrane space. The accumulation of protons in this intermembrane space results in proton concentration gradient across the inner membrane, from a high proton concentration in the intermembrane space to a low proton concentration in the matrix of the mitochondrion. The proton concentration gradient represents a stored form of energy, much like the capacitor in an electronic device.

ATP Synthesis

Finally, in the fourth and final step of oxidative phosphorylation, the proton gradient is used to drive the synthesis of ATP. The energy from the proton gradient is used by an ATP-synthesizing enzyme, also found in the inner membrane of the mitochondrion called *ATP synthase*. ATP synthase is a very large molecule. At very high magnification, the ATP synthase molecule looks like a lollipop sticking out from the inner membrane into the matrix of the mitochrondrion.

In 1961 Peter Mitchell proposed that the stepwise transfer of electrons by the electron transport system and the proton gradient worked together to synthesize ATP. His proposal, called the *chemiosmotic hypothesis*, represented a radical departure from other ideas at the time. At first the chemiosmotic hypothesis found little support among scientists, but the chemiosmotic hypothesis has stood the test of time. Although some details remain to be worked out, the experimental evidence accumulated by Mitchell, as well as by many other investigators since 1961, overwhelmingly supports this model. Mitchell received the Nobel Prize in Chemistry in 1978 for his proposal of the chemiosmotic hypothesis and the elegant research he performed in its support.

Work from the laboratory of scientist Efraim Racker has demonstrated that there are different functions for the two parts of the lollipop. The spherical part of the lollipop can be removed by mechanical shaking and has been found still to be able to synthesize ATP. Racker called the sphere F_1, or coupling factor 1. The "stick" portion of the lollipop is embedded in the inner mitochondrial membrane. This part of the enzyme acts as a tunnel for protons to travel back into the mitochondrion. The stick portion of the enzyme can be inactivated by the antibiotic oligomycin, so it is called the F_o, or oligomycin-sensitive factor. Another name for the ATP synthase is thus F_oF_1ATPase.

The fine details of how the movement of protons through the channel in the stick drives ATP synthesis have not been completely worked out. The spherical portion of the molecule can make ATP without the proton gradient. Once synthesized, however, the ATP remains tightly attached, so no additional ATP can be made by the enzyme. The large concentration of protons in the intermembrane space causes a net flux of protons back into the matrix through the tunnel provided by the ATP synthase. Paul Boyer suggested that when protons move through the lollipop stick into the mitochondrion, the sphere changes its shape. This shape change, in turn, causes the enzyme to release newly synthesized ATP. The ATP can now be used by the cell, and the enzyme can now make more ATP. Boyer's proposal, which has gradually gained acceptance among researchers, is reasonable. Many biological enzyme reactions work by using similar shape changes to attach and release the substrates. A molecule of ATP is released for every three hydrogen ions that are returned to the matrix.

Alina C. Lopo

See also: ATP and other energetic molecules; Carbohydrates; Cell-to-cell communication; Energy flow in plant cells; Exergonic and endergonic reactions; Glycolysis and fermentation; Krebs cycle; Lipids; Mitochondria; Proteins and amino acids.

Sources for Further Study

Berg, Jeremy Mark, John Tymoczko, and Lubert Stryer. *Biochemistry*. 5th ed. New York: W. H. Freeman, 2001. Includes descriptions of oxidative phosphorylation and the electron transport system. Full-color diagrams aid in understanding what is known about the structure and function of the components of the electron transport system. This is a college-level biochemistry textbook, and background in chemistry will be useful; however, many of the figures provide useful illustrative material for all levels.

James, D. C., and G. S. Matthews. *Understanding the Biochemistry of Respiration*. New York: Cambridge University Press, 1991. In-depth coverage of respiration is presented in a clear, step-by-step format designed to be used by undergraduate students.

Levings, Charles S., III, and Indra K. Vasil, eds. *The Molecular Biology of Plant Mitochondria*. Boston: Kluwer Academic, 1995. From the Advances in Cellular and Molecular Biology of Plants series. Includes bibliographical references and index.

Mitchell, Peter. "Keilin's Respiratory Chain Concept and Its Chemiosmotic Consequences." *Science* 206 (1979): 1148-1159. This is the lecture that Peter Mitchell delivered at the Nobel Prize ceremonies. Mitchell summarizes the work that led him to propose the hypothesis as well as subsequent work in its support.

Wolfe, Stephen L. *Biology: The Foundations*. Belmont, Calif.: Wadsworth, 1983. This is an introductory biology text. Chapter 7 includes an excellent description of oxidative phosphorylation and of the chemical reactions that provide energy to drive the process. Contains numerous simple diagrams of the process as well as several information boxes and supplements relating oxidative phosphorylation to the whole organism.

OZONE LAYER AND OZONE HOLE DEBATE

Categories: Environmental issues; pollution

Ozone, a form of the element oxygen, forms naturally in the stratosphere and provides the earth with a filter from ultraviolet radiation. Some human activities have been linked to a decrease in the amount of ozone present, an effect that has been described as a hole in the ozone layer. The size of the hole located over Antarctica fluctuates with the seasons but over the past several decades has steadily increased in average size.

Ozone is a highly reactive form of oxygen. It is composed of three oxygen atoms in a molecule (O_3) rather than the more usual two atoms (O_2). Ozone is formed from the more common diatomic oxygen where high energy is present. Near the earth, ozone forms in high-temperature combustion processes, such as in automobile engines and in electrical sparks.

In the stratosphere, ozone forms because of high-energy ultraviolet radiation. Once formed, it is quick to react with other molecules. Near the earth there are many molecules with which to react, and the ozone concentration remains low. In the stratosphere there are few molecules present, so ozone concentration builds up and forms what is known as the ozone layer. Ozone also disappears naturally by decomposing to ordinary oxygen, so there is a natural limit to the concentration that accumulates, and equilibrium, or a "steady state," occurs. The ozone layer is actually quite diffuse, and the ozone concentration is never very high.

Changes in the Ozone Layer

Since the mid-1950's, measurements of ozone concentrations in the atmosphere have been made regularly. In the early 1970's analysis of the measurements suggested that something new was causing a reduction in the concentration of ozone in the stratosphere, particularly in the region over the South Pole. Continued measurements established a similar change above the North Pole and a spreading of the effect over a larger area. Laboratory experiments show that molecular fragments containing unpaired electrons are very effective in speeding the decomposition of ozone. This catalytic effect is particularly strong in the presence of small ice crystals, as are present in the stratosphere in the polar regions in winter.

Chlorofluorocarbons

Chlorofluorocarbons (CFCs) are a class of chemicals that have found wide use as propellants in aerosol cans, cleaning solvents for electronic circuit boards, and working fluids in air conditioning and refrigeration. These molecules are very stable, which is a prime factor in their utility, but this property also allows the molecules to drift into the stratosphere after they are released. Most other es-

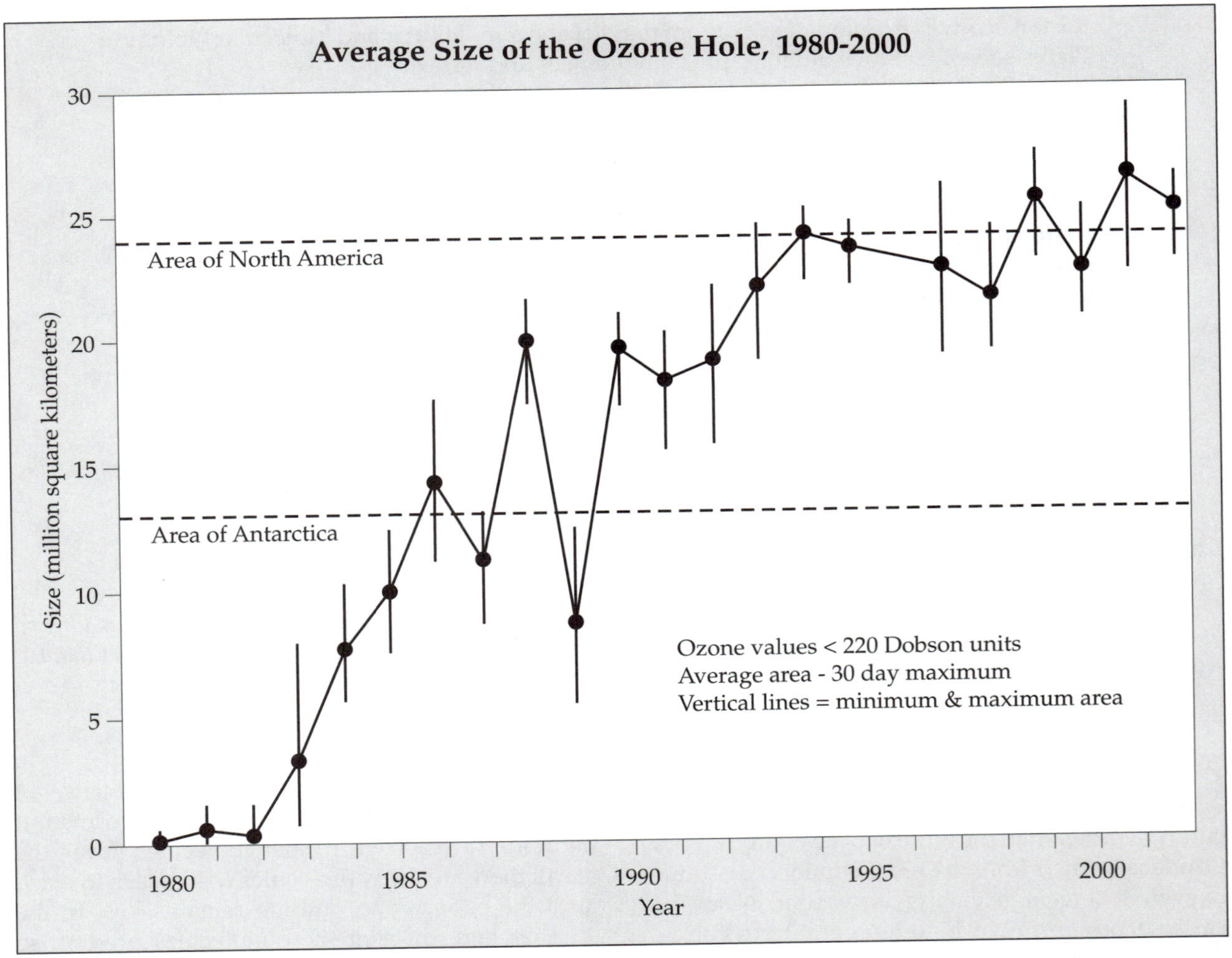

Source: Goddard Space Flight Center, National Aeronautics and Space Administration, http:\\toms.gsfc.nasa.gov\multi\oz_hole_area.jpg; accessed April 15, 2002.

caping molecules react or are washed out by precipitation before they gain much height in the atmosphere.

In the stratosphere, CFCs decompose by irradiation and form molecular fragments to which ozone is sensitive. CFCs are not the only artificial cause of ozone depletion, but they have been recognized as a major contributor. Much of what is known about the way that the ozone layer forms and decomposes comes from the work of Paul Crutzen, Mario J. Molina, and F. Sherwood Rowland, who received the 1995 Nobel Prize in Chemistry for their work on this subject.

Halons are compounds that have been used extensively in fire extinguishers and have been shown to be ozone-depleting. In some cases halons have the potential to destroy ten times as much ozone as chlorofluorocarbons. Limits on their use have been recommended.

HULTON ARCHIVE

The Importance of Ozone

Ozone is decomposed when the energy available in part of the ultraviolet region of the spectrum is absorbed by the molecule. When the energy is used in such a fashion it is no longer present in the sunlight that comes through the stratosphere to the earth. Thus, the ozone layer acts as a filter to limit the earth's exposure to high-energy light. This type of energy, if it does make it to the earth, is capable of causing the reaction of other molecules, including those of biological importance. The evidence is overwhelming that the primary cause of nonmelanoma skin cancers is chronic long-term exposure to ultraviolet light. Increased ultraviolet levels also cause cellular modifications in plants, including food crops, which may lead to their death. Of particular concern is the inhibition of photosynthesis in the phytoplankton, the photosynthetic organisms that form the base of the ocean food chain. With a diminishing level of filtering, one would expect that there would be a global increase in the effects of overexposure to ultraviolet radiation.

The Ozone Debate

Some scientists believe that ozone depletion is part of a natural cycle related to sunspot activity. Knowledge of what has happened in the distant past is circumstantial and not easy to interpret, but most scientists agree that human activities play a significant role in the current decrease in the ozone layer. In terms of the human contribution, CFCs have received the major attention, and their production was severely limited by international agreement in the 1987 Montreal Protocol and later revisions.

CFCs are no longer used as propellants, and their role as cleaners is all but over. However, their use as refrigerant fluids continues while economically viable, safe substitutes are being sought. People in developed countries have become extremely dependent on air conditioning; nearly all large buildings are designed to be air-conditioned rather than open to the outside. Part of the controversy concerning banning CFCs is based on ethical considerations. The developed countries used CFCs to help build their economies, so prohibiting CFC use in developing countries is considered inequitable by some. The search for substitutes has proved difficult, with economic, safety, and environmental concerns all placing limits on what is acceptable.

Kenneth H. Brown

See also: Air pollution; Greenhouse effect; Photosynthesis; Phytoplankton; Rain forests and the atmosphere.

Sources for Further Study

Asimov, Isaac. *What's Happening to the Ozone Layer?* Milwaukee, Wis.: Gareth Stevens, 1992. Good source for general readers; explains related concepts very clearly.

Brune, William. "There's Safety in Numbers." *Nature* 379, no. 6565 (February 8, 1996): 486-487. Presents the evidence of the chlorofluorocarbon-ozone depletion link.

Maduro, Rogelio A. *The Holes in the Ozone Scare: The Scientific Evidence That the Sky Isn't Falling*. Washington, D.C.: 21st Century Science Associates, 1992. Argues that scientific evidence does not support the theory of ozone depletion. Suggests a conspiracy theory.

Prather, Michael, et al. "The Ozone Layer: The Road Not Taken." *Nature* 381, no. 6583 (June 13, 1996): 551-555. Predictions stated indicate that if chlorofluorocarbons had not been linked to ozone depletion as early as 1974, the current state of the ozone layer would be more greatly compromised.

Roan, Sharon. *Ozone Crisis: The Fifteen-Year Evolution of a Sudden Global Emergency*. New York: Wiley, 1990. Good coverage of the early days of the controversy.

PACIFIC ISLAND AGRICULTURE

Categories: Agriculture; economic botany and plant uses; food; world regions

Agricultural practices in the Pacific Islands have an unsettled history. Typhoons, storms, drought, and volcanic activity can cause havoc to these islands' agriculture. For some, the rise of sea level can be devastating.

Throughout history, attempts by foreigners to encourage commercial agriculture in the Pacific Islands through *monoculture* (single cropping) have threatened fragile island environments. Large-scale land clearing and the use of fertilizers and pesticides have caused erosion, pollution, loss of biodiversity, and depletion of precious soils. Introducing new crops to these communities has resulted in the loss of native crops, harm to native species, and the elimination of traditional mixed farming methods.

The Pacific Islands have unique and fragile ecosystems, many of which are endangered by modern agricultural practices. The islands can be divided into two main types: the high islands, which are generally volcanic islands, and the *atolls*, or low islands. Volcanic lava wears down rapidly and provides fertile soil for cultivation. Atolls are low-lying coral reefs, which generally have an inadequate supply of fresh water and poor soils.

Coconuts, which can grow in poor soil and ripen throughout the year, are one of the few crops that thrive on the low islands. Coconuts are an important source of nutritious food and are easy to store. The meat of the coconut can be dried into a product called copra, which is pressed to make multipurpose oils.

Islands vary greatly in the amounts of rainfall they receive, the steepness of their slopes, and their varieties of plant life. Differences in rainfall and vegetation also exist on different parts of the same island. For example, one side of the island of Hawaii has one of the driest deserts on earth, whereas a few miles away on the other side of the island is a tropical rain forest.

Melanesian Islands

These fairly large islands are located to the northeast of Australia. They are quite damp, have a hot climate, and display a mountainous terrain covered with dense vegetation. Papua New Guinea is the largest island in Melanesia, slightly larger in size than California. It has a mountainous interior with rolling foothills that are surrounded by lowlands along the coastal areas. Its highest point is Mount Wilhelm, which rises to 14,795 feet (4,509 meters). Permanent crops occupy only 1 percent of the land. Crops are terraced in areas having steep slopes and extreme vegetation. Irrigation water is often brought to the crops through bamboo pipes. Approximately 64 percent of the labor force is involved in subsistence agriculture. Products grown include coffee, cocoa, coconuts, palm kernels, tea, rubber, fruit, sweet potatoes, vegetables, poultry, and pork. Palm oil, coffee, and cocoa are exported. In 1997 droughts brought on by the El Niño weather cycle caused extreme damage to coffee, cocoa, and coconut production.

Vanuatu, which includes eighty islands in the South Pacific due east of Australia's Cape York Peninsula, covers a total area a bit larger than the state of Connecticut. Its mostly mountainous terrain provides minimal arable land. Approximately 2 percent of the land is arable, and another 2 percent is used for pasture. About two-thirds of the population is involved in subsistence or small-scale agriculture. The main agricultural products are coconuts, cocoa, yams, coffee, fruits, vegetables, fish, and beef. Copra, beef, cocoa, and coffee are exported.

New Caledonia, located east of Australia in the South Pacific, is almost the size of New Jersey. It consists of coastal plains with interior mountains that range up to 5,340 feet (1,628 meters) in height. New Caledonia, which is known for its nickel resources, imports much of its food supply. A few vegetables are grown, but raising livestock is more common. Of New Caledonia's land, 12 percent is in permanent pasture, used for raising beef cattle.

The Fiji Islands include 332 islands, 110 of which are inhabited, located east of Vanuatu in the South Pacific. These islands are volcanic in origin. Approximately 10 percent of the land is arable, and 10 percent is in permanent pasture. About 67 percent of the labor force is involved in subsistence agriculture. Sugarcane is an important crop in Fiji and constitutes 32 percent of Fiji's exports. Other products grown in Fiji are coconuts, cassava, rice, sweet potatoes, cattle, pigs, and goats.

The Solomon Islands are a cluster of small islands that collectively cover an area almost the size of Maryland. They are located in the Solomon Sea between Papua New Guinea and Vanuatu. Some of the islands have rugged mountainous terrain; others are low coral atolls. Only 1 percent of the land is arable and 1 percent devoted to pastures. Approximately 24 percent of the working population is involved in agriculture, forestry, or fishing. Beans, cocoa, coconuts, palm kernels, rice, potatoes, fruit, and vegetables are grown on the islands. Cattle and pigs are the primary livestock raised there. Palm oil, cocoa, copra, and tuna are exported.

PhotoDisc

Tropical fruits like these on display at a market are a major export from many Pacific islands.

Micronesian Islands

Micronesia comprises thousands of relatively small islands located along the equator in the central Pacific Ocean and up to 1,200 miles (2,000 kilometers) north of the equator in the western Pacific Ocean. The region covers 1.54 million square miles (4 million square kilometers) and is subdivided into four areas: the Kiribati group, which lies around the intersection of the equator (0 degrees latitude) and the international date line (longitude 180 degrees); the Marshall Islands, which are about 900 miles (1,500 kilometers) to the northwest of Kiribati; the Federated States of Micronesia, which extend westward from the Marshall Islands for another 900 miles; and Guam, a U.S. territory located within the western cluster of islands of the Federated States of Micronesia.

The Kiribati group includes the Gilbert, Line, and Phoenix Islands, which are primarily low-lying atolls encircled by extensive living reefs. Of the thirty-three islands in the group, twenty are inhabited. Agriculture is mainly subsistence, with copra being one of Kiribati's few exports. Grown on the island for local consumption are taro, breadfruit, sweet potatoes, vegetables, and coconuts. Taro is one of the oldest cultivated plants in Pacific Island history and was once a staple food of the island people. This starchy, edible tuber can be cultivated by clearing or partially clearing a patch in the tropical rain forest and planting the taro in the moist ground.

The Marshall Islands contain two island chains of 30 atolls and 1,152 islands. Agriculture exists as small farms that provide commercial crops of tomatoes, melons, coconuts, and breadfruit. Coconuts, cacao, taro, breadfruit, pigs, and chickens are produced for local consumption.

Leading Agricultural Crops of Pacific Island Nations

Country	*Products*	*Percent Arable Land*
Fiji	Sugarcane, coconuts, cassava, rice, sweet potatoes, bananas	10
Micronesia	Black pepper, tropical fruits and vegetables, coconuts, cassava, sweet potatoes	—
New Zealand	Wheat, barley, potatoes, pulses, fruits, vegetables	9
Palau	Coconuts, copra, cassava, sweet potatoes	—
Papua New Guinea	Coffee, tea, cocoa, coconuts, palm kernels, tea, rubber, sweet potatoes, fruit, vegetables	—
The Philippines	Rice, coconuts, corn, sugarcane, bananas, pineapples, mangoes	19
Polynesia	Coconuts, vegetables, fruits, vanilla	—
Solomon Islands	Coconuts, palm oil, rice, cocoa, yams, vegetables, timber, beans, potatoes	1
Vanuatu	Copra, cocoa, coffee, coconut, taro, yams, fruits, vegetables	2

Source: Data are from *Time Almanac 2000*. Boston: Infoplease, 1999.

The Federated States of Micronesia include Pohnpei, the Truk Islands, the Yap Islands, and Kosrae. There are a total of 607 diverse islands, some high and mountainous, others low-lying atolls. Volcanic outcroppings are found on Kosrae, Pohnpei, and Truk. Agriculture on the islands is mainly subsistence farming. Products grown include black pepper, coconuts, tropical fruits and vegetables, cassava, and sweet potatoes. Pigs and chickens are the main livestock raised. Bananas and black pepper are exported.

Guam, the largest island in the Mariana archipelago, is of volcanic origin and surrounded by coral reefs. It has a flat coral limestone plateau that serves as a source of fresh water for the islands. In Guam, 15 percent of the land is used as permanent pasture. and another 11 percent is arable. Although fruits, vegetable, copra, eggs, poultry, pork, and beef are raised on the island, much of its food is imported because the economy relies heavily on U.S. military spending and the tourist trade.

Polynesia

Polynesia comprises a diverse set of islands lying within a triangular area having corners at New Zealand, the Hawaiian Islands, and Easter Island. French Polynesia is at the center of the triangle, 15 degrees south latitude and longitude 140 degrees west, and includes 118 islands and atolls.

The five archipelagoes of French Polynesia include four volcanic island chains (the Society Islands, the Marquesas, the Gambiers, and the Australs) and the low-lying atolls of the Tuamotus. The mountainous volcanic islands contain fertile soils along their narrow coastal strips. The atolls have little soil and lack a permanent water supply. Permanent pasture covers 5 percent of French Polynesia, and 6 percent of the land is used for permanent crops. Agricultural products raised include coconuts, vegetables, fruits, vanilla, poultry, beef, and dairy products. Coconut products and vanilla are exported. Thirteen percent of the population is involved in agriculture. On Tahiti, the largest island in French Polynesia, less than 10 percent of the land is arable. Exports include vanilla and coffee, but the main export from French Polynesia is pearls.

Samoa is a chain of seven islands lying several thousand miles to the west of Tahiti. Collectively, the islands are almost the size of Rhode Island, and they are covered with rugged mountains with a narrow coastal plain. Approximately 24 percent of the land sustains crops of bananas, taro, yams, and coconuts. Taro is a crop that can be used in land reclamation by building mud ridges or mounds in swampy ponds between the beach rampart and the foothills. Coconuts grown on the island are processed into creams and copra for export.

Tuvalu is a group of nine coral atolls located almost halfway between Hawaii and Australia. The soil is poor on these islands, and there are no streams or rivers. Water is captured in catchment systems and put into storage. Islanders live by subsistence farming and fishing. Coconut farming allows the islanders to export copra.

The Cook Islands comprise a combined area almost the size of Washington, D.C. Located halfway between Hawaii and New Zealand, the northern islands are primarily low coral atolls, and the southern Cook Islands are hilly volcanic islands. Permanent crops occupy about 13 percent of the land; 29 percent of the labor force is involved in agriculture. The Cook Islands produce a wide diversity of crops, including pineapple, tomatoes, beans, papayas, bananas, yams, taro, coffee, and citrus fruits. Agriculture is an important part of the economy, and copra, fresh and canned citrus fruit, and coffee are major exports.

The Hawaiian Islands are a volcanic chain of more than 130 islands, centered near 25 degrees north latitude and longitude 160 degrees west. Hawaii, with more than one million people, has the largest population in the Polynesian Island group. Lanai, one of the seven largest islands in Hawaii, is privately owned, and almost all its cultivated land is planted in pineapples. There are more than fifty-five hundred farms in Hawaii, and more than forty crops are grown commercially.

Because of the absence of adequate water on some sides of the Hawaiian Islands, water is brought through aqueducts from the wet sides of the islands to the dry sides, then kept in lined reservoirs to be used for irrigation, ranching, and tourism. Sugarcane requires enormous amounts of water to grow, and most pineapple crops are grown under irrigation. Hawaii is the second-largest producer of macadamia nuts in the world. The islands of Hawaii, Kauai, Maui, Molokai, and Oahu produce 7.6 million pounds of green coffee a year. Hawaii is also a prime producer of pineapple, cane from sugar, greenhouse and nursery plants, and dairy products. The main exports are fruits, coffee, and nuts.

New Zealand

About the size of Colorado, New Zealand is divided into two large islands and numerous smaller islands. About 50 percent of its land is in permanent pasture. A mountainous country with large coastal plains, 9 percent of the land in New Zealand is arable, and 5 percent is planted in permanent crops. Agriculture accounts for 9 percent of the gross national product and employs approximately 10 percent of the labor force.

In New Zealand, livestock outnumber people twenty-three to one. Wool, lamb, mutton, beef, and dairy products account for more than one-third of New Zealand's exports, and New Zealand is one of the world's top exporters of these products. New Zealand exports 90 percent of its dairy products. Other exports include cheese, fruits, and vegetables. Industry in New Zealand is primarily based on food products, including processed fruit, wines, and textiles.

Toby R. Stewart

See also: Caribbean agriculture; Monoculture; Pacific Island flora.

Sources for Further Study

Campbell, Ian C. *A History of the Pacific Islands*. Berkeley: University of California Press, 1996. Describes the development of various cultures in the Pacific Islands and their experience with intensive European contact during the last two hundred years.

Clarke, W. C., and R. R. Thaman, eds. *Agroforestry in the Pacific Islands: Systems for Sustainability*. New York: United Nations University Press, 1993. Covers agroforestry in Melanesia, Polynesia, and Micronesia as well as Pacific Island urban agroforestry. Includes appendix: One hundred Pacific Island agroforestry trees.

Gibson, Arrell Morgan. *Yankees in Paradise: The Pacific Basin Frontier*. Albuquerque: University of New Mexico Press, 1993. The story of Americans searching for whales and marketable commodities. Includes a chapter titled "The Agrarian Frontier."

Mackenzie, John M., and Herman Hiery, eds. *European Impact and Pacific Influence: British and German Colonial Policy in the Pacific Islands and the Indigenous Response*. New York: Taurus Academic Studies, 1997. Studies the effects of British and German ambitions on the Pacific Islanders. Part 2 examines the ecological effect of European intervention.

Thawley, John. *Austalasia and South Pacific Islands Bibliography*. Lanham, Md.: Scarecrow Press, 1997. Directed toward the student who needs specific information about Australasia or the South Pacific. Each section is defined by key topics including flora and fauna.

PACIFIC ISLAND FLORA

Category: World regions

Pacific Island ecosystems are unique and fragile, but with extensive wildlife management, many native species can be saved from extinction.

The vast region of the Pacific Ocean collectively called Oceania holds thousands of islands. Oceania spreads across the Pacific from 20 degrees north latitude to 50 degrees south latitude and from longitude 125 degrees east to 130 degrees west. The major groupings of islands are Melanesia, Micronesia, Polynesia, and New Zealand. Melanesia ("black islands") is a group of large islands immediately north and east of Australia, from New Guinea to New Caledonia. Micronesia ("little islands") is made up of hundreds of tiny atolls in the western Pacific. Polynesia ("many islands") covers

Digital Stock

Hawaiian palm trees.

a huge region in the central Pacific. New Zealand lies east and south of Australia. For botanical purposes, these islands can be categorized by climate and formation type. Climates range from tropical to sub-Antarctic, dry to very rainy. Types include volcanic (Fiji, Guam, and Hawaii), tectonic (New Zealand and New Guinea), and low coral atolls (nearly all of Micronesia's islands).

Unique Ecosystems

Organisms have a hard time reaching islands across the broad expanses of the Pacific Ocean. The islands' isolation leads to trends in the number of species on any given island. Bigger islands have more species; those farthest from continents have fewer species. To reach the islands, plants must be carried by animals or rely on wind or water currents. Birds are usually the first visitors, bringing with them hitchhiking insects and plant seeds in their digestive tracts.

Island plants and animals evolve together, affected by difficult conditions. Soil is often poor and food limited. Harsh environments and isolation contribute to the formation of new and unique species. Island ecosystems are sensitive to disturbances, whether from natural causes, such as severe storms, or human activities, such as construction, agriculture, logging, and introduced species.

Introduced species (exotics), both accidental and deliberate, are a serious problem. Rats and feral animals (domestic animals that have gone wild) can devastate island ecosystems. Pigs, rats, and goats are particularly devastating to vegetation. Introduced plants may overgrow native ones. Exotics also bring diseases or other problems to which native plants and animals have no resistance. For example, a Hawaiian bee crawls headfirst into native, barrel-shaped flowers, gathers the nectar, and then backs out. A plant that was introduced in Hawaii by landscapers attracted bees, but the flowers were smaller than those of the native plants. Once a bee crawled in, it became stuck like a cork in a bottle. This led to thousands of these plants, each with hundreds of flowers, becoming stoppered with dead bees.

Many tropical and temperate islands have coastal wetlands and mangrove swamps growing at the edge of the sea. Mangroves are low-growing, salt-tolerant trees that form dense tangles virtually impenetrable to humans. Wetlands and mangrove swamps are important breeding grounds for many types of fish and crabs and also trap sediment, stabilize shorelines, and protect coastlines from storms. When humans fill in the wetlands and cut down the mangroves, it causes coastal erosion and the loss of food fish.

Fiji Islands

The Fiji Islands are mostly volcanic in origin and lie in the South Pacific Ocean between longitudes 175 degrees east and 178 degrees west and 15 degrees and 22 degrees south latitudes, about 1,300 miles (2,100 kilometers) north of Auckland, New Zealand. Some parts of the islands receive up to 13 feet (4 meters) of rain per year, while other parts remain dry. A range of volcanic peaks divides the islands. The differences in weather and elevation create a variety of habitats—dense rain forests, grassy savanna, and mangrove swamps—and a large diversity of species.

Human disruption on Fiji has been moderate. About half the total area is still forested, and less than one-fourth of the land is suitable for agriculture. Trees include mahogany, pine, pandanus, coconut palms, mangoes, guava, and figs. Banyan figs are difficult to cut down and are responsible for some of the lack of forest clearing. The figs are an important food for many birds and animals. Other rain-forest plants include orchids, ferns, and epiphytes (plants that grow upon other plants).

There are nearly fifteen hundred endemic Fijian plant species, including ten species of palm tree on the island of Viti Levu. Grassy savannas are found higher on the volcanic slopes and in the dry zones. They are often planted with coconut palms and taro, a plant with potato-like tubers that grows on many Pacific Islands.

There are twelve reserve areas in the Fijian islands, but several are being logged and provide little sanctuary to native plants and animals. The government is interested in increased logging of mahogany and pine. The Fiji Pine Commission hopes to encourage the development of pine forests. Pines grow quickly and could form a sustainable logging industry, unlike the valuable but slow-growing mahogany trees. Increased world interest in herbal remedies has created a market for Fiji's traditional crop, kava root, and for ginger processing. The University of the South Pacific is located in Fiji and is a center of serious research into South Pacific species. Tourism is important to Fiji's economy and, with management, could be a source of income to Fijians while preserving native wildlife.

New Guinea

The world's largest tropical island, New Guinea is located north of Australia, just south of the equator. It is tectonic in origin, with large changes in elevation and many different habitats. Because of its size and varied terrain, New Guinea has a greater variety of habitats than any similar-sized land area in the world. In fact, New Guinea is so rugged that it is one of the least explored or developed places on earth. It provides the best remaining example of the types of organisms that can develop in island isolation.

New Guinea habitats include cold tundra, tropical rain forests, grassy savannas, coastal zones, montane rain forests, cloud forests, and bogs. There are at least twenty thousand species of flowering plants, including more than twenty-five hundred species of orchids, and hundreds of birds and animals. Many New Guinea species are unusual. Endemic Klinki pines are the tallest tropical trees in the world, reaching 295 feet (90 meters) tall. Many forests host "ant-plants," warty looking epiphytes that have hollow mazes inside their tissues. Ants live in the maze, safe from predators. The ants provide nutrition for the plant in the form of droppings, scraps of food, and dead ants.

Even though New Guinea is rugged and isolated, human impact is increasing. The population is rising, which means that more forests are being logged and grasslands plowed for agriculture, roads, and development. Humans have brought in food plants such as the sago palm, which can be cultivated in areas where traditional crops will not survive. They also cultivate pandanus trees, several varieties of fruiting vine, breadfruit, fungi, tubers, sugarcane, bananas, taro, and yams. Gold, silver, and copper have been discovered, which encourages destructive mining.

It has proven difficult to develop New Guinea economically without destroying the unique life of the island. It is hoped that lessons learned on other islands, such as Guam and New Zealand, may be applied to New Guinea. The government has tried incentives to keep wild areas wild, including encouraging ecologically friendly businesses, such as crocodile and butterfly farms and ecotourism. The National Park reserve that includes Mount Jaya is the only place in the world where it is possible to visit a glacier and a coral reef in the same park.

New Zealand

Located off the eastern edge of Australia, New Zealand has a fairly moderate climate that comes from conflicting warm, humid Pacific and colder Antarctic weather. It is similar to New Guinea, with rugged terrain, high mountains, and habitats from grassy open plains to dense forests, wet areas to near-deserts. Unlike New Guinea, however, New Zealand has been occupied and developed by humans for hundreds of years. Before large-scale agriculture, about half of New Zealand was covered with forests and one-third with grassland communities. Now half is pasture for grazing, and one-quarter is forest, mostly introduced species. Much of the remaining native forest is maintained as national parks and reserves. Pastureland usually consists of a single species of grass and does not support the wide variety of bird and animal life of the original grassland communities. There are more than four thousand species of beetles, two thousand species of flies, and fifteen hundred species of butterflies and moths. In a reversal of the usual ecological concerns, some native insects are destroying introduced pasture grasses.

Coral Atolls

The Federated States of Micronesia consist mainly of small atolls. Coral atolls are found only in tropical latitudes because coral (small, colonial animals) grow only in warm water. Coral reefs support a tremendous variety of fish, crabs, and mollusks. Atolls tend to have porous, infertile soil and to be very low in elevation. The inhabited state of Tokelau, three small islands located at 9 degrees south latitude, longitude 172 degrees west, has a maximum elevation of 16 feet (5 meters). Due to the low profile, poor soil, and occasional scouring by typhoons, flora are mostly limited to hardy root crops and fast-growing trees such as coconut and pandanus. Other vegetation may include native and introduced species such as papaya, banana, arrowroot, taro, lime, breadfruit, and pumpkin. Common fauna are lizards, rodents, crabs, and other small creatures. Pigs, ducks, and chickens are raised for food.

Human disturbances on coral atolls often have been particularly violent; several nuclear test bombs were exploded on Bikini Atoll and other islands in the 1940's and 1950's. Kwajalein, the largest atoll in the world, is used by the U.S. military for intercontinental ballistic missile target practice. Johnston Atoll, about 820 miles (1,320 kilometers) southwest of Honolulu, is a U.S. military base and storage facility for radioactive and toxic substances. It is also designated as a protected area and bird-breeding ground.

Future Prospects

Island ecosystems are unique and fragile. Some, like those in Guam and New Zealand, can never be returned to their original state, but with extensive wildlife management, many native species can be saved from extinction. New Guinea, Fiji, and many smaller islands are in earlier stages of development, and wildlife destruction can still be controlled. Conservationists sometimes do not realize that islands are not small, geographical zoos and gardens; people live there and want to improve their lives. Development cannot be stopped, but it can be managed so that the humans can improve their standard of living as they wish and the original, amazing island dwellers can still survive.

Kelly Howard

See also: Biological invasions; Caribbean flora; Invasive plants; Pacific Island agriculture.

Sources for Further Study

Hargreaves, Dorothy. *Tropical Blossoms of the Pacific*. Kailua, Hawaii: Hargreaves, 1970. Illustrated identification guide to flora.

Kirch, Patrick V., and Terry L. Hunt, eds. *Historical Ecology in the Pacific Islands: Prehistoric Environmental and Landscape Change*. New Haven, Conn.: Yale University Press, 1997. Research findings challenge earlier views that the Pacific Islands underwent dramatic environmental change only after Western contact.

Stuessy, Tod F., and Mikio Ono, eds. *Evolution and Speciation of Island Plants*. New York: Cambridge University Press, 1998. Covers issues in oceanic island plant biology, focusing on patterns and processes in the Pacific.

Whistler, W. Arthur. *Flowers of the Pacific Island Seashore*. Honolulu: Isle Botanica, 1992. Illustrated book on tropical flowers identifies not only cultivated ornamental plants but also plants found growing all over the islands of Polynesia, Melanesia, and Micronesia.

_______. *Wayside Plants of the Islands: A Guide to the Lowland Flora of the Pacific Islands*. Honolulu: Isle Botanica, 1995. Guidebook to the identification of the most common plant species encountered in the lowlands of the Pacific Islands.

PACLITAXEL

Categories: Economic botany and plant uses; medicine and health

Also known by the brand name Taxol, paclitaxel is a potent cancer-fighting drug originally derived from the bark of the Pacific yew tree, a small- to medium-sized understory tree that occupies Pacific coastal forests from southwestern Alaska to California.

Development of paclitaxel as a drug began in 1962 with the collection in Washington State of the reddish-purple bark of the Pacific yew tree (*Taxus brevifolia nutt*) by Kurt Blum, then a technician with the National Cancer Institute (NCI). The NCI was employing a "shotgun" approach to cancer research: A wide variety of plant parts of various species were being screened for anticancer activity. Thereafter, several scientists, including Monroe Wall and M. C. Wani at Research Triangle Institute in North Carolina and Susan Horwitz and Peter Schiff of the Albert Einstein College of Medicine in New York, recognized the potential of paclitaxel and became intensely interested.

The cancer-fighting compound paclitaxel is derived from the bark of the Pacific yew tree. The large amount of bark required to manufacture the drug raised fears among conservationists that continued harvest could threaten the species.

After years of delay, the pharmaceutical company Bristol-Myers Squibb continued tests and production on a larger scale. Paclitaxel was found to arrest the growth of cancer cells by attaching to their microtubules, thus preventing cell division. By the late 1980's Taxol had become the drug of choice, despite its high cost, for the treatment of a wide range of cancers, especially ovarian and breast cancer.

In spite of Taxol's prominence as a success story in the "herbal renaissance" of the twentieth century, several problems involved in production and medicinal use have persisted. For one, the cost of Taxol treatment has been prohibitive for many who desperately need it. The large amount of bark required (all the bark from a one-century-old tree yields only enough paclitaxel for a 300-milligram dose) raised fears among conservationists that continued harvest could threaten the species. While occurring over a wide area, the Pacific yew tree exists only in relatively small numbers. Furthermore, it is a slow-growing species that rarely reaches a height of more than 18 meters (60 feet); stripping the bark kills the tree. Plantations of the Pacific yew tree could be established, but it would take years for them to become productive.

Several means of producing paclitaxel without the destruction of wild yew trees have been proposed. Attempts have been made to produce paclitaxel from tissue cultures. Efforts to identify other *Taxus* species that may contain paclitaxel have been only marginally successful. In 1993 Bristol-Myers Squibb announced that it had found a semisynthetic method of producing paclitaxel that does not require yew bark. Paclitaxel-like compounds have been found in extracts from needles of the European yew tree (*Taxus baccata*) and those of several yew shrub species. An important advantage is that needles can be harvested without killing the trees or shrubs. Similar compounds have also been found in a fungus that grows on *Taxus* species.

Thomas E. Hemmerly

See also: Endangered species; Medicinal plants; Mitosis and meiosis.

Sources for Further Study

Georg, Gunda, ed. *Taxane Anticancer Agents: Basic Science and Current Status*. Washington, D.C.: American Chemical Society, 1995. Presents an overview of the development of paclitaxel as an anticancer drug, including its discovery, initial development, supply issues, and approaches to its total synthesis.

Goodman, Jordan, and Vivien Walsh. *The Story of Taxol: Nature and Politics in the Pursuit of an Anti-Cancer Drug*. New York: Cambridge University Press, 2001. Recounts a longstanding argument of environmentalists for preserving threatened habitats: that the plants within them may contain cures for ailments. This proved to be true with the discovery of the compound in Pacific yew bark, which pitted environmentalists against pharmaceutical companies.

Suffness, Matthew, eds. *Taxol: Science and Applications, Pharmacy and Toxicology, Basic and Clinical Aspects*. Boca Raton, Fla.: CRC Press, 1995. Brings together all aspects of Taxol research, development, and clinical use.

PALEOBOTANY

Categories: Classification and systematics; disciplines; evolution; paleobotany

The study of plants in the fossil record, in order to understand both the evolution of plant life and the ecology of ancient eras, is known as paleobotany.

Only a small percentage of the plants that ever lived left a record of their existence, surviving as *fossils*: mineralized wood, flowers in amber, leaf imprints in coal, or other indicators of life in an earlier era. Paleobotanists document this fossil record and use it to interpret the past evolution of plants.

Importance of Plant Fossils

Paleontology (or paleobiology) is the science concerned with fossils, the physical evidence of prehistoric life—including plants, animals, and microorganisms—on the earth. Paleobotany focuses on plant fossils, including algae, fungi, and related organisms, as well as mosses, ferns, and seed plants. As most organisms decompose rapidly after death, their preservation in nature is a rare event. Most individuals are not represented in the fossil record, and even many species that must have existed have vanished without a trace.

As a branch of botany, paleobotany is of importance primarily because the record of fossil plants helps scientists understand the long process of plant evolution. Especially since the 1940's, fossil evidence has helped to explain the origin of major classes of organisms, such as *algae* and *fungi*. Researchers now also have evidence for the origin of the earliest vascular plants and the formation of reproductive structures, such as cones of gymnosperms (evergreen trees and relatives) and flowers of angiosperms (flowering plants). The location of fossils, including both their temporal (age) and their spatial (geographical) arrangement, is used to determine past climates.

The climates of the world have changed continuously as continents have shifted over the earth's surface. For example, the location of coal deposits (which are the remains of giant tree ferns) in what is now Pennsylvania indicates the warmer climate that must have existed then. Although perhaps most of the contributions to paleobotany have been made by professionally trained scientists with a solid background in geology, botany, and related sciences, amateurs have also made significant discoveries. Many valuable specimens of university and museum collections were made by people interested in paleobotany as a hobby.

How Fossils Form

The formation of a fossil is an exceptional event, one that requires a special combination of favorable environmental conditions. In the most common fossilization process, the plant becomes covered by a soft sediment that then hardens to form a sedimentary rock. This type of rock forms gradually, over long periods of time, as particles produced by erosion are compacted on the bottom of the body of water.

The large-scale process by which plant parts become impregnated with minerals produces what has traditionally been called *petrified wood*. The modern term for this process is *permineralization*. Soluble carbonates, silicates, and other compounds infiltrate plant cells and the spaces between them. Eventually, the mineral deposits may completely replace the naturally occurring organic matter, preserving the details of the plant's microscopic architecture. Well known are the petrified forests of western United States, many of which are protected within national parks, such as Petrified Forest National Park in Arizona.

Being trapped in a sedimentary rock does not automatically guarantee that the organism will be preserved. The environment must be an *anaerobic* one—that is, one in which oxygen is excluded—thus preventing the decay that would otherwise result. The process may be interrupted by the action of waves or other erosive forces which re-expose the developing fossil before the process of fossilization is completed. Even after the process is completed, the well-preserved specimen may become distorted or altered in appearance because of the combined effects of time, pressure, and high temperatures that convert sedimentary rocks into metamorphic rocks.

As one would expect, the harder cells and tissues of plants are more likely to be preserved as fossils than are softer ones. For example, the thick-walled cells of wood and bark (called xylem) are more often preserved than are those of the pith (center of a stem) or cortex (found in stems and roots beneath the bark or outer covering). Other cells that are often fossilized are pollen grains and spores, both of which have outer shells that are highly resistant to decay.

Limestone and dolomite are among the most common types of rocks that form in such a way that they trap plants and form fossils. Coal, a combustible sedimentary rock, is formed in much the same way as other rocks but is distinctive because the sediment involved is of plant, rather than mineral, origin. Within this matrix of plant-derived material is often embedded a variety of plant parts.

Special Types of Fossils

Two special kinds of rock that may contain plant and animal fossils are diatomite and amber. Diatomite is a rock that forms from the silica cell walls of a group of unicellular algae known as *diatoms*. Because silica is the same material that sand and quartz are composed of, it is unusually permanent. Diatoms are found in both fresh water and salt water in great numbers and diversity. When they die, their cell walls accumulate underneath the water and become compacted over time into diatomite. The rock, itself formed by fossilization, may have fossil remains of various kinds of plants and animals preserved within it.

Amber, considered a semiprecious stone by gemologists and valued because of its beauty and distinctive appearance, is also of interest to paleobotanists. Amber is basically the fossilized resin produced by ancient cone-bearing evergreen trees. Sticky resins ooze from trees in response to injuries. Before such resins harden, various small animals, floral parts, pollen grains, fungal spores, and other plant parts may become trapped and be preserved intact. Deposits of amber valued for their use as jewelry and as fossils are recovered mainly from two world areas: the Baltic region of northern Europe and the Dominican Republic in the Caribbean Sea.

Paleobotanists are sometimes challenged by puzzling specimens. Outright fakes are sometimes presented by pranksters, but more common are various mineral structures that bear a superficial resemblance to a plant. Such specimens are called *pseudofossils*. Mineral deposits called dendrites found in rock crevices bear a resemblance to fern leaves. A coprolite (fossilized feces) from the upper Cretaceous period in Alabama was initially mistaken for the cone of a conifer (cone-bearing evergreen tree); these specimens may be referred to as pseudo-plant fossils, as they are true fossils of animals. During the formation of flint, bands are sometimes formed that suggest fossil mollusks or coral. Suspicious specimens require careful analysis by a specialist. In general, plants and animals, and therefore their remains, possess details and a characteristic regularity of form absent in pseudofossils.

Naming and Classifying Fossils

In order to prevent confusion, fossils, like living species, need to be named and classified in a consistent, systematic fashion recognized by paleobiol-

The Geologic Time Scale

MYA	*Eon*	*Era*	*Period*	*Epoch*	*Developments*
0.01	Phanerozoic Eon (544 mya-today)	Cenozoic (65 mya-today)	Quaternary (1.8 mya-today)	Holocene (11,000 ya-today)	Ice Age ends. Human activities begin to impact biosphere.
1.8				Pleistocene (1.8 mya-11,000 ya)	Ice Age in northwest Europe, Siberia, and North America. Plants migrate in response. New speciations lead to modern plants. Modern humans evolve.
5			Tertiary (65-1.8 mya)	Pliocene (5-1.8 mya)	Cooling period leads to Ice Age.
23				Miocene (23-5 mya)	Erect-walking human ancestors
38				Oligocene (38-23 mya)	Primate ancestors of humans
54				Eocene (54-38 mya)	Intense mountain building occurs (Alps, Himalaya, Rockies); modern mammals appear (rodents, hoofed animals); diverse conferous forests
65				Paleocene (65-54 mya)	Cretaceous-Tertiary event leads to dinosaurs' extinction c. 65 mya; seed ferns, bennettites and caytonias die off; ginkgoes decline; deciduous plants rise.
146		Mesozoic (245-65 mya)	Cretaceous (146-65 mya)		Breakup of supercontinents into today's form; birds appear. Cycads, other gymnosperms still widespread, but angiosperms dominate by 90 mya; animal-aided pollination begins to evolve.
208			Jurassic (208-146 mya)		Earliest mammals; ginkgoes thrive in moister areas; drier climates in North and South America, parts of Africa, central Asia; rise of modern gymnosperms, such as junipers, pine trees. Earliest angiosperm fossil (from China) dates from end of this period.
245			Triassic (245-208 mya)		Diminished land vegetation, with lack of variation reflecting global frost-free climate; gymnosperms dominate, bennettites and gnetophytes appear; dinosaurs develop.
286		Paleozoic (544-245 mya)	Permian (286-245 mya)		Permian extinction event initiates drier, colder period; supercontinent Pangaea has formed; tree-sized lycophytes, club mosses, cordaites die off; horsetails, peltasperms, cycads, conifers dominate.
325			Carboniferous (360-286 mya)	Pennsylvanian (325-286 mya)	Gymnosperms, ferns, calamites, lycopods thrive (first seed plants); reptiles appear. Plant life diverse, from small creeping forms to tall forest trees. Coal beds form in swamp forests from the dominant seedless vascular plants.
360				Mississippian (360-325 mya)	
410			Devonian (410-360 mya)		Club mosses, early ferns, lycophytes, progymnosperms; amphibians, diverse insects; horsetails, gymnosperms present by end of period.
440			Silurian (440-410 mya)		Early land plants: nonvascular bryophytes (mosses, hornworts), followed by seedless vascular plants in now-extinct phyla *Rhyniophyta*, *Zosterophyllophyta*, *Trimerophytophyta*.
505			Ordovician (505-440 mya)		Life colonizes land; earliest vertebrates appear in fossil record.
544			Cambrian (544-505 mya)	Tommotian (530-527 mya)	Cambrian diversification of life
900	Precambrian Time (4,500-544 mya)	Proterozoic (2500-544 mya)	Neoproterozoic (900-544 mya)	Vendian (650-544 mya)	Age of algae, earliest invertebrates
1600			Mesoproterozoic (1600-900 mya)		Eukaryotic life established.
2500			Paleoproterozoic (2500-1600 mya)		Transition from prokaryotic to eukaryotic life leads to multicellular organisms.
3800		Archaean (3800-2500 mya)			Microbial life (anaerobic and cyanobacteria) as early as 3.5 bya
4500		Hadean (4500-3800 mya)			Earth forms 4.5 bya.

Notes: bya = billions of years ago; mya = millions of years ago; ya = years ago.
Source: Data on time periods in this version of the geologic time scale are based on new findings in the last decade of the twentieth century as presented by the Geologic Society of America, which notably moves the transition between the Precambrian and Cambrian times from 570 mya to 544 mya.

ogists throughout the world. The branch of biology devoted to the naming and classification of organisms is called *taxonomy*. The same system is applied to both living and fossilized plants.

According to the system of *binomial nomenclature*, each species is given a scientific name consisting of two parts: the genus name followed by the species name. The former is capitalized, whereas the latter is written in lower case; both are italicized. Often a name follows that belongs to the person who assigned that name. As example, the scientific name of an extinct redwood tree is *Sequoia dakotensis* Brown, while that of a stemless palm is *Nipa burtinii* Brongniart. The ginkgo tree, *Ginkgo biloba* L., is considered a "living fossil." Known from the fossil record, it persists as a commonly planted shade tree. The initial following its scientific name is that of Carolus Linnaeus, the Swedish botanist who established this binomial system of nomenclature in 1753.

Species, living or dead, are classified using a hierarchical system (also by Linnaeus), which reflects degrees of similarity or dissimilarity to other species. All three trees already mentioned, because they are (or were) vascular plants, are assigned to the division *Tracheophyta*. Within that division the redwood and ginkgo, because they produce uncovered seeds, are placed into the class *Gymnospermopsida* (naked seeded plants), whereas the palm, a flowering plant, is assigned to the class *Angiospermophytina*.

The plant fossil record is often used to establish natural relationships among various extant plant species and other taxa at higher levels. This is especially true of the vascular plants. In fact, the division *Tracheophyta* was established by A. J. Eames in 1936 to show the natural relationship between seed plants and ferns. The basis for this new category was the discovery, earlier in the twentieth century, of Devonian fossils of a group of primitive vascular plants known as *psilotophytes*. They were recognized as ancestral to both ferns and seed plants. Previously, ferns and seed plants had been assigned to a separate division of the plant kingdom.

As the plant fossil record becomes more complete, further revision of the classification system becomes necessary to allow the system to more nearly reflect the true or natural relationships among the various categories. This is, at least, the goal of both paleobotanists and those who study modern plants.

Thomas E. Hemmerly

See also: Angiosperm evolution; Cladistics; Coal; Cycads and palms; Dendrochronology; Diatoms; Evolution of plants; Ferns; Fossil plants; Ginkgos; Gymnosperms; Paleoecology; Petrified wood; Psilotophytes; Seedless vascular plants; Species and speciation; Systematics and taxonomy; Systematics: overview; *Tracheobionta*.

Sources for Further Study

Agashe, Shripad N. *Paleobotany: Plants of the Past, Their Evolution, Paleoenvironment, and Application in Exploration of Fossil Fuels*. Enfield, N.H.: Science Publishers, 1997. Describes the geologic history of the earth and the origin and evolution of different groups of plants on the earth's surface. Covers principles and applications of paleobotany. With illustrations.

Cleal, Christopher J., and Barry A. Thomas. *Plant Fossils: The History of Land Vegetation*. New York: Boydell Press, 1999. For palaeobotany students or anyone interested in the origin and evolution of land plants. Describes methods of study and reviews major figures in paleobotany since the field's inception in the early 1800's. With illustrations.

Crane, Peter R. "Paleobotany: Back to the Future." *American Journal of Botany* 87, no. 6 (June, 2000): S2. Twentieth century advances in paleobotany have emphasized thorough integration of data with information of living plants, meaning scientists know more about the past than they did in the past.

Gensel, Patricia G. "Plants, Fossils, and Evolution: Lessons from the Fossil Hunters." *American Journal of Botany* 87, no. 6 (June, 2000): S1. Discusses the importance of and provides a template for recording and sharing information about plant biology and diversity and the patterns and processes of evolution.

Gensel, Patricia G., and Dianne Edwards, eds. *Plants Invade the Land: Perspectives in Paleobiology and Earth History*. New York: Columbia University Press, 2001. Contributors pre-

sent fossil record-based understanding of the evolution of plants on land. With summary chart of time period covered (500 million to 360 million years ago).
Pigg, Kathleen B. "Paleobotany." *Geotimes* 45, no. 7 (July, 2000): 36-37. A description of the advances in the field of paleobotany in 1999.

PALEOECOLOGY

Categories: Disciplines; ecology; evolution; paleobotany

Paleoecology is the study of ancient environments. As a field of science, paleoecology is most closely related to paleontology, the study of fossils. It is also related to paleobotany and a number of other areas of study dealing with the distant past.

Because paleoecology and its related disciplines (paleontology, paleobotany, paleoclimatology, paleogeography, and others) deal with the past, scientists are unable to apply the usual scientific criteria of direct observation and measurement of phenomena. In order to make any conclusions about the past, scientists must assume at least one statement to be true without direct observation: The processes that exist in the modern universe and on the modern earth existed in the past.

Uniformitarianism and Catastrophism

Paleontologists must assume that ancient plants and animals had tolerances to temperature, moisture, and other environmental parameters similar to those of modern organisms. The belief that the present is the key to the past is called *uniformitarianism*. It has been a key concept of the biological and earth sciences since the early nineteenth century. Uniformitarianism does not include the belief that the ancient earth was like the modern earth in its life-forms or geography.

During the early part of the nineteenth century, another worldview dominated: *catastrophism*, the belief that the earth is relatively young and was formed by violent upheavals, floods, and other catastrophes at an intensity unlike those of modern earth. Many catastrophists explained the presence of fossils at high elevations by the biblical flood of Noah. The uniformitarian viewpoint prevailed and, although admitting that local catastrophes may be important, their long-term, earthwide importance was denied. Catastrophism was revived in the 1980's to explain certain important events. The rapid extinction of the large dinosaurs at the close of the Mesozoic era has been attributed to the climatic changes associated with an alleged encounter between the earth and a comet—certainly a catastrophic event.

Climatic Cycles

One of the most intensively investigated paleoecological problems has been the changing environments associated with the ice ages of the past million years. Analysis of pollen from bogs in many parts of the world indicates that there have been at least four advances and retreats of glaciers during that period. Evidence for this is the changing proportions of pollen from tree species found at the various depths of bogs. In North America, for example, spruces (indicators of cool climate) formerly lived much farther south than they do now. They were largely replaced almost eight thousand years ago by other tree species, such as oaks, which are indicative of warmer climates. This warming trend was a result of the latest glacial retreat.

Dendrochronology (tree-ring analysis) not only enables paleoecologists to date past events such as forest fires and droughts but also allows them to study longer-term cycles of weather and climate, especially those of precipitation and temperature. In addition, trees serve as accumulators of past mineral levels in the atmosphere and soil. Lead levels of tree wood showed a sharp increase as the automobile became common in the first half of the twentieth century because of lead additives in gasoline. Tree rings formed since the 1970's have shown a decrease in lead because of the decline in use of leaded fuels. Tree-ring analysis has also been a valuable tool for archaeologists' study of climatic changes responsible for shifting patterns of population and agriculture among American Indians of the southwestern United States.

Traces of Ancient Environments

Fossil evidence is the chief source of paleoecological information. A fossil bed of intact clamshells with both valves (halves) present in most individuals usually indicates that the clams were preserved in the site in which they lived (*autochthonous deposition*). Had they been transported by currents or tides to another site of deposition (*allochthonous deposition*), the valves would have been separated, broken, and worn. Similarly, many coal beds have yielded plant fossils that indicate that their ancient environments were low-lying swamp forests with sluggish drainage periodically flooded by water carrying a heavy load of sand. The resulting fossils may include buried tree stumps and trunks with roots still embedded in their original substrate and numerous fragments of twigs, leaves, and bark within the sediment.

Certain dome- or mushroom-shaped structures called *stromatolites* are found in some of the most ancient of earth's sedimentary rocks. These structures may be several meters in diameter and consist of layers of material trapped by blue-green algae (*cyanobacteria*). Such structures are currently being formed in shallow, warm waters. Uniformitarian interpretation of the three-billion-year-old stromatolites is that they were formed under similar conditions. Their frequent association with mud cracks and other shallow- and above-water features leads to the interpretation that they were formed in shallow inshore environments subject to frequent exposure to the air.

Relative oceanic temperature can be estimated by observing the direction in which the shells of certain planktonic organisms coil. The shell of *Globigerina pachyderma* coils to the left in cool water and to the right in warmer water. *Globigerina menardii* shells coil in an opposite fashion—to the right in cool water and to the left in warmer water. Uniformitarian theory leads one to believe that ancient *Globigerina* populations responded to water temperature in a similar manner. Sea-bottom core samples showing fossils with left- or right-coiling shells may be used to determine the relative water temperature at certain periods. Eighteen-thousand-year-old sediments taken from the Atlantic Ocean show a high frequency of left-handed *pachyderma* and right-handed *menardii* shells. Such observations indicate that colder water was much farther south about eighteen thousand years ago, a date that corresponds to the maximum development of the last ice age.

Interpreting Clues

Fossil arrangement and position can be clues to the environments in which the organisms lived or in which they were preserved. Sea-floor currents can align objects such as small fish and shells. Not only can the existence of the current be inferred, but also its direction and velocity can be determined. Currents and tides can create other features in sediments which are sometimes indicators of environment. If a mixture of gravel, sand, silt, and clay is being transported by a moving body of water such as a stream, tide, or current, the sediments will often become sorted by the current and be deposited as conglomerates—sandstones, siltstones, and shales. Such graded bedding can be used to determine the direction and velocity of currents. Larger particles, such as gravel, would tend to be deposited nearer the sediment source than smaller particles such as clay. Similarly, preserved ripple marks indicate current direction. Mud cracks in a rock layer indicate that the original muddy sediment was exposed to the atmosphere at least for a time after its deposition.

Certain minerals within fossil beds or within the fossil remains themselves can sometimes be used to interpret the paleoenvironment. The presence of pyrite in a sediment almost always indicates that the sedimentary environment was deficient in oxygen, and this, in turn, often indicates deep, still water. Such conditions exist today in the Black Sea and even in some deep lakes, with great accumulations of dead organic matter.

The method of preservation of the remains of the fossilized organism can be an indication of the environment in which the creature lived (or died). Amber, a fossilized resin, frequently contains the embedded bodies of ancient insects trapped in the resin like flies on flypaper. This ancient environment probably contained resin-bearing plants (mostly conifers) and broken limbs and stumps that oozed resin to trap these insects. Mummified remains in desert areas and frozen carcasses in the northern tundra indicate the environments in which the remains were preserved thousands of years ago.

Fossils and Fuels

One of the most immediately important applications of paleoecological data is in the search for increasingly scarce *fossil fuels*: petroleum, natural gas, and coal. Reconstruction of ancient environments and their paleogeographical distributions is probably the most accurate way to predict the presence of reservoirs of these natural resources. Because these substances are formed in environments that are biologically highly productive, with abundant plant and animal life, they are commonly found associated with reefs (petroleum, gas) and swamps (coal).

Index fossils and fossil assemblages can indicate such environments with a high degree of accuracy. Petroleum and gas reservoirs must be porous and permeable in order for the material to accumulate in high concentrations. Reef material and porous sandstones formed from ancient sandbars meet such criteria. Most of the historically famous oil-producing region of Pennsylvania lies within an area 40 to 100 kilometers west of a former shoreline near present-day Pittsburgh. Within this area, most oil pools are within the ancient offshore sandbar belt. Sediments forming this "Pocono formation" came from the east, from an area near present Atlantic City, New Jersey. These sediments, all deposited within the same time frame, grade from coarse, nonmarine conglomerates and sandstones in the east near their source to fine-grained sandstones and shales to the west in the oil-producing region.

P. E. Bostick

See also: Anaerobic photosynthesis; *Archaea*; Bacteria; Coal; Dendrochronology; Evolution: convergent and divergent; Evolution: gradualism vs. punctuated equilibrium; Evolution: historical perspective; Evolution of cells; Evolution of plants; Fossil plants; Paleobotany; Petrified wood; Prokaryotes; Stromatolites.

Sources for Further Study

Agashe, Shripad N. *Paleobotany: Plants of the Past, Their Evolution, Paleoenvironment, and Application in Exploration of Fossil Fuels*. Enfield, N.H.: Science Publishers, 1997. Describes the geologic history of the earth and the origin and evolution of different groups of plants on the earth's surface. Covers principles and applications of paleobotany. With illustrations.

Arduini, Paolo, and Giorgio Teruzzi. *Simon and Schuster's Guide to Fossils*. New York: Simon & Schuster, 1986. An excellent introduction to the subject of fossils and fossilization. Good discussion of paleoecology. Well illustrated with color photographs and symbols indicating the time periods and habitats of the organisms.

Bennett, K. D. *Evolution and Ecology: The Pace of Life*. New York: Cambridge University Press, 1997. Presents paleoecological evidence that evolutionary processes visible in ecological time cannot be used to predict macroeveolutionary trends, contrary to Charles Darwin's thesis.

Cowen, Richard. *History of Life*. 3d ed. Malden, Md.: Blackwell Science, 2000. A brief, easily understood introduction to evolution and paleontology. Good description of the paleogeography of various continents and regions. Includes a brief list of further readings.

Davis, Richard A. *Depositional Systems: An Introduction to Sedimentology and Stratigraphy*. Englewood Cliffs, N.J.: Prentice-Hall, 1992. This is an advanced but understandable textbook about sediments and their environmental aspects of deposition. Includes glossary, extensive bibliography.

Dodd, J. Robert, and Robert J. Stanton, Jr. *Paleoecology, Concepts and Applications*. 2d ed. New York: Wiley, 1990. This revised edition reflects developments and changing emphasis in the field of paleoecology; includes references on all topics.

National Research Council. Commission on Geosciences, Environment, and Resources. Board on Earth Sciences and Resources. *Effects of Past Global Change on Life*. Washington, D.C.: National Academy Press, 1995. Recalls how particular environmental changes have caused certain species to become extinct or give rise to other species and shows that some systems are more fragile than others. Provides a scientific framework for policymakers.

Shipman, Pat. *Life History of a Fossil: An Introduction to Taphonomy and Paleoecology.* Cambridge, Mass.: Harvard University Press, 1993. Perhaps the best reference available on the fossilization process (taphonomy). Written at the advanced college level, this is a well-illustrated book with graphs, maps, line drawings, and photographs. Includes glossary and further references.

PARASITIC PLANTS

Categories: Fungi; poisonous, toxic, and invasive plants

Parasitic plants are those symbiotic life-forms that form mutualistic relationships with other plants at their hosts' expense.

The more is learned about plants and other organisms, the more apparent it is that many, if not most, plants are involved in *symbiotic associations* in which one or more other organisms live in close interaction with the plant—almost as if they are one. Symbiosis is frequently thought of as being a *mutualistic* relationship in which both the plant and its symbiont benefit. Typical examples include the alga and fungus that make up a *lichen* and the bacteria living in a legume root nodule. *Commensalistic* relationships, such as Spanish moss hanging from a live oak, are less commonly thought of; in this relationship, one plant obviously benefits, while the other seems to be little affected.

Parasitism is the third symbiotic condition, in which one plant benefits, but the other is harmed. Fungi are by far the most common parasites of plants; the study of these fungi is a major component of the science of plant pathology. Less common, but frequently well known, are some flowering plants, such as mistletoe, that parasitize other plants.

Haustorium

The defining feature of parasitic plants is the presence of a *haustorium*. The concept of a haustorium is broadly defined as a structure formed by the parasite to connect it physiologically to its host—a morphological/physiological bridge. The haustorial organ may be a simple outgrowth and proliferation of tissue, or it may be an elaborate, highly organized structure. Its appearance is often a useful diagnostic feature. Typically, the haustorium serves three functions: to attach the parasite to the host (*adhesion*), to penetrate into the host tissue or even into individual host cells (*intrusion*), and to conduct water and nutrients from the host to the parasite (*conduction*).

Fungal Parasites

Fungal parasites are among the most destructive diseases of plants. (Some, such as athlete's foot, can affect animals.) Parasitic fungi occur in each of the fungal divisions. They can have tremendous economic impacts. For instance, rust fungi, members of the *Basidiomycetes*, infect every species of cereal grain and can reduce yields by as much as 25 percent.

In addition to their direct economic impact, parasitic fungi have historically had profound social effects. In the 1800's outbreaks of *Phytophthora infestans*, late blight, decimated the potato crops of Europe just prior to harvest. This *oomycete* first turned the potato leaves and stems to mush and then infected the tubers. As a result, millions of people who depended on potatoes as their primary source of food starved to death. Ireland was particularly hard-hit, and millions emigrated from there.

Some fungal parasites have a mixed impact on humans. The genus *Claviceps*, an *ascomycete*, infects cereal grains, producing structures called *ergots*. A variety of alkaloids are produced in the ergot tissue that causes ergotism in animals that eat infected grain. Symptoms include convulsions, psychotic delusions, and gangrene. It has been suggested that the Salem "witches" executed in colonial Massachusetts or their "victims" were suffering from the effects of eating ergotted grain. Today ergot is the source of vasoconstricting drugs used in medicine to control bleeding and relieve migraine headaches.

Angiosperm Parasites

Among vascular plants, parasites are limited to about twelve families of dicots. The best-known examples are the Christmas mistletoes, *Viscum album* in Europe and *Phoradendron serotinum* in North America. While most mistletoes are tropical, the two mentioned above are common in the temperate regions. Unlike many parasitic flowering plants, mistletoes are green and photosynthetic and can thus produce their own food. Many species, including the Christmas mistletoes, produce fleshy, rigid mature leaves, but in other species the leaves are reduced to scales. Most species are *epiphytic* and grow in the branches of a host tree. Some tropical species are terrestrial, and at least one forms a 30-foot-tall tree. Even these tree species form a characteristic *haustorium* connecting the parasite to the host. Mistletoes cause severe economic losses in many areas. For instance, dwarf mistletoe (*Anceuthobium*) attacks many gymnosperms in the southwestern United States, particularly ponderosa pine.

Another well-known flowering plant parasite is dodder (*Cuscuta*). Dodder is distributed worldwide. It is easily recognizable because its rapid growth can quickly cover a host plant with a network of yellowish stems and scalelike, yellow leaves. Although dodder seeds germinate on the ground, the root disintegrates as soon as haustoria make connections with a host, at which time the dodder becomes completely dependent on the host for nutrients. Dodder is designated as a noxious weed throughout the continental United States.

AP/Wide World Photos

Common in temperate regions of North America, Christmas mistletoe (Phoradendron serotinum) *is a well-known holiday decoration and a plant parasite. The green leafy plant at the top of this hackberry tree can do considerable damage to its host.*

Origin of Parasitism

It is generally believed that parasites evolved from "normal" plants. As haustoria developed and became more specialized, structural parts and physiological processes normally associated with support and nutrient assimilation were reduced until they were no longer capable of supporting the plant, which became dependent on its parasitic habit. For instance, within the mistletoe family there is a complete gradient from terrestrial root-parasites, in which the roots of the tree-mistletoe parasitize the roots of host plants, to epiphytic shoot parasites, which are completely dependent on their host. This gradient suggests that the epiphytic forms evolved from terrestrial species through increasing specialization of the haustorial organs and reduction of roots and leaves.

The visible effects of a parasite on its host ranges from spectacular malformations, such as "witches' brooms" (a proliferation of short branches by the host at the site of parasite attachment), to no discernable effect. The physiological effects are also variable. While all parasites weaken their host to some degree, in some cases the effect is hardly discernable. This makes sense because it would not be advantageous for the parasite to so weaken its host that the host dies. Nevertheless, this extreme is also evident, as described above for late blight of potato.

Marshall D. Sundberg

See also: Adaptations; Ascomycetes; Bacteria; *Basidiomycetes*; Basidiosporic fungi; Chytrids; Coevolution; Deuteromycetes; Diseases and disorders; Evolution of plants; Fungi; Mycorrhizae; Oomycetes; *Protista*; Rusts; *Ustomycetes*; Zygomycetes.

Sources for Further Study

Bhandari, N. N., and K. G. Mukerji. *The Haustorium*. New York: Wiley, 1993. From the research studies in botany series, a close look at the defining feature of parasites. Includes references, index.

Kuijt, Job. *The Biology of Parasitic Flowering Plants*. Berkeley: University of California Press, 1969. Technical but comprehensive treatment of parasitic angiosperms.

Large, E. C. *Advance of the Fungi*. New York: Dover, 1962. This classic book describes the social and economic importance of a number of interesting fungi.

Raven, Peter H., Ray F. Evert, and Susan E. Eichhorn. *Biology of Plants*. 6th ed. New York: W. H. Freeman/Worth, 1999. Brief introduction to various aspects of plant parasitism in an introductory-level college text.

PEAT

Category: Economic botany and plant uses

Peat is an unconsolidated accumulation of partly decomposed plant material, used as a fuel source, as a mulch, and for other uses.

Peat has been burned for heating and cooking or used as soil since the New Stone (Neolithic) Age, at least fifty-five hundred years ago. Peat has many uses in agriculture, industry, and energy generation because of its organic chemical content and combustion properties. Although abundant in the middle latitudes of the Northern Hemisphere, it has been exploited as fuel primarily in northwestern Europe.

Like crude oil and coal, peat is composed of the remains of dead organisms compressed underground, and it can be burned in home stoves and fireplaces or in factories and public power plants. Because peat is lightly compressed plant matter, it also works well as soil for agriculture and horticulture. Worldwide, reserves of peat are comparable to those of other fossil fuels. For example, according to some estimates, resources in the United States surpass the combined potential energy yield of the nation's petroleum and natural gas.

Peat forms in bogs, fens, sedge meadows, and some swamps as the debris of peat mosses (sphagnum), grasses, and sedges falls to the wet earth and becomes water-soaked. In the absence of oxygen underwater, the plant matter and microorganisms compact without completely decomposing, forming soft, usually fibrous soils that are tan to black in color. The organic component, which includes cellulose, lignin, and some humus, is greater than 20 percent, and in most peat soils plant fragments are visible. The ash content is less than 50 percent, usually as low as 10 percent. Although the rate varies widely, in general a peat field increases in depth about three centimeters yearly. The bottoms of large peat fields are typically about ten thousand years old and can be as much as 50 meters below the surface, although 3-meter to 6-meter fields are common.

Geographical Distribution

Most deposits lie between 40 and 65 degrees latitude of the Northern Hemisphere. World reserves of exploitable peat exceed 200 billion tons, of which more than half is in Russia. Canada, the United States, Great Britain, Ireland, Finland, Norway, Sweden, Germany, Iceland, France, and Poland also have substantial peat fields. In the United States, Alaska contains more than half the reserves, but peat is also abundant in Minnesota, Washington, Michigan, Wisconsin, Maine, New York, North Carolina, Florida, and Louisiana. Some countries well below the fortieth meridian have exploitable peat reserves, especially Indonesia, Cuba, and Israel.

Energy Potential and Uses

In northern Europe, peat has fueled fires since the New Stone Age. It provides one-half to two-

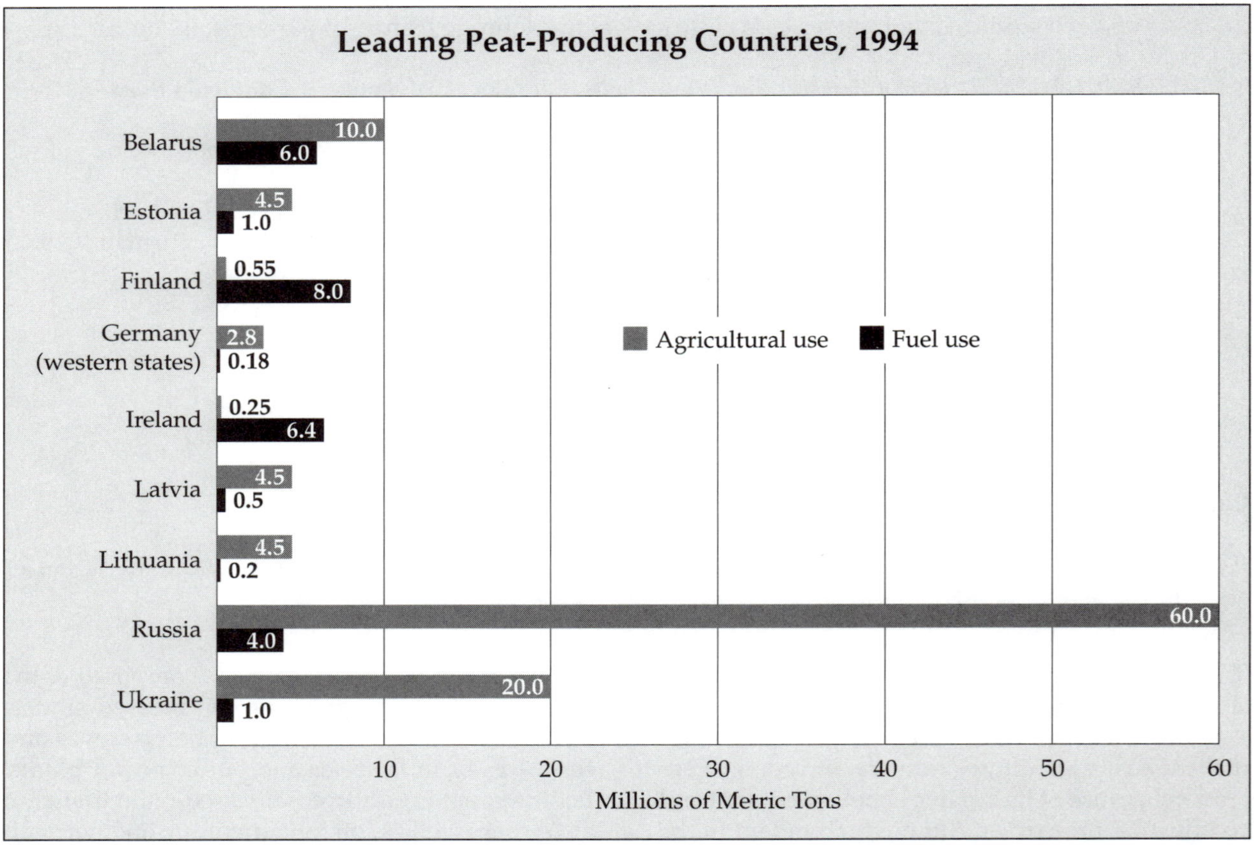

Note: Venezuela and the eastern German states are also major peat producers, but output is not reported, so no reliable estimates can be made concerning production. Reported world 1994 peat production was about 139 million metric tons.
Source: U.S. Bureau of Mines, *Minerals Yearbook, 1994*. U.S. Government Printing Office, 1996.

thirds as much energy as coal, or about 3.8 megajoules per dry kilogram, yet gives off far fewer pollutants such as sulfur and ash. It can be converted into coke, charcoal, or a synthetic natural gas.

Only in Ireland, Russia, Finland, and Great Britain is peat employed primarily as a fuel, where in fact it is a traditional domestic resource. Dried and pressed into briquettes, peat burns easily in fireplaces, stoves, and braziers. During the twentieth century the four countries burned increasing amounts of peat to generate electricity. Because it has very limited wood and fossil fuel resources, Ireland has consumed about three times as much peat for power generation as for domestic heating, whereas the other countries primarily rely on coal for that purpose.

AP/Wide World Photos

Layers of dark-colored peat can be seen at the center rear of this trench, in between layers of light-colored rock and gravel.

Agricultural and Horticultural Uses

The United States and Canada, as well as some European countries, process most of their peat as potting soil, lawn dressing, and soil conditioners. Because they are much lighter and fluffier than mineral soils, peat preparations let water and oxygen penetrate easily and increase water retention, and so can be soil supplements or mulch. Throughout the United States commercial nurseries and homeowners apply such products to gardens and tree beds. Farmers have raised grasses, clover, wild rice, cranberries, blueberries, strawberries, Christmas trees, and root and leafy vegetables on peat fields, and ranchers have used them for hay and grazing. However, peat fields are difficult to drain and clear and often remain wet, promoting rot and disease. They can be low in nutrients.

Other Uses

During the energy crisis of the 1970's, researchers investigated peat as an alternative to petroleum, although few of the efforts resulted in commercial products because oil again became cheaper than peat for industrial chemicals in the 1980's. Peat yields such mineral and organic substances as dyes, paraffin, naphtha, ammonium sulfate, acetic acid, ethyl and methyl alcohol, waxes, and phenols. Combined with clay, it forms lightweight blocks for construction. It can remove heavy metals from industrial waste and can be turned into coke for iron processing or into charcoal for purifying water. With its mildly antibiotic properties, peat served as a lightweight surgical dressing during World War I. Another of peat's well-known functions—and one of its oldest—is giving the smoky flavor to Scotch and Irish whiskeys as their malts slowly dry over open peat fires.

Because peat fields, once harvested, regenerate only after thousands of years, peat is not a renewable resource in any practical sense. Accordingly, intensive peat "mining" has caused concern among environmentalists. They worry that the rapid exploitation of peat fields, especially in Ireland and Great Britain, may permanently destroy bogs and fens and thereby threaten the wildlife dependent upon wetland habitats.

Roger Smith

See also: Coal; Wetlands.

Sources for Further Study

Charman, Daniel J., and Barry G. Warner. *Peatlands and Environmental Change*. New York: J. Wiley, 2001. Examines peatland conservation and rehabilitation. Focuses on peatlands as a system.

Crum, Howard. *A Focus on Peatlands and Peat Mosses*. Ann Arbor: University of Michigan Press, 1992. Surveys the types, ecology, and uses of peat in detail with many maps, drawings, and photographs.

Fuchsman, Charles H. *Peat, Industrial Chemistry, and Technology*. New York: Academic Press, 1980. Explains the chemicals contained in peat that have commercial value.

Godwin, Harry. *The Archives of the Peat Bogs*. New York: Cambridge University Press, 1981. Examines the biochemistry, ecology, stratification, and resource availability of peat worldwide. Also discusses the importance of peat to Stone Age cultures.

McQueen, Cyrus B. *Field Guide to the Peat Mosses of Boreal North America*. Hanover, N.H.: University Press of New England, 1990. Color photographs accompany these descriptions of sphagnum species.

Moore, P. D., and D. J. Bellamy. *Peatlands*. New York: Springer-Verlag, 1974. Examines the biochemistry, ecology, stratification, and resource availability of peat worldwide.

Parkyn, L., H. A. Ingram, and R. E. Stoneman, eds. *Conserving Peatlands*. New York: CAB International, 1997. Describes processes and influences involved in sustaining the remaining examples of these fragile ecosystems.

Wright, H. E., Barbara Coffin, and Norman E. Aaseng, eds. *The Patterned Peatlands of Minnesota*. Minneapolis: University of Minnesota Press, 1992. Minnesota has more peat than any other U.S. state except Alaska, and these peatlands exhibit a remarkable display of the complex adjustment of living organisms to their environment.

PEROXISOMES

Categories: Anatomy; cellular biology; physiology

Peroxisomes are distinct membrane-bound organelles within plant cells. A versatile microbody, the peroxisome plays key roles in photorespiration and in the conversion of stored fats to sucrose during the germination process in many seeds.

Peroxisomes are small, vesicle-like organelles that are found in virtually all eukaryotic plant cells. They are surrounded by a single lipid bilayer membrane and are spherically shaped microbodies with a granular interior containing proteins that usually function as enzymes. Many of these enzymes are important in the process of glycolic acid metabolism. Because peroxisomes contain no genome or ribosomes, their protein is imported from the cytosol. Peroxisomes reproduce by cycles of growth and fission and are not members of the endomembrane system.

Typically, peroxisomes vary in diameter from 0.2 to 2 micrometers. They are generally distributed throughout the cytosol, usually with more abundance around the nucleus and where the cell walls of two cells abut. Many researchers believe that peroxisomes are of primitive origin, predating the appearance of the mitochondria, and developed in response to the increased levels of oxygen in the environment that was generated by cyanobacteria.

Functions

In plants, peroxisomes perform many important roles in plant tissues and development. Their various functions are dictated by their enzyme content, which is constantly modified throughout the development of a plant. One function of peroxisomes is to remove hydrogen atoms from organic substrates by using oxygen, resulting in the production of hydrogen peroxide, a strong oxidizing agent.

In green leaves, peroxisomes play a vital role in photosynthesis by oxidizing glycolate to glyoxylate, a critical reaction that fixes carbon dioxide from the air into carbohydrate. The process is known as *photorespiration*. It provides a pathway for energy transfer to the bonds of sugar molecules. Photorespiration takes place during daylight, and, like normal cellular respiration, it requires oxygen and produces carbon dioxide and water.

In addition to photorespiration, a second function of peroxisomes that is unique to plants is the conversion of fatty acids to sugars by oxidation. It provides a vital pathway for young germinating seeds, particularly sunflower, watermelon, castor bean, and soy bean seeds, for the use of fatty acids as a source of food energy for germination and growth until they can carry out photosynthesis.

In seedlings and cotyledons, the specialized peroxisomes that carry out this function are called *glyoxysomes*. During this oxidation process glyoxysomes generate jasmonic acid, an important signaling molecule in plants. In leguminous plants, peroxisomes are vital in the formation of ureides, compounds that transport nitrogen to cellular locations, where it can be used in organic combination. Peroxisomes also function as sites of defense against activated oxygen species that produce oxidative stress in plants.

Alvin K. Benson

See also: Cell theory; Cytosol; Eukaryotic cells; Microbodies; Plant cells: molecular level; Proteins and amino acids; Respiration.

Sources for Further Study

Gunning, Brian E. S., and Martin W. Steer. *Plant Cell Biology: Structure and Function*. Boston: Jones and Bartlett, 1996. Comprehensive overview of plant cells. Includes illustrations, index.

Hopkins, William G. *Introduction to Plant Physiology*. 2d ed. New York: John Wiley and Sons, 1999. Develops the foundations of plant physiology, covering cellular organelles and the endoplasmic reticulum. Includes illustrations, bibliographical references, and index.

Salisbury, Frank B., and Cleon W. Ross. *Plant Physiology*. 4th ed. Pacific Grove, Calif.: Brooks/Cole, 1999. A fundamental textbook that covers plant cell biology. Includes illustrations, bibliographical references, and index.

PESTICIDES

Categories: Agriculture; environmental issues; pests and pest control

Pesticides are substances designed to kill unwanted plants, fungi, or animals that interfere, directly or indirectly, with human activities.

While the use of pesticides has mushroomed since the introduction of *monoculture* (the agricultural practice of growing only one crop on a large amount of acreage), the application of toxins to control pests is by no means new. The use of sulfur as an insecticide dates back before 500 B.C.E. Salts from heavy metals such as arsenic, lead, and mercury were used as insecticides from the fifteenth century until the early part of the twentieth century, and residues of these toxic compounds are still being accumulated in plants that are grown in soil where these materials were used. In the seventeenth and eighteenth centuries, natural plant extracts, such as nicotine sulfate from tobacco leaves and rotenone from tropical legumes, were used as insecticides. Other natural products, such as pyrethrum from the chrysanthemum flower, garlic oil, lemon oil, and red pepper, have long been used to control insects. *Biopesticides* are beneficial microbes, fungi, insects, or animals that kill pests.

The major types of pesticides in common use are *insecticides* (to kill insects), *nematocides* (to kill nematodes), *fungicides* (to kill fungi), *herbicides* (to kill weeds), and *rodenticides* (to kill rodents). Herbicides and insecticides make up the majority of the pesticides applied in the environment. In 1939 the discovery of dichloro-diphenyl-trichlorethane (DDT) as a strong insecticide opened the door for the synthesis of a wide array of synthetic organic compounds to be used as pesticides. *Chlorinated hydrocarbons* such as DDT were the first group of synthetic pesticides. Other commonly used chlorinated hydrocarbons have in the past included aldrin, endrin, lindane, chlordane, and mirex. Because of their low *biodegradability* and persistence in the environment, the use of these compounds was banned or severely restricted in the United States after years of use.

Organophosphates such as malathion, parathion, and methamidophos have replaced the chlorinated hydrocarbons. These compounds biodegrade in a fairly short time but are generally much more toxic to humans and other animals than the compounds they replaced. In addition, they are water-soluble and therefore more likely to contaminate water supplies. Carbamates such as carbaryl, maneb, and aldicarb have also been used in place of chlorinated hydrocarbons. These compounds rapidly biodegrade and are less toxic to humans than organophosphates, but they are less effective in killing insects.

Herbicides are classified according to their method of killing rather than their chemical composition. As their name suggests, contact herbicides such as atrazine and paraquat kill when they come in contact with a plant's leaf surface. Contact herbicides generally disrupt the photosynthetic mechanism. Systemic herbicides such as diuron and fenuron circulate throughout the plant after being absorbed. They generally mimic the plant hor-

PhotoDisc

Chemical pesticides are sprayed on this field of tulips.

mones and cause abnormal growth to the extent that the plant can no longer supply sufficient nutrients to support growth. Soil sterilants such as triflurain, diphenamid, and daiapon kill microorganisms necessary for plant growth and also act as systemic herbicides.

Pesticide Use

In the United States, approximately 55,000 different pesticide formulations are available, and Americans apply about 500 million kilograms (1.1 billion pounds) of pesticides each year. Fungicides account for 12 percent of all pesticides used by farmers, insecticides account for 19 percent, and herbicides account for 69 percent. These pesticides have been used primarily on four crops: soybeans, wheat, cotton, and corn. Approximately $5 billion is spent each year on pesticides in the United States, and about 20 percent of this is for nonfarm use. On a per-unit-of-land basis, homeowners apply approximately five times as much pesticide as do farmers. On a worldwide basis, approximately 2.5 tons (2,270 kilograms) of pesticides are applied each year. Most of these chemicals are applied in developed countries, but the amount of pesticide used in developing countries is rapidly increasing. Approximately $20 billion is spent worldwide each year, and this expenditure is expected to increase in the future, particularly in the developing countries.

Pesticide use has had a beneficial impact on the lives of humans by increasing food production and reducing food costs. Even with pesticides, pests reduce the world's potential food supply by as much as 55 percent. Without pesticides, this loss would be much higher, resulting in increased starvation and higher food costs. Pesticides also increase the profit margin for farmers. It has been estimated that for every dollar spent on pesticides, farmers experience an increase in yield worth three to five dollars.

Pesticides appear to work better and faster than alternative methods of controlling pests. These chemicals can rapidly control most pests, are cost-effective, can be easily shipped and applied, and have a long shelf life compared to alternative methods. In addition, farmers can quickly switch to another pesticide if *genetic resistance* to a given pesticide develops.

Perhaps the most compelling argument for the use of pesticides is the fact that pesticides have saved lives. It has been suggested that since the introduction of DDT, the use of pesticides has prevented approximately seven million premature human deaths from insect-transmitted diseases such as sleeping sickness, bubonic plague, typhus, and malaria. Perhaps even more lives have been saved from starvation because of the increased food production resulting from the use of pesticides. It has been argued that this one benefit far outweighs the potential health risks of pesticides. In addition, new pesticides are continually being developed, and safer and more effective pest control may be available in the future. In spite of all the advantages of using pesticides, their benefit must be balanced against the potential environmental damage they may cause.

Environmental Concerns

An ideal pesticide should have the following characteristics: It should not kill any organism other than the target pest; it should in no way affect the health of nontarget organisms; it should degrade into nontoxic chemicals in a relatively short time; it should prevent the development of resistance in the organism it is designed to kill; and it should be cost-effective. Since no currently available pesticide meets all of these criteria, a number of environmental problems have developed, one of which is broad-spectrum poisoning. Most, if not all, chemical pesticides are not selective; they kill a wide range of organisms rather than just the target pest. Killing beneficial insects, such as bees, lady bird beetles, and wasps, may result in a range of problems. For example, reduced pollination and explosions in the populations of unaffected insects can occur.

When DDT was first used as an insecticide, many people believed that it was the final solution for controlling many insect pests. Initially, DDT dramatically reduced the number of problem insects; within a few years, however, a number of species had developed genetic resistance to the chemical and could no longer be controlled with it. By the 1990's there were approximately two hundred insect species with genetic resistance to DDT. Other chemicals were designed to replace DDT, but many insects also developed resistance to these newer insecticides. As a result, although many synthetic chemicals have been introduced to the environment, the pest problem is still as great as it ever was.

Depending on the type of chemical used, pesticides remain in the environment for varying lengths of time. Chlorinated hydrocarbons, for example, can persist in the environment for up to fifteen years. From an economic standpoint, this can be beneficial because the pesticide has to be applied less frequently, but from an environmental standpoint, it is detrimental. In addition, when many pesticides are degraded, their breakdown products, which may also persist in the environment for long periods of time, can be toxic to other organisms.

Pesticides may concentrate as they move up the *food chain*. All organisms are integral components of at least one food pyramid. While a given pesticide may not be toxic to species at the base, it may have detrimental effects on organisms that feed at the apex because the concentration increases at each higher level of the pyramid, a phenomenon known as *biomagnification*. With DDT, for example, some birds can be sprayed with the chemical without any apparent effect, but if these same birds eat fish that have eaten insects that contain DDT, they lose the ability to metabolize calcium properly. As a result, they lay soft-shelled eggs, which causes deaths of most of the offspring.

Pesticides can be hazardous to human health. Many pesticides, particularly insecticides, are toxic to humans, and thousands of people have been killed by direct exposure to high concentrations of these chemicals. Many of these deaths have been children who were accidentally exposed to toxic pesticides because of careless packaging or storage. Numerous agricultural laborers, particularly in developing countries where there are no stringent guidelines for handling pesticides, have also been killed as a result of direct exposure to these chemicals. Workers in pesticide factories are also a high-risk group, and many of them have been poisoned through job-related contact with the chemicals. Pesticides have been suspected of causing long-term health problems such as cancer. Some of the pesticides have been shown to cause cancer in laboratory animals, but there is currently no direct evidence to show a cause-and-effect relationship between pesticides and cancer in humans.

D. R. Gossett

See also: Agriculture: modern problems; Agriculture: world food supplies; Biopesticides; Fertilizers; Green Revolution; Herbicides; Hormones; Monoculture.

Sources for Further Study

Altieri, Miguel A. *Agroecology: The Science of Sustainable Agriculture*. 2d ed. Boulder, Colo.: Westview Press, 1995. Describes alternative methods for restoring soil fertility, soil conservation, and the biological control of pests.

Carson, Rachel. *Silent Spring*. 1962. Reprint. Thorndike, Maine: G. K. Hall, 1997. Somewhat outdated but excellent dramatization of the potential dangers of pesticides. The book is historically significant: It served as a major catalyst for bringing the hazards of using pesticides to the attention of the general public.

Mannion, Antionette M., and Sophia R. Bowlby. *Environmental Issues in the 1990's*. Chichester, N.Y.: John Wiley & Sons, 1992. Another good discussion of the environmental impact of agricultural chemicals.

Milne, George W. A., ed. *Ashgate Handbook of Pesticides and Agricultural Chemicals*. Burlington, Vt.: Ashgate, 2000. A guide to the eighteen hundred most commonly used agricultural chemicals, organized by function and covering the common, chemical, and trade names, molecular formulas, physical properties and uses, acute toxicity, and manufactures and suppliers.

Nadakavukaren, Anne. *Man and Environment: A Health Perspective*. 2d ed. Prospect Heights, Ill.: Waveland Press, 1990. A very good discussion on human health issues associated with the use of pesticides.

Pierce, Christine, and Donald VanDeVeer. *People, Penguins, and Plastic Trees*. 2d ed. Belmont, Calif.: Wadsworth, 1995. Good discussions of environmental ethics, including the ethics of using agricultural chemicals.

PETRIFIED WOOD

Categories: Evolution; paleobotany

Petrified wood is formed by the fossilized remains of ancient trees that were saturated with mineral-filled water which, over time, converted the woody tissues into stone.

Petrified wood is studied by scientists interested in prehistoric plants and their environments. The Latin word *petros*, meaning rock, is the source for the scientific term *petrification*. Petrified wood is actually the stone remnant of a prehistoric tree.

Formation

During the Triassic period, *gymnosperms*—seed-producing trees without flowers, such as gingkos and conifers—grew over much of the earth's landmass. Volcanic eruptions triggered tremors, lightning, and heavy rains, which washed trees from higher elevations down to swampy valleys. As they were pushed downhill, the trees were stripped of their bark, branches, and roots from the force of the water's impact and broke into pieces. Under normal circumstances, trees soaking in deep, muddy water would decay, but silt rapidly and completely covered these trees, preventing exposure to oxygen and inhibiting aerobic decomposition. Volcanic ash in the floodwater consisted of inorganic compounds such as magnesium carbonate and iron sulphide, and the trees also absorbed silicon dioxide (silica) that had dissolved in groundwater. The minerals filled the spaces between cells in the tree trunks and branches. Molecules of these inorganic materials replaced molecules of organic tissues. During the next millions of years, wood gradually became stone in the process of silicification. Assisted by extreme pressure and temperatures, the silica that was lodged in the wood was transformed into quartz. Plants that have undergone petrification are also referred to as being *permineralized*.

The trees remained preserved under the soil for millions of years until soil erosion and shifting plates exposed them. Manganese, lithium, copper, and iron created patterns of bright colors as the wood *fossilized*. Some petrified wood displays varying rings of vivid colors, resembling agates. Other pieces are brown and look like driftwood. More significant than its beauty, information about the history of plants on earth is revealed by petrified wood. Unlike other fossils that are seen as an impression or compression, petrified wood is a three-dimensional representation of its organic material that preserves its external shape and internal structure. The preserved tree trunk sections also indicate the size of the Triassic forest. Scientists have even seen chromosomes and stages of nuclear division in petrified cells. A termite nest was discovered in one petrified log, offering clues about that insect's communal evolution.

Identifying Petrified Wood

Geologists and paleobotanists analyze petrified wood samples to specify the type of ancient tree that became fossilized. The cell structure of hardwood fossils is more diverse than that of softwood trees, causing its source to be more easily identified. Scientists choose pieces that exhibit an intact cell structure and examine the specimen with a microscope to scrutinize how the wood's bands and pores are arranged, both of which are crucial identifying characteristics. Softwood samples require greater magnification of thin slivers only one or two cells thick in order to permit enough light to shine through when mounted on a slide.

Wood anatomists then describe the sample's cell structure, which they compare to records of previously identified petrified wood and existing trees. North Carolina State University and the International Association of Wood Anatomists created a computer database, the General Unknown Entry and Search System (GUESS), which can identify matching cell patterns. Other databases contain information about existing hardwood and softwood trees; one has information on at least 1,356 types of trees of more than twelve hundred species.

Three types of petrified wood are found in the Tertiary strata. *Nondescript silicified wood* has under-

Petrified wood in the Crystal Forest at Petrified Forest National Park in Arizona.

gone silicification but still appears to be woody structurally. Difficult to identify because of its generic structure, nondescript silicified wood requires expert authentication for accurate labeling. *Petrified palm wood* has rod structures that reinforced the tissue strength before the wood grain became silicified and that look like spots or lines when the wood is cut. Popular among rock collectors, this type of petrified wood is both Arizona's state fossil and state rock. *Massive silicified wood* is difficult to recognize because the tree's grain was destroyed during silicification, thus making identification reliant on awareness of the area in which the tree was located and comparison to other petrified wood in adjacent territory.

Petrified Wood in the United States

Wood in Arizona's Petrified Forest National Park originated from a forest of giant conifer-like trees that grew from Texas to Utah. The park's 93,533 acres are home to one of the world's largest assortments of petrified wood. This desert area is dotted with stone log fragments. Although some visitors expect to see rock trees standing in clumps similar to a natural forest, these petrified trees rest where they fell individually or in groups.

Because the ancient forest lived simultaneously with the dinosaurs, archaeologists look for dinosaur fossils near petrified wood in the Arizona park. Although collection and thefts have greatly reduced the number of petrified logs, authorities believe that some areas of the park may shelter petrified wood buried as much as 100 meters beneath the surface.

By the twentieth century, prospectors and tourists had taken the most beautiful pieces; as a result, local residents sought government protection against further theft. In 1906 the petrified wood site was declared a national monument and was named a national park in 1962. The U.S. National Park Service tries to protect petrified wood by preventing the nearly one million yearly visitors from seizing samples. Rangers patrol sites and ask tourists to report any thefts they witness. Despite these precautions, several tons of petrified wood disappear annually from the park. Privately owned sites adjacent to the park offer collectors opportunities to search for petrified wood without any restrictions. Stores sell small pieces.

Petrified wood has been discovered in many other regions of the United States, especially in western areas where volcanic activity occurred, such as Yellowstone National Park, and in areas where rivers and streams deposited large amounts of sand, such as Louisiana and Texas. Washington State is home to the Ginkgo Petrified Forest State Park, which contains petrified logs that began fossilizing during the Miocene epoch. This petrified wood is unique because it includes petrified ginkgo, an indigenous tree that no longer grows naturally there. An unusual type of petrified wood that resembles pebbles and that originated in the Chehalis Valley is sometimes seen on the state's beaches.

The Calistoga, California, petrified forest is considered one of the best sources of Pliocene fossils similar to existing redwoods. Measuring more than 2 meters in diameter, the fossil logs reveal gray stone veins of quartz. Petrified wood in New Mexico's Bisti Badlands is not as colorful as neighboring Arizona's fossil wood. In Utah, petrified wood has been found near the Escalante River and the Coyote Buttes region near the Paria River. Petrified wood has also been found in Mississippi and Alabama.

Petrified Wood Worldwide

A petrified forest on the Greek island of Lesbos was named a protected site by presidential decree. Scientists have determined that the petrified wood began fossilizing during the Late Oligocene to Lower-Middle Miocene epochs. Unlike other sites of petrified forests, the Lesbos fossil trunks are erect and still have intact roots penetrating into fossilized soil, branches, leaves, cones, and seeds.

These trees were fossilized where they grew, offering insight into the environment and climate of ancient Lesbos. The ancient trees were well preserved during the petrification process, and such details as rings indicating age and growth patterns are visible. Both gymnosperms and angiosperms (flowering plants) grew in the ancient forests, along with pteridophytes such as ferns. Many of these ancient species no longer grow in the Mediterranean. Instead, they are found in Asian and American tropical and subtropical regions, indicating that Lesbos's petrified forest presents information about how life on earth evolved as the continents moved apart.

Other sites of fossilized wood also reveal details about the planet's development. Argentina's Petrified Forest on the Central Steppes was created after the formation of the Andes Mountains. Larger than American wood fossils, Argentinean petrified wood includes pieces 27 meters long and 3 meters in diameter. The petrified forest in Namibia contains giant tree trunks as long as 30 meters. Petrified wood has also been discovered in Australia, India, England, Turkey, and Switzerland.

Petrified Wood's Legacy

Paleobotanists research petrified wood to determine how the earth and the plants that have grown on its surface have changed since ancient times. They study how plants are related and descended from similar ancestors. Throughout history, many more tree species have existed than are documented, and many species, existing and extinct, await discovery. Pieces of fossil wood are often the oldest known specimens of a tree species and might be the predecessors of living trees.

Petrified wood helps researchers comprehend how trees have evolved to adapt to environmental and climatic conditions. For example, researchers have hypothesized that the cell structure of tree xylem has not changed as much as fruit and leaves. Fossil wood sometimes reveals the reason for the demise of extinct trees. The study of petrified wood has altered how scientists perceive both ancient geological ecosystems and modern environments by offering perspective on which plants lived on the earth and survived, adapted, or died according to changes in the atmosphere and crust.

Elizabeth D. Schafer

See also: Fossil plants; Paleobotany; Paleoecology; Plant tissues; Wood.

Sources for Further Study

Daniels, Frank, Brooks B. Britt, and Richard D. Dayvault. *Petrified Wood: The World of Fossilized Wood, Cones, Ferns, and Cycads*. Grand Junction, Colo.: Western Colorado, 1998. Beautiful photographic presentation of petrified wood.

Long, Robert A., Rose Houk, and Doug Henderson. *Dawn of the Dinosaurs: The Triassic in Petrified Forest*. Petrified Forest, Ariz.: Petrified Forest Museum Association, 1988. Written by former Petrified Forest National Park paleontologist Long and veteran natural history writer Houk, with moody, atmospheric paintings by Henderson.

Lubick, George M. *Petrified Forest National Park: A Wilderness Bound in Time*. Tucson: University of Arizona Press, 1996. Text speeds the reader on an ancient ecological journey, from the time of the forest's early life to the discovery of its fossils, and through the political manuevering to create a place for the forest in American history.

PHEROMONES

Categories: Cellular biology; physiology

Pheromones are any chemical or chemical mixture that, when released by one member of a species, affects the physiology or behavior of another member of the same species.

Pheromones are semiochemicals that carry information between members of a single species. To do this, the pheromone must be released into the atmosphere or placed on some structure in the organisms' environment. It is thus made available to other members of the species for interpretation and response. It is also available to members of other species, however, so it is a potential *allelochemic*.

Types of Pheromones

In complex interactions, a pheromone may also be acting as a *kairomone*, passing messages between species to the benefit of the recipient; an *allomone*, passing messages between species to the benefit of the sender; or a *hormone*, passing messages within a single organism. One possible example of a hormone as a pheromone is the plant hormone ethylene, which is produced by an individual plant to stimulate ripening of fruit, loss of leaves, and other physiological changes. Some evidence suggests that ethylene, produced in response to damage caused by insects feeding on the plant, stimulates production of chemicals that are detrimental to the insects, thus acting as a hormone. It also passes through the atmosphere to surrounding plants and stimulates their production of defensive chemicals, thus acting as a pheromone. Not everyone is convinced by the evidence that has been presented for this phenomenon, but the possibility is intriguing.

There are two general types of pheromone: those that elicit an immediate and predictable behavioral response, called *releaser* or *signal pheromones*, and those that bring about a less obvious physiological response, called *primer pheromones* because they prime the system for a possible behavioral response. Pheromones are also categorized according to the messages they carry. There are trail, marker, aggregation, attractant, repellant, arrestant, stimulant, alarm, and other pheromones. Their functions are suggested by the terms used to name them.

The chemical compounds that act as pheromones are numerous and diverse. Most are lipids or chemical relatives of the lipids, including many steroids. Even a single pheromonal message may require a number of different compounds, each present in the proper proportion, so that the active pheromone is actually a mixture of chemical compounds.

Functions and Sources

Different physical and chemical characteristics are required for pheromones with different functions. Attractant pheromones must generally be volatile to permit atmospheric dispersal to their targets. On the other hand, many marking pheromones need not be especially volatile because they are placed at stations which are checked periodically by the target individuals. Some pheromones are exchanged by direct contact, and these need not have any appreciable volatile component.

Pheromones are widespread in nature, occurring in most, if not all, species. Most are poorly understood. The best-known are those found in insects, partly because of their potential use in the control of pest populations and partly because the relative simplicity of insect behavior allowed for rapid progress in the identification of pheromones and their actions. Despite these advantages, much remains to be learned even about insect pheromones.

Pheromones and Pest Control

Pheromones and other semiochemicals are of interest from the standpoint of understanding communication among living things. In addition, they have the potential to provide effective, safe agents for pest control. The possibilities include sex-attractant pheromones to draw insects of a particular species to a trap (or to confuse the males and keep them from finding females) and repellant pheromones to drive a species of insect away from a valuable crop. One reason for the enthusiasm generated by pheromones in this role is their specificity. Whereas insecticides generally kill valuable insects as well as pests, pheromones may target one or a few species.

These chemicals were presented as a panacea for insect and other pest problems in the 1970's, but most actual attempts to control pest populations failed. Lack of understanding of the particular pest and its ecological context was called the most common cause of failure. Some maintain that pest-control applications must be made with extensive knowledge and careful consideration of pest characteristics and the ecological system. In this context, pheromones have become a part of integrated pest management (IPM) strategies, in which they are used along with the pest's parasites and predators, resistant crop varieties, insecticides, and other weapons to control pests. In this role, pheromones have shown great promise.

Carl W. Hoagstrom

See also: Allelopathy; Biopesticides; Hormones; Integrated pest management; Metabolites: primary vs. secondary.

Sources for Further Study

Agosta, William. "Using Chemicals to Communicate." *Journal of Chemical Communication* 71, no. 3 (March, 1994): 242-247. Discusses biologists' exploration of plant pheromone concentrations through observations of reactions produced when the pheromone is released.

Rhoades, David F. "Pheromonal Communication Between Plants." In *Chemically Mediated Interactions Between Plants and Other Organisms*, edited by Gillian A. Cooper-Driver, Tony Swain, and Eric E. Conn. New York: Plenum Press, 1985. Written for botanists but mostly accessible to nonspecialists. Presents arguments for an alarm pheromone in plants. Includes many references; one index for the book.

Schiestl, Florian P., et al. "Orchid Pollination by Sexual Swindle." *Nature*, June 3, 1999, 421-422. The odor of *Orphrys sphegodes* is almost identical to the sex pheromone of female bees and thus prompts almost equally intense reactions. Discusses innovation of this orchid for attracting and deceiving pollinators.

PHOSPHORUS CYCLE

Categories: Biogeochemical cycles; cellular biology; ecology; nutrients and nutrition

The constant exchange of a mineral or elemental nutrient between organisms and the physical environment is called a biogeochemical cycle. Along with the carbon cycle and the oxygen cycle, one of the most important biogeochemical cycles is that of the element phosphorus. The phosphorus cycle involves the movement of the element phosphorus as it circulates through the living and nonliving portions of the biosphere.

Many of the chemical elements found on the earth are vital to the processes and systems of living organisms. Unlike oxygen and carbon, phosphorus follows complex pathways. It circulates

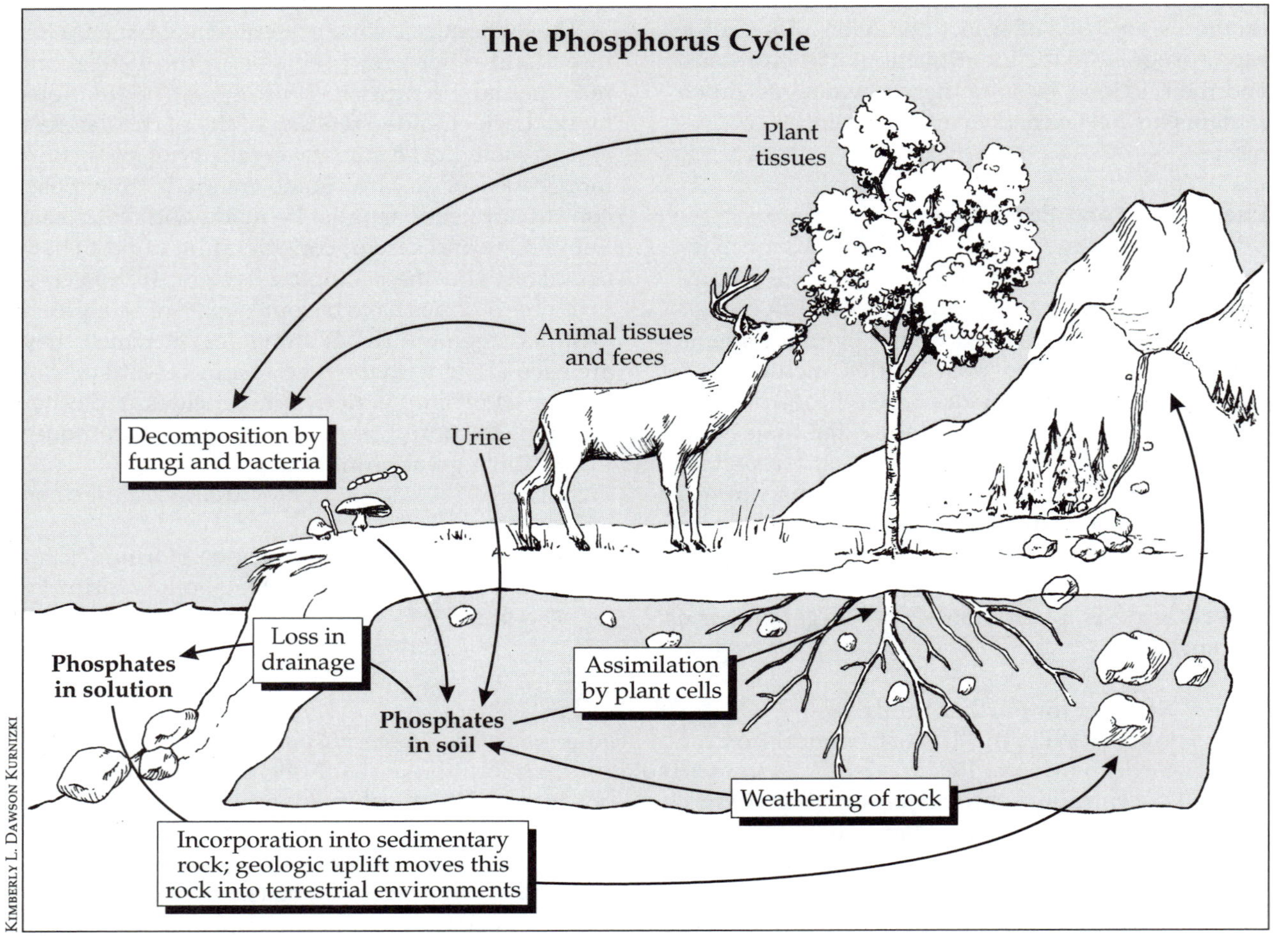

The biogeochemical phosphorus cycle is the movement of the essential element phosphorus through the earth's ecosystems. Released largely from eroding rocks, as well as from dead plant and animal tissues by decomposers such as bacteria and fungi, phosphorus migrates into the soil, where it is picked up by plant cells and is assimilated into plant tissues. The plant tissues are then eaten by animals and released back into the soil via urination, defecation, and decomposition of dead animals. In marine and freshwater aquatic environments, phosphorus is a large component of shells, from which it sediments back into rock and can return to the land environment as a result of seismic uplift.

through the earth's soils, rocks, waters, and atmosphere and through the organisms that inhabit these many ecosystems.

Elements or minerals are stored in discrete parts of the earth's ecosystems called *compartments*. Examples of compartments include all the plants in a forest, a certain species of tree, or even the leaves or needles of a tree. Chemical elements reside within the compartments in certain amounts, or *pools*. A basic description of biogeochemical cycles involves following nutrients in the form of minerals or elements from pool to pool through the multitudes of ecosystem compartments.

Phosphorus and Plants

Phosphorus compounds reside primarily in rocks. Phosphorus does not go through an atmospheric phase, but rather, phosphorus-laden rocks release *phosphate* (PO_4^{-3}) into the ecosystem as the result of weathering and erosion. To plants, phosphorus is a vital nutrient (second only to nitrogen). Plants absorb phosphates through their root hairs. Phosphorus then passes on through the food chain when the plants are consumed by other organisms. Phosphorus is an essential component of many biological molecules, including deoxyribonucleic acid (DNA) and ribonucleic acid (RNA). Adenosine triphosphate (ATP), one of the nucleotides that make up DNA and RNA, is also the main energy transfer molecule in the multitude of chemical reactions taking place within organisms.

Because phosphorus is a major plant nutrient, massive amounts of phosphate-based fertilizers are either derived from natural sources (in the form of bat or bird guano) or chemically manufactured for use by agriculture. As late as the early 1970's, phosphates were a major constituent of household detergents, until it was discovered that large amounts of phosphates were being released into the environment. In aquatic systems such as rivers and lakes, where such runoff eventually appears, an infusion of phosphates can cause *algal blooms* (rapidly forming, dense populations of algae). When the algae die, they are consumed by bacteria. Decomposition by bacteria requires large amounts of oxygen, which soon depletes the available oxygen in the water. If the process is allowed to continue unchecked, fish and other organisms die from lack of oxygen. Both phosphates and nitrates contribute to cultural *eutrophication*.

Phosphates not taken up by plants go into the sedimentary phase, where they are very chemically reactive with other minerals. Some of these reactions produce compounds that effectively remove phosphates from the active nutrient pool. This sedimentary phase is characterized by its long residence time compared to the rapid cycling through the biological phase. Phosphates can remain locked up in rocks for millions of years before being exposed and broken down by weathering, which once again makes them available to plants.

Phosphorus and the Environment

Because the phosphorus cycle is so complex, its interactions with other biogeochemical cycles are not completely understood. The study of these interactions is emerging as a vital field among the environmental sciences. Excessive phosphates in a eutrophic lake disrupt the carbon cycle by reacting with bicarbonates, thus increasing the pH. Many freshwater organisms depend on a neutral pH level for their survival. The presence of phosphorus under these oxygen-depleted conditions can also indirectly affect the sulfur cycle, leading to the conversion of sulfate to sulfide. When sulfide combines with hydrogen to form the gas hydrogen sulfide, it takes on the familiar "rotten egg" smell.

One of the keys to preventing environmental degradation through the altering of global chemical cycles lies in recognizing the effects of such alterations. With the perception of an environmental crisis in the early 1970's, more attention was paid to the role of human activity in these cycles. Test lakes were studied to determine why freshwater fisheries were becoming oxygen-depleted at accelerated rates. Dramatic progress has been made in eliminating the problem of algal blooms and oxygen depletion by limiting the phosphorus-laden effluents being discharged into lakes.

David M. Schlom, updated by Bryan Ness

See also: Eutrophication; Fertilizers; Nutrients.

Sources for Further Study

Krebs, Charles J. *Ecology: The Experimental Analysis of Distribution and Abundance*. 5th ed. San Francisco: Benjamin/Cummings, 2001. Extensive coverage of the biogeochemical cycles in chapter dealing with nutrient cycling. Referenced, with an index including many entries on phosphorus. The nutrient cycle is discussed throughout the book. This is an excellent college-level text with some mathematical models of nutrient cycles and algae growth. Numerous illustrations, figures, graphs, and diagrams.

Lasserre, P., and J. M. Martin, eds. *Biogeochemical Processes at the Land-Sea Boundary*. New York: Elsevier, 1986. A collection of very accessible technical papers dealing with nutrient cycling interactions at the continental-ocean boundary. Contains excellent discussions of ecosystem and nutrient-cycle modeling. Suitable for the advanced student or college-level reader with an interest in land-sea interactions. No index.

Pomeroy, Lawrence, ed. *Cycles of Essential Elements*. Stroudsburg, Pa.: Dowden, Hutchinson & Ross, 1974. A historical perspective of the topic, this volume is a collection of classic papers on nutrient cycling from 1855 to 1973. Part of the "Benchmark Papers in Ecology" series, this book offers insight into the evolving science of ecology. Chapter 4 is a classic paper by Juday, Birge, and Kemmerer titled "Phosphorus Content of Lake Waters in

Northeast Wisconsin." Other articles include discussions of particle analysis and cycles in various bodies of water.

Tiessen, Holm, ed. *Phosphorus in the Global Environment: Transfers, Cycles, and Management*. New York: Wiley, 1995. Traces the global phosphorus cycle and discusses policy issues concerning phosphorus management for global food security and pollution control.

PHOTOPERIODISM

Categories: Physiology; reproduction

The reproductive cycles of many organisms, both plant and animal, are regulated by the length of the light and dark period, called the photoperiod. In flowering plants (angiosperms), flowers are organs for sexual reproduction, and photoperiodism refers to the process by which these plants flower in response to the relative lengths of day and night.

Along the equator, the lengths of day and night remain constant because the sun rises and sets at the same time throughout the year. The lengths of the day and night are also equal (each is six months long) at the exact North and South Poles due to the fact that the sun remains below the horizon for six months each year and above the horizon for the other six months. Anyplace else in the world, the days become longer in the summer and shorter in the winter. The reproductive cycles of many organisms, both plant and animal, are regulated by the length of the light and dark period, called the photoperiod. In flowering plants (angiosperms), flowers are organs for sexual reproduction, and photoperiodism refers to the process by which these plants flower in response to the relative lengths of day and night.

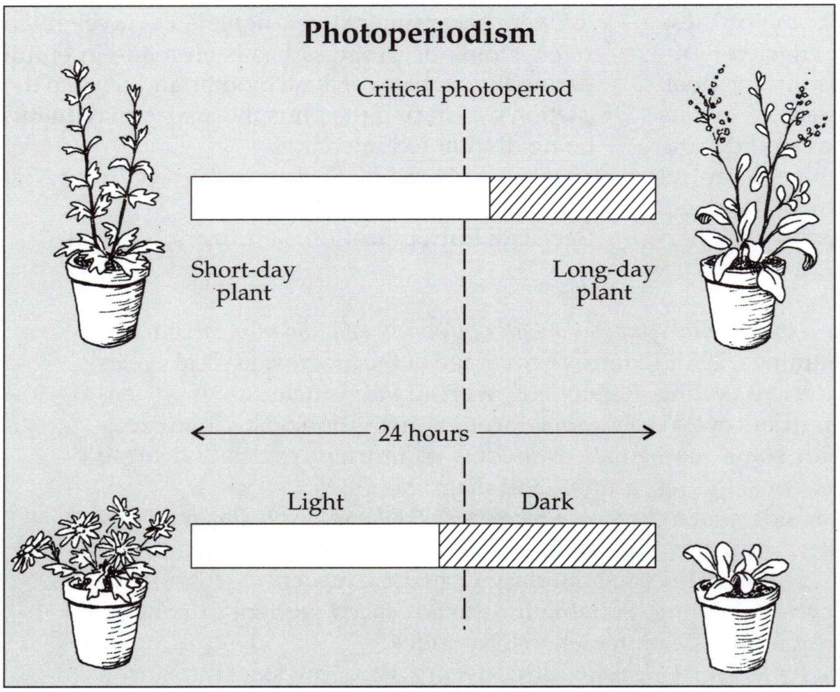

Kimberly L. Dawson Kurnizki

Short-day plants flower when the length of the day is less than the critical photoperiod—that is, when days are relatively short, as in the late fall or early winter. Long-day plants flower when the days are longer than the critical photoperiod—as in late spring and early summer.

The synchronization of reproduction with seasonal time is a very important aspect of plant physiology. Reproduction in many angiosperms is dependent on *cross-pollination*, the process of pollen being transferred from one flower to another. Hence, it is important for all of the plants of the same species in a given region to flower at the same time. Even in nonflowering plants such as mosses, ferns, and some algae, it is usually beneficial for reproductive structures to be formed in a given season. The ability to detect the length of the day or night or both makes it possible to synchronize the reproductive event to a particular time of year. While there have been hundreds of studies which show that many plants respond to changes in the photoperiod, there have been no broad sweeping generalities to provide a better understanding of this phenomenon. Each species, and often each cultivar or variety within a species, appears to have its own photoperiodic response.

Photoperiodic Classification

The photoperiodic classification of plants is usually made on the basis of flowering, but other aspects of their development may also be affected by day length. Based on their flowering response, plants are classified as short-day plants (SDPs), long-day plants (LDPs), intermediate-day plants, ambiphotoperiodic, or day-neutral plants.

Short-day plants flower when the days are relatively short (generally nine hours or less), such as in the late fall or early winter. In some SDP species flowering is *qualitative*, meaning that short days are absolutely required, while in other SDP species flowering is *quantitative*, which means flowering is accelerated under short days, but short days are not an absolute requirement. Some examples of SDPs include rice, cocklebur, and soybean.

Long-day plants flower when the days are relatively long (generally fifteen hours or greater), as would occur in late spring and early summer. As with SDPs, there are qualitative and quantitative species of LDPs.

Intermediate-day plants require quite narrow day lengths (between twelve and fourteen hours) in order to flower, and flowering is inhibited by either short or long days. Sugarcane is an example of an intermediate-day plant.

Ambiphotoperiodic plants are a specialized group of plants that will flower in either short days or long days, but flowering is inhibited by intermediate day lengths.

In *day-neutral plants*, flowering is not regulated by day lengths. In other words, day-neutral plants flower regardless of the day length. There are also many interesting interactions between photoperiod and temperature. A plant may respond to a certain day length at one temperature but exhibit a different response at another temperature. For example, both the poinsettia and morning glory are absolute SDPs at high temperature; however, they are absolute LDPs at low temperature and day-neutral at intermediate temperatures.

Chemical Control

Flowering is regulated by chemicals produced in the plant, and a variety of plant hormones, including *auxins*, *ethylene*, *gibberellins*, *cytokinins*, and *abscisic acid*, have been shown to influence flowering in different species. The critical aspect of photoperiodism, however, is the measurement of seasonal time by detecting the lengths of day and night. The discovery of the *night break phenomenon*, which showed that interruption of the night period with light inhibited flowering in SDPs, established that the length of the dark period is the most critical for initiating a photoperiodic response.

The chemical *phytochrome* is responsible for measuring the dark period. Phytochrome, found in the leaves of plants, exists in two forms, P_r and P_{fr}. P_r absorbs red light during the day and is converted to P_{fr}. P_{fr} absorbs far-red light during the night and is converted to P_r. The prevailing hypothesis is that P_{fr} inhibits flowering, and the length of the dark period has to be sufficient for the P_{fr} to fall below some critical level. When the P_{fr} falls below this level, chemical messages are sent to the floral regions, and flowering is initiated.

While phytochrome definitely has been shown to trigger the flowering response, it is not the only chemical involved. It has been shown that a blue-light photoreceptor may also play a role in photoperiodism. In addition, phytochrome is not translocated in the plant. It remains in the leaves. Hence, other chemicals which have not been positively identified are responsible for signaling the photoperiodic response.

D. R. Gossett

See also: Circadian rhythms; Flower structure; Flowering regulation; Hormones; Pollination; Reproduction in plants.

Sources for Further Study

Halevy, A. H., ed. *CRC Handbook of Flowering*. Boca Raton, Fla.: CRC Press, 1985. An excellent source of information dealing with all aspects of flowering, for the more advanced student. Includes bibliographical references.

Hopkins, William G. *Introduction to Plant Physiology*. 2d ed. New York: John Wiley and Sons, 1999. This general text for the beginning plant physiology student contains a very clear discussion of photoperiodism. Includes diagrams, illustrations, and index.

Raven, Peter H., Ray F. Evert, and Susan E. Eichhorn. *Biology of Plants*. 6th ed. New York: W. H. Freeman/Worth, 1999. A very good general botany text with a simplified discussion of photoperiodism. Includes diagrams, illustrations, and index.

Salisbury, Frank B., and Cleon W. Ross. *Plant Physiology*. 4th ed. Belmont, Calif.: Wadsworth, 1992. One of the standard texts for the more advanced plant physiology student, with an excellent discussion on photoperiodism. Includes biographical references, diagrams, illustrations, and index.

Taiz, Lincoln, and Eduardo Zeiger. *Plant Physiology*. 2d ed. Sunderland, Mass.: Sinauer Associates, 1998. A text written for the advanced plant physiology student, with an outstanding discussion of photoperiodism. Includes biographical references, diagrams, illustrations, and index.

PHOTORESPIRATION

Categories: Cellular biology; photosynthesis and respiration; physiology

Photorespiration is a biochemical process in plants in which, especially under conditions of water stress, oxygen inhibits the Calvin cycle, the carbon fixation portion of photosynthesis.

Photorespiration results in the light-dependent uptake of oxygen and release of carbon dioxide and is associated with the synthesis and metabolism of a small molecule called *glycolate*. Photorespiration takes place in green plants at the same time that photosynthesis does. Because in photosynthesis carbon dioxide is taken in, and in photorespiration carbon dioxide is given off, these two processes work against each other. The end result is that photorespiration decreases the net amount of carbon dioxide which is converted into sugars by a photosynthesizing plant. By interfering with photosynthesis in this way, photorespiration may significantly limit the growth rate of some plants.

Photosynthesis

In green plants, photosynthesis takes place in the special energy-storing molecules called *chloroplasts*. Photosynthesis can be divided into two parts: the *light reactions* and the *dark reactions*. In the light reactions, light energy from the sun is captured by the plant and converted into chemical energy in the form of chloroplasts. An additional feature of the light reactions is that a molecule of water is split so that its oxygen is released. In the dark reactions, a series of steps called the *Calvin cycle* converts carbon dioxide from the air into organic molecules such as sugars and starch. The Calvin cycle requires energy in order to operate, and this is provided by the energy-storing molecules (such as adenosine triphosphate) formed in the light reactions. The carbohydrates thus formed can serve as food for the plant or for an animal that eats the plant.

The overall pattern of *gas exchange* in photosynthesis, therefore, is the release of oxygen and the uptake of carbon dioxide. It has been found that, to a lesser extent, light can also cause plants to do just the opposite—that is, to consume oxygen and release carbon dioxide. This phenomenon was discovered in the 1950's and is termed photorespiration. If net photosynthesis is defined as the total amount of carbon dioxide taken in minus the amount given off, it is apparent that increasing the rate of photorespiration will decrease net photosynthesis. In terms of agricultural plants, this translates into a decrease in the productivity of the crop.

Rubisco and Glycolate

Each of the reactions of the Calvin cycle must be catalyzed by an enzyme. The first reaction of the cycle, in which carbon dioxide is taken up, utilizes an enzyme popularly known as *Rubisco*, an abbreviation for ribulose bisphosphate carboxylase/oxygenase. The normal function of Rubisco is to take carbon dioxide from the atmosphere and combine it with another molecule in the chloroplast, ribulose bisphosphate (RuBP). The resulting compound is then acted upon by other enzymes which eventually convert it into the simple sugar glyceraldehydes 3-phosphate, which is used in the synthesis of more complex sugars and other compounds.

Rubisco, however, does not always behave in its normal fashion. It is sometimes unable to distinguish between molecules of carbon dioxide and oxygen. Rubisco will sometimes "mistakenly" incorporate an oxygen molecule into RuBP rather than the carbon dioxide that would normally have been used. The oxygen may come from the atmosphere, or it may originate from the oxygen that is continually being produced by the splitting of water during the light reactions of photosynthesis.

The result of this metabolic error is that, rather than forming compounds that can be converted into sugar, the plant forms a substance known as *glycolate*. Understanding what happens to glycolate is the key to understanding the process of photorespiration. The utilization of oxygen in the formation of glycolate accounts for part of the oxygen uptake that is observed during photorespiration. Instead of being used in sugar synthesis, the glycolate enters a different metabolic pathway, where it is acted upon by a different series of enzymes with different consequences. These sequential reactions, referred to as the *glycolate pathway*, result in the conversion of glycolate into a series of different compounds. This pathway constitutes the remainder of the process of photorespiration.

The Glycolate Pathway

An unusual feature of photorespiration is that it involves three separate cellular organelles: the chloroplast, the *peroxisome*, and the *mitochondrion*. The first stage of photorespiration involves the formation of glycolate in the chloroplast. The glycolate does not undergo further reactions in the chloroplast but instead is transported to the peroxisome.

Once inside the peroxisome, the glycolate enters a series of reactions, one of which causes oxygen to be converted into hydrogen peroxide. This represents a second point in the process at which oxygen is consumed. If hydrogen peroxide were present in large quantities it could have a toxic effect upon the cell, so the peroxisome also contains the enzyme catalase, which destroys most of the hydrogen peroxide thus formed.

In subsequent steps of the glycolate pathway, one of the compounds formed is glycine, which enters the mitochondrion and loses a carbon atom in the form of carbon dioxide. This, then, accounts for the observed release of carbon dioxide during photorespiration. If the carbon dioxide is lost to the atmosphere, it represents a decrease in the net amount of carbon taken up by the plant and, therefore, a decrease in net photosynthesis.

Further reactions of the glycolate pathway occur in the mitochondrion and peroxisome, and eventually a compound is formed which is returned to the chloroplast, where the process began. This compound is capable of reentering the Calvin cycle and can actually be used for the synthesis of sugars. The critical point, however, is that not all the carbon atoms which left the chloroplast in the form of glycolate are returning to it. Part of the carbon was lost in the form of the carbon dioxide that was released in the mitochondrion. Furthermore, certain steps in the glycolate pathway require the expenditure of energy. The process is, therefore, doubly wasteful in that it results in the loss of both carbon dioxide and energy storage molecules.

To summarize the process, oxygen is utilized in the chloroplast to form the two-carbon compound glycolate. The glycolate then enters a series of reactions that occur in the peroxisome and mitochondrion and that take up additional oxygen and release a portion of the carbon in the form of carbon dioxide. The remaining carbon is converted into 3-phosphoglycerate, which can be returned to the chloroplast and reenter the Calvin cycle, where it is one of the normal intermediate compounds. This accounts for the light-dependent uptake of oxygen and release of carbon dioxide, which constitute photorespiration.

Factors That Increase Photorespiration

Although some amount of photorespiration occurs in many plants regardless of conditions, photorespiratory rates increase any time that carbon dioxide levels are low and oxygen levels are high. Such conditions occur whenever stomata (specialized pores for gas exchange) remain closed, or partially closed, while photosynthesis is under way. Under most conditions plants are able to keep their stomata open, so photorespiratory rates remain low. When plants become water stressed, they close their stomata to prevent further water loss by transpiration. Water stress is most likely under hot, dry conditions. Under these conditions, the stomata close as far as needed to conserve water, thus restricting normal gas exchange. Carbon dioxide levels slowly rise as water is split during the light reactions. Consequently, photorespiratory rates accelerate, and photosynthetic efficiency drops to as low as 50 percent of normal.

In dry tropical and desert environments water stress, and thus photorespiration, can significantly reduce plant growth potential. Some plants have evolved solutions to this problem by modifying the way they carry out photosynthesis. One common adaptation is called C_4 metabolism. This modification involves a different leaf anatomy, called Kranz anatomy, as well as a different enzyme pathway for initially fixing carbon dioxide that is not prone to problems with oxygen. The Calvin cycle still functions in C_4 plants, as they are called, but it is protected from photorespiration by the C_4 adaptations. Many tropical grasses, including corn and sugarcane, use this approach. Unfortunately, many crop plants, including wheat, soybeans, spinach, and tomatoes, do not possess this adaptation.

The other major adaptation is called *CAM metabolism*, which is short for crassulacean acid metabolism. This adaptation is common in succulents (plants that store excess water in their stems and leaves, making them very juicy), such as pineapples, cacti, and stonecrops. Stonecrops are in the

PhotoDisc

In dry tropical and desert environments water stress, and thus photorespiration, can significantly reduce plant growth potential. Some plants have evolved solutions to this problem by modifying the way they carry out photosynthesis. One common adaptation is called C_4 metabolism. The Calvin cycle still functions in C_4 plants, but it is protected from photorespiration by the C_4 adaptations. Many tropical grasses, including sugarcane, use this approach.

family *Crassulaceae*, which is the source of the name for this adaptation. CAM plants, as they are called, only open their stomata at night, when transpiration rates are low and water loss is minimal. Carbon dioxide enters the leaf at night and is attached to organic molecules in a different pathway that does not require light as an energy source. Then during the day, when the stomata are closed, this carbon dioxide is released and enters the Calvin cycle. Photorespiration is prevented because carbon dioxide levels can be maintained at appropriate levels, even though the stomata are closed.

Why Photorespiration?

Several theories have been proposed to explain why plants photorespire. One possibility is that when plants first evolved, conditions on the primitive earth were very different from what they are now. The early atmosphere contained little oxygen, so the inability of Rubisco to distinguish between oxygen and carbon dioxide was not a problem. As the oxygen level in the atmosphere gradually increased, the formation of glycolate during photosynthesis began to occur, and this led to the problem of photorespiration. The glycolate pathway then developed as a mechanism for salvaging some of the material that leaves the Calvin cycle in the form of glycolate, ultimately returning a portion of it to the cycle.

Seen in this context, the real culprit in photorespiration is the formation of glycolate by Rubisco, while the glycolate pathway is an evolutionary adaptation for making the best of a bad situation. Perhaps, millions of years in the future, plants will evolve a form of Rubisco that can more effectively distinguish between these two gases, and photorespiration will diminish or cease.

An alternative theory about why plants photorespire is that the process does, in fact, perform an important function: protecting the plant from the harmful effects of very high internal concentrations of oxygen or energy storage molecules. This high concentration could occur when the plant is exposed to high light intensities, causing photosynthesis to generate these substances very rapidly. Photorespiration would then consume some of the excess oxygen and energetic molecules, depleting them to levels that would not be harmful to the plant. It has not yet been conclusively shown, however, that photorespiration really does play such a protective role. Further research will be required before scientists know whether photorespiration is beneficial to the plant.

Thomas M. Brennan, updated by Bryan Ness

See also: ATP and other energetic molecules; C_4 and CAM photosynthesis; Calvin cycle; Gas exchange in plants; Glycolysis and fermentation; Photosynthesis; Photosynthetic light absorption; Photosynthetic light reactions; Respiration.

Sources for Further Study

Dennis, David T., et al., eds. *Plant Metabolism*. 2d ed. New York: Longman, 1997. A well-balanced overview. Revised edition contains new chapters on protein synthesis and the molecular biology of plant development and on the manipulation of carbon allocation in plants.

Foyer, Christine H. *Photosynthesis*. New York: John Wiley & Sons, 1984. A college-level monograph dealing with all aspects of photosynthesis. Chapter 9 gives a very readable discussion of photorespiration and the glycolate pathway. Earlier chapters of the book provide a detailed discussion of chloroplast structure and the light and dark reactions of photosynthesis. Technical references provided at the end of each chapter.

Kaufman, Peter B., and Michael L. Evans. *Plants: Their Biology and Importance*. New York: Harper & Row, 1989. An introductory-level college textbook suitable for the nonscience major interested in plant biology. Chapter 8 includes a very lucid discussion of photorespiration, its effect on plant productivity, and the relationship of these processes to various environmental factors. Chapter 9 deals briefly with some of the interconnections between respiration, photorespiration, and photosynthesis.

Kozaki, Akiko, and Go Takeba. "Photorespiration Protects C_3 Plants from Photooxidation." *Nature* 384, no. 6609 (December, 1996): 557-561. Discusses photorespiration as a defense mechanism against the harmful effects of light in C_3 plants, such as wheat and rice, and ways in which the protective role of photorespiration is supported.

Salisbury, Frank B., and Cleon W. Ross. *Plant Physiology*. 4th ed. Belmont, Calif.: Wadsworth, 1992. An intermediate-level college textbook which presupposes a moderate knowledge of chemistry. The chapter on photorespiration and photosynthesis includes some background information on how the process of photorespiration was discovered. Also includes a good discussion of the anatomical and physiological features found in plants that do not photorespire.

Stumpf, Paul, and E. E. Conn, eds. *Metabolism and Respiration*. Vol. 2 in *The Biochemistry of Plants: A Comprehensive Treatise*. New York: Academic Press, 1980. Chapter 12 of this volume contains a comprehensive and detailed but clearly written discussion of all aspects of photorespiration, including the methodology used in studying the process. The author takes a strongly chemical approach to the material; intended for readers with some prior knowledge of plant physiology and biochemistry.

PHOTOSYNTHESIS

Category: Cellular biology

Photosynthesis is the process by which organic sugar molecules are synthesized from an inorganic carbon source (carbon dioxide or bicarbonate), using sunlight as the energy source to drive the process. Although most often associated with plants (in which the reactions of photosynthesis occur within compartments called chloroplasts), algae and certain types of bacteria are also capable of photosynthesis.

From an ecological perspective, photosynthesis is significant because the conversion of inorganic carbon to organic carbon represents the entry point of carbon atoms into biological systems. Photosynthesis is also significant because it is the means whereby oxygen is released into the atmosphere. The atmospheric concentration of oxygen is approximately 21 percent, and most of this oxygen originates from photosynthesis. In addition, solar energy absorbed during photosynthesis serves as the ultimate source of energy for almost all nonphotosynthetic organisms.

Nature of Light

Light from the sun is composed of various types of radiation. Only a portion of this solar radiation can be used by plants for photosynthesis. This *photosynthetically active radiation* (PAR) ranges in wavelength from 400 to 700 nanometers and corresponds approximately to the visible light perceived by the human eye. The energy content of light depends on its wavelength, with shorter wavelengths having a higher energy content than longer wavelengths.

Role of Pigments

For light energy to drive photosynthesis, it first must be absorbed. Several types of photosynthetic pigments are found in plants. When these pigments absorb light, some of the pigments' electrons are elevated to a high energy level. These high-energy electrons are used to drive the reactions of photosynthesis, thus converting light energy into chemical energy. The most common photosynthetic pigment in plants is the green-colored *chlorophyll*. Two types of chlorophyll are found in plants, chlorophyll *a* and chlorophyll *b*, with other types of chlorophyll found in various types of algae and photosynthetic bacteria.

Additional plant accessory pigments, such as *carotenoids*, which are yellow or orange, play a minor role in the absorption of wavelengths of light not absorbed by chlorophyll. Carotenoids also help protect chlorophyll from damage that may occur as a result of absorbing excess light energy. As the most abundant plant pigment, chlorophyll gives plants their green color and usually masks the other colored pigments. In deciduous trees and shrubs, however, chlorophyll is degraded during the au-

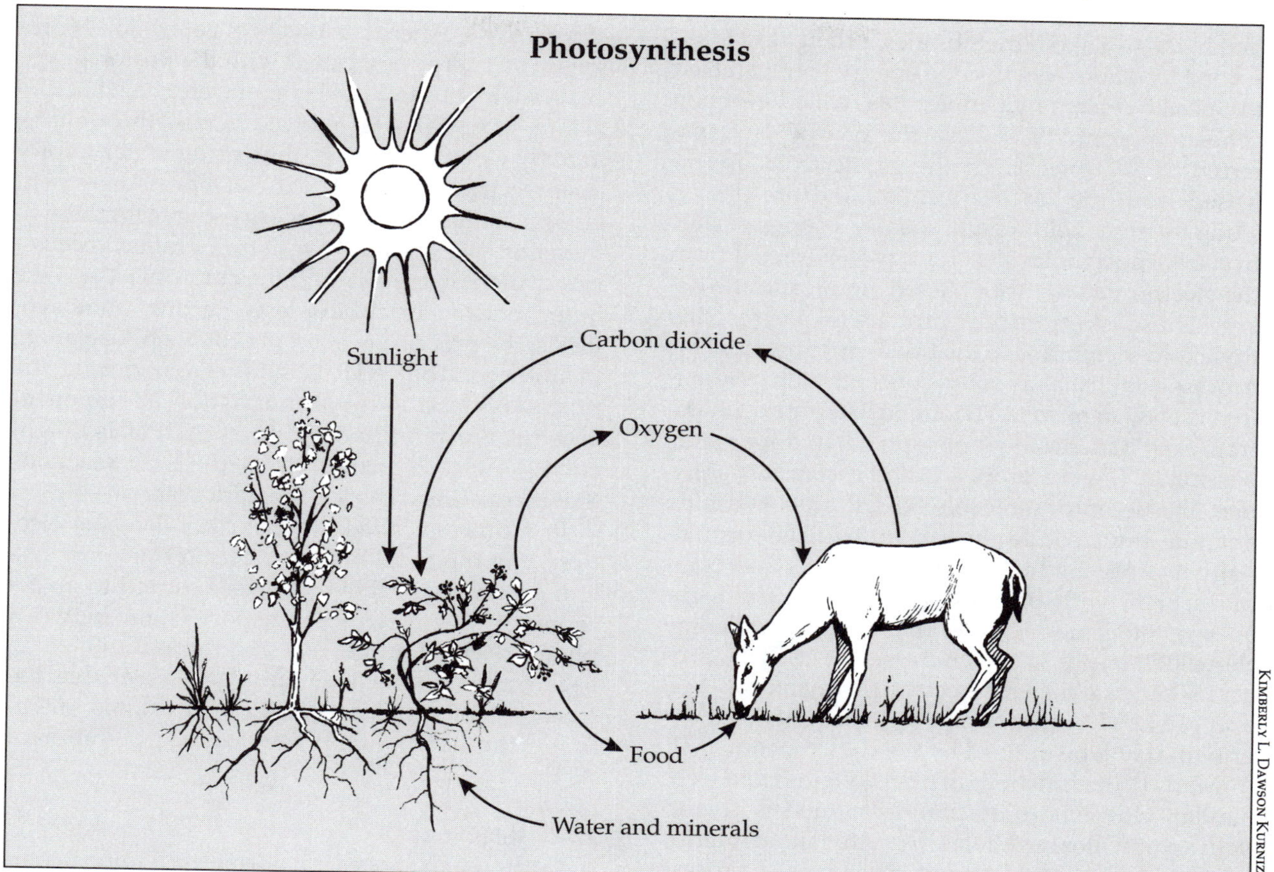

KIMBERLY L. DAWSON KURNIZKI

The processes of photosynthesis and cellular respiration are complementary: Oxygen released into the atmosphere, a by-product of photosynthesis, is breathed in by animals, which in turn breathe out carbon dioxide, essential to photosynthesis. Likewise, photosynthesis is at the beginning of the carbon cycle and the food chain: Algae and plants generate organic (biochemical) energy from nonorganic sunlight, water, and minerals; animals eat plants and one another; decomposers (fungi and bacteria) return the essential elements to the ecosystem; and the cycle begins again.

tumn, revealing a spectacular display of colors from carotenoids and other pigments.

Reactions of Photosynthesis

The process of photosynthesis is complex, involving many biochemical reactions. Historically, the reactions of photosynthesis have been divided into the *light reactions* and the *dark reactions*. The light reactions include the absorption of light and the conversion of light energy to chemical energy. The dark reactions use the chemical energy produced in the light reactions to incorporate (or fix) carbon dioxide molecules into organic molecules (sugars). Within the chloroplast, the light reactions are localized in the internal network of membranes called *thylakoid membranes*.

The dark reactions occur in the aqueous region of the chloroplast called the *stroma*. The term "dark reactions" is somewhat misleading because several photosynthetic enzymes are not active in the dark, so these reactions will not occur without light. Although it is common to separate the light and dark reactions when describing photosynthesis, it should be noted that these reactions are tightly coupled and occur simultaneously in the plant.

Light Reactions

Chlorophyll and other accessory pigments that absorb light energy, along with certain proteins, are organized into structures called *photosystems*. Two types of photosystems occur in plants, photosystem I and photosystem II, and both are embedded in the thylakoid membranes. When light is absorbed by photosystems, the energy is transferred to special chlorophyll molecules, called reaction center chlorophylls, where the energy is transferred to electrons. High-energy electrons are released from the reaction centers and are passed along the thylakoid membranes by a series of electron transport molecules. Energy is extracted from the electrons as they are passed along, and the energy is used to transport protons (H^+) across the thylakoid membrane to the thylakoid interior. This process establishes a proton concentration gradient that is used to make ATP (adenosine triphosphate) from ADP (adenosine diphosphate) and inorganic phosphate (P_i) in a process called *photophosphorylation*. The acceptor molecule $NADP^+$ (nicotinamide adenine dinucleotide phosphate, oxidized form) finally accepts the high-energy electrons and combines them with protons (H^+) to form the high-energy molecule NADPH (nicotinamide adenine dinucleotide phosphate, reduced form). This process is called noncyclic electron flow, and the ATP and NADPH produced are forms of chemical energy that will be utilized by the dark reactions.

Many of the functions of photosytems I and II are similar. However, only photosystem II is able to split apart water molecules in the thylakoid interior into electrons, protons, and oxygen in a process called *photolysis*. The electrons released from water during photolysis replace electrons lost by chlorophyll molecules during the electron transport reactions. The protons released from water accumulate in the thylakoid interior, adding to the concentration gradient established by noncyclic electron flow. Oxygen produced from the photolysis of water is released as a gas to the atmosphere. Therefore, oxygen gas may be considered a by-product of plant photosynthesis. Algae and some bacteria (cyanobacteria) also release oxygen during photosynthesis, but other photosynthetic bacteria do not split water molecules and thus do not release oxygen.

The proton gradient created across the thylakoid membranes represents a source of potential energy used in making ATP. Protons are unable to diffuse across the thylakoid membranes unless permitted to do so by a special enzyme complex called the ATP synthase. The energy required to generate ATP is provided by protons as they move through the ATP synthase from the thylakoid interior, where there is a high concentration of protons to the stroma, where there is a lower concentration of protons. The energy associated with the proton gradient is analogous to a reservoir of water held back by a dam. Water may be allowed to pass through the dam by way of a turbine, thus using water to produce electrical power. The use of a proton gradient across a membrane as the energy source for the synthesis of ATP by ATP synthase is called *chemiosmosis*, and it occurs in both the chloroplast and the mitochondria. In chloroplasts, during photosynthesis, the process is called photophosphorylation. In mitochondria, ATP is synthesized during the process of oxidative phosphorylation, a component of cellular respiration. ATP, like NADPH, is a high-energy molecule produced by the light reactions that is consumed during the dark reactions.

In a process called *cyclic electron flow*, the electrons can travel within the electron transport system as described above but are diverted to an acceptor in the electron transport chain between photosystems I and II. Passing through the chain back to photosystem I, the electrons enable the transport of protons across the thykaloid membrane, thus supplying power for the generation of ATP.

Dark Reactions

Carbon dioxide gas is a normal, but minor, component of the atmosphere and enters leaves when air diffuses through stomata, small pores on the plant surfaces. The first reaction in converting carbon dioxide to sugar molecules occurs when the enzyme *Rubisco* (also known as ribulose bisphosphate carboxylase/oxygenase, and reportedly the most abundant protein on earth) combines ribulose bisphosphate (RuBP) containing five carbon atoms with a carbon dioxide molecule to produce two identical molecules of a simple sugar, each containing three carbon atoms. These three-carbon sugar molecules are then subsequently metabolized through a series of reactions leading to the production of a three-carbon sugar called glyceraldehyde 3-phosphate (G3P). Some of the G3P is used to make glucose and other organic molecules, and the remaining G3P is used to regenerate RuBP so the process can continue. This cyclic pathway is known as the *Calvin cycle* (or the Calvin-Benson cycle).

Sugar products may be stored within the chloroplast as starch, or they may be transported as sucrose to other parts of the plant as needed. ATP and NADPH from the light reactions are required for several of the reactions of the Calvin cycle. Because the first molecule produced by this pathway is a simple sugar with three carbon atoms, the pathway is known as the C_3 pathway. Plants using this pathway are known as C_3 plants, and common examples include many trees and the majority of agricultural crops.

The Rubisco enzyme, in addition to catalyzing the uptake of carbon dioxide during the Calvin cycle, can take up oxygen, initiating another metabolic pathway called *photorespiration*. In photorespiration, when oxygen is attached to RuBP instead of carbon dioxide, a product results that cannot be used in the Calvin cycle, and that product must go through a different set of complex reactions. For this reason photorespiration is often described as a wasteful process that competes with photosynthesis. The relative amounts of carbon dioxide and oxygen gases inside the chloroplast determine the relative rates of photosynthesis and photorespiration. Experiments in which the carbon dioxide concentration of air has been altered have demonstrated that the rate of photosynthesis increases and the rate of photorespiration decreases when the concentration of carbon dioxide is increased. In some agricultural and horticultural greenhouse operations, carbon dioxide amounts in the atmosphere are elevated to stimulate photosynthesis, leading to increases in plant production and yield.

Some plants have an adaptation whereby carbon dioxide is initially fixed by a pathway other than the Calvin cycle. This adaptation involves the enzyme phosphoenolpyruvate carboxylase (*PEP carboxylase*), an enzyme that lacks the oxygenase activity of Rubisco. PEP carboxylase actually attaches bicarbonate to phosphoenolpyruvate (PEP) in mesophyll cells that are in contact with air spaces in the leaf. PEP is then converted into a series of organic acids and, in the process, is transported into a specialized set of cells called *bundle sheath cells* that are separated from the air spaces in the leaf. In the bundle sheath cells carbon dioxide is released from the last organic acid in the series, which raises the carbon dioxide level in these cells where the carbon dioxide is used in the C_3 cycle. Raising the carbon dioxide concentration within the chloroplast increases photosynthesis while reducing photorespiration. The initial product of the pathway in the mesophyll cells is an organic acid with four carbon atoms, and thus the pathway is called the C_4 pathway. Plants possessing this pathway are known as C_4 plants. Examples include most grasses and a few crops, including corn and sugarcane.

A second adaptation that circumvents photorespiration is the CAM (crassulacean acid metabolism) pathway, named after the family of plants in it was first observed, *Crassulaceae*, or the stonecrop family. *CAM photosynthesis* is similar to C_4 photosynthesis in that it is another adaptation for raising the concentration of carbon dioxide inside the chloroplast. CAM plants accomplish photosynthesis using a biochemical process essentially the same as that of C_4 plants, but instead of carrying out these reactions in separate cells, they carry out certain reactions at night. CAM plants open their stomata only at night, and the carbon dioxide is transferred to PEP, which is converted into another organic acid that is stored throughout the night. During the day, the stomata remain closed, and the carbon dioxide needed for the C_3 cycle is supplied by releasing carbon dioxide from the last organic acid in the CAM cycle. Examples of CAM plants include cactus and pineapple. Both C_4 and CAM plants typically require less water than do C_3 plants and may be found in warmer and drier environments. C_4 plants tend to have high rates of photosynthesis, whereas CAM plants have low photosynthetic rates because the CAM cycle is less efficient than the C_4 cycle. C_3 plants typically have intermediate photosynthetic rates under optimal conditions.

Photosynthesis and the Environment

Several environmental factors affect the rate of photosynthesis. For example, temperature extremes and water stress inhibit photosynthesis. As light intensity increases, so do photosynthetic rates. However, when photosynthesis becomes light-saturated, further increases in light intensity will not result in greater rates of photosynthesis. Leaves of plants that grow in full-sun conditions are smaller and thicker, with more extensive vascular systems than those found in shade plants. Although so-called sun leaves and shade leaves have similar photosynthetic rates in low light, shade leaves have much lower rates of photosynthesis at high light intensities and can be damaged when exposed to such conditions.

As mentioned above, atmospheric carbon dioxide concentrations can also regulate photosynthesis. At the present time the concentration of carbon dioxide in the atmosphere is less than 0.04 percent, but scientific data show that this concentration is increasing. Higher concentrations of atmospheric carbon dioxide may stimulate plant photosynthesis and plant growth but may have other undesirable climatic effects.

William J. Campbell

See also: Anaerobic photosynthesis; C_4 and CAM photosynthesis; Chemotaxis; Calvin cycle; Chloroplasts and other plastids; Energy flow in plant cells; Eukaryotic cells; Membrane structure; Photosynthetic light absorption; Photosynthetic light reactions; Pigments in plants; Plant cells: molecular level; Respiration.

Sources for Further Study

Buchanan, Bob B., et al. *Biochemistry and Molecular Biology of Plants*. Rockville, Md.: American Society of Plant Physiologists, 2000. A comprehensive and up-to-date book presenting the current understanding of the functioning of plants. Includes diagrams, photographs, bibliographical references, and index.

Hopkins, William G. *Introduction to Plant Physiology*. 2d ed. New York: John Wiley & Sons, 1999. An overview of the processes and functions of plants. Includes diagrams, photographs, bibliographical references, and index.

Rao, K. K., and D. O. Hall. *Photosynthesis*. 6th ed. Cambridge, England: Cambridge University Press, 1999. Covers all aspects of photosynthesis. Includes diagrams, photographs, bibliographical references, and index.

Taiz, Lincoln, and Eduardo Zeiger. *Plant Physiology*. 2d ed. Sunderland, Mass.: Sinauer Associates, 1998. Thorough coverage of the physiology of flowering plants. Includes diagrams, photographs, bibliographical references, and index.

PHOTOSYNTHETIC LIGHT ABSORPTION

Categories: Cellular biology; photosynthesis and respiration; physiology

Photosynthetic light absorption involves plants' use of pigments to facilitate the conversion of light energy into chemical energy.

Photosynthesis occurs in green plants, algae, and certain types of bacteria. There is considerable variation among the types of pigments found in these different groups of organisms, but the basic mechanisms by which they absorb light are similar. Photosynthetic pigments are always attached to membranes within a cell. In algae and higher plants, the photosynthetic pigments are located in the *chloroplast*, where photosynthesis takes place. The pigment molecules are not dispersed randomly within the chloroplast but are arrayed on the surface of the *thylakoid membranes*. The chloroplasts become oriented in such a way as to present a large surface area to the sun or other light source, thereby maximizing the ability of the pigments to absorb light energy. In photosynthetic bacteria, the light-absorbing pigments are not organized in chloroplasts but are located on membranes that are dispersed throughout the cell.

Chlorophylls and Carotenoids

The two primary types of pigments utilized by most photosynthetic organisms are the green *chlorophylls* and the yellow to orange *carotenoids*. There are several forms of chlorophyll that differ from one another in small details of their molecular structures. The forms are designated chlorophyll *a*, *b*, *c*, and so on. Chlorophyll *a* is found in all plants and algae, while the other forms are dispersed among various taxonomic groups. The chlorophyll molecule consists of two parts: an elaborate ring structure that actually absorbs the light, and a long tail-like section that anchors the molecule in the membrane. Photosynthetic bacteria contain similar pigments called *bacteriochlorophyll a* and *b*.

Carotenoids, the other major group of photosynthetic pigments, also occur in various forms and are found in all types of photosynthetic plants, algae, and bacteria. The *xanthophylls*, which are oxygenated carotenoids, form another widespread and diverse subgroup of pigments. Carotenoid molecules have elongated structures and, like the chlorophylls, are embedded in the photosynthetic membranes. It is a popular misconception that the change in leaf color that takes place in the fall is the result of the formation of new yellow or orange carotenoid pigments. Actually, the carotenoids are present all the time but are masked by the presence of the green chlorophyll. In fall, the chlorophyll begins to decompose more rapidly than the carotenoids, whose yellow colors are then exposed.

Chlorophyll *a*, or something very similar to it, occurs almost universally in photosynthetic organisms, from bacteria to higher plants, because it is an essential component of photosynthetic reaction centers. All of the other chlorophylls and carotenoids involved in light absorption are referred to as *accessory pigments*. Another kind of accessory pigment found in some groups of algae and photosynthetic bacteria are the *phycobiliproteins*, which may impart a red or blue color to the cells in which they occur. These molecules consist of a light-absorbing portion bound to a protein. In fact, all types of pigment molecules seem to be bound to proteins within the photosynthetic membranes. These pigment-protein associations are sometimes referred to as *light-harvesting complexes*, a term that accurately describes their function.

Properties of Light

To understand the functioning of photosynthetic pigments, it is necessary to consider first the physical nature of light. Visible light is only a small portion of the electromagnetic spectrum, which ranges from very short wavelength radiation, such as X rays, to extremely long wavelength radiation, such as radio waves. The visible portion of the spectrum is intermediate in wavelength and ranges from blue (at the short end) to red (at the long end). Sunlight contains a mixture of all the visible wavelengths, which humans perceive as white light. The energy of light is inversely proportional to its wavelength; blue light has more energy than an equivalent amount of red light. Light may be thought of as consisting of either waves or particles. For purposes of studying light absorption by pigments, it is easier to think of light as particles, referred to as *photons* or *quanta*.

When a photon is absorbed by a pigment molecule, the photon's energy is transferred to one of the electrons of the pigment. The electron is thus said to enter an excited state and contains the energy originally associated with the photon of light. A specific kind of pigment is not capable of absorbing all the photons it encounters. Only photons of certain energy (and therefore wavelength) can be absorbed by each pigment. For example, chlorophyll primarily absorbs light in the blue and red wavelengths but not in the green portion of the spectrum. Consequently, the green light to which chlorophyll is exposed is either transmitted through it or reflected from it, with the result being that the pigment appears green. The color of the pigment results from the wavelengths of light that are not absorbed. Carotenoids do not absorb light in the yellow to orange portion of the spectrum and, therefore, are seen as being that color.

The process of light absorption begins when a photon of appropriate energy strikes a chlorophyll or carotenoid molecule located on a thylakoid or other photosynthetic membrane, thus causing an electron in the pigment to be raised to an excited state. If two pigment molecules are situated adjacent to each other in exactly the right orientation and are separated by a very small distance, it is possible for the energy of excitation to be transferred from one molecule to the next. This transfer process (referred to as *Forster resonance*) enables the excitation energy to migrate throughout the array of pigment molecules that are attached to the photosynthetic membrane.

In addition to pigment molecules, the membranes also contain a smaller number of special structures called *reaction centers*, which consist of special chlorophyll and protein molecules arranged in a very specific fashion. The excitation energy migrating throughout the pigment array will eventually find its way to one of the reaction centers, and there it is utilized to form new energy-containing molecules. All of this occurs with a large number of photons and pigment molecules simultaneously. The array of pigment molecules feeding excitation energy into the reaction centers contains both chlorophylls and carotenoids and is sometimes referred to as an *antenna*, to indicate its role in light absorption. This overall process constitutes the light reactions of photosynthesis. The energy-containing molecules thus formed will then be used in the *Calvin cycle* to convert carbon dioxide into carbohydrates: Light energy has been converted into chemical energy.

Pigment Functions

Chlorophyll *a*, as well as the other chlorophylls, makes up a major portion of the antenna pigments that absorb light energy and transfer it to the reaction centers. The carotenoids, which are also part of the antenna, seem to contribute in two ways to the effectiveness of the light absorption process. First, they increase the range of wavelengths that can be absorbed. Chlorophyll absorbs mainly in the blue and red portions of the spectrum but is not effective at absorbing other wavelengths. Carotenoids are able to absorb some of the green light that would be unusable if chlorophyll were the only pigment present, so having a combination of different pigments makes the organism more effective at using the various wavelengths that occur in sunlight.

A second function of the carotenoids has to do with their ability to protect chlorophyll from damage by intense light. Under conditions of high light intensity, chlorophyll has a tendency to decompose through a process called photooxidation. The presence of carotenoids prevents this decomposition from occurring and enables the chlorophyll to continue to function effectively at light intensities that would otherwise cause damage.

Although virtually all photosynthetic organisms utilize some form of chlorophyll in light absorption, an exception is found in the halobacteria. These bacteria live in conditions of very high salt concentration, such as the Dead Sea or the Great Salt Lake. In these bacteria is found a purple membrane containing molecules of a pigment called *bacteriorhodopsin* which, remarkably, is very similar to rhodopsin, the pigment found in the visual systems of higher animals. When bacteriorhodopsin molecules absorb light, they cause a hydrogen ion (a proton of H^+) to be ejected across the cell membrane, and this leads to the formation of energy-containing molecules that the bacterium can utilize for its various metabolic requirements. Research with these unique photosynthetic organisms may also lead to a better understanding of the molecular basis and evolution of vision in animals.

Thomas M. Brennan

See also: Anaerobic photosynthesis; Calvin cycle; Chloroplasts and other plastids; Photosynthesis; Photosynthetic light reactions; Pigments in plants.

Sources for Further Study

Curtis, Helena, and N. Sue Barnes. *Biology*. 5th ed. New York: Worth, 1989. An introductory college or advanced high-school-level textbook. Chapter 10, "Photosynthesis, Light, and Life," covers the absorption of light by photosynthetic pigments and its biological significance. Includes suggestions for further reading.

Häder, Donat-Peter, and Manfred Tevini. *General Photobiology*. Elmsford, N.Y.: Pergamon Press, 1987. A college-level reference book that introduces the reader to the general principles of photobiology as well as a number of specific processes, including photosynthesis. Contains sections on the physical nature of light, absorption and energy transfer, and methods of study.

Hall, David O., and Krishna Rao. *Photosynthesis*. 6th ed. New York: Cambridge University Press, 1999. Provides a clear, concise, and vivid account of the process of photosynthesis, including details at the macro and molecular levels.

Salisbury, Frank B., and Cleon W. Ross. *Plant Physiology*. 4th ed. Belmont, Calif.: Wadsworth, 1992. An intermediate-level college textbook which presupposes a moderate knowledge of chemistry. The chapter "Photosynthesis: Chloroplasts and Light" includes a good discussion of photosynthetic pigments and the concepts of absorption and action spectra.

Wilkins, Malcolm. *Plantwatching: How Plants Remember, Tell Time, Form Partnerships, and More*. New York: Facts on File, 1988. This book is intended to introduce the nonscientist to various aspects of plant biology. It is clearly written and well illustrated. Chapter 9, "Photosynthesis: The Capture of Solar Energy," deals with light absorption by photosynthetic pigments.; photosynthesis and respiration; physiology.

PHOTOSYNTHETIC LIGHT REACTIONS

Categories: Cellular biology; photosynthesis and respiration; physiology

Photosynthetic light reactions involve the absorption of light energy by plant pigments and the conversion of light energy into adenosine triphosphate (ATP).

Photosynthesis is the process by which plants, algae, and certain types of bacteria use the energy of sunlight to manufacture organic molecules from carbon dioxide and water. The process may be divided into two parts: the *light reactions* and the *dark reactions*. In the light reactions of photosynthesis, light energy coming from the sun or from an artificial light source is absorbed by pigments and used to boost electrons into higher energy levels so they can be used to do cellular work. In the dark reactions (also called the *Calvin cycle*), energy-containing molecules from the light reactions are used to convert carbon dioxide into carbohydrates. All living organisms ultimately depend upon this process as their source of food.

In algae and higher plants, the light reactions of photosynthesis take place on *thylakoid membranes* located within *chloroplasts*. The surfaces of the thylakoids are covered with molecules of the green pigment *chlorophyll* as well as yellow *carotenoid* pigments. Also located on the thylakoids, though fewer in number, are special structures called *reaction centers*. The process begins when a unit of light energy (referred to as a photon or quantum) strikes a pigment molecule and causes one of its electrons to be raised to a higher energy level, or an excited state. Many chlorophyll and carotenoid molecules are located adjacent to one another on the thylakoids, and the energy of the excited electrons may be transferred from one to the next until it reaches a reaction center. Overall, a very large number of pigment molecules are absorbing light, becoming excited, and passing the excitation energy to reaction centers. It is in the reaction centers that the central events of the photosynthetic light reactions take place.

Photosystem I

Higher plants contain two different types of reaction centers, referred to as *photosystem I* and *photosystem II*. (The numbers I and II have no functional significance; they simply reflect the order in which they were discovered.) What the two types of reaction centers actually consist of are groups of special proteins complexed with several chlorophyll molecules and structured in a very specific arrangement within the thylakoid membrane. Also embedded in the thylakoid membrane are a series of proteins and other molecules that are capable of transporting electrons from photosystem II to photosystem I. This process is referred to as the *electron transport system*. The two photosystems and the electron transport system work together to form the energy-containing molecules produced in the photosynthetic light reactions.

The process is best understood by looking first at what happens in photosystem I. As described above, light energy is absorbed by a pigment molecule and transferred to the reaction center, causing two photosystem I electrons to be raised to an excited state. The excited electrons are then passed from the reaction center to a primary electron acceptor. The primary electron acceptor passes the electrons to the first of a series of electron transport proteins in the inner thykaloid membrane, the last of these proteins being ferredoxin, an iron-containing protein. Ferredoxin passes the electrons to a coenzyme called *nicotinamide adenine dinucleotide phosphate* ($NADP^+$), which then becomes reduced and, after joining with a free proton (H^+), becomes NADPH. The important point is that some of the energy of the excited electrons is incorporated into the molecule of NADPH, thereby converting light energy into chemical energy. The NADPH is not attached to the thylakoid membrane but remains in the *stroma* (the region inside both the outer membranes of the chloroplast and surrounding the thykaloids) of the chloroplast, where it will be used in the Calvin cycle.

Photosystem II

However, photosystem I, by itself, could not continue to function in the manner described above without having some way of replacing the electrons which are being removed. This is where photosystem II comes in. The basic operation of photosystem II is similar to photosystem I insofar as light energy is absorbed by pigment molecules and transferred to the reaction center. However, the excited electron is not used to form NADPH; instead it enters the electron transport system and is passed from one molecule to the next until it eventually reaches photosystem I, where it replaces the electron previously lost in the formation of NADPH.

The electron transport system does much more than merely replace electrons in photosystem I. As an electron moves through the electron transport system, its energy is used to transfer protons (hydrogen ions) from the stroma to the thykaloid space (the region inside the thykaloids). The thylakoids are not simply flat sheets of membrane but are folded in such a way as to form numerous saclike compartments within the chloroplast. The result is that a proton gradient is established across the thykaloid membrane.

Many electrons are carried through the electron transport system, resulting in the accumulation of a high concentration of protons within the thylakoid compartments. This high concentration of protons on one side of the membrane and relatively low concentration on the other represents a source of *potential energy*, somewhat analogous to the energy contained in a body of water held back by a dam. The protons are then allowed to move back across the membrane through special proteins called *ATP synthase*. ATP synthase is an enzyme capable of catalyzing the joining of adenosine diphosphate (ADP) with inorganic phosphate (P_i) in the stroma to form adenosine triphosphate (ATP). The passage of protons through ATP synthase results in a change in shape in the protein, which brings about the reaction between ADP and P_i. The formation of ATP by this mechanism is called *photophosphorylation*. Once again, light energy has been transformed into chemical energy.

Electron Replacement

If electrons from photosystem II are used to replace those from photosystem I, then the electrons from photosystem II must be replaced as well. This problem is solved by the use of water as a source of electrons. Photosystem II is capable of splitting apart molecules of water to extract electrons, using the electrons to replace the ones used by the electron transport system. Water is always available in a functioning photosynthetic plant cell, and this represents a virtually limitless source of electrons. Water molecules contain hydrogen and oxygen, and when they are split apart in this manner, the protons left over from hydrogen simply go into solution, but the oxygen forms oxygen gas and is released into the atmosphere. Although the oxygen is really no more than a by-product as far as photosynthesis is concerned, it is of profound significance to all higher organisms.

The overall flow of electrons in the light reactions is from water to photosystem II, through the electron transport system where ATP is formed, to photosystem I and finally to NADPH. This arrangement was hypothesized by Robin Hill and Fay Bendall in 1960. Because of the way in which the process was diagrammed in their paper, it is often referred to as the *Z scheme*. In energy terms, what has been accomplished is the conversion of light energy to chemical energy in the form of ATP and NADPH. These energy-rich molecules are essential to the Calvin cycle as energy for the conversion of carbon dioxide into carbohydrates. The process is known as *noncyclic electron flow*.

Alternatively, in *cyclic electron flow*, the electrons can travel within the electron transport system as described above but are diverted to an acceptor in the electron transport chain between photosystems I and II. Passing through the chain back to photosystem I, the electrons enable the transport of protons across the thykaloid membrane, thus supplying power for the generation of ATP. This process is called *cyclic photophosphorylation*, because it involves a cyclic flow of electrons. In this way, photosystem I can work independently of photosystem II. Apparently, this is the manner in which some bacteria carry out photosynthesis.

Bacterial Photosynthesis

Certain types of bacteria also carry on photosynthesis. These organisms do not contain chloroplasts, so the light-absorbing pigments and reaction centers are located on membranes spread throughout the cell. In one group, the cyanobacteria, the photosynthetic process is similar to that described for algae and higher plants in that there are two photosystems, water serves as the primary electron donor, and oxygen is released.

The other types of photosynthetic bacteria, however, are more primitive. They have only one photosystem and use inorganic compounds such as hydrogen sulfide instead of water as a source of electrons. The cyanobacteria possess chlorophyll, and the other photosynthetic bacteria contain a similar pigment known as *bacteriochlorophyll*. In addition, these organisms contain a variety of accessory pigments that also are involved in the photosynthetic light reactions.

Thomas M. Brennan, updated by Bryan Ness

See also: Active transport; Anaerobic photosynthesis; Calvin cycle; Photosynthesis; Photosynthetic light absorption; Pigments in plants.

Sources for Further Study

Curtis, Helena, and N. Sue Barnes. *Biology*. 5th ed. New York: Worth, 1989. An introductory college or advanced high-school-level textbook. Chapter 10, "Photosynthesis, Light, and Life," includes a well-written and illustrated discussion of both the light and dark reactions of photosynthesis, with suggestions for further reading. Chapter 54, "Ecosystems," addresses the role of photosynthesis in energy flow through ecosystems.

Hall, David O., and Krishna Rao. *Photosynthesis*. 6th ed. New York: Cambridge University Press, 1999. Provides a clear, concise, and vivid account of the process of photosynthesis, including details of the processes at the macro and molecular levels.

Nakatani, Herbert Y. *Photosynthesis*. Burlington, N.C.: Carolina Biological Supply Company, 1988. A brief discussion of the ecological significance of photosynthesis and the mechanisms of both the light and dark reactions. Well-written and illustrated, with suggestions for further reading.

Ort, Donald R., Charles F. Yocum, and Iris F. Heichel, eds. *Oxygenic Photosynthesis: The Light Reactions*. Boston: Kluwer, 1996. Describes reactions in eukaryotic thykaloid membranes; topics include photosystems, plastocyanin, ferredoxin, light-harvesting complexes, and the coupling factor.

Salisbury, Frank B., and Cleon W. Ross. *Plant Physiology*. 4th ed. Belmont, Calif.: Wadsworth, 1992. An intermediate-level college textbook which presupposes a moderate knowledge of chemistry. The chapter "Photosynthesis: Chloroplasts and Light" covers all aspects of the light reactions and chloroplast structure and also includes an interesting summary of the history of research on photosynthesis.

Wilkins, Malcolm. *Plantwatching: How Plants Remember, Tell Times, Form Partnerships, and More*. New York: Facts on File, 1988. A unusually interesting and well-written book dealing with many aspects of plant life at a level understandable by the nonscientist. Chapter 9, "Photosynthesis: The Capture of Solar Energy," covers the light reactions clearly and effectively, with good illustrations.

PHYTOPLANKTON

Categories: Algae; microorganisms; water-related life

The term "plankton," from Greek planktos *for "wandering," is applied to any organism that floats or drifts with the movement of the ocean water. Most plankton are microscopic and are usually single-celled, a chain of cells, or a loose group of cells. Algal and cyanobacterial plankton are referred to as phytoplankton. The heterotrophic crustaceans and larvae of animals are referred to as zooplankton.*

The group of organisms known as phytoplankton (literally, "plant" plankton) do not constitute a taxonomic group but rather refer to a collection of diverse, largely algal and cyanobacterial, microorganisms that live in water and are at the base of the food chain. The phytoplankton, including diatoms, unicellular cyanobacteria and coccolithophorids in nutrient-poor waters, and cryptomonads, manufacture organic material from carbon dioxide, usually through photosynthesis. Phytoplankton are responsible for one-half of the world's primary photosynthesis and produce one-half of the oxygen in the atmosphere.

Eighty to ninety percent of the weight of phytoplankton is water, with the rest made up of protein, fat, salt, carbohydrates, and minerals. Some species have compounds of calcium or silica that make up their shells or skeletons. Phytoplankton include many of the algal phyla: *Chrysophyta* (chrysophytes), *Phaeophyta* (golden-brown algae), coccolithophores, silicoflagellates, and diatoms. The most common type of phytoplankton is the *diatom* (phylum *Bacillariophyta*), a single-celled organism that can form complex chains. *Dinoflagellates* (phylum *Dinophyta*) are the most complex of the phytoplankton. They are unicellular and mobile. Green algae (phylum *Chlorophyta*) are usually found in estuaries or lagoons in the late summer and fall. Some species can cause toxic *algal blooms* associated with coastal pollution and eutrophication. Cyanobacteria (often called blue-green algae but not true algae) are prominent near shore waters with limited circulation and brackish waters.

Photosynthesis

Phytoplankton are primary *producers*, responsible for a half the world's primary photosynthesis: the conversion of light energy and inorganic matter into bioenergy and organic matter. Each year, 28 billion tons of carbon and 250 billion to 300 billion tons

HULTON ARCHIVE

Massive fish kills can result when large blooms of phytoplankton die. When a blooming phytoplankton population crashes, bacterial decomposition depletes the oxygen in the water, which reduces water quality and conditions, and fish and other marine animals suffocate.

of photosynthetically produced materials are generated in the oceans by phytoplankton. All animal organisms eliminate carbon dioxide into the atmosphere, and plants remove carbon dioxide from the air through photosynthesis. In the oceans' *carbon cycle,* carbon dioxide from the atmosphere dissolves in the ocean. Photosynthesis by marine plants, mainly phytoplankton, converts the carbon dioxide into organic matter. Carbon dioxide is later released by plants and animals during respiration, while carbon is also excreted as waste or in the dead bodies of organisms. Bacteria decompose organic matter and release the carbon dioxide back into the water. Carbon may be deposited as calcium carbonate in biogenous sediments and coral reefs (made of skeletons and shells of marine organisms).

Food Chain

Because they are primary producers of organic matter through photosynthesis, phytoplankton play a key role in the world's food chain: They are its very beginning. Sunlight usually penetrates only 200 to 300 feet deep into ocean waters, a region called the photic zone. Most marine plant and animal life and feeding take place in this zone. Phytoplankton, the first level in the marine food chain, are the primary food source for zooplankton and larger organisms. These microscopic plants use the sun's energy to absorb minerals to make basic nutrients and are eaten by *herbivores,* or plant eaters. Herbivores are a food source for *carnivores,* the meat eaters. In temperate zones, phytoplankton increase greatly in the spring, decline in the summer, and increase again in the fall. Zooplankton (animal plankton) are at their maximum abundance after the spring increase, and their grazing on the phytoplankton causes a decrease in phytoplankton population in the summer. Fish and invertebrates that eat zooplankton become more abundant and so on, up the food chain. Krill, planktonic crustaceans, and larvae commonly eaten by whales, fish, seals, penguins, and seabirds feed on diatom phytoplankton.

Red Tides

The term *red tide* is applied to red, orange, brown, or bright-green phytoplankton blooms, or even to blooms that do not discolor the water. Red tides are poorly understood and unpredictable. No one is certain what causes the rapid growth of a single species of phytoplankton, although they can blossom where sunlight, dissolved nutrient salts, and carbon dioxide are available to trigger photosynthesis. Dense phytoplankton blooms occur in stable water where lots of nutrients from sewage and runoff are available. Natural events, such as storms and hurricanes, may remobilize populations buried in the sediment. These nuisance blooms, usually caused by dinoflagellates, which turn the water a reddish brown, and cyanobacteria, are becoming more frequent in coastal waters, possibly because of increased human populations and sewage. In shallower bodies of water, such as bays and estuaries, nutrients from winter snow runoffs, spring rains, tributaries, and sewage bring about spring and summer blooms.

Some of the poisons produced during red tides are the most powerful toxins known. The release of toxins by dinoflagellates may poison the higher levels of the food chain as well as suppress other phytoplankton species. These toxins cause high mortality in fish and other marine vertebrates. They can kill the whales and seabirds that eat contaminated fish. Dinoflagellates produce a deadly neurotoxin called saxitoxin, which is fifty times more lethal than strychnine or curare. Commercial shellfish, such as mussels, clams, and crabs, can store certain levels of the toxin in their bodies.

People who eat contaminated shellfish may experience minor symptoms, such as nausea, diarrhea, and vomiting, or more severe symptoms such as loss of balance, coordination, and memory, tingling, numbness, slurred speech, shooting pains, and paralysis. In severe cases, death results from cardiac arrest. When the toxins are blown ashore in sea spray, they can cause sore throats or eye and skin irritations.

Toxic blooms costs millions of dollars in economic losses, especially for fisheries which cannot harvest some species of shellfish. Smaller fish farms can be devastated. Additionally, coastal fish deaths foul beaches and shore water with decaying bodies, which can cripple tourism in the coastal regions.

Not all blooms are harmful, but they do affect the marine environment. Even when no toxins are released, massive fish kills can result when the large blooms of phytoplankton die. When the blooming phytoplankton population crashes, bacterial decomposition depletes the oxygen in the water, which in turn reduces water quality and conditions, and fish and other marine animals suffocate.

Virginia L. Hodges

See also: Algae; Aquatic plants; Bacteria; Brown algae; *Chlorophyceae*; Chrysophytes; Diatoms; Dinoflagellates; Eutrophication; Food chain; Green algae; Invasive plants; Marine plants; Photosynthesis; Red algae.

Sources for Further Study

Castro, Peter, and Michael Huber. *Marine Biology*. 3d ed. Boston: McGraw-Hill, 2000. Well organized, basic textbook with summary and bibliography with each chapter. Includes illustrations, glossary, and index.

Cousteau, Jacques. *The Ocean World*. New York: Harry N. Abrams, 1985. Discusses all marine life. Includes color photographs, glossary, bibliography, and index.

Levinton, Jeffrey S. *Marine Biology: Function, Biodiversity, Ecology*. New York: Oxford University Press, 1995. Well-organized, advanced textbook, with a good section on phytoplankton. Bibliography and review questions with each chapter. Includes glossary, index.

Sumich, James L. *An Introduction to the Biology of Marine Life*. 5th ed. Dubuque, Iowa: William C. Brown, 1992. Textbook format with basic explanations, review and discussion questions, and bibliography with each chapter. Includes glossary and taxonomic and subject indexes.

PIGMENTS IN PLANTS

Categories: Animal-plant interactions; economic botany and plant uses; photosynthesis and respiration; physiology

Photosynthetic pigments color plants and participate in photosynthesis. Other plant pigments are important in flowers and fruits to attract pollinators and seed dispersers. Humans use plant pigments in vitamins and dyes.

Plant pigments can be classified as either *nitrogenous* or *non-nitrogenous*, that is, either nitrogen-containing or non-nitrogen-containing.

Non-nitrogenous pigments

Non-nitrogenous forms are widely distributed and include the *carotenoids* and the *quinones*. Carotenoids are yellow, orange, or red pigments often involved as accessory pigments in photosynthesis. They are insoluble in water but soluble in a variety of nonpolar solvents. They are easily bleached by light or oxygen. Carotenoids occur as hydrocarbon carotenes and oxygenated xanthophylls. They are made by bacteria, fungi, algae, and other plants. Variations of their occurrence and their multiplicity make carotenoids useful in taxonomic differentiation of these organisms.

The quinones include benzoquinones, naphthoquinones, and anthraquinones. Benzoquinones occur in fungi and higher plants as yellow, orange, red, or violet pigments. Yellow coenzyme Q variants (ubiquinones) and plastoquinones are found in most plants. Because of their low tissue levels, they do not affect plant color. Naphthoquinones of bacteria and leaves, seeds, and woody parts of higher plants have yellow, orange, red, or purple pigments, soluble in nonpolar solvents. A familiar example is vitamin K. Brightly colored anthraquinones also occur in many plants.

Many other complex plant quinones are water-soluble *flavonoids*, all containing a common fifteen-carbon skeleton, flavone (2-phenylben-zopyrone). The various flavonoids differ in how many hydroxyl or methoxyl groups they contain. They occur as sugar-containing substances called *glycosides*, which is the basis of their water solubility. The members of one plentiful group of flavonoids are the anthoxanthins, which are yellow pigments. Another important flavonoid group, anthocyanins, are orange, red, crimson, or blue.

Carotenoids are yellow, orange, or red pigments often involved as accessory pigments in photosynthesis. In the fall, destruction of the green pigment chlorophyll is accompanied by increased visibility of the yellow or orange pigments of the carotenoids, which were previously masked by chlorophyll in these maple leaves.

Nitrogenous Pigments

The nitrogenous (nitrogen-containing) pigments are *tetrapyrrole porphyrins* and their derivatives, indigoids and flavins. The porphyrins are water-soluble, cyclic, nitrogen-containing substances. The basic structural unit of all porphyrins is a large tetrapyrrole ring made up of four smaller connected pyrrole rings. Porphyrins combine with metal ions and proteins. In plants they are represented by green *chlorophylls*. The chlorophylls contain magnesium ions and are associated with water-insoluble proteins. Related *bilins* are a group of yellow, green, red, or brown compounds that have linear structures composed of four connected pyrrole rings. The bilins include red bilirubin, green biliverdin, and the phycobilins of red algae or green plants. Examples are blue phycocyanobilin and phytochrome as well as red phycoerythrin. Plant bilins bind water-soluble and water-insoluble proteins.

Indigoids, a group of indole pigments, derive from the amino acid tryptophan. They are red, blue, or purple. The flavins (lyochromes) are pale yellow, water-soluble pigments widely distributed in plants. The most plentiful flavin is riboflavin (vitamin B_2). Flavins are made by bacteria, yeasts, and green plants.

Functions

Non-nitrogenous benzoquinones and ubiquinones are involved in electron transport processes important in respiration. Naphthoquinones have similar functions in photosynthesis and are also represented by K vitamins. Anthroquinones are the brilliantly colored compounds seen in the colors of flowers. Similarly, flavonoid glycosides color many flowers. For example, the anthoxanthins produce the yellow color of buttercups.

Anthocyanins make flower petals orange-red, crimson, and blue. They cause much of the red color seen in some plant buds and shoots, the colors of autumn leaves, fruits (such as berries), or roots (such as beets). A typical anthocyanin is red in acid solution, violet in neutral solution, and blue in alkaline solution. Blue or red cornflowers and violet dahlias contain the same anthocyanin, their color differences simply being the result of pH (acidity or alkalinity) differences of the cell sap.

The function of flavonoids is still unclear. However, it is believed that they are essential in flowers for attracting bees and other pollinators, encouraging cross-pollination. They are also thought to attract larger animals to bright-colored, edible fruits, enhancing seed dissemination, and to protect plants from damage by ultraviolet light.

Nitrogen-containing tetrapyrrole plant porphyrins are best known because of the key role they play in photosynthesis. Chlorophylls enable photosynthetic conversion of sunlight to chemical energy, which is used by plant cells to use carbon dioxide to make organic molecules. Chlorophylls include chlorophylls *a* and *b* of higher plants and green algae and the bacteriochlorophylls in photosynthetic bacteria. The various chlorophylls differ in only minor ways as a result of side chain groups attached to their pyrroles. In higher plants, chlorophylls bind to proteins and lipids in the thylakoid membranes of chloroplasts, where they function in photosynthesis. Some of the phycobilins of blue algae, red algae, and green plants act as accessory photosynthetic pigments. Another function of bilins is as phytochromes, essential to photoperiodic processes.

Economic Uses

Many plant pigments meet human nutritional needs. The carotenes, derived from carotenoids, are used in the biosynthesis of vitamin A, essential to vision and growth. Furthermore, napthoquinone photosynthetic pigments lead to another important vitamin group: K vitamins, essential to blood clotting. Yet another vitamin derived from plant pigments is a nitrogen-containing flavin called riboflavin, better known as vitamin B_2.

Many plant pigments are used as dyes or as model compounds from which other dyes have been synthesized. The naphthoquinones from leaves, seeds, and woody parts of higher plants are isolated as yellow, orange, red, or purple materials soluble in organic solvents and used as fabric dyes. Indigo, a blue indigoid which occurs in many plants of Asia, Africa, and South America, has been used as a blue dye and the model for many industrially synthesized dyes. Similarly, anthraquinones, brightly colored plant pigments, are widely used. Moreover, because their colors change in acidic, basic, and neutral solutions, anthraquinones are used as acid-base indicators. Flavonoid anthocyanins (as was mentioned) are also used as acid-base indicators. A variety of edible plant pigments are also used to add color to foods.

Sanford S. Singer

See also: Algae; *Archaea*; Bacteria; Angiosperm evolution; Angiosperms; Animal-plant interactions; Biochemical coevolution in angiosperms; Chloroplasts and other plastids; Chromatography; Coevolution; Flower types; Flowering regulation; Fruit: structure and types; Leaf abscission; Metabolites: primary vs. secondary; Photoperiodism; Photosynthesis; Photosynthetic light absorption; Photosynthetic light reactions; Vacuoles.

Sources for Further Study

Gross, Jeana. *Pigments in Vegetables: Chlorophylls and Carotenoids*. New York: Van Nostrand Reinhold, 1991. Provides information on chlorophylls and carotenoids in vegetation. Includes illustrations, bibliography.

Mazza, G., and Enrico Miniati. *Anthocyanins in Fruits, Vegetables, and Grains*. Boca Raton, Fla.: CRC Press, 1993. Describes anthocyanin types, their structures, pH effects, biosynthesis, presumed functions, distribution in numerous vegetables, fruits, and grains. Includes an excellent bibliography and illustrations.

Proctor, John, and Susan Proctor. *Color in Plants and Flowers*. New York: Everest House, 1978. An interesting, brief, nicely illustrated book on plant and flower color.

Raven, Peter H., Ray F. Evert, and Susan E. Eichhorn. *Biology of Plants*. 6th ed. New York: W. H. Freeman/Worth, 1999. This well-illustrated text presents many aspects of flower structure, also identifying flower structure, color, and means of pollination.

Singhal, G. S, ed. *Concepts in Photobiology: Photosynthesis and Photomorphogenesis*. Boston: Kluwer Academic, 1999. Topics related to pigments include photobiology, photosynthesis principles, chlorophylls, phycobiliins, and phytochromes. It has a fine bibliography.

Young, Andrew, and George Britton, eds. *Carotenoids in Photosynthesis*. London: Chapman and Hall, 1993. The illustrated text discusses carotenoid distribution in plants, their protein complexes, biosynthesis, and photosynthetic functions. Includes bibliography, illustrations.

PLANT BIOTECHNOLOGY

Categories: Biotechnology; economic botany and plant uses; environmental issues; genetics

Plant biotechnology may be defined as the application of knowledge obtained from study of the life sciences to create technological improvements in plant species. By this very broad definition, plant biotechnology has been conducted for more than ten thousand years.

The roots of plant biotechnology can be traced back to the time when humans started collecting seeds from their favorite wild plants and began cultivating them in tended fields. It appears that when the plants were harvested, the seeds of the most desirable plants were retained and replanted the next growing season. While these primitive agriculturists did not have extensive knowledge of the life sciences, they evidently did understand the basic principles of collecting and replanting the seeds of any naturally occurring variant plants with improved qualities, such as those with the largest fruits or the highest yield, in a process that we call *artificial selection*. This domestication and controlled improvement of plant species was the beginning of plant biotechnology. This very simple process of selectively breeding naturally occurring variants with observably improved qualities served as the basis of agriculture for thousands of years and resulted in thousands of domesticated plant cultivars that no longer resembled the wild plants from which they descended.

The second era of plant biotechnology began in the late 1800's as the base of knowledge derived from the study of the life sciences increased dramatically. In the 1860's Gregor Mendel, using data obtained from controlled pea breeding experiments, deduced some basic principles of genetics and presented these in a short monograph modestly titled "Versuche über Pflanzenhybriden" (in *Verhandlungen des naturforschenden Vereins*, 1866; *Experiments with Plant-Hybridisation*, 1910). In this publication, Mendel proposed that heritable genetic factors segregate during sexual reproduction of plants and that factors for different traits assort independently of each other. Mendel's work suggested a mechanism of heritable factors that could be manipulated by controlled breeding of plants through selective fertilization and also suggested that the pattern of inheritance for these factors could be analyzed or, in some cases, predicted by the use of mathematical statistics.

These findings complemented the work of Charles Darwin, who expounded the principles of descent with modification and selection as the chief factor of evolutionary change in his 1859 book *On the Origin of Species by Means of Natural Selection*. The application of these principles to agriculture resulted in deliberately produced hybrid varieties for a large number of cultivated plants via selective fertilization. These artificially selected hybrids soon began to benefit humankind with tremendous increases in both the productivity and the quality of food crops.

Genetic Engineering

The third era of plant biotechnology involves a drastic change in the way crop improvement may be accomplished, by direct manipulation of genetic elements (genes). This process is known as *genetic engineering* and results in plants that are called genetically modified organisms (GMOs), to distinguish them from plants that are produced by conventional plant-breeding methods. Genetically modified plants can contribute desirable genes from outside traditional breeding boundaries.

Even genes from outside the plant kingdom can now be brought into plants. For example, animal genes, including human genes, have been transferred into plants, a feat not replicated in nature.

Public Concern

It is perhaps this lack of natural boundaries for genetic exchange that seems so foreign to conventional scientific thought and that makes plant genetic engineering controversial. The thought of taking genes from animals, bacteria, viruses, or any other organism and putting them into plants, especially plants consumed for food, has raised a host of questions among concerned scientists and public alike. Negative public perception of genetically modified crops has affected the development and commercialization of many plant biotechnology products, especially food plants. While there are dozens of genetically engineered plants ready for field production, public pressure has delayed the release of some of these plants and has caused the withdrawal of others from the marketplace.

This public concern also appears to be driving increased government review of products and decreased government funding for plant biotechnology projects in Europe. Negative public perceptions do not seem to be as strong in Asia, since the pressures of feeding large populations tend to outweigh the perceived risks. The social climate of the United States toward biotechnology, although guarded, appears to be less apprehensive than that of most European countries. Therefore, many agricultural biotechnology projects have moved from European countries to U.S. laboratories.

Economic Goals

To what end are humans genetically engineering plants? This is an essential question for researchers,

Time Line of Plant Biotechnology

Year	*Event*
1838	German scientists Matthias Schleiden and Theodor Schwann presented their cell theory: that all life-forms are made up of cells.
1858	Biologist Rudolf Virchow adds to the cell theory, proposing that, "where a cell exists, there must have been a preexisting cell."
1902	Austrian botanist Gottlieb Haberlandt completes the cell theory with his idea of totipotency: Cells must contain all the genetic information necessary to create an entire, multicellular organism. Therefore, every plant cell is capable of developing into an entire plant.
1939	R. J. Gautheret demonstrates the first successful culture of isolated plant tissues as a continuously dividing callus tissue.
1953	James Watson and Francis Crick make their landmark proposal for the double-helical structure of deoxyribonucleic acid (DNA), the molecule that carries genetic material.
1954	The first whole plant is regenerated, or cloned, from a single adult plant cell by W. H. Muir and colleagues.
1967	DNA ligase, the enzyme that joins DNA molecules, is discovered.
1970	Daniel Nathans, Hamilton Smith, and others discover restriction endonucleases.
1972	Researchers at Stanford University construct the first recombinant DNA molecules, and the following year DNA is inserted into *Escherichia coli* cells.
1980	The U.S. Supreme Court rules that genetically altered life-forms can be patented, allowing Exxon to patent a microorganism that eats oil.
1983	The National Institutes of Health permit scientists at the University of California at Berkeley to release genetically engineered bacteria designed to retard frost formation on crop plants.
1983	The U.S. Patent Office begins to issue a series of patents for genetically modified plants.

executives of biotechnology companies, and consumers at large. Before addressing technical questions about how to apply biotechnology, the desired goals must be clearly defined. The general goals of plant biotechnology appear to be (1) economic improvement of existing products, (2) improvement of human nutrition, and (3) development of novel products from plants.

Economic improvements include increases in yield, quality, pest resistance, nutritional value, harvestability, or any other change that adds value to an established agricultural product. Examples of this category include insect-protected tomatoes, potatoes, cotton, and corn; herbicide-resistant canola, corn, cotton, flax, and soybeans; canola and soybeans with genetically altered oil compositions; virus-resistant squash and papayas; and improved-ripening tomatoes. All these examples were introduced to agriculture in the later half of the 1990's.

Nutritional Goals

Additionally, some products appearing in the scientific literature but awaiting commercialization have the potential to dramatically improve human nutritional deficiencies, which are especially prevalent in developing countries. These products include "golden rice," genetically modified rice that produces carotenoids, a dietary source of vitamin A. Golden rice has the potential to prevent vitamin A deficiency in developing countries, where this vitamin deficiency is a leading cause of blindness.

Researchers are also using genetic engineering to increase the amount of the iron-storing protein ferritin in seed crops such as legumes. Iron deficiency, which affects 30 percent of the human population, can impair cognitive development and cause other other health problems. This proposed enhancement of iron content in consumable plant

Year	*Event*
1984	The Plant Gene Expression Center, a collaborative effort between academia and the U.S. Department of Agriculture, is established to research plant molecular biology, sequence plant genomes, and develop genetically modified plants.
1985	Field testing of plants genetically modified to resist plant pathogens and disease vectors begins.
1986	The first release of a genetically modified crop, genetically engineered tobacco plants, is approved by the Environmental Protection Agency.
1987	Calgene receives a patent for a DNA sequence that extends the shelf life of tomatoes.
1987	Advanced Genetic Sciences, Inc., field-tests a recombinant organism designed to inhibit frost in strawberries.
1990	Calgene conducts field experiments with herbicide-resistant cotton plants.
1990	At the Plant Gene Expression Center, biologist Michael Fromm announces the use of a high-speed "gene gun" to transform corn. Gene guns are used to shoot genetic material directly into cells via DNA-coated microparticles.
1993	Kary Mullis wins the Nobel Prize in Chemistry for development of polymerase chain reaction technology, a technique he invented in 1981 for quickly multiplying DNA sequences in vitro.
1993	The U.S. Food and Drug Administration (FDA) announces its finding that genetically modified foods are not "inherently dangerous" and not in need of regulation. The following year, the FDA approves the first genetically modified whole food crop, Calgene's Flavr Savr tomato.
1996	The genomes of *Saccharomyces cerevisiae* (baker's yeast), with 12 million base pairs, and of ancient archaea cells, which live near thermal vents at the ocean bottom, advance biologists' understanding of the evolution of life.
2000	Researchers complete the full genomic sequence for the model flowering plant *Arabidopsis thaliana*.
2001	Researchers complete the genomic sequence for rice, *Oryza sativa*.

products could help more than a billion people who suffer from chronic iron deficiency.

Novel Products

Novel products include those not traditionally associated with plants and are limited only by imagination and currently available techniques. These include the production of plastics, vaccines, antibodies, human blood proteins, and new pharmaceuticals. One project has involved the production of hepatitis B vaccine in transgenic tomatoes. This project, which underwent clinical trials in the late 1990's, has the potential to provide a simple and inexpensive means of vaccinating people against hepatitis B. By oral administration of tomato juice containing the vaccine protein, humans are thought to develop an immune response that may protect them from infection by the hepatitis B virus. Hepatitis B is epidemic in Asia and increasing at an alarming rate in the rest of the world. The disease ultimately causes liver disease, cancer, and death in millions of infected people.

Plant Tissue Cultures

Central to plant biotechnology is the use of in vitro methods. Researchers use *plant tissue cultures*, for example, to grow plant cells on sterile nutrient media. Countless recipes for these nutrient media exist. The choice of which one to use is based on the plant species and the tissue type to be grown. All such media contain at least some of the important nutitional elements, such as nitrogen, potassium, calcium, magnesium, sulfur, phosphorus, iron, boron, manganese, zinc, iodine, molybdenum, copper, and cobalt, usually in the form of inorganic salts or as metal chelates, and an organic energy source, such as sucrose. The media may also contain vitamins, hormones, and other ingredients, depending on the intended use.

To initiate plant tissue culture, a piece of a living plant is excised and disinfected using a chemical disinfectant. This piece of plant tissue, called an *explant*, is placed on a sterile plant tissue culture medium to grow. Many plant tissues may be used to obtain explants for plant tissue culture, including those from leaves, petioles, shoots, tubers, roots, and meristematic regions. When an explant is placed in the sterile tissue culture medium, cells that are not terminally differentiated will grow and divide. If plant hormones are included in the recipe, the plant cells can be coaxed to develop into different types of tissues or organs. By using a succession of media containing different hormones, it is possible to regenerate whole plants from single cells. The choice of tissue used for the explant and the choice of hormones included in the tissue culture medium depend on the desired result.

Micropropagation

Micropropagation, another biotechnology technique, is the production of many clonal plants using tissue culture methods. By means of micropropagation, it is possible to generate many thousands of plant clones using tissue explants obtained from a single parent plant. The main advantage to micropropagation is the potential of producing thousands of exact copies of a plant with desirable traits. Micropropagation is especially important for rare plants, genetically engineered plants, and plants that have sexual reproductive problems. Many plant species are now routinely propagated by micropropagation methods, including orchids, ferns, many flowering ornamentals, and vegetable plants.

Steps in Genetic Engineering

The first genetically engineered plants, tobacco plants, were reported in the scientific literature in 1984. Since 1984 there have been thousands of genetically engineered plants produced in laboratories worldwide. The process of genetically engineering a plant involves several key steps:

- isolating the genetic sequence (gene) to be placed from its biological source
- placing the gene in an appropriate vehicle to facilitate insertion into plant cells
- inserting the gene into the plant in a process known as *plant transformation*
- selecting the few plant cells that contain the new gene (transformed cells) out of all the plant cells in the explant
- multiplying the transformed cells in sterile tissue culture
- regenerating the transformed cells into a whole plant that can grow outside the tissue culture vessel

The gene or genes to be placed in the plant may be obtained from virtually any biological source:

AP/Wide World Photos

The practice of taking genes from animals, bacteria, viruses, or other organisms and putting them into plants consumed for food has raised many questions among concerned scientists and the public alike. In Mauvezin, France, activists slash genetically modified plants in this cornfield. The poster reads: Monsanto . . . no; My health . . . yes!

animals, bacteria, fungi, viruses, or other plants. Placing genes into an appropriate vehicle for transfer into a plant involves using various molecular biology techniques, such as restriction enzymes and ligation, to essentially "cut and paste" the gene or genes of interest into another DNA molecule, which serves as the transfer vehicle (vector).

Plant Transformation Methods

Currently *plant transformation* with foreign genes may be accomplished by several proven methods, including bacteria-mediated transfer, microparticle bombardment, electroporation, microinjection, sonication, and chemical treatment.

By far, the most often utilized method of plant transformation involves the use of naturally occuring plant pathogenic bacteria from the genus *Agrobacterium*. In nature, this bacterium infects plants and transfers some of its own bacterial DNA into the plant. Through the action of proteins produced by the bacteria, bacterial DNA is made to integrate permanently into the plant's own genomic DNA. Expression of the bacterial DNA in the plant causes the plant to produce unusual quantities of plant hormones and other compounds, called opines, which provide food for the bacteria. The unusual quantities of plant hormones around the infection site cause the plant cells to grow abnormally, producing characteristic tumors. Scientists have harnessed this pathogenic bacterium to insert genes into plants by deleting the bacterial genes that cause tumors in the plant and then inserting desirable genes in their place. When the modified *Agrobacterium* infects a plant, it transfers the desirable genes into the plant genome instead of causing tumors. The desirable genes become a permanent part of the plant genome, and expression of these genes in plant cells produces desirable products.

One major drawback of the *Agrobacterium* method is that insertion of bacterial DNA into the plant genome is essentially random. The gene may not be efficiently transcribed at its location, or the insertion of bacterial DNA may knock out an important plant gene by inserting in the middle of it—or both may occur. Therefore, the fact that a cell is genetically transformed does not guarantee that it will perform as desired.

Microparticle bombardment is the introduction of foreign DNA constructs into plant cells by attaching the DNA to small metal particles and blasting the particles into plant cells using either a compressed air gun or a gun powered by a 0.22 caliber gun cartridge. This is truly a "brute force" method of introducing DNA into a cell that inadvertently causes many lethal casualties among the bombarded plant cells. However, some plant cells blasted with the DNA-containing metal particles will recover and survive. The plant cells may express the DNA for only a short time (transient expression), because the DNA does not readily integrate into the plant genome, but occasionally the foreign DNA may spontaneously recombine into the plant genome and become permanent.

Other ways of introducing foreign DNA into plant cells include electroporation, microinjection, sonication, and chemical treatment. These methods are not used extensively, because they generally require the production of *protoplasts* (plant cells that lack their cell walls) from plant cells before transformation. To create protoplasts, the plant cell wall is removed by digestion with the enzymes cellulase and pectinase. Protoplasts are fragile structures, but the absence of a cell wall is desirable because it leaves only the plasma membrane as a barrier to foreign DNA entering a plant cell.

Electroporation uses very brief pulses of high-voltage electrical energy to create temporary holes in the plasma membrane through which the foreign DNA can pass. *Microinjection* involves physically injecting a small amount of DNA into a plant cell using a microscope and an extremely fine needle. *Sonication* uses ultrasonic waves to punch temporary holes in the plasma membrane; this method is therefore similar to electroporation. *Chemical treatment* involves the use of polyethylene glycol to render the plasma membrane permeable to foreign DNA.

All the transformation procedures produce only a few transformed cells out of the millions of cells in an explant, so selection of transformed cells is essential.

Selection of Transformed Plant Cells

Selecting the few transformed plant cells out of all the plant cells in an explant requires some advance planning. Most foreign DNA constructs introduced into a plant are designed and built to contain additional genes that function as selectable markers or reporter genes. *Selectable markers* include genes for resistance to antibiotics or herbicides. Plant cells containing and expressing these genes will be tolerant of antibiotics or herbicides added to the plant tissue culture media, while the nontransformed plant cells will be killed off. The surviving cells in the tissue culture media are mostly transformed.

Instead of selectable markers, *reporter genes* may be used. Reporter genes induce an easily observable trait to transformed plant cells that facilitates the physical isolation of these cells. Reporter genes include beta-glucuronidase, luciferase, and plant pigment genes. Beta-glucuronidase (commonly known as GUS) allows the plant cells expressing this gene to metabolize colorigenic substrates while nontransformed plant cells cannot. To use this test, researchers treat a small amount of plant tissue with the colorigenic chemical substrate. If the cell turns color (blue) it is known to be transformed and expressing the GUS gene. If the cell does not turn color, it probably is not transformed. Another reporter gene is luciferase, an enzyme isolated from fireflies. Luciferase makes plant cells glow in the presence of certain chemicals if the gene is present; hence, transformed cells glow, whereas nontransformed cells do not glow. Plant pigment genes, such as anthocyanin pigment genes, occur naturally in plants and produce pigments that impart color to flowers. Inclusion of these pigment genes as reporter genes will allow transformed plant cells to be selected by their color. Transformed cells have color, while nontransformed cells remain colorless. Both selectable markers and reporter genes allow selection of cells into which genes have been successfully inserted and are operating properly.

Regenerating Whole Transformed Plants

After successfully getting a gene construct into a plant cell and selecting the transformed cells, it is possible to get the plant cells to multiply in tissue culture. Also, by treating the plant cells with combinations of plant hormones, the cells are made to differentiate into various plant organs or whole plants.

For example, treating transformed plant cells with a high concentration of the plant hormone cytokinin causes shoots to develop. Transferring these shoots to another medium, one that is high in the plant hormone auxin, will cause roots to develop on the shoots. In this way a whole *transgenic plant* may be regenerated from transformed plant cells. Once a transformed plant is regenerated in tissue culture, the plant may be transferred to a climate-controlled greenhouse, where it can grow to maturity.

Future generations of transgenic plants may then be propagated sexually via seeds or asexually via vegetative propagation methods. Often transgenic plants must be grown in containment greenhouses to prevent accidental release into the environment. In such high-tech greenhouses, all factors contributing to optimal plant growth—lighting, temperature, humidity, nutrients, and other environmental conditions—are tightly controlled. Often hydroponic systems, which use a solution of plant nutrients as a growth medium in place of soil, are employed to control all aspects of plant nutrition.

Robert A. Sinnott

See also: Biotechnology; Cloning of plants; DNA: recombinant technology; Environmental biotechnology; Genetically modified bacteria; Genetically modified foods.

Sources for Further Study

Dixon, Richard A., and Robert A. Gonzales, eds. *Plant Cell Culture: A Practical Approach*. 2d ed. New York: IRL Press at Oxford University Press, 1994. A handbook for researchers, illustrated, with bibliography and index.

Gamborg O. L., and G. C. Phillips, eds. *Plant Cell, Tissue and Organ Culture: Fundamental Methods*. New York: Springer Verlag, 1995. A substantial (358-page) laboratory manual for researchers. Bibliography, index.

Kyte, Lydiane, and John G. Kleyn. *Plants from Test Tubes: An Introduction to Micropropagation*. 3d ed. Portland, Oreg.: Timber Press, 1996. A laboratory manual with illustrations, bibliography, index.

Owen, Meran R. L., and Jan Pen, eds. *Transgenic Plants: A Production System for Industrial and Pharmaceutical Proteins*. New York: John Wiley, 1996. Geared for commercial purposes. Illustrations, bibliography, index.

PLANT CELLS: MOLECULAR LEVEL

Categories: Cellular biology; physiology

Water, ions, salts, and gases all are types of inorganic molecules that are essential to cellular function. The chemical properties of water make it an ideal solvent and buffer for the chemistry that occurs inside cells. The capillary action that helps water travel up plant tissues from the roots is a direct consequence of the polarity of the water molecule.

The chemistry of life on earth is carbon and water chemistry. Water is the most abundant compound in living cells and makes up as much as 90 percent of the weight of most plant tissues. Many of the molecules that are part of larger macromolecules in cells are linked together chemically by *dehydration synthesis*, or the loss of water. These macromolecules are broken up into their compo-

Parts of a Plant Cell

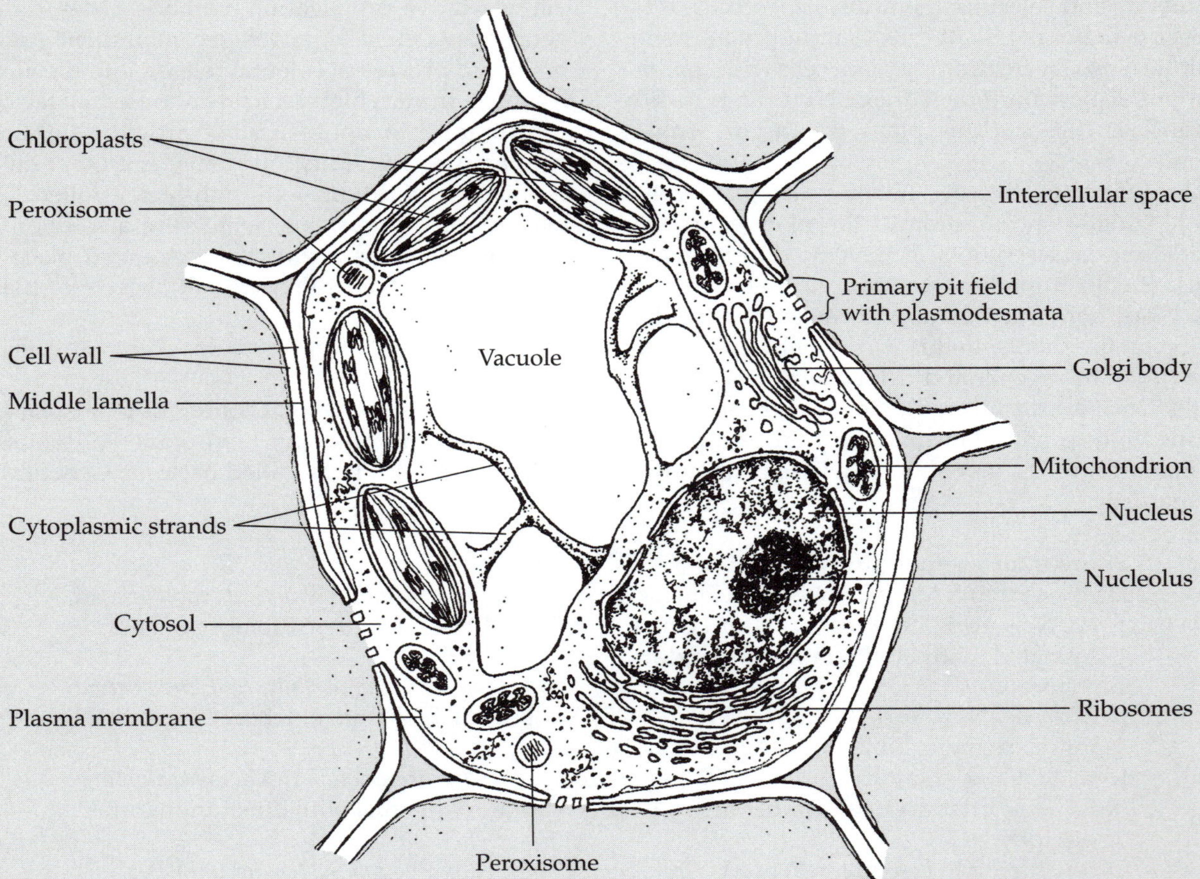

nent units by the addition of a water molecule between the units, a process known as *hydrolysis*. The chemical properties of water make it an ideal solvent and buffer for the chemistry that occurs inside cells.

Because the electrons of the covalent bonds within the water molecule are more often orbiting the oxygen atom, the oxygen atom gains a slightly negative charge. The hydrogen atoms are slightly positive. This separation of charge across the water molecule is said to make it polar. Because of its polar nature, water is able to dissolve, or ionize, a variety of molecules. This gives water its buffering capacity.

Water molecules are attracted to one another because of this polarity. This weak attraction, which occurs in the form of hydrogen bonds, has great chemical consequences when many molecules of water are involved. Hydrogen bonding allows water to have surface tension. The capillary action that helps water travel up plant tissues from the roots is a direct consequence of the polarity of the water molecule. Water is also able to absorb heat without vaporizing (changing from a liquid to a gas state) quickly. Therefore, physiological temperatures can be maintained as water molecules absorb the heat from metabolic reactions.

Water, ions, salts, and gases all are types of *inorganic molecules* that are essential to cellular function. Inorganic molecules are chemical molecules that do not contain carbon. The remainder of the molecules within cells are built around the unique properties of the carbon atom and are called organic molecules.

Organic Macromolecules

There are four major classes of organic molecules in cells: *carbohydrates*, *lipids*, *nucleic acids*, and *proteins*. All of these molecules contain carbon backbones, and almost all of them contain oxygen and hydrogen as well as other elements. Some or all of the members of each class of organic molecules occur as very large molecules, called macromolecules, that are polymers of smaller molecules joined together by covalent bonds. For example, starch and cellulose are carbohydrate polymers of simpler carbohydrates called sugars. Likewise, fats and oils are lipid polymers composed of smaller lipids called fatty acids and the sugar alcohol called glycerol.

Carbohydrates

Carbohydrates are molecules that consist of primarily carbon, hydrogen, and oxygen atoms. Carbohydrates are the primary source of stored energy in most living organisms. They can also serve as structural molecules in cell walls and as markers on some cell membranes, identifying different types of cells.

Simple sugars, or *monosaccharides*, are sugars that are small molecules composed of a chain of covalently bonded carbon atoms with associated hydrogen and oxygen atoms. These molecules always have a ratio of one carbon atom to two hydrogen atoms to one oxygen atom (CH_2O). The monosaccharide glucose is the primary sugar produced from simpler sugars made in photosynthesis.

When two simple sugars are covalently linked together, they form a *disaccharide*. In plants, the disaccharide sucrose, which is composed of one fructose molecule and one glucose molecule, is the most common sugar. Sucrose is the same thing as so-called table sugar, which is harvested from sugar cane or sugar beets.

Many sugars can be linked together to form a carbohydrate polymer, or *polysaccharide*. Starch is composed of many glucose molecules linked together and is the major form of carbohydrate storage in plants. When energy is required, the individual sugars of the polysaccharides are hydrolyzed (broken down to simpler molecules), and the glucose that is released is used by the mitochondria to generate energy. Polysaccharides are also important structural molecules in plants. The most abundant polysaccharide in nature is cellulose, another polymer of glucose and a major component of plant cell walls.

Lipids

Lipids are diverse group of unrelated molecules which includes fats, oils, steroids and sterols, waxes, and other water-insoluble molecules. Lipids are characterized by their hydrophobic, or "water-fearing," chemical behavior, which is what makes them insoluble in water. Unlike other molecules that ionize and are dissolved by water, lipid molecules are nonpolar. They are repelled by the polar nature of water and tend to aggregate in aqueous solutions. Lipids also are used to store energy and are especially abundant in seeds because lipids contain more energy by weight than carbohydrates.

Examples of lipids commonly found in biological systems include fats and oils that are storage molecules known as *triglycerides*. A triglyceride consists of glycerol (a three-carbon molecule) and three fatty acid molecules, long-chain hydrocarbon molecules that are attached to each of the three glycerol-carbon atoms by ester linkages.

The long chain of carbon atoms of the fatty acid can be saturated or unsaturated with respect to hydrogen content. *Saturated fatty acids* contain as many hydrogen atoms as allowed bonded to each carbon atom. Saturated fats tend to be solid at room temperature and include substances such as butter and lard. *Unsaturated fatty acids* do not have the maximum number of hydrogen atoms because some of the carbon atoms form double bonds with adjacent carbon atoms in the chain. Unsaturated fats tend to be liquid at room temperature and include substances such as corn oil and olive oil.

Plants have many lipids that are unique to them. For instance, cutin and suberin are two lipid polymers that form structural components of many plant cell walls. These two molecules form a meshwork that secures another type of lipid polymer found in plants, wax. *Waxes* are long-chain lipid compounds that are integrated into the cutin and suberin meshwork and are important in preventing water loss for plants. Waxes give apple peels their characteristic shiny appearance.

Phospholipids are a type of lipid molecule that is found in all living organisms. They are structurally similar to triglycerides, except instead of having three fatty acids attached to glycerol, they have only two. Replacing the third fatty acid is a charged phosphate group. This unique structure results in one end of the molecule being hydrophilic (the phosphate end, often called the head) and the other being hydrophobic (the end with the two fatty acids, often called the tail). Consequently, phospholipids will spontaneously form an oily layer at the water surface, orienting their charged phosphate heads toward the water and their fatty acid tails away from the water and toward the air. This is the basis for the phospholipid bilayer structure that underlies the formation of all cellular membranes. In the case of a lipid bilayer, because there is water on both sides, the two layers are tail to tail, with their heads oriented to the inside and outside of the membrane, where they come into contact with water.

Nucleic Acids

The information that directs all cellular activity is contained within the chemical structure of the *nucleic acids*. Nucleic acids are polymers of smaller molecules called *nucleotides*. Nucleotides, in turn, are composed of three types of covalently linked molecules: a ribose sugar, a phosphate group, and a nitrogen-containing base. The two major nucleotides that are found in cells are *deoxyribonucleic acid* (DNA) and *ribonucleic acid* (RNA).

DNA contains the genetic information that directs the development and activity of the organism. In eukaryotic cells DNA resides in the nucleus in linear molecules of repeating nucleotide units, although there are circular molecules of DNA found in the mitochondria and chloroplasts of eukaryotic cells. DNA nucleotides are composed of a five-carbon deoxyribose sugar, a phosphate group, and one of four possible bases: adenine (A), thymine (T), cytosine (C), and guanine (G). The information of the DNA molecule is found in the sequence of the nitrogenous bases along its length. Any region of DNA that directs a cellular function or encodes another molecule is called a *gene*. Not all DNA regions encode proteins. Some regions encode the instructions for RNA molecules that are used as catalysts and for protein synthesis reactions. Some genes are regulatory, controlling the time and place where certain genes are expressed. In many eukaryotes, genes only account for 10 percent of the DNA. Although some of the remaining 90 percent carries various structural functions, most of it is of uncertain function.

In 1953 Francis Crick and James Watson constructed a molecular structure for the DNA molecule, relying heavily on the experimental data generated by Rosalind Franklin. The structure they proposed, which has since been supported by additional experimental data, was that of a *double helix*. The DNA molecule can be envisioned as a ladder. The sugars and phosphates of the nucleotides alternate with each other to form the backbone, the outside vertical support, and the bases form the individual rungs of the ladder. The ladder is twisted to create a helical structure. DNA can exist as single strands and in other confirmations in the cell, but the "B-form" of the DNA double helix is the most common form in the cell.

RNA molecules are also polymers of nucleotides, but the nucleotides of the RNA molecule differ slightly from those of the DNA molecule. RNA nucleotides contain a five-carbon ribose sugar, a phosphate group, and one of four bases. Three of the four bases are the same as found in DNA: adenine, guanine, and cytosine. Instead of thymine, RNA uses the base uracil. RNA bases can pair in essentially the same way as DNA bases, but most often RNA exists as single-stranded molecules in cells. These long strands of RNA can often pair with other bases in short regions, causing the RNA to fold up into highly complex, three-dimensional structures important for RNA function.

RNA is found throughout cells. Messenger RNA (mRNA) is made by the cell using the DNA sequence in genes as a template for making a complementary strand of RNA in a process called transcription. In After being transcribed and modified in certain complex ways, most mRNA is transported to the cytoplasm where it is used to direct the synthesis of proteins. Ribosomal RNA (rRNA) is a major component of ribosomes, which are responsible for coordinating protein synthesis, along with transfer RNA (tRNA). Some RNA molecules, like protein molecules, can also catalyze chemical reactions. Catalytic RNA molecules are called *ribozymes*, and they play roles in gene expression and protein synthesis.

Single nucleotides and compounds that are made from them are involved in many cellular processes. The universal unit of "energy currency" in the cell is *adenosine triphosphate* (ATP). Guanosine triphosphate (GTP) is a molecule that is involved in relaying signals received at the cell membrane to the nucleus of the cell. Compounds, such as NADH and NADPH, that are involved in metabolic reactions in the mitochondria and in energy capture reactions in the chloroplasts also contain nucleotides.

Proteins

Protein molecules are large, complex molecules with a huge variety of structures and functions within cells. Most chemical reactions in cells are catalyzed by proteins called *enzymes*. Proteins form the basis of the cytoskeleton of cells, providing structure and motility. Proteins are also essential for the communication between cells and within cells. In plants, the largest concentration of proteins can be found in some seeds.

Proteins are polymers of nitrogen-containing molecules called *amino acids*. The amino acids are much simpler molecules than the nitrogenous bases found in nucleic acids. The same twenty amino acids is are used in the manufacture of proteins in the cells of all living organisms. An amino acid is built around a single carbon atom called the *alpha carbon*. Bonded to the alpha carbon are a hydrogen atom (H), a carboxyl group (COOH), and an amino group that contains nitrogen (NH_2). A specialized "R" group is attached at the last site. The R-groups are different for each of the twenty amino acids, and their chemical properties, such as charge, hydrophilic or hydrophobic nature, and size, dictate protein function and shape.

The order and number of amino acids that are linked together to form a protein are determined by the order of the codons in the DNA that encode that protein. The order and number of the amino acids in a protein is called the *primary structure*, and it ultimately determines the shape of the protein. Proteins can have *secondary structures* formed by hydrogen bonding between the peptide bonds that link the amino acids together. The two common secondary structures in proteins are the *alpha helix* and the *beta pleated sheet*. The amino acid chain (also called a peptide chain) can fold up on itself to form globular structures. This is known as *tertiary structure*. Tertiary structure is determined by the number and order of amino acids in the protein and is formed when molecules in the R-groups of the amino acids interact with one another. When two or more peptide chains interact to form a single functional molecule, the protein is said to have *quaternary structure*.

Michele Arduengo

See also: ATP and other energetic molecules; Calvin cycle; Carbohydrates; Cell-to-cell communication; Cell wall; Cells and diffusion; Chloroplast DNA; Chloroplasts and other plastids; Chromatin; Chromosomes; Cytoplasm; Cytoskeleton; Cytosol; DNA: historical overview; DNA in plants; DNA replication; Eukaryotic cells; Fluorescent staining of cytoskeletal elements; Gene regulation; Genetic code; Genetics: mutations; Glycolysis and fermentation; Hormones; Krebs cycle; Lipids; Liquid transport systems; Membrane structure; Metabolites: primary vs. secondary; Microbodies; Mitochondria; Mitochondrial DNA; Mitosis and meiosis; Molecular systematics; Nuclear envelope; Nucleic acids; Nucleolus; Nucleus; Nutrients; Oil bodies; Osmosis, simple diffusion, and facilitated diffusion; Oxidative phosphorylation; Peroxisomes; Pheromones; Plasma membranes; Prokaryotes; Proteins and amino acids; Ribosomes; RNA; Sugars; Vacuoles; Vesicle-mediated transport; Viruses and viroids.

Sources for Further Study

American Society for Cell Biology. *Exploring the Cell*. Available on the World Wide Web at http://www.ascb.org/pubs/exploring.pdf. Features excellent images of cells and cellular processes.

Lodish, Harvey, et al. *Molecular Cell Biology*. 4th ed. New York: W. H. Freeman, 2000. One of the most authoritative cell biology reference texts available. Provides excellent graphics, CD-ROM animations of cellular processes, and primary literature references.

Raven, Peter H., Ray F. Evert, and Susan E. Eichhorn. *Biology of Plants*. 6th ed. New York: W. H. Freeman/Worth, 1999. Introductory botany text provides a chapter introducing eukaryotic cells in general and plant cells specifically.

PLANT DOMESTICATION AND BREEDING

Categories: Agriculture; economic botany and plant uses

Plant domestication and breeding are the processes by which wild plants are intentionally raised to meet human food, fiber, shelter, medicinal, or aesthetic needs.

No one knows exactly when the first crop was cultivated, but most authorities believe that it occurred at some time between eight and ten thousand years ago. For centuries prior to that time, humans had known that some wild plants and plant parts (such as fruits, leaves, and roots) were edible. These plants appeared periodically (usually annually) and randomly throughout a given region. Eventually humans discovered that these wild plants grew from seeds and that the seeds from certain wild plants could be collected, planted, and later gathered for food. This most likely occurred at about the same time in both the Sumerian region between the Tigris and Euphrates Rivers and in Mexico and Central America. While the earliest attempts at domesticating plants were primarily to supplement the food supply provided by hunting and gathering, people soon improved their ability to domesticate and breed plants to the point that they could depend on an annual supply of food. This food supply allowed the development of permanent settlements.

Early Crop Domestication

By six thousand years ago, agriculture was firmly established in Asia, India, Mesopotamia, Egypt, Mexico, Central America, and South America. Before recorded history, these areas had domesticated some of the world's most important food (corn, rice, and wheat) and *fiber* (cotton, flax, and hemp) crops. The place of origin of wheat is unknown, but many authorities believe that it may have grown wild in the Tigris and Euphrates Valleys and spread from there to the rest of the Old World. Wheat was grown by Stone Age Europeans and was reportedly produced in China as far back as 2700 B.C.E. Wheat is now the major staple for about 35 percent of the people of the world. The earliest traces of the human utilization of corn date back to about 5200 B.C.E. It was probably first cultivated in the high plateau region of central or southern Mexico and represented the basic food plant of all pre-Columbian advanced cultures and civilizations, including the Inca of South America and the Maya of Central America.

Botanists believe that rice originated in Southeast Asia. Rice was being cultivated in India as early as 3000 B.C.E. and spread from there throughout Asia and Malaysia. Today rice is one of the world's most important cereal grains and is the principal food crop of almost half of the world's people. Hemp, most likely the first plant cultivated for its fiber, was cultivated for the purpose of making cloth in China as early as the twenty-eighth century B.C.E. It was used as the cordage or rope on almost all ancient sailing vessels. Linen, made from flax, is one of the oldest fabrics. Traces of flax plants have been identified in archaeological sites dating back to the Stone Age, and flax was cultivated in Mesopotamia and Egypt five thousand years ago. Cotton has been known and highly valued by people throughout the world for more than three thousand years. From India, where a vigorous cotton industry began as early as 1500 B.C.E., the cultivation of cotton spread to Egypt and then to Spain and Italy. In the West Indies and South America, a different species of cotton was grown long before the Europeans arrived. Other important plants that have been under domestic cultivation since antiquity include dates, figs, olives, onions, grapes, bananas, lemons, cucumbers, lentils, garlic, lettuce, mint, radishes, and various melons.

Modern Plant Breeding

Genetic variability is prevalent in plants and other organisms that reproduce sexually and thereby produce spontaneous mutants. Throughout most of history, plant domestication and breeding were primarily based on the propagation of mutants. When a grower observed a plant with a potentially desirable *mutation* (such as a change that produced bigger fruit, brighter flowers, or increased insect resistance), the grower would collect seeds or take cuttings and produce additional plants with the desirable characteristic. Advances in the understanding of genetics in the early part of the twentieth century made it possible to breed some of the desirable characteristics resulting from mutation into plants that previously had lacked the characteristic.

The obvious advantages of producing plants with improved characteristics such as higher yield made plant breeding very desirable. As human populations continued to grow, there was a need to select and produce higher-yielding crops. The development and widespread use of new high-yield varieties of crop plants in the 1960's is often referred to as the *Green Revolution*. Basic information supplied by biological scientists allowed plant breeders to fuse a variety of characteristics from different plants to produce new, higher-yielding varieties of numerous crops, particularly seed grains.

When a plant characteristic is identified as desirable, it is studied both morphologically and biochemically to determine the mechanism of inheritance. If it is determined that the mechanism is transferable, attempts are made to incorporate the trait into the target plant. If the plants are closely related, traditional breeding techniques are used to crossbreed the plant with the desirable trait with the plant that lacks the characteristic. Although this process is often tedious, it is based on a fairly simple concept. Basically, pollen from one of the plant types is used to fertilize the other plant type. This

AP/Wide World Photos

A scientist prepares peach flowers for cross-pollination. Hand-pollinating is a technique to produce hybrid fruit that combines properties such as heat resistance, large fruit size, and good flavor.

process often requires specialized handling techniques to ensure that only the pollen from the plant with the desired characteristic is allowed to fertilize the eggs of the recipient plant.

Sometimes this process involves the use of bags or other materials to isolate the recipient flowers, which are then pollinated by hand. Another technique involves the introduction of a gene for male sterility into the recipient plant. In these cases, only pollen from another plant can be used to fertilize the egg. Once plants with the desirable characteristics are developed, the lines are often inbred to maintain large numbers of progeny with the desired traits. In many cases, inbred lines will lose vigor after several generations. When this occurs, two inbred lines may be crossed to produce hybrids. A majority of the hybrid offspring will still contain the desired characteristics but will be more vigorous.

Recombinant Technology

Until recently, the use of traditional breeding techniques between two very closely related species was the only means of transferring heritable characteristics from one to the other. The advent of *recombinant technologies* in the manipulation of deoxyribonucleic acid (DNA), however, made it possible to transfer genetic characteristics from any plant (or from any organism) to any other. The simplest method for accomplishing this transfer involves the use of a vector, usually a piece of circular DNA called a *plasmid*. The plasmid is removed from a microorganism such a bacterium and cut open by an enzyme called a restriction endonuclease, or restriction enzyme. A section of DNA from the plant donor cell that contains the gene for an identified desirable trait is cut from the donor cell DNA by the same restriction endonuclease. The section of plant donor cell DNA with the gene for the characteristic of interest is then combined with the open plasmid DNA, and the plasmid closes with the new gene as part of its structure. The recombinant plasmid (DNA from two sources) is placed back into the bacterium, where it will replicate and code for protein just as it did in the donor cell. The bacterium is then used as a vector to transfer the gene to another plant, where it will also be transcribed and translated.

D. R. Gossett

See also: Agricultural revolution; Agriculture: traditional; Agronomy; DNA: recombinant technology; Genetically modified foods; Green Revolution; High-yield crops; Hybridization; Plant fibers.

Sources for Further Study

American Chemical Society. *Genetically Modified Foods: Safety Issues*. Washington, D.C.: American Chemical Society, 1995. A thorough overview of the technology, applications, and risks associated with genetically engineered foods.

Fennema, O. R., ed. *Principles of Food Science*. New York: Dekker, 1975- . Covers topics in food science, including the harvesting, preservation, and marketing of a variety of horticultural food crops.

Metcalfe, Darrel S., and Donald M. Elkins. *Crop Production: Principles and Practices*. 4th ed. New York: Macmillan, 1980. One of the most valuable sources available on the practical aspects of crop production.

Rissler, Jane. *The Ecological Risks of Engineered Crops*. Cambridge, Mass.: MIT Press, 1996. A carefully reasoned science and policy assessment showing that the commercialization and release of transgenic crops on millions of acres of farmland can pose serious environmental risks. The authors propose a practical, feasible method of conducting precommercialization evaluations that will balance the needs of ecological safety with those of agriculture and business, and that will assist governments seeking to identify and protect against two of the most significant risks.

PLANT FIBERS

Categories: Agriculture; economic botany and plant uses

Plants are the natural sources of many raw materials used to produce textiles, ropes, twine, and similar products.

The major fiber crops are cotton, flax, and hemp, although less important plants, such as ramie, jute, and sisal, are grown in small amounts. With a total annual production of more than 13 million tons, cotton is by far the most important fiber crop in the world. Because humans heavily rely on cotton for clothing and other textiles, it enters the daily lives of more people than any other product except salt.

Cotton

Cotton (*Gossypium*) fiber has been known and highly valued by people throughout the world for more than three thousand years. The early history of cotton is obscure. A vigorous cotton industry was present in India as early as 1500 B.C.E. From India, the cultivation of cotton spread to Egypt and then to Spain and Italy. In the New World, a different species of cotton was being grown in the West Indies and South America long before Europeans arrived. In the United States, cotton is grown from the East Coast to the West Coast in the nineteen southernmost states.

Botanically, cotton is in the mallow family, which also includes okra, hollyhock, hibiscus, and althea. Cotton has a taproot and branching stems. Flowers form at the tips of fruiting branches, and the ovary within each flower develops into a *boll*, which contains the *seed*, *fiber*, and *fuzz*. The fiber, most commonly referred to as *lint*, develops from epidermal cells in the seed coat of the cottonseed. The fiber reaches its maximum length in twenty to twenty-five days, and an additional twenty-five days are required for the fiber to thicken. Fiber length from 2.0 to 2.4 centimeters is referred to as short-staple cotton, and fiber length from 2.4 to 3.8 centimeters is called long-staple cotton.

The boll normally opens forty-five to sixty-five days after flowering. Cotton is native to tropical regions but has adapted to the humid, subtropical climate, where there are warm days (30 degrees Celsius), relatively warm nights, and a frost-free season of at least 200 to 210 days. There are eight species of cotton in the genus *Gossypium*, but only three species are of commercial importance. *Gossypium hirsutum*, also known as upland cotton, has a variable staple length and is produced primarily in North and Central America. *Gossypium barbadense*, a long-staple cotton, is primarily produced in South America and Africa. *Gossypium herbaceum* is a shorter-staple cotton native to India and eastern Asia.

Cotton is one of the more labor-intensive and expensive crops to produce. The most opportune time to plant cotton is at least two weeks after the last killing-frost date of the region. Prior to seeding, the field is prepared by plowing to a depth of 2.5 centimeters. Fertilizer, which is applied before seeding or at the same time the seeds are planted, is placed to the side and below the cotton seed. Once the seeds germinate and emerge from the soil, they often have to be thinned, and shortly afterward the producer begins to apply irrigation water as needed. After the plants have developed a stand, weed control becomes crucial. Weeds are controlled both by cultivation and herbicides.

Cotton plants are subject to invasion by a variety of insect pests, such as the boll worm and boll weevil; therefore considerable attention is given to insect control, typically using a number of different insecticides.

When the bolls ripen with mature fiber, the leaves of the plant are removed by the application of a chemical defoliant, and the fiber is harvested. Harvesting was once done almost entirely by hand, but today mechanical pickers harvest almost all the cotton produced in the United States. The picked cotton is ginned to remove the seed and compressed into bales. The bales are transported to a cotton mill, where the cotton is cleaned and spun into yarn, which is then woven into fabric. One pound of fiber is sufficient to produce up to 6 square yards of fabric.

*Hemp (*Cannabis sativa*), a term used to identify both the plant and the fiber it produces, is used to make strong and durable commercial fibers.*

Flax

Flax (*Linum usitatissimum*) is the fiber used to make linen. While some flax is still grown for the purpose of producing this fabric, much of the flax, particularly that grown in the United States, is used to produce the flaxseed, from which linseed can be extracted. Linen made from flax is one of the oldest fabrics. Flax was cultivated in Mesopotamia and Egypt five thousand years ago, and traces of flax plants have been identified in archaeological sites dating back to the Stone Age. Flax was one of the first crops brought to North America by European settlers. Today, most of the flax produced in the United States is grown in the north-central states.

An annual plant, flax grows to a height of 60 to 100 centimeters and bears five-celled bolls or capsules with ten seeds each at the ends of fertile branches. Because the flax fiber is found in the stems from the ground to the lowest branches, varieties that are long-stemmed with little branching are grown for fiber production. Selection of quality, disease-free seed is essential in flax production. Flax fields are usually prepared in the fall to allow the soil to settle before planting. Flax is usually sown in early spring, two to three weeks prior to the date of the last killing frost of the region. Considerable attention is given to controlling weeds in a flax field. When the crop is harvested for fiber, the plants are pulled from the soil, the seeds are removed, and the flax straw is "retted" to separate the fiber from the woody part of the stem. When the straw is completely retted, it is dried and then broken apart to remove the 50-centimeter fibers which can be woven into fabrics.

Hemp

Hemp (*Cannabis sativa*), a term used to identify both the plant and the fiber it produces, is used to make the strongest and most durable commercial fibers available. Hemp was most likely the first plant cultivated for its fiber. It was cultivated for the purpose of making cloth in China as early as the twenty-eighth century B.C.E. It was also used as a drug by the ancient Persians as early as 1400 B.C.E. and was used as the cordage or rope on almost all ancient sailing vessels. Today hemp is commercially produced for heavy textiles in numerous countries, but less than 1,000 acres is devoted to commercial hemp production in the United States. Hemp production is problematic in the United States because it is illegal to grow *Cannabis sativa*, the source of marijuana.

Hemp is an annual plant in the mulberry family. The plant is dioecious, meaning that it has staminate or "male" flowers and pistillate or "female" flowers. It has a rigid stalk, which can reach a thickness of more than 2.5 centimeters in diameter, and a height of 5 meters. The plant has a hollow stem, and the bark or "bast" located outside the woody shell is used to make the bast fiber, which is then used to make hemp twine, ropes, and other textiles where strength and durability are desired.

Humid climates with moderate temperatures and a period of at least 120 frost-free days are necessary for hemp production. Unlike flax, hemp requires that the soil be plowed and thoroughly disked or harrowed prior to planting. The entire aboveground portion of the plant is harvested when the male plants are in full flower. After two to three days the plants are tied in bundles and set in shocks. Hemp fiber is retted and prepared for the mills in a manner very similar to that of flax except that heavier machines are used to handle the stronger hemp stalks.

Minor Crops

Ramie (*Boehmeria nivea*) is produced primarily in Asia and is used to make strong cloth, such as Chinese linen. Jute (*Corchorus capsularis*) is grown primarily in India and Pakistan and is used to manufacture burlap for bags and sacks. Sisal (*Agave sisalana*) is produced in East Africa and the West Indies and is used to make different types of cordage, such as baler twine.

D. R. Gossett

See also: Plant domestication and breeding; Textiles and fabrics.

Sources for Further Study

Dillman, A. C. "Flaxseed Production." In *Farmer's Bulletin Number 1747*. Washington, D.C.: U.S. Department of Agriculture, 1935. Old but very useful publication on flax production.

Kipps, Michael S. *The Production of Field Crops*. 6th ed. New York: McGraw-Hill, 1970. Excellent discussions of the importance and production of cotton, hemp, and flax.

Lester, D. H. "Hemp," In *Yearbook*. Washington, D.C.: U.S. Department of Agriculture, 1913. Old but very useful publication on hemp.

Mauney, J. R., and J. M. Stewart, eds. *Cotton Physiology*. Memphis, Tenn.: Cotton Foundation, 1986. Contains excellent discussions of the physiology of the cotton plant.

Metcalfe, Darrel S., and Donald M. Elkins. *Crop Production: Principles and Practices*. 4th ed. New York: Macmillan, 1980. One of the most valuable sources available on the practical aspects of cotton production.

PLANT LIFE SPANS

Category: Reproduction and life cycles

The cycle of a plant's life, from seed germination to death, is referred to as its life span. Some plants have short life spans (less than one year), whereas others have life spans that are measured in centuries.

The longest-lived organisms are plants. For example, one bristlecone pine tree in eastern California is forty-nine hundred years old, and some creosote bushes, also in California, are estimated to be about twelve thousand years old. People have long recognized this variation in plant longevity, but the understanding of plant life spans improved greatly after research during the 1960's.

Types of Life Spans

The life span of an individual plant depends upon two factors. The first is the innate, genetically determined potential for longevity. The second is the effects of the environment, including soil and weather conditions, competing plants, disease-causing microbes, and herbivores.

Historically, people have classified the life spans of plants into three categories: *annuals*, *biennials*, and *perennials*. Annual plants live for up to one year. Biennials live for approximately two years. Perennials live for more than two years, often for several decades, even centuries.

While this categorization is useful in many ways, botanists have come to recognize that it is inaccurate, especially for plants that grow under natural conditions. Plant life histories are now classified mainly according to the number of times that each individual normally reproduces before it dies. Two main categories are recognized using this system: *monocarpic* plants and *polycarpic* plants. Monocarpic plants reproduce once before they die (*mono* means "one"; *carpic* means "fruits"). Polycarpic plants reproduce several or many times before they die (*poly* means "many"). Some botanists have defined a third group, the *paucicarpic* plants, that are intermediate between the two. Paucicarpic plants reproduce up to five times (*pauci* means "few").

Monocarpic Plants

Monocarpic plants have a general life history that involves four separate stages: germination, vegetative growth, reproduction, and death. The period of vegetative growth is very important to the monocarpic plant because during this time the plant manufactures and stores starch, which is rich in energy. When reproduction occurs, all that stored energy is devoted to producing flowers, fruits, and seeds; none is saved for the following year. The plant literally reproduces itself to death. Monocarpic plants vary greatly in their longevity, and it is possible to recognize several subcategories: *ephemerals*, annuals, *obligate biennials*, *facultative biennials*, and *long-lived monocarpic perennials*.

Ephemerals are plants that germinate, grow, reproduce, and die within a few weeks or months. They are typically found in environments in which conditions favor active plant growth for only a short period of time during the year, such as a desert. Desert ephemerals spend most of the year as seeds. When a heavy rainstorm occurs, the seeds germinate, and the new plants grow quickly and reproduce before the soil dries out. One species in the Sahara Desert can complete its life cycle in as little as ten days.

Annuals are plants that progress from germination to true seed within a six- to twelve-month period. Botanists recognize two major subcategories of annual plants. One is the summer annual, which germinates in the spring: The plant grows vegetatively during the summer and reproduces during the autumn. Examples of summer annuals include touch-me-not, common ragweed, and goosefoot. The second subcategory is the winter annual, in which germination occurs in the fall, vegetative growth occurs in the winter, and reproduction occurs in the spring. Daisy fleabane and winter wheat are examples of winter annuals.

Obligate biennials are monocarpic plants that germinate and grow vegetatively over the course of one year and throughout much of a second year. At the end of the second year, the plant always reproduces, sets seed, and dies (hence the designation "obligate"). During the 1960's and 1970's, some botanists doubted that obligate biennials existed in nature. Studies conducted during the 1970's and 1980's demonstrated that some plants, such as the white and yellow sweet clovers, are indeed obligate biennials.

Facultative biennials are monocarpic plants that have the ability to germinate, grow, and reproduce within two years. They can behave as biennials only when they grow under favorable conditions, with adequate moisture, light, and soil nutrients. More commonly, these plants grow under stressful conditions—either infertile soils or high competition. On such sites, they grow vegetatively for three, four, or even five years before they reproduce. Examples of facultative biennials include wild carrot, foxglove, burdock, teasel, and thistle.

Long-lived monocarpic perennials are able to live for many years or a few decades before they reproduce—once—and then die. Well-known examples include bamboo and plants from the arid southwestern United States, such as species of *Yucca* and the century plant *Agave*. These may reproduce only after they attain an age of sixty, eighty, or even one hundred years.

Polycarpic and Paucicarpic Plants

Paucicarpic and polycarpic plants normally reproduce more than once before they die. They are able to survive for at least one year following reproduction and hence are true perennials. Paucicarpic plants are short-lived herbs that may die after reproduction but more commonly live to reproduce two, three, or four times before dying. Paucicarpic plants are therefore intermediate between the true monocarps and the true polycarps. Examples include the common and English plantains, which are weeds found in lawns and fields throughout temperate North America and Europe.

True polycarpic plants survive to reproduce many times during their lifetimes and usually remain alive for at least ten years. Unlike the monocarps, polycarps do not expend all of their energy in reproduction. They save some of their energy and maintain part of the plant for the post-reproductive period. In seasonal climates, some of that energy must be directed to forming structures that allow the plant to survive the unfavorable season—a cold winter or a rainless period. These structures are called *perennating buds*, and they differ from plant to plant in their location relative to the ground surface. In some plants, called *cryptophytes* (the prefix *crypto* means "hidden"), the perennating buds are buried several centimeters under the ground. Examples of cryptophytes include milkweed, iris, and onion. Conversely, *hemicryptophytes* (*hemi* means "partial") have their perennating buds at the soil surface; a good example is the dandelion. Both cryptophytes and hemicryptophytes are herbaceous plants, never producing an aboveground woody structure.

Phanerophytes are polycarpic plants that do produce an aboveground woody structure—the perennating buds are borne above the ground surface. Some phanerophytes are shrubs that have several shoots. Examples of shrubs include lilac, blueberry,

Digital Stock

Extreme examples of seed longevity can be seen in species of lotus which have survived for more than fifteen hundred years.

hawthorn, hydrangea, rhododendron, and many dogwoods and willows. A second category of phanerophytes is the trees, which typically have a single woody stem emerging from the rootstock.

In theory, most species of polycarpic plants can live for decades, if not centuries, under ideal conditions. Many do not appear to have a maximum life span because they rejuvenate their tissues with each reproductive period, as in some polycarpic herbs or because the tissues that they accumulate do not put much of an added strain on the plant, as in many phanerophytes.

In nature, such polycarpic plants are not killed by old age. External factors such as *herbivory* (consumption by animals), fire, severe weather, disease, and competition from other plants contribute heavily to die-off among individuals. Other polycarpic plants form senescent tissue that hastens their death.

Potential and Real Life Spans

There have not been many studies of the longevity of most polycarpic plant species. The logistic problems involved and the consideration of mortality are more closely related to the size of the plant than to its age. Knowledge of the longevity in many species, particularly polycarpic herbs, is very poor.

Many herbs, such as buttercups and clovers, live for five to twenty-five years. Other herbs, such as blazing star, milkweed, and some goldenrods, may live for twenty-five to fifty years. Some shrubs, including blueberries and sumacs, can live for thirty to seventy years. Trees such as gray birch, pin cherry, and trembling aspen live for fifty to one hundred fifty years. Conifers such as hemlock, white pine, and red spruce have longevity in the range of two hundred to three hundred years, with some trees living five hundred to six hundred years. Hardwood trees such as sugar maple, white oak, sycamore, and beech can live for a similar duration. The oldest trees are the redwoods, at thirteen hundred years, and the bristlecone pines, at three thousand to five thousand years.

Most plants do not live to their maximal potential, succumbing to environmental factors. The average age that plants attain is well below the maximum life span. Beginning in the 1960's, ecologists began to examine the length of time that individual plants in a population remain alive. Although the actual patterns differ greatly from one species to another, plants typically suffer heavy mortality shortly after germination. For most species, fewer than 10 percent of newly emergent seedlings survive for two months. After that point, there are additional losses, although the death rate slows.

Some plants produce seeds that can remain dormant for many years. For example, seeds of many weed species, including monocarpic and polycarpic herbs, can remain dormant for seventy years in abandoned farm soil. Under experimental conditions, seeds of some of these species were found to be capable of germinating after one hundred years. Extreme examples of seed longevity can be seen in species of goosefoot and lotus, both of which have survived for more than fifteen hundred years. The variety of plant life cycles appears to be related to the earth's widely divergent habitats.

Kenneth M. Klemow

See also: Angiosperm life cycle; Dendrochronology; Growth and growth control; Hormones; Leaf abscission; Seeds.

Sources for Further Study

Barbour, Michael G., Jack H. Burk, and Wanna D. Pitts. *Terrestrial Plant Ecology*. 3d ed. Menlo Park, Calif.: Benjamin/Cummings, 1999. A well-written text intended for undergraduate students. Excellent overview of most phases of plant ecology, including physiological ecology, population biology, and community ecology. Chapter 4 focuses on the patterns of survival found in representative plants. Chapter 5 addresses plant life histories and includes a good explanation of the interrelationships among different life spans and the environments in which they are found.

Dennis, David T., and David H. Turpin, eds. *Plant Physiology, Biochemistry, and Molecular Biology*. New York: John Wiley, 1990. Examines the assimilation and metabolism of carbon and nitrogen in an integrative fashion, assessing the physiology, biochemistry, and molecular biology of each topic discussed, for beginning or advanced students.

Hardin, James W., Donald J. Leopold, and F. M. White. *Harlow and Harrar's Textbook of Dendrology*. 9th ed. Boston: McGraw-Hill, 2001. This well-illustrated book describes the biological features of nearly all species of evergreen and deciduous trees in North America. It includes information on the life spans of most of the species as well as descriptions of vegetative and reproductive features, habitat preferences, and forestry.

PLANT SCIENCE

Categories: Disciplines; history of plant science

Botany is the study of plants, stationary organisms with chlorophyll that are able to make their own food. Major categories of the plant kingdom include algae, mosses, ferns, and seed plants. Plant science includes many subdisciplines of both botanical (nonapplied science) and applied studies of plants, especially agriculture.

Taxonomy

One of the basic subdisciplines of plant science and life science in general, *taxonomy* (also known as *systematics*) is the study of relationships and organization of plant species. The great diversity within the plant kingdom requires a system by which plant species are named and classified. The modern system is a modification of the system first established in the eighteenth century by the Swedish botanist Carolus Linnaeus. Each species is placed into a hierarchy of groups that indicate its similarity and dissimilarity to other species. These categories (taxa) are from the most to the least inclusive: domain, kingdom, phylum, class, order, family, genus, species.

The species is the most natural and fundamental unit. Similar species are grouped into a genus, similar genera into a family, families into orders, orders into classes, classes into phyla, and phyla into the kingdoms typically studied in plant science courses: true plants, or *Plantae*, *Fungi*, and *Protista* (which include many unicellular organisms and algae). Each species is given a scientific name which includes the genus name followed by the species name. An example is the scientific name of the dwarf crested iris: *Iris cristata*.

Morphology

Morphology includes the study of the general structure of plants. Morphologists study the parts of a plant and how they are arranged and function. For example, when a seed of an angiosperm (flowering plant) germinates, the radicle of the seed embryo develops downward to form a root system. Growth in the length of the root occurs within the meristem (region of cell division). Branch roots form due to the activity of pericycle cells within the root. Some epidermal cells develop root hairs as extensions of the cells. The shoot system, which includes the stem and leaves, develops from the epicotyl of the seed embryo. Stems are often branched, allowing for the attachment of leaves in such a manner as to permit their maximum exposure to sunlight.

Also included in the study of morphology are the reproductive parts of plants. The pollination of flowers causes the ovary of the flower to mature into a fruit. At the same time, the one or more ovules inside the ovary become seeds.

Anatomy

Anatomy is the study of plant tissues. New plant cells are formed within meristems. There, the cells begin the process of becoming specialized (differentiated) for a particular function. As a result, three categories of plant tissues are formed: dermal, vascular, and ground.

Dermal tissues, which form protective coverings, include the epidermis, which covers all parts of a young plant, and others that develop as a plant matures. The periderm commonly replaces the epidermis and includes tissues found in bark.

Vascular tissues, derived from cambium cells within the meristem, conduct water and dissolved compounds within a plant. They include xylem and phloem.

Ground tissues are the less specialized tissues. Among their functions are storage, support, and photosynthesis. A common type is parynchema.

Cytology

Cytology is the study life at the cellular level; another name for this discipline is cell biology. Plant cells share many features with animal cells. Both forms of life are composed of eukaryotic cells, which have a distinct nucleus surrounded by a nuclear envelope which separates it from the cytoplasm. (By contrast, bacteria have nucleus-free cells, called prokaryotic cells.) Inside the nucleus is chromatin, which becomes organized into chromosomes as a cell divides. Chromosomes are composed of functional units called genes, which serve as the control center of the cell. Genes are composed of nucleoprotein. The nucleus also contains a nucleolus.

The cytoplasm is differentiated into numerous organelles, each specializing in a particular activity. Among those which plant cells share with animal cells are mitochondria (which conduct cellular respiration), ribosomes (which conduct protein synthesis), endoplasmic reticulum (for strengthening), Golgi apparatus (for packaging), and a plasma membrane (which functions as the cell's outer boundary). Not found in animal cells are chloroplasts (which conduct photosynthesis). Surrounding each plant cell are several layers of compounds (especially cellulose) that form the cell wall.

AP/Wide World Photos

By counting the flowers on these lantana plants, this scientist can determine how many fruits and seeds the plants will produce.

Physiology

Physiology is the study of the various functions performed in and by living organisms. Physiological processes of plants include the flow of energy, movement of solutes, and control by hormones. Chemical reactions are mediated (their rate is controlled) by enzymes.

For example, those studying plant physiology are concerned with the way that plants trap light energy as light is absorbed by chlorophyll. As a result of a series of chemical reactions, glucose, a six-carbon carbohydrate, is formed (preceding glucose are molecules of PGAL, a three-carbon sugar, which pair to form glucose). The glucose may be oxidized within the same cell or within another cell of the same plant (or utilized by an animal). This oxidation process produces adenosine triphosphate (ATP). As this compound is converted to adenosine diphosphate (ADP), energy is released, allowing organisms to perform other essential energy-requiring life activities.

Plant physiologist would also be concerned with the transport of nutrients and water throughout a plant. By means of their roots, plants absorb water and dissolved materials from the soil, after which they are conducted upward, by means of xylem tissue, to all parts of the plant. This upward movement is called transpiration. The glucose formed in leaves is dissolved in water and transported, by means of phloem tissue, to all parts of the plant. This movement, called translocation, is commonly downward, but also may be upward. By these processes, water, minerals, and sugars are transported to all parts of a plant.

Another area of concern for plant physiologists is the function of plant hormones (phytohormones); in fact, those specializing in this area have their own discipline, *endocrinology*. Phytohormones are compounds produced within a plant. They are transported to other parts of plant, where they regulate growth and development. Early in the twentieth century, auxin was the first phytohormone to be discovered. It promotes growth by causing cells to elongate but was found also to inhibit growth of lateral buds. Gibberellins, a second group of hormones, also stimulate growth by causing cell elongation. Among their activities is the promotion of seed germination. Cytokinins are abundant in dividing tissues, where they stimulate cell division. Abscisic acid is a growth-inhibiting hormone that maintains dormancy in buds and fruits and also is associated with the falling of leaves in autumn. Ethylene causes fruits to ripen. Several hormones are used in agriculture for increasing growth rates of crops.

Genetics

Genetics, the study of heredity and the mechanisms that control it, is an outgrowth of the studies of Gregor Mendel. In the 1860's, he performed experiments with garden peas which resulted in a new way of explaining how traits are passed from generation to generation. Mendel's ideas were revived in 1900 as other European investigators confirmed his basic tenets. Heredity is due to discrete hereditary particles which soon came to be called genes, which are located on located on paired chromosomes. The application of the principles of genetics, begun in the first few decades of the twentieth century, has resulted in the development of greatly improved varieties of crop plants.

Molecular Biology

Genetics today is in many ways the concern of another discipline of plant science, *molecular biology*. The basic chemical nature of genes and how they express themselves remained in question until the 1950's, when James Watson and Francis Crick developed the double-helix model of deoxyribonucleic acid (DNA), explaining how genes occur in great variety, replicate (duplicate) themselves, and produce phenotypes (observable traits). Today, genetics is largely concerned with studying the chemical reactions that control DNA replication. Many researchers are also at work mapping the genomes (identifying the genes responsible for expressed characteristics) of various organisms. In 2001, researchers completed a map of the genome of the model plant *Arabidopsis thaliana* as well as the more complex genome of the rice plant, *Oryza sativa*.

Molecular biology has many practical applications. A knowledge of the genetic control of cells has already resulted in new crop plants. The term "genetic engineering" indicates that plants can be designed for specific purposes.

Ecology

The early Greek scientist Theophrastus, in the third century B.C.E., recognized environmental effects on plants. Much later, naturalists documented the geographical distribution of plants as determined by various climatic factors. Such studies were the roots of the scientific discipline *ecology*, which emerged in the late nineteenth century. Plant ecologists of the early twentieth century were concerned largely with describing the nature and distribution of world plant communities and developing a "successional theory" as a means of understanding the dynamics of changing plant communities. Now, ecologists study plants as integral parts of ecosystems that also include animals and microorganisms. They are concerned with countering the threat of loss of species as a result of human activities such as pollution and habitat destruction.

Paleobotany

Fossils have long been recognized as remnants of plants and animals that lived and died many millennia ago. The animal fossil record was an important factor in the development of Charles Darwin's theory of evolution in the 1800's. *Paleobotany* as a subdiscipline can be traced to the efforts of Albert Seward of Cambridge University of England in the late nineteenth and early twentieth centuries. Studies of plant fossils have resulted in a clearer understanding of plant evolution.

Economic Botany

People have always relied on plants to provide basic necessities of life: food, shelter, and clothing.

Economic botany developed as a specialty within botany to acquaint botanists with plant uses. Topics considered in economic botany include plant domestication, food and beverage plants, essential oils, oils and waxes, latexes and resins, medicines, fibers, tannins and dyes, wood products, and ornamental plants.

Related Disciplines

Courses in botany and plant science often address organisms that are not, in the strict sense, plants but that nevertheless are appropriately studied in the same context. Hence, although *bacteria* differ from plants primarily because of their cells, which are prokaryotic (lacking a nucleus and most cytoplasmic organelles), they are often studied in botany courses. Bacteria are early and evolutionarily significant organisms. Some are closely related to the protists known as algae and therefore important in the study of photosynthesis.

Fungi, too, are often studied in the context of plant science. These are mostly multicellular filamentous eukaryotic organisms lacking chlorophyll. Because they do not make their own food but live in or on the food provided by plant and animal tissues, fungi are heterotrophs (rather than autotrophs, like plants, which make their own food through photosynthesis). In some ways, therefore, fungi are more similar to animals than they are to plants. Nevertheless, they are traditionally studied in the context of plant courses because they were once considered to be plants, given their lack of movement and other gross similarities. Their world significance parallels that of bacteria: They, too, are important as *decomposers*, returning nutrients and other elements to the environment. The study of fungi is *mycology*.

Protists are unicellular eukaryotes, forming one of the four kingdoms of *Eukarya*, the others being fungi, plants, and animals. Included among the protists are slime molds and protozoans which, lacking chlorophyll, are said to be heterotrophic protists, obtaining their food from other sources, generally other organisms. Also included are the algae, which are autotrophic eukaryotes—many using photosynthesis, like plants, to generate their own food. For this reason, protists are often studied in the context of plant science, and algae are almost always included in such studies. The study of algae is *phycology*.

Viruses, the study of which is *virology*, are noncellular entities that can reproduce only inside specific host cells. Not generally considered to be living, they are nevertheless important because of the infections they cause. To the extent that they cause infection in plants, they are important in the study of plant science, particularly *plant pathology*.

Thomas E. Hemmerly

See also: Agriculture: history and overview; Agronomy; Algae; Biotechnology; Botany; Cell theory; Cladistics; Dendrochronology; Ecology: history; Environmental biotechnology; Evolution: historical perspective; Fungi; Genetics: Mendelian; Genetics: post-Mendelian; History of plant science; Hormones; Hydroponics; Molecular systematics; Paleobotany; Paleoecology; Plant biotechnology; Population genetics; *Protista*; Systematics and taxonomy; Systematics: overview; Viruses and viroids.

Sources for Further Study

Madigan, Michael T., et al. *Brock Biology of Microorganisms*. 9th ed. Upper Saddle River, N.J.: Prentice Hall, 2000. This microbiology textbook serves as a reference for other microorganisms.

Magner, Lois N. *A History of the Life Sciences*. 2d ed. New York: Marcel Dekker, 1994. A concise treatment of the history of biology.

Raven, Peter H., Ray F. Evert, and Susan E. Eichhorn. *Biology of Plants*. 6th ed. New York: W. H. Freeman, 1999. An excellent textbook that gives an overview of plants and related groups and provides rationales for the inclusion of various subdisciplines under the umbrella of plant science.

PLANT TISSUES

Categories: Anatomy; physiology

Plant tissues are the distinctive structural and functional units of a plant that carry out all its basic life functions, including growth, reproduction, support, metabolism, circulation, and protection from the environment.

The body plan of a plant is very different from that of most animals. Terrestrial plant bodies are anchored in a growing medium, which has an enormous influence over the form and behavior of plant tissues.

Growth and Protective Tissue

Meristematic tissues in plant bodies are responsible for the growth that results from an increase in cell number. In the meristems, individual cells divide to produce pairs of daughter cells which have the ability to divide further or to enlarge and differentiate. The meristems are located at the ends of branches and roots (shoot apical meristems and root apical meristems, respectively) and within the *cambia* of woody plants, which grow in girth. The shoot and root apical meristematic tissues produce cells that account for the lengthening of the shoots and roots.

The primary developmental tissues are in a region called the *zone of elongation*. These developmental tissues are distinguished from meristematic tissues by the larger size of their cells and by their locations. Three primary developmental tissues are produced by the shoot and root apical meristems. They are the *protoderm*, the *ground tissues*, and the *procambium*. As these primary developmental tissues mature, they will ultimately differentiate into the metabolically more active portions of the plant.

In a region called the *zone of maturation*, the cells begin to take on the characteristics of mature, functioning tissues. The protoderm differentiates to form the *epidermis*, a mature tissue protecting the surfaces of plant parts which do not have secondary vascular tissues. The epidermis is made of cells which have one side in contact with the environment (air, water, or soil). The other side is in contact with other cells in the plant body.

Epidermal tissue in contact with air is usually protected by a layer of wax called the *cuticle*. It may also be covered with hairs, water-filled cells, poison-filled barbs, or even digestive glands. These specialized structures provide protection from particular environmental conditions and may even serve as paths for the absorption of nutrients in the case of carnivorous plants.

The underlying tissues must have access to atmospheric gases for their metabolic activities. To accomplish this, the epidermal tissues are punctuated by pores which open and close (*stomata*) or are permanently open (*lenticels*). Epidermal tissues in contact with the ground require a different kind of protection. These tissues may secrete mucus, which protects growing underground structures. There are epidermal cells that fall off the plant body to provide a lubricating barrier between the rest of the plant and the soil. Finally, the epidermal tissues nearest the root tip may be covered with long subcellular hairs that contribute significantly to the root's ability to absorb water and minerals. Epidermal tissues of plant organs that normally grow in water are less likely to bear the specialized structures of epidermal tissues from aerial or subterranean parts. These cells are often more like parenchyma cells of ground tissues than they are like epidermal cells of subterranean or aerial structures.

Ground Tissues

The ground tissues, the second of the primary developmental tissues, differentiate in the zone of maturation to form tissues called *parenchyma*, *collenchyma*, or *sclerenchyma*. The parenchymous tissues are the primary site of cellular metabolism. The organelles of parenchyma cells in different parts of the plant vary so that they can accommodate differences in metabolic functions. Cells of leaf parenchyma and some stem parenchyma have large numbers of chloroplasts to carry out photosynthesis. Stem and root parenchyma cells have amyloplasts, organelles that store starch. *Chromoplasts* in the pa-

Parts of a Plant

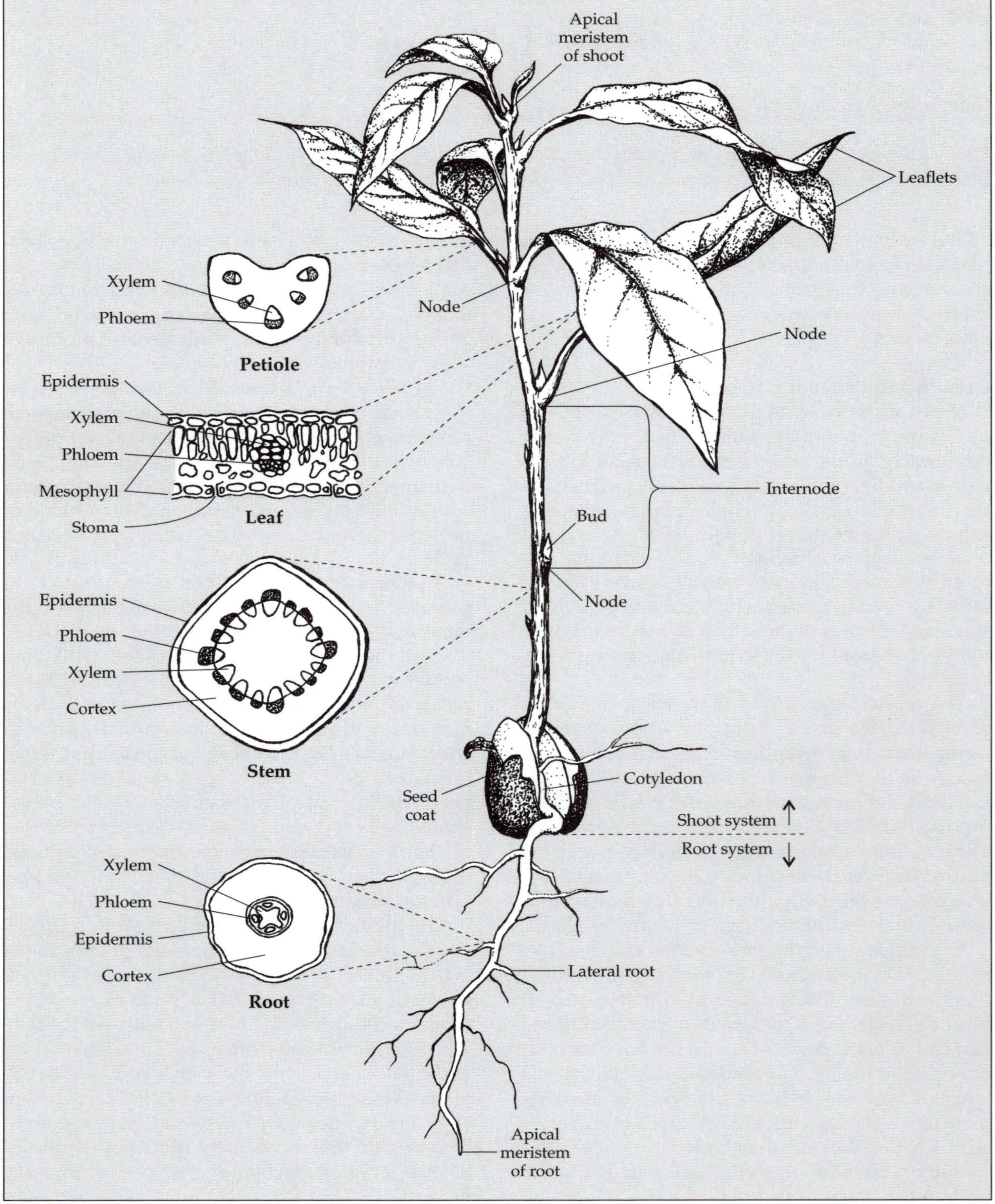

Kimberly L. Dawson Kurnizki

renchyma of flower petals contribute to the color of the flower petals. Parenchyma cells producing large quantities of protein have more ribosomes than those specialized for starch storage. Reproductive parenchyma cells may have unusual nuclear characteristics that prevent these tissues from competing with the developing embryos for nutrients or space.

Parenchyma fill the inner parts of leaves, stems, and roots. These cells have large, water-filled *vacuoles*. The water pressure from these vacuoles provides much of the rigidity of the body of nonwoody plants. When a leaf is limp, its parenchyma cells are usually depleted of water. Many of the chemicals that give plants their unique tastes or pharmaceutical characteristics are produced and stored in parenchyma. For example, the bulk of a carrot root (especially outside the central core), the mass of a potato tuber (which is actually a unique form of stem), and much of a lettuce leaf are all made of parenchyma.

Collenchyma cells are similar to parenchyma cells in many ways. They use water pressure to provide support. However, they are normally found near the surface of stems and leaves. Collenchyma cells have a unique pattern of cell-wall thickening that allows expansion in diameter but not in length. This makes collenchyma especially suited to providing support for soft-bodied plant parts that have completed much of their longitudinal growth. Collenchyma cells rarely provide bulk to plant structures. Instead, they form thin sheets just below the epidermis and outside much of the parenchyma. Because collenchyma is thin, it has a smaller volume than parenchyma and contributes less to the metabolism of the plant organs. It may nevertheless support some of the photosynthesis of the plant, and it provides textures to the organs as well.

Sclerenchyma cells occur throughout the body of the plant and include three types of cells: elongated *fibers*; branched *sclereids*, resembling a three-dimensional jigsaw puzzle piece; and globular *stone cells*. All three cell types have heavy, secondary cell walls and have lost many organelles. Sclerenchyma is a type of differentiated tissue that functions when its cells are dead.

Fibers support plant organs in the same way as does collenchyma, but because the secondary cell walls of sclerenchyma cells resist longitudinal and latitudinal expansion, they are not common in growing tissues. Their rigidity helps to supply support even when tissues are water-stressed, but it also limits the potential for the organs to expand in girth or length. Fibers, sclereids, and stone cells all provide protection against predation. The gritty texture of a ripe pear, the shell of a nut, and the strings of a coconut husk are all composed of sclerenchyma cells and promote the wear and breakage of predators' teeth and other chewing structures.

Procambium

The procambium, the third of the primary developmental tissues, differentiates to form primary *xylem* and primary *phloem* as well as the vascular cambium. The *vascular cambium* produces cells that differentiate into secondary xylem and secondary phloem. It also regenerates the supply of cells in the vascular cambium.

An example of xylem is the woody tissue at the center of most trees. (Palm trees are a notable exception.) Smaller bundles of xylem are found in the roots, stems, and leaves of most plants, even when they are not woody. Xylem tissues are made of four cell types: fibers and parenchyma cells (which also occur in sclerenchyma and parenchyma) and xylem vessel elements and *tracheids* (which are found only in the xylem). These cells work in concert to move water upward through the plant. The xylem vessel elements and tracheids provide the actual channels for the movement, and the fibers serve largely as physical supporting structures.

The parenchyma cells are responsible for some lateral movement in the xylem tissues. These parenchyma cells also have the ability to revert to a meristematic condition, providing a mechanism for the xylem to replace damaged cells. Tracheids and vessels have unusual, patterned secondary cell walls that resist the physical stresses involved in moving xylem sap. The cell organelles are lost before the vessels and tracheids are functional. The sap moves through a channel where the body of the cell had been before it was lost. Xylem is another tissue that contains cell types that function when they are dead.

An example of phloem is the tissue on the inside of the bark of most trees (again, palm trees are an exception). Smaller bundles of phloem are found in the roots, stems, and leaves of most plants, even when they are not woody. Phloem tissues are made of fibers, parenchyma cells, *sieve tube elements*, and *companion cells*. These four cell types work in concert to move sugars, other organic molecules,

Plant Tissues

Tissue Type	*Location in Plant*	*Functions and Characteristics*
Meristems		
Apical meristems	Roots and shoot tips	Site of primary growth; these cells eventually differentiate into the plant's primary tissues: dermal, vascular, and ground.
Lateral meristems	Stems	Secondary growth of vascular and cork cambia, lateral budding and branching limited by apical dominance unless the apical meristem is cut as in pruning.
Dermal tissue (epidermis)	Outer layer	Retention and absorption of water and minerals, protection against herbivores, control of gas exchange. Includes stomata, trichomes, root hairs.
Vascular tissue		
Xylem	Throughout	Conducts water through plant. Contains two kinds of conducting cells: tracheids and vessel elements.
Phloem	Throughout	Transports dissolved organic materials throughout plant. Contains sieve cells, sieve-tube elements, other cells responsible for conducting nutrients and information.
Ground tissue		
Parenchyma cells	Throughout	Food and water storage, sites for metabolism (respiration, photosynthesis), healing.
Chlorenchyma cells	Throughout	Chloroplast-containing parenchyma cells specialized for photosynthesis.
Collenchyma cells	Beneath epidermis	Support growing regions of shoots; common in petioles, elongating stems, expanding leaves.
Sclerenchyma cells	Mature regions	Rigid, producing thick secondary walls; usually dead at maturity. Support and strengthen leaves, stems, roots. Two types: sclereids (short, compact, forming cores, seed coats, other tough, gritty tissue) and fibers (long, slender, occuring in bundles).

and some ions throughout the body of the plant.

The sieve tube elements (or *sieve cells*, in some plants) provide the actual channels for the movement. The fibers serve largely as physical supporting structures. The parenchyma cells are responsible for some lateral movement and also provide a mechanism to replace damaged cells. Sieve tube elements and sieve cells have unusual, perforated cell walls whose appearance indeed resembles a sieve. Many of the cell organelles are lost before the sieve tube elements and sieve cells are functional. The phloem sap moves through living cells, but they resemble no other cells in the plant.

The companion cells (which in some plants are called *albuminous cells*, to indicate a different developmental origin) are similar to parenchyma cells, but they provide substantial metabolic support to the sieve tube elements and sieve cells. These cells function together: The companion cells could live independently, while the sieve tube element could not, but the important function is carried out by the sieve tube element.

Most species that grow in girth are woody, and the wood of woody plants is composed almost entirely of secondary xylem. The bark of woody plants is made of phloem and corky layers. There are two principal cambia, the vascular cambium and the *cork cambium*. Both contribute to the increase in girth. The vascular cambial tissues produce the cells that will differentiate to form the secondary xylem and phloem of woody species. The cork cambium produces the corky cells on the outside of the bark.

Craig R. Landgren

See also: Angiosperm cells and tissues; Flower structure; Growth and growth control; Leaf anatomy; Roots; Shoots; Stems; Wood.

Sources for Further Study

Burgess, Jeremy. *An Introduction to Plant Cell Development*. New York: Cambridge University Press, 1985. A text for the advanced student, with difficult and technical language on occasion. Assumes significant knowledge of botany. Illustrated with both line drawings and photographs.

Dennis, David T., and David H. Turpin, eds. *Plant Physiology, Biochemistry, and Molecular Biology*. New York: John Wiley, 1990. Examines the assimilation and metabolism of carbon and nitrogen in an integrative fashion, assessing the physiology, biochemistry, and molecular biology of each topic discussed, for beginning or advanced students.

Hartmann, Hudson T., Anton M. Kofranek, and Vincent E. Rubatzky. *Plant Science: Growth, Development, and Utilization of Cultivated Plants*. 2d ed. Englewood Cliffs, N.J.: Prentice Hall, 1988. An easily read introductory college botany text which places plant tissues in context with the biology and economical uses of plants. Chapter 2, "Structure of Higher Plants," describes the relationship between plant tissues and cell types and between plant tissues and organs. Heavily illustrated (in black and white) with both line drawings and photographs.

Kaufman, Peter B., et al. *Plants: Their Biology and Importance*. New York: Harper & Row, 1989. Illustrated with black-and-white and color drawings and photographs, this college-level botany text is easily understood. Chapter 2, "Plant Tissues and Their Functions," provides a good review of plant tissues but is not as well illustrated as it might be.

Raven, Peter H., and George B. Johnson. *Biology*. 5th ed. Boston: WCB\McGraw-Hill, 1999. Beautifully illustrated with many color drawings and photographs, this college-level biology text places plant tissues in the context of the biology of plants and places the plants in the context of the living and nonliving world. The chapter "Vascular Plant Structure" provides a simple review of plant cell types, tissues, and organs.

Taiz, Lincoln, and Eduardo Zeiger. 2d ed. *Plant Physiology*. Sunderland, Mass.: Sinauer Associates, 1998. Clear, comprehensive textbook covers the transport and translocation of water and solutes, biochemistry and metabolism, and growth and development.

PLANTAE

Categories: *Plantae*; Taxonomic groups

Life on earth is dependent on the ability of plants to capture the sun's energy. Directly or indirectly, members of the green kingdom, Plantae, *provide food and shelter for nearly all other organisms, including humans. Plants also generate much of the earth's oxygen. The biosphere would not exist without plants.*

Most plants are multicellular, autotrophic organisms, that is, able to produce their own food from inorganic elements by converting water and carbon dioxide to sugar. Plants are sessile, stationed in one spot throughout their lives. Most plants have a complex life cycle called *alternation of generations* between diploid and haploid forms. The diploid generation, in which the plant body is made up of diploid cells, is called the *sporophyte*. The sporophyte produces haploid spores by meiosis.

Phyla of Kingdom *Plantae*

Phylum	*Common Name*	*Living Species*
Bryophytes (nonvascular plants)		
Anthocerotophyta	Hornworts	100
Bryophyta	Mosses	8,000
Hepatophyta	Liverworts	6,000
Seedless vascular plants		
Lycopophyta	Lycopods (Lycophytes)	1,200
Psilotophyta	Whiskferns	2 genera
Pterophyta	Ferns	12,000
Sphenophyta	Horsetails	15
Gymnosperms		
Coniferophyta	Conifers	550
Cycadophyta	Cycads	150-200
Ginkgophyta	Ginkgo	1
Gnetophyta	Gnetophytes	90
Angiosperms		
Anthophyta	Flowering plants	235,000

Source: Data on species are from Peter H. Raven et al., *Biology of Plants*, 6th ed. (New York: W. H. Freeman/Worth, 1999).

These haploid spores then grow mitotically to produce the haploid generation, called the *gametophyte*, which produces haploid gametes. Gametes fuse to form a diploid zygote, a fertilized reproductive cell that marks the beginning of a new sporophyte generation.

Various methods may be used to group members of *Plantae*, which comprises about 300,000 species. Based upon where plants live, they can be divided into terrestrial (land) and aquatic plants. Members of these two groups vary widely in their size, body structure, and level of complexity as a result of their interactions with their environments. Plants that live in water usually lack true roots, stems, and leaves as well as complex reproductive structures, such as flowers. Because they are surrounded by water, aquatic plants also lack the rigid supporting substances required by terrestrial plants. Terrestrial plants are more complex and may be broadly grouped into *vascular* (plants with vessels) and *nonvascular* (plants without vessels). Vascular plants are divided into various phyla, based upon their adaptive features and complexity of structure.

Algae vs. *Plantae*

Those plants known as algae are not members of kingdom *Plantae* but instead are primarily members of another group of eukaryotic life, *Protista*, which also produce their own food by photosynthesis. Algal pigments are red or brown, absorbing the green, violet, and blue lights that most readily penetrate the water. The combination of various pigments with chlorophyll resulted in the distinctive coloration of algae of the three phyla: red algae, brown algae, and green algae.

Within the kingdom *Protista*, three main algal phyla exist. Red algae (phylum *Rhodophyta*) are multicellular with red pigment (phycobilins) masking their green chlorophyll. Brown algae (phylum *Phaeophyta*) are multicellular and include the largest and most complex of the marine algae. Green algae (phylum *Chlorophyta*) are mostly multicellular inhabitants of freshwater environments, with about seven thousand species. The algae are thought to be primitive evolutionary precursors to the species which now make up the kingdom *Plantae*.

Transition from Water to Land

The move from water to land requires adaptations and some new features. The advantages to life on land seem obvious, with plentiful access to carbon dioxide and sunlight, reduced competition, fewer predators, and increased nutrient concentrations. The challenges are also plentiful: The supportive buoyancy of water is missing, and the air tends to dry things out.

Conditions on land favored the evolution of structures that support the plant body, vessels that transport water and nutrients throughout the body, and structures that conserve water. Adaptations to dry land called conducting vessels (collectively known as the vascular system) emerged to transport water and minerals as well as products of photosynthesis. Roots or rootlike structures evolved to help anchor the plant and absorb nutrients and water from the soil. A stiffening substance called lignin, made up of rigid polymer, enables plants to stand in wind, hence exposing maximal surface area to sunlight. Numerous small pores called stomata in the leaves and stems open to allow gas exchange and close to conserve water when necessary. A waxy cuticle covering of the surfaces of leaves and stems also reduces the loss of water.

Based upon their structure, complexity, and distribution over the globe, land plants can be classified into three phyla of *bryophytes* and nine phyla of vascular plants, which include *seedless plants*, *gymnosperms*, and *angiosperms*.

Bryophytes

Three phyla of plants—the liverworts, hornworts, and mosses—have been commonly known as bryophytes. The sixteen thousand species of bryophytes are among the least complex terrestrial plants. They are the plant equivalent of amphibians. Although they have rootlike anchoring structures (rhizoids), they lack true roots, leaves, and stems. They are also nonvascular, lacking well-developed structures for conducting water and nutrients. Because they must rely upon slow diffusion or poorly developed tissues for distribution of water and nutrients, their body size is limited. Most bryophytes are less than 1 inch (2.5 centimeters) tall.

The liverworts, phylum *Hepatophyta*, comprise six thousand species of small, inconspicuous plants forming large colonies in moist, shaded soil or rocks, tree trunks, or branches. Due to the liver-shaped gametophyte in some genera and the fiction that these plants might be useful in treating liver-related diseases, they were named liverworts many centuries ago. Liverworts are the simplest of all living land plants, lacking cuticle, stomata, and vascular tissue.

The hornworts, phylum *Anthocerophyta*, are a small phylum of plants consisting of about one hundred species. By appearance, many species of hornworts have a remarkable resemblance to green algae. Hornworts, however, have stomata, an important structure for land plants. Like liverworts, hornworts lack specialized conducting tissue.

The mosses, phylum *Bryophyta*, constitute a diverse group of some ninety-five hundred species of small plants. Many species have both stomata and specialized conducting tissue, resembling the remaining phyla of land plants. Mosses usually thrive in relatively moist areas, where a variety of species can be found. Some mosses are used to monitor air pollution because of their acute sensitivity to air pollutants such as sulfur dioxide. There are three classes of mosses: *Bryidae* (the "true" mosses), *Sphagnidae* (the peat mosses), and *Andreaeidae* (the granite mosses), each with distinctive features.

Seedless Vascular Plants

The overall pattern of plant diversification may be explained in terms of the successive rise to dominance of each of four major plant groups. Early vascular plants, including *Rhyniophyta*, *Zosterophyllophyta*, and *Trimerophyta*, were primitive in morphology yet dominant during a period from about 420 million to 370 million years ago. Ferns, lycophytes, sphenophytes, and progymnosperms were dominant from about 380 million to 290 million years ago. Seed plants arose about 360 million years ago, with gymnosperms dominating the globe until 100 million years ago. Finally, the angiosperms, or flowering plants (phylum *Anthophyta*) appeared about 127 million years ago and have been the most dominant and diverse group for the past 100 million years.

The seedless vascular plants dominated the landscape during the Carboniferous period. They are the primary source of coal, which was formed through gradual transformation of plant bodies under high pressure and heat. The three dominant phyla—the *Rhyniophyta*, *Zosterophyllophyta*, and *Trimerophyta*—had become extinct by about 360 million years ago. The modern representatives of seedless vascular plants have become reduced in size and importance. The landscape once dominated by seedless plants has largely been replaced by the more versatile seed plants. Five phyla of seedless plants have living representatives: *Psilotophyta*, *Lycophyta*, *Sphenophyta*, *Pterophyta*, and *Progymnospermophyta*.

Psilotophytes

Commonly called the whiskferns, the phylum *Psilotophyta* includes two living genera, *Psilotum* and *Tmesipteris*. Both are very simple plants. *Psilotum* is widely known as a greenhouse weed that prefers tropical and subtropical habitats. In the United States, it is found in Arizona, Florida, Hawaii, Louisiana, Puerto Rico, and Texas. *Tmesipteris* is restricted to South Pacific regions such as Australia, New Caledonia, and New Zealand.

Psilotum is unique among vascular plants in that it lacks both true roots and leaves. The underground portion of *Psilotum* forms a system of rhizomes with many rhizoids that are a result of a symbiotic relationship between fungi and *Psilotum*. *Psilotum* is homosporous, meaning the male and female gametophytes are produced through the germination of the spores of same origin. *Psilotum* sperm require water to swim to and fertilize the egg.

Tmesipteris grows as an epiphyte on tree ferns and other plants and in rock crevices. The leaflike appendages of *Tmesipteris* are larger than those of *Psilotum*. Otherwise, *Tmesipteris* is very similar to *Psilotum*.

Lycophytes

There are about one thousand living species of phylum *Lycophyta* that belong to three orders of ten to fifteen genera. At least three orders of *Lycophyta* have become extinct; these include small and large trees, the dominant plants of the coal-forming forest of the Carboniferous period. The three orders of living *Lycophyta* consist of herbs, each including a single family: *Lycopodiaceae*, *Selaginellaceae*, and *Isoetaceae*.

Lycopodiaceae are commonly known as club mosses. All except two genera of *Lycophyta* belong to this family. Most of the estimated four hundred species of *Lycopodiaceae* are tropical. They rarely form conspicuous elements in any plant community, except in some temperate forests where several species may form distinct mats on the forest floor. Because they are evergreen, they are most noticeable during winter months. *Lycopodiaceae* are homosporous and require water for fertilization. Among the genera that grow in the United States and Canada are *Huperzia* (the fir mosses, seven species), *Lycopodium* (tree club mosses, five species), *Diphasiastrum* (club mosses and ground pines, eleven species), and *Lycopodiella* (six species).

Selaginella is the only living genus of the family *Selaginellaceae*. Among its seven hundred species, most are tropical, growing in moist habitats. A few species occur in deserts, becoming dormant during the driest season. The well-known resurrection plant, *S. lepidophylla*, grows in Mexico, New Mexico, and Texas. *Selaginella* are heterosporous, having separate male and female gametophytes.

The only member of the family *Isoetaceae* is *Isoetes*, commonly known as quillwort. *Isoetes* may be aquatic or grow in pools that have alternate dry and wet seasons. *Isoetes* is also heterosporous. The unique feature of some species of *Isoetes* is their ability to acquire carbon dioxide for photosynthesis from soil where they grow rather than from the atmosphere. Their leaves have thick cuticles and lack stomata.

Horsetails

Only one genus, *Equisetum*, consisting of fifteen living species, makes up the phylum *Sphenophyta*. Members of this phylum are known as the horsetails, believed to be the oldest surviving plants on earth. They are widely distributed in moist or damp places, by streams, and along the edges of woods and roadsides. The leaves are reduced to tiny scales on the branches. The ribs are tough and strengthened with silicon deposits. Due to its abrasive texture, *Equisetum* was used in past times to scour pots, pans, and floors. Hence, they were also called "scouring rushes."

The roots of *Equisetum* are adventitious, emerging at the nodes of the rhizomes. The aerial stems of *Equisetum* arise from branching underground rhizomes. Although the plants may die back during the dry season, the rhizomes are perennial. *Equisetum* is homosporous. Its gametophytes are green and free-living, most being about the size of a pinhead. They become established mainly in mud that has recently been flooded and is rich in nutrients. The gametophytes, which reach sexual maturity in three to five weeks, are either bisexual or male. The sperm require water to swim to the eggs. A fertilized egg develops into an embryo or young sporophyte.

Pterophyta

Members of the phylum *Pterophyta* are commonly called ferns, the largest group of plants other than the flowering plants. About eleven thousand living species of ferns are widely distributed on the earth, among which three-fourths of the species are found in the tropics. There the greatest diversity of fern species exists, and they are abundant in many plant communities. In the small tropical country of Costa Rica, 1,000 species of ferns have been identified, whereas only 380 species of fern occur in the United States and Canada combined. In both form and habitat, ferns exhibit amazing diversity. Some are small and have undivided leaves, while others can reach up to 30 meters in height, with a trunk of more than 30 centimeters in diameter. Most living ferns are homosporous, except for two orders of water ferns.

Two genera of the order *Ophioglossales*, the grape ferns (*Botrychium*), and the adder's-tongues (*Ophioglossum*), are widespread in the north temperate region. A single leaf is usually produced each year from the rhizome. Each leaf consists of two parts:

the blade (the vegetative portion) and a fertile segment that typically bears two rows of eusporangia, hence the name eusporangiate ferns.

The order *Filicales* consists of 35 families and 320 genera, with more than 10,500 species. Most the familiar ferns are members of this order, such as the garden and woodland ferns of temperate regions. They have rhizomes that produce new sets of leaves each year. The root system is primarily adventitious, arising from the rhizomes near the bases of the leaves. They are homosporous. With a high surface-to-volume ratio, their bodies capture sunlight much more effectively than those of the lycophytes.

The water ferns, *Marsileales* and *Salviniales*, are the only living heterosporous ferns. Members of *Marsileales* grow in mud, on damp soil, or often with their four-leaf-clover-like leaves floating on the surface of water. Their unique, drought-resistant, bean-shaped reproductive structures are able to germinate even after one hundred years of dry storage. Members of *Salviniales*, genera *Azolla* and *Salvinia*, are small plants that float on the surface of water. They are harvested and used as feed or fertilizer in some Asian countries.

Gymnosperms

Gymnosperms, literally "naked seeds," are the earliest-evolved plants that produce seeds. Living gymnosperms comprise four phyla: *Cycadophyta*, *Ginkgophyta*, *Coniferophyta*, and *Gnetophyta*. The life cycles of all bear a remarkable resemblance: an alteration of heteromorphic generations, with large, independent sporophytes and greatly reduced gametophytes. The ovules are exposed on the surfaces of the megasporophylls. At maturity the female gametophyte of most gymnosperms is multicellular in structure, with several archegonia. The male gametophytes develop as pollen grains. Except for the ginkgo and cycads, the sperm cells of seed plants are nonmotile.

In seed plants, water is not required for transfer of sperm or fertilization of an egg. Pollen grains that usually contain two sperm nuclei may be transferred to the egg via various means, such as wind,

Digital Stock

The smallest of the modern flowering plants, or angiosperms, is duckweed, a few millimeters in diameter, which floats on ponds.

insects, and animals. A pollen grain then germinates and sends one nucleus to fuse with the egg, which in turn develops into embryo. After fertilization, each ovule develops into a seed.

Among the four phyla, the conifers (*Coniferophyta*) are the largest and most widespread gymnosperms, with about 50 genera and 550 species. They still dominate many of the earth's plant communities, with pines, firs, spruces, and other familiar evergreen trees over wide stretches of the north. Living cycads (*Cycadophyta*) constitute 11 genera, and some 140 species grow primarily in tropical and warm regions. Cycads are palmlike plants, with trunks and sluggish secondary growth. Only one living species of ginkgo (*Ginkgophyta*) exists. The phylum *Gnetophyta* consists of three genera that are close relatives of the angiosperms, with which they share many characteristics.

Angiosperms

As plants adapted to terrestrial environments, more effective means of reproduction and distribution emerged. Flowers, fruits, and seeds resulted in the dominance by *angiosperms*: flowering plants, the most diverse group of plants. Modern flowering plants, or angiosperms, constitute the phylum *Anthophyta*, which are incredibly diverse, with more than 230,000 species. They range in size from a few millimeters in diameter, such as the duckweed that floats on ponds, to more than 320 feet (about 100 meters) tall, such as the eucalyptus. From desert cacti to tropical orchids to grasses to major food crops, angiosperms dominate the plant kingdom.

Anthophyta is divided into two large classes: *Monocotyledons* (65,000 species) and *Eudicotyledones*, or "true" dicots (170,000 species).

In addition to some features shared with gymnosperms, angiosperms have a few unique characteristics. Within seeds, nutrients and food are usually stored in a triploid tissue called an *endosperm*. The presence of *carpels* makes flowers shiny and more attractive for pollinators, enhancing their reproductive success. The nutritious fruit-encasing seeds offer protection and ensure the wide distribution of angiosperms as various animals eat fruits and disperse the seeds throughout the ecosystem. Angiosperms have broad leaves, giving them the advantage of collecting more sunlight for photosynthesis, especially in warm, moist climates. The extra energy gained in spring and summer allows trees to drop their leaves and enter a dormant period, which reduces water evaporation when water is in short supply.

Pollination in angiosperms takes place by the transfer of pollen from anther to stigma. Each pollen grain contains sperm, typically with two or three cells. One cell grows, sending pollen tubes to the ovule, where one sperm nucleus unites with the egg and produces a diploid zygote. The other cell or cells fuse with two polar nuclei, giving rise to primary endosperm nucleus. This phenomenon, called *double fertilization*, is a unique characteristic for angiosperms. The zygote then develops into an embryo (sporophyte). The primary endosperm grows and matures into a nutritive endosperm. Both *self-pollination* and *cross-pollination* occur in flowering plants. The angiosperm domination of earth began about 100 million years ago and has persisted to the present.

Ming Y. Zheng

See also: Angiosperms; Angiosperm evolution; Aquatic plants; Bryophytes; Conifers; Cycads and palms; Eudicots; Evolution of plants; Ferns; Ginkgos; Gnetophytes; Gymnosperms; Hornworts; Horsetails; Liverworts; Lycophytes; Monocots vs. dicots; *Monocotyledones*; Mosses; Plant life spans; Psilotophytes; Reproduction in plants; *Rhyniophyta*; Seedless vascular plants; *Spermatophyta*; *Tracheobionta*; *Trimerophytophyta*; *Zosterophyllophyta*.

Sources for Further Study

Beck, C. B., ed. *Origin and Evolution of Gymnosperms*. New York: Columbia University Press, 1988. An excellent collection of essays on every aspect of gymnosperms and their evolution.

Buckles, M. P. *The Flowers Around Us: A Photographic Essay on Their Reproductive Structures*. Columbia: University of Missouri Press, 1985. Description and discussion on the diversity of flowers and fruits. Well illustrated with diagrams and photos.

Gifford, E. M., and A. S. Foster. *Morphology and Evolution of Vascular Plants*. 3d ed. New York: W. H. Freeman, 1989. A well-written, general discussion on vascular plants. Illustrated by photos.

Schofield, W. B. *Introduction to Bryology*. New York: Macmillan, 1985. An excellent analysis of all three phyla of mosses, dealing with their morphology, physiology, ecology, and geography.

South, G. R., and A. Whittick. *Introduction to Phycology*. Oxford, England: Blackwell Scientific, 1987. Excellent account of algae, well written and illustrated.

PLANTS WITH POTENTIAL

Categories: Agriculture; economic botany and plant uses

One of the primary reasons humans cultivate plants is to satisfy an economic need for natural resources. In order for a plant to realize its full economic potential, it must not only fill an economic need but also do so in a cost-efficient manner.

There are several examples of plants which, because of their unique products, appear to fulfill an economic need. However, these plants may not do so in a cost-effective manner. With development of improved agronomic practices and plant-processing methods, which lower the cost of production, some plants may eventually realize their full economic potential.

Another factor that may affect a plant's economic potential includes the availability and price of competing products on the world market. This factor is beyond the control of domestic agronomists. Drastic changes in world conditions, such as in times of natural disaster, severe economic recession, or wartime, can have a dramatic impact on the economic feasibility of natural resources. Crops with economic potential may move in and out of economically favorable conditions as world markets change and new markets develop.

Guayule

Guayule (*Parthenium argentatum*) is a shrubby member of the *Compositae* family that is native to the desert regions of the southwestern United States and northern Mexico. Other species of this genus are found in all regions of the Americas. Guayule is one of more than two thousand plant species that can potentially be used to produce latex and rubber. Only guayule and its chief competitor, *Hevea brasiliensis*, have been used to produce commercial quantities of rubber.

Although commercial production of guayule-derived rubber dates back to the 1920's, when Continental Rubber Company produced small quantities of latex and rubber from guayule plants grown in Arizona and California, it was not until the Emergency Rubber Project during World War II that large-scale production of guayule rubber commenced. With the U.S. economy on a wartime footing and with supplies of imported Hevea rubber becoming uncertain, U.S. Code Title 7, Section 171, authorized the U.S. Department of Agriculture to acquire the technology of the Continental Rubber Company, plant up to 500,000 acres of guayule, and develop factories for the production of guayule-based rubber. With the end of World War II and reestablishment of the Hevea rubber supply from Asia, guayule-based rubber was no longer economically competitive. The unfavorable price difference between guayule-based rubber and Hevea-based rubber has remained unchanged, even though many agronomic improvements have been made for guayule.

However, new life may be developing for guayule-based rubber in a large niche market: medical products. A method for producing hypoallergenic latex derived from guayule has been developed and patented by the U.S. Department of Agriculture. A private company, Yulex Corporation, has licensed this technology and intends to use it to manufacture medical items, such as surgical gloves. The gloves produced from guayule-based latex do not contain the allergenic proteins that Hevea-based latex contains, so they will not cause allergic reactions in the estimated twenty million Americans who are allergic to Hevea rubber products. In this case, the cost disparity between Hevea rubber and guayule rubber is offset by the technical

*Guayule (*Parthenium argentatum*) is a shrub native to the desert regions of the southwestern United States and northern Mexico. The plant is one of more than two thousand species that can potentially be used to produce latex and rubber. Here, a worker in the U.S. Department of Agriculture inspects guayule plantlet tissue cultures.*

improvement that guayule latex brings to the high-value market for medical devices.

Jojoba

Jojoba (*Simmondsia chinensis*) is a woody, evergreen desert shrub that, despite its misleading species epitaph, *chinensis*, is native to the southwestern United States. In cultivation, jojoba may be irrigated during its two- to three-year establishment period after which, assuming that the roots find groundwater, the plants do not require irrigation. During the plant's initial production period of three to ten years, the female plants may produce 350 kilograms of seeds per hectare. After ten years of growth, the plants may yield 500 to 800 kilograms per hectare for many decades.

Jojoba oil is extracted from jojoba seeds and comprises approximately 40 to 60 percent of the mass of the seeds. Jojoba oil is not really an oil per se, as it is not a triglyceride; jojoba oil is a plant wax similar to plant cuticular waxes, being composed of long-chain alcohols and fatty acids. The value of jojoba oil comes from its desirable stability. Jojoba oil is stable up to 300 degrees Celsius and does not become rancid even after decades of storage. Also, jojoba oil is very similar chemically to the highly prized sperm whale oil, so it is useful in cosmetics.

Jojoba oil was first produced in commercially important quantities during World War II, as a high-temperature lubricant and an extender for petroleum-based lubricants. These jojoba-based products were used for engines, machinery, vehicles, and guns. As seen with guayule, the economics of jojoba oil production did not compare favorably with abundant petroleum products after World War II, so production of jojoba oil decreased sharply after the war.

In recent decades, the economics of jojoba oil production has gone through many fluctuations. In the 1970's the Green Revolution reignited interest in renewable, natural resources, especially products that could replace petrochemicals and animal-derived products. Growers took advantage of tax incentives to start farming jojoba. Then the disappearance of tax incentives and the decade-long production time to achieve commercially useful quantities of jojoba seed proved to be economically disastrous for many growers. Many jojoba farms shut down operations. In the 1990's the price of jojoba oil ranged from $40 per gallon to $200 per gallon, an unacceptable fluctuation in price that discouraged many industries from becoming dependent on jojoba oil. The price of jojoba oil must stabilize before industries can once again explore adding jojoba to their lines. Additionally, some unique, value-added products containing chemically modified jojoba oil are becoming common in many upscale health and beauty products.

Hesperaloe

Hesperaloe (*Hesperaloe funifera*) is a plant in the agave family that produces long, thin fibers that may be processed into exceptionally light and strong paper. Long-term biomass production studies began on hesperaloe in 1988, and since then many agronomic and processing improvements have been made. The fibers of hesperaloe seem suitable for the production of high-value products, such as ultralight coated papers.

Kenaf

Kenaf (*Hibiscus cannabinus*) is another fast-growing fiber crop that is finding utility in niche markets. Kenaf may be used to produce bright white paper, building materials, and absorbent materials. Additionally, the black lignin liquor, a by-product of kenaf processing, may add value to the crop by functioning as a binder for animal feeds, a fertilizer, or a termite-resistant coating.

Lesquerella

Lesquerella (*Lesquerella fendleri*) is an industrial crop under development for its unique seed oils. Initial research indicates that domestically produced lesquerella may eventually replace imported castor oil in many cosmetics, pharmaceuticals, and industrial products. Enhanced agronomic techniques, processing methods, conventional breeding, genetic engineering, and emerging niche markets may one day push lesquerella and other plants with potential into the realm of commercial viability.

Robert A. Sinnott

See also: Agricultural crops: experimental; Culturally significant plants; Green Revolution; Medicinal plants; Rubber.

Sources for Further Study

Ricketts, Cliff, and Omri Rawlins. *Introduction to Agribusiness*. Albany, N.Y.: Delmar, 1999. Introduction to economic aspects of agricultural industries in the United States.

Simpson, Beryl, and Molly Ogorzaly. *Economic Botany: Plants in Our World*. 3d ed. New York: McGraw-Hill, 2000. Introductory-level text for economic botany.

Wiersema, John H., and Leon Blanca. *World Economic Plants: A Standard Reference*. Boca Raton, Fla.: CRC Press, 1999. A comprehensive review of ten thousand vascular plants, relating them to the global economy.

PLASMA MEMBRANES

Categories: Anatomy; cellular biology

The plasma membrane is a structure of the plant cell that forms a semipermeable, or selective, barrier between the interior of the cell and the external environment; they also function in transport of molecules into and out of the cell.

In addition to forming the structural barrier between the internal contents of a cell and the external environment, plasma membranes contain proteins involved in the transport of molecules and other substances into and out of the cell, and they contain proteins and other molecules that are essential for receiving signals from the environment and from plant hormones that direct growth and division. Carbohydrates associated with the plasma membrane are markers of cell type. In plants, the plasma membrane is the site of cellulose synthesis.

Lipid molecules provide the structure for the plasma membrane, which is described by the *fluid mosaic model* as a dynamic ocean of lipids in which other molecules float. *Phospholipids* are the most abundant lipid of plasma membranes, and they are organized in a fluid phospholipid bilayer in which

Mechanisms for Transport Across Cell Membranes

Type of Transport	*Mechanisms and Functions*	
Diffusion	Passive movement of substances into and out of the cell.	
	Simple diffusion	Dissolved gases (oxygen, carbon dioxide) and lipid-soluble substances move down the concentration gradient through the lipid portion of the membrane.
	Facilitated diffusion	Specialized membrane proteins allow substances with low lipid-solubility to move down the concentration gradient by binding to carrier (transport) proteins.
	Osmosis	Net movement of water molecules down the concentration gradient across the membrane, driven by the difference in solute concentrations on either side.
Active transport	Movement of substances against the concentration gradient ("uphill"), requiring an expenditure of cellular energy, usually from adenosine triphosphate.	
	Cotransport (symport)	Active transport involving a specialized protein molecule called a symport carrier.
	Countertransport (antiport)	Active transport that uses a specialized protein molecule called an antiport carrier.
Endocytosis (vesicle-mediated)	Movement of materials into the cell: The cell's plasma membrane engulfs material and packs it into saclike vesicles that detach from the plasma membrane and move into the cell.	
	Phagocytosis	"Cell eating": cellular intake of large and generally insoluble molecules and macromolecules (and little fluid) that cannot be taken into the cell using other membrane transport mechanisms.
	Pinocytosis	"Cell drinking": cellular intake of fluids and dissolved solutes by the formation of membrane-bound vesicles at the cell membrane surface, which are then taken into the cell interior and released.
	Receptor-mediated	Cellular intake of nutrients and other essential macromolecules at specific receptor sites. Receptor-ligand complexes combine to form clusters around which a portion of the cell membrane encircles, producing a vesicle that invaginates inward to pinch off as a coated vesicle. Once inside, the vesicle dissolves as a result of changes in the acidity of the cytoplasm.
Exocytosis (vesicle-mediated)	Transport of macromolecules and large particles outside the cell, the reverse of endocytosis. Materials inside the cell are packed in a vesicle, which fuses to the plasma membrane and expels its contents into the extracellular medium.	

sterols, proteins, and other molecules are interspersed. Phospholipids are amphipathic molecules, containing water-loving (hydrophilic) regions and water-fearing (hydrophobic) regions.

Each phospholipid consists of a three-carbon glycerol backbone; two of the carbons are attached to long-chain fatty acid molecules, and the third carbon is attached to a phosphate-containing group. Because the fatty acids are nonpolar and hydrophobic, they tend to aggregate and exclude water. This aggregation allows the phospholipids to form a bilayer structure that has the fatty acids of both layers in the middle and the charged, phosphate-containing groups toward the outside. This bilayer structure allows one surface of the plasma membrane bilayer to interact with the aqueous external environment, while the other interacts with the aqueous internal cellular environment.

Sterols are also found within the plasma membranes of plant cells. The major sterol found in plant cell plasma membranes is stigmasterol (as opposed to cholesterol, which is found in animal cell plasma membranes). Sterols found in plant cells are important economically as the starting material for steroid-based drugs such as birth control pills.

Membrane Proteins and Carbohydrates

Some membrane proteins span the entire length of the phospholipid bilayer and are called *transmembrane proteins*. Transmembrane proteins are sometimes referred to as integral membrane proteins and have varied structures and functions. They may pass through the lipid bilayer only once, or they may be "multiple pass" transmembrane proteins, weaving into and out of the membrane many times.

The portion of a transmembrane protein that passes through the interior of the membrane often consists of amino acids that have nonpolar side chains (R-groups) and is known as the transmembrane domain. The portion of the transmembrane protein that is on the external surface of the membrane and interacts with the aqueous environment often contains charged, or polar, amino acids in its sequence.

Membrane proteins are often important for receiving signals from the external environment as membrane receptors. For instance, protein or peptide hormones interact with transmembrane protein receptors on the plasma membrane. Membrane proteins are also involved in receiving signals such as light photons. Membrane proteins form pores that allow ions (charged particles) to pass through the interior of the membrane. Membrane proteins called carriers are essential for bringing nutrient molecules such as simple sugars into the cell.

Not all proteins within the membrane are transmembrane proteins. Some are only loosely associated with the membrane, attached to other proteins, or anchored in the membrane by a lipid tail. These proteins, which do not span both sides of the membrane, are often called *peripheral membrane proteins*.

In addition to proteins, the plasma membrane contains carbohydrate molecules. Carbohydrate molecules are usually attached to membrane proteins or to lipid molecules within the bilayer. Carbohydrates provide important information about cell type and identity.

Transport Across Membranes

Transport of molecules into and out of the cells is an important function of the plasma membrane. Hydrophobic molecules, such as oxygen, and small, uncharged molecules, such as carbon dioxide, cross the membrane by *simple diffusion*. These molecules use the potential energy of a chemical gradient to drive their movement from an area of higher concentration on one side of the membrane to an area of lower concentration on the other side. Diffusion works best when this concentration gradient is steep. For example, in cells that do not have the ability to carry out photosynthesis, oxygen is used almost as quickly as it enters the cell. This maintains a sharp gradient of oxygen molecules across the membrane, so that molecules continually flow from the area of greater oxygen concentration outside the cell to the area of lower concentration inside the cell.

Molecules that are polar are excluded from the hydrophobic area of the bilayer. Two factors influence the transport of these kinds of molecules: the concentration gradient and the electrical gradient. Lipid bilayers separate differences in electrical charge from one side of the membrane to the other, acting as a kind of biological capacitor. If the inside of the cell is more negative than the outside of the cell, negatively charged ions would have to move from the inside to the outside of the cell to travel with the electrical gradient. The combination of the concentration and electrical gradients is called the *electrochemical gradient*.

Transport of charged or polar molecules requires the assistance of proteins within the membrane, known as transporters. Channel proteins form pores within the membrane and allow small, charged molecules, usually inorganic ions, to flow across the membrane from one side to the other. If the direction of travel of the ion is down its electrochemical gradient, the process does not require additional energy and is called *passive transport*. Carrier proteins change shape to deposit a small molecule, such as a sugar, from one side of a membrane to the other. Pumps are proteins within the membrane that use energy from adenosine triphosphate (ATP) or light to transport molecules across the membrane. When energy is used in transport, the process is called *active transport*.

Cellulose Biosynthesis

In plants, the plasma membrane is the site for the synthesis of cellulose, the most abundant biopolymer on earth. Electron microscope studies suggest that the plant cell membrane contains rosette structures that are complexes of many proteins and are the sites of cellulose synthesis. Studies in bacteria, cotton plants, and the weed *Arabidoposis thaliana* have allowed scientists to isolate the gene that actually carries out the chemical reactions linking glucose molecules together into the long cellulose microfibril structure. This gene encodes a protein called glycosyl transferase. Antibodies against the catalytic, or active, subunit of glycosyl transferase specifically label these rosette structures. Two of these transferase molecules act simultaneously from opposite sides to add two glucoses at a time to the growing microfibril, accounting for the rotation of alternating glucoses in cellulose molecules.

Michele Arduengo

See also: Active transport; ATP and other energetic molecules; Carbohydrates; Cell wall; Cells and diffusion; Endocytosis and exocytosis; Lipids; Liquid transport systems; Membrane structure; Osmosis, simple diffusion, and facilitated diffusion; Plant cells: molecular level; Proteins and amino acids.

Sources for Further Study

Carpita, Nicholas, and Claudia Vergar. "Enhanced: A Recipe for Cellulose." *Science* 279 (1998): 672. A short review of a research article describing cellulose biosynthesis. Includes useful graphics with review.

Lodish, Harvey, et al. *Molecular Cell Biology*. 4th ed. New York: W. H. Freeman, 2000. This is one of the most authoritative cell biology reference texts available. Provides excellent graphics, CD-ROM animations of cellular processes, and primary literature references.

Raven, Peter H., Ray F. Evert, and Susan E. Eichhorn. *Biology of Plants*. 6th ed. New York: W. H. Freeman/Worth, 1999. Introductory botany text provides a chapter introducing eukaryotic cells in general and plant cells specifically.

PLASMODIAL SLIME MOLDS

Categories: Molds; *Protista*; taxonomic groups

The plasmodial slime molds, or myxomycetes, phylum Myxomycota, *are a group of funguslike organisms usually present and sometimes abundant in terrestrial ecosystems. However, this group comprises about eight hundred species, related neither to cellular slime molds nor to fungi.*

The plasmodial slime molds have no cell walls and exist as thin masses of protoplasm, which appear to be streaming in a fanlike shape, under favorable conditions. As these masses, called *plasmodia*, travel, they absorb small particles of decaying plant and animal matter as well as bacteria, fungi, and yeasts. When mature, a plasmodium may weigh 20-30 grams and take up an area of 1 meter or more.

Myxomycetes have been known from their *fruit-*

ing bodies (often a sporangium sprouting from a small mound that forms when the plasmodium stops moving) since at least the middle of the seventeenth century. Their life cycle has been understood for more than a century. The reproductive, spore-producing stage in the life cycle can achieve macroscopic dimensions and be collected and preserved for study in much the same way as mushrooms and other fungi or even specimens of bryophytes, lichens, and vascular plants. However, most species of myxomycetes tend to be rather inconspicuous or sporadic in their occurrence and thus not always easy to detect in the field. Moreover, fruiting bodies of most species are relatively ephemeral and do not persist in nature for very long. Myxomycetes also spend a portion of their life cycle as true eukaryotic microorganisms, when their very presence in a given habitat can be exceedingly difficult, if not impossible, to determine.

Life Cycle

The life cycle of a myxomycete involves two very different trophic (or feeding) stages, one consisting of uninucleate (single-nucleus) *amoeboid cells*, with or without flagella, and the other consisting of a distinctive multinucleate structure, the *plasmodium*. Under favorable conditions, the plasmodium gives rise to one or more fruiting bodies containing spores. The fruiting bodies produced by myxomycetes are somewhat suggestive of those produced by some fungi, although they are considerably smaller (usually no more than 1-2 millimeters tall). The spores of myxomycetes are for most species apparently wind-dispersed and complete the life cycle by germinating to produce the uninucleate amoeboid cells. These cells feed and divide by binary fission to build up large populations in the various habitats in which these organisms occur.

The transformation from one trophic stage to the other in the myxomycete life cycle is in most cases the result of fusion between compatible amoeboid cells, which thus function as gametes. The fusion of the two cells produces a diploid zygote that feeds, grows, and undergoes repeated mitotic nuclear divisions to develop into the plasmodium. Bacteria represent the primary food resource for both trophic stages, but plasmodia are also known to feed upon yeasts, cyanobacteria, and fungal spores.

Myxomycete plasmodia usually occur in situations in which they are relatively inconspicuous, but careful examination of the inner surface of dead bark on a fallen log or the lower surface of a piece of coarse woody debris on the ground in a forest, especially after a period of rainy weather, often will turn up an example or two. Most of the plasmodia encountered in nature are relatively small, but some species are capable of producing a plasmodium that can reach a size of more than 1 meter across. Under adverse conditions, such as drying out of the immediate environment or low temperatures, a plasmodium may convert into a hardened, resistant structure called a *sclerotium*, which is capable of reforming the plasmodium upon the return of favorable conditions. Moreover, the amoeboid cells can undergo a reversible transformation to dormant structures called *microcysts*. Both sclerotia and microcysts can remain viable for long periods of time and are probably very important in the continued survival of myxomycetes in some habitats, such as deserts.

Structure of Fruiting Bodies

Identification of myxomycetes is based almost entirely upon features of the fruiting bodies produced by these organisms. Fruiting bodies (also sometimes referred to as "sporophores" or "sporocarps") occur in four generally distinguishable forms or types, although there are a number of species that regularly produce what appears to be a combination of two types. The most common type of fruiting body is the *sporangium*, which may be sessile or stalked, with wide variations in color and shape. The actual spore-containing part of the sporangium (as opposed to the entire structure, which also includes a stalk in those forms characterized by this feature) is referred to as a *sporotheca*. Sporangia usually occur in groups, because they are derived from separate portions of the same plasmodium.

A second type of fruiting body, an *aethalium*, is a cushion-shaped, sessile structure. Aethalia are presumed to be masses of completely fused sporangia and are relatively large, sometimes exceeding several centimeters in extent. A third type is the *pseudoaethalium* (literally, a false aethalium). This type of fruiting body, which is comparatively uncommon, is composed of sporangia closely crowded together. Pseudoaethalia are usually sessile, although a few examples are stalked. The fourth type of fruiting body is called a *plasmodiocarp*. Almost always sessile, plasmodiocarps take the form of the main veins of the plasmodium from which they were derived.

A typical fruiting body consists of as many as six major parts: hypothallus, stalk, columella, peridium, capillitium, and spores. Not all of these parts are present in all types of fruiting bodies. The *hypothallus* is a remnant of the plasmodium sometimes found at the base of a fruiting body. The *stalk* (also called a stipe) is the structure that lifts the sporotheca above the substrate. As already noted, some fruiting bodies are sessile and thus lack a stalk. The *peridium* is a covering over the outside of the sporotheca that encloses the actual mass of spores. It may or may not be evident in a mature fruiting body. The peridium may split open along clearly discernible lines of dehiscence, as a preformed lid, or in an irregular pattern. In an aethalium, the relatively thick covering over the spore mass is referred to as a *cortex* rather than a peridium. The *columella* is an extension of the stalk into the sporotheca, although it may not resemble the stalk.

The *capillitium* consists of threadlike elements within the spore mass of a fruiting body. Many species of myxomycetes have a capillitium, either as a single connected network or as many free elements called *elaters*. The elements of the capillitium may be smooth, sculptured, or spiny, or they may appear to consist of several interwoven strands. Some elements may be elastic, allowing for expansion when the peridium opens, while other types are hygroscopic and capable of dispersing spores by a twisting motion. Spores of myxomycetes are quite small and range in size from slightly less than 5 to occasionally more than 15 micrometers. Nearly all of them appear to be round, and most are ornamented to some degree. Spore size and also color are very important in identification. Spores can be dark or light to brightly colored.

Occurrence in Nature

There are approximately eight hundred recognized species of myxomycetes, and these have been placed in six different taxonomic orders: *Ceratiomyxales*, *Echinosteliales*, *Liceales*, *Physarales*, *Stemonitales*, and *Trichiales*. However, members of the *Ceratiomyxales* are distinctly different from members of the other orders, and many modern biologists have removed these organisms from the myxomycetes and reassigned them to another group of slime molds, the *protostelids*. The majority of species of myxomycetes are probably cosmopolitan, and at least some species apparently occur in any terrestrial ecosystem with plants (and thus plant detritus) present. However, a few species do appear to be confined to the tropics or subtropics, and others have been collected only in temperate regions. Compared to most other organisms, myxomycetes show very little evidence of endemism, with the same species likely to be encountered in any habitat on earth where the environmental conditions suitable for its growth and development apparently exist.

Although the ability of a plasmodium to migrate some distance from the substrate upon or within which it developed has the potential of obscuring myxomycete-substrate relationships, fruiting bodies of particular species of myxomycetes tend to be rather consistently associated with certain types of substrates. For example, some species almost always occur on decaying wood or bark, whereas others are more often found on dead leaves and other plant debris and only rarely occur on wood or bark. In addition to these substrates, myxomycetes also are known to occur on the bark surface of living trees, on the dung of herbivorous animals, in soil, and on aerial portions of dead but still-standing herbaceous plants. The myxomycetes associated with decaying wood are the best known, because the species typically occurring on this substrate tend to be among those characteristically producing fruiting bodies of sufficient size to be detected in the field. Many of the more common and widely known myxomycete taxa, including various species of *Arcyria*, *Lycogala*, *Stemonitis*, and *Trichia*, are predominantly associated with decaying wood.

Steven L. Stephenson

See also: Bacteria; *Protista*.

Sources for Further Study

Farr, Marie L. *How to Know the True Slime Molds*. Dubuque, Iowa: Wm. C. Brown, 1981. A good general introduction to myxomycetes, with illustrated keys to many of the species found in North America.

Ing, Bruce. *The Myxomycetes of Britain and Ireland*. Slough, England: Richmond, 1999. A largely taxonomic treatment of the species known from Britain and Ireland but with information on ecology and the techniques used in collecting and studying myxomycetes.

Katsaros, Peter. *Illustrated Guide to Common Slime Molds*. Eureka, Calif.: Mad River Press, 1989. Essentially a collection of photographs of some of the more common species, with some general information on structure and ecology.
Keller, Harold W., and Karl L. Braun. *Myxomycetes of Ohio: Their Systematics, Biology, and Use in Teaching*. Columbus: Ohio Biological Survey, 1999. A primarily taxonomic account of the species known to occur in Ohio.
Martin, George W., and Constantine J. Alexopoulos. *The Myxomycetes*. Iowa City: University of Iowa Press, 1969. The standard monograph on myxomycete taxonomy, now out of print.
Stephenson, Steven L., and Henry Stempen. *Myxomycetes: A Handbook of Slime Molds*. Portland, Oreg.: Timber Press, 2000. A relatively nontechnical but fairly comprehensive introduction to the biology, ecology, and taxonomy of the myxomycetes.

POISONOUS AND NOXIOUS PLANTS

Categories: Diseases and conditions; medicine and health; poisonous, toxic, and invasive plants

Poisonous plants have evolved toxic substances that function to defend them against herbivores and thereby better adapt them for survival.

After evolving adaptations that facilitated colonization of terrestrial habitats, plants were confronted with a different type of problem. This was the problem of *herbivory*, or the inclination of many different types of organisms, from bacteria to insects to four-legged herbivores, to eat plants. Pressures from herbivory drove many different types of plants, from many different families, to evolve defenses. Some of these defenses included changes in form, such as the evolution of thorns, spikes, or thicker, tougher leaves. Other plants evolved to produce chemical compounds that make them taste bad, interrupt the growth and life cycles of the herbivores, make the herbivores sick, or kill them outright.

Phytochemicals

One of the most interesting aspects of plants, especially prevalent in the angiosperms (flowering plants), is their evolution of substances called *secondary metabolites*, sometimes referred to as *phytochemicals*. Once considered waste products, these substances include an array of chemical compounds: alkaloids, quinones, essential oils, terpenoids, glycosides (including cyanogenic, cardioactive, anthraquinone, coumarin, and saponin glycosides), flavonoids, raphides (also called oxalates, which contain needle-like crystals of calcium oxalate), resins, and phytotoxins (highly toxic protein molecules). The presence of many of these compounds can characterize whole families, or even genera, of flowering plants.

Effects on Humans

The phytochemicals listed above have a wide range of effects. In humans, some of these compounds will cause mild to severe skin irritation, or contact dermatitis; others cause mild to severe gastric distress. Some cause hallucinations or psychoactive symptoms. The ingestion of many other types of phytochemicals proves fatal. Interestingly, many of these phytochemicals also have important medical uses. The effects of the phytochemicals are dependent on dosage: At low doses, some phytochemicals are therapeutic; at higher doses, some can kill.

Alkaloids

Alkaloids are nitrogenous, bitter-tasting compounds of plant origin. More than three thousand alkaloids have been identified from about four thousand plant species. Their greatest effects are mainly on the nervous system, producing either physiological or psychological results. Plant families producing alkaloids include the *Apocynaceae*, *Berberidaceae*, *Fabaceae*, *Papaveraceae*, *Ranunculaceae*, *Rubiaceae*, and *Solanaceae*. Some well-known alkaloids include caffeine, cocaine, ephedrine, morphine, nicotine, and quinine.

Glycosides

Glycosides are compounds that combine a sugar, usually glucose, with an active component. While there are many types of glycosides, some of the most important groups of potentially poisonous glycosides include the cyanogenic, cardioactive, anthraquinone, coumarin, and saponin glycosides.

Cyanogenic glycosides are found in many members of the *Rosaceae* and are found in the seeds, pits, and bark of almonds, apples, apricots, cherries, peaches, pears, and plums. When cyanogenic glycosides break down, they release a compound called hydrogen cyanide.

Two other types of glycosides, cardioactive glycosides and saponins, feature a steroid molecule as part of their chemical structure. Digitalis, a cardioactive glycoside, in the right amounts can strengthen and slow the heart rate, helping patients who suffer from congestive heart failure. Other cardioactive glycosides from plants such as milkweed and oleander are highly toxic. Saponins can cause severe irritation of the digestive system and hemolytic anemia. Anthraquinone glycosides exhibit purgative activities. Plants containing anthraquinone glycosides include rhubarb (*Rheum* species) and senna (*Cassia senna*).

Household Plants

Many common household plants are poisonous to both humans and animals. One family of popular household plants that can cause problems is the

Common Poisonous Plants and Fungi

Common Name	*Scientific Name*	*Common Name*	*Scientific Name*
Aconite	*Aconitum* spp.	Choke cherry	*Prunus* spp.
Alfalfa	*Medicago sativa*	Christmas rose	*Helleborus niger*
Amaryllis (bulbs)	*Hippeastrum puniceum*	Clovers (alsike, red, white)	*Trifolium* spp.
Anemone	*Anemone tuberosum*	Cocklebur	*Xanthium strumarium*
Angel's trumpet	*Datura* spp.	Corn cockle	*Agrostemma githago*
Apple (seeds, leaves)	*Malus sylvestris*	Corn lily	*Veratrum californicum*
Apricot (seeds, leaves)	*Prunus armeniacea*	Cow cockle	*Saponaria* spp.
Arrowgrass	*Triglochin maritima*	Creeping charlie	*Glechoma* spp.
Asparagus (berries)	*Asparagus officinalis*	Croton	*Croton* spp.
Azalea	*Rhododendron* spp.	Crowfoot	*Ranunculus* spp.
Baneberry	*Actaea* spp.	Crown-of-thorns	*Euphorbia milii*
Belladonna	*Atropa belladonna*	Crown vetch	*Coronilla varia*
Birdsfoot trefoil	*Lotus corniculatus*	Daffodil (bulbs)	*Narcissus pseudonarcissus*
Bitter cherry	*Prunus* spp.	Daphne	*Daphne* spp.
Black cherry	*Prunus* spp.	Datura	*Datura* spp.
Black locust	*Robinia pseudoacacia*	Deadly nightshade	*Atropa belladonna*
Bleeding heart	*Dicentra* spp.	Death angel mushroom	*Amanita* spp.
Bloodroot	*Sanquinaria canadensis*	Death camas	*Zigadenus* spp.
Bouncing bet	*Saponaria* spp.	Death cap mushroom	*Amanita* spp.
Bracken fern	*Pteridium aquilinium*	Delphiniums and larkspurs	*Delphinium* spp.
Broad beans	*Vicia* spp.	Destroying angels	*Amanita verna*
Buckeye	*Aesculus* spp.	Devil's trumpet	*Datura* spp.
Buckwheat	*Fagoypyrum esculentum*	Dock	*Rumex* spp.
Buffalo bur	*Solanum* spp.	Dogbane	*Apocynum* spp.
Buttercups	*Ranunculus* spp.	Dolls eyes	*Actaea* spp.
Caladium	*Caladium bicolor*	Downy thornapple	*Datura* spp.
Caley pea	*Lathyrus* spp.	Drooping leucothoe	*Leucothoe axillaris*
Cardinal flower	*Lobelia* spp.	Dutchman's breeches	*Dicentra* spp.
Castor bean	*Ricinus communis*		
Celandine	*Chelidonium majus*		

Araceae, the philodendron family, including plants such as philodendron and dieffenbachia. All members of this family, including these plants, contain needlelike crystals of calcium oxalate that, when ingested, cause painful burning and swelling of the lips, tongue, mouth, and throat. This burning and swelling can last for several days, making talking and even breathing difficult. *Dieffenbachia* is often referred to by the common name of dumb cane, because eating it makes people unable to talk for a few days.

Landscape Plants

Many landscape plants are also poisonous. For example, the yew (genus *Taxus*), commonly planted as a landscape plant, is deadly poisonous. Children who eat the bright red aril, which contains the seed, are poisoned by the potent alkaloid taxine. Yews are poisonous to livestock as well, causing death to horses and cattle. Death results from cardiac or respiratory failure.

Other poisonous landscape and garden plants include oleander, rhododendrons, azaleas, hyacinths, lily of the valley, daffodils, tulips, and star-of-Bethlehem. Many legumes are also toxic, including rosary pea, lupines, and wisteria. Castor bean plant, a member of the family *Euphorbiaceae*, produces seeds that are so toxic that one seed will kill a child and three seeds are fatal to adults. The toxin produced by the seeds is called ricin, which many scientists consider to be the most potent natural toxin known.

Common Name	*Scientific Name*	*Common Name*	*Scientific Name*
Eastern skunk cabbage	*Symplocarpus foetidus*	Jimsonweed	*Datura* spp.
Eggplant (leaves, stems)	*Solanum melongena*	Johnson grass	*Sorghum* spp.
Elderberry	*Sambucus canadensis*	Klamath weed	*Hypericum perforatum*
Ergot	*Claviceps* spp.	Laburnum	*Laburnum anagyroides*
Everlasting pea	*Lathyrus* spp.	Lamb's quarters	*Chenopodium album*
False hellbore	*Veratrum californicum*	Lantana	*Lantana camara*
Fiddleneck	*Amsinckia intermedia*	Larkspur	*Delphinium* spp.
Flax	*Linum usitatissimum*	Lily of the valley	*Convallaria majalis*
Fly agaric	*Amanita muscaria*	Lobelia	*Lobelia cardinalis*
Foxglove	*Digitalis purpurea*	Locoweed	*Astragalus, Oxytropis* spp.
Gill over the ground	*Glechoma* spp.	Lucerne	*Medicago sativa*
Gloriosa lily	*Gloriosa* spp.	Lupine	*Lupinus* spp.
Golden chain	*Laburnum anagyroides*	Mandrake	*Podophyllum peltatum*
Great lobelia	*Lobelia* spp.	Marijuana	*Cannabis sativa*
Ground ivy	*Glechoma* spp.	Marsh marigold or cowslip	*Caltha palustris*
Groundsels	*Senecio* spp.	Mayapple	*Podophyllum peltatum*
Halogeton	*Halogeton glomeratus*	Milkweed	*Asclepias* spp.
Henbane	*Hyoscyanamus niger*	Mistletoe	*Phoradendron* spp.
Holly (berries)	*Ilex* spp.	Monkey agaric mushroom	*Amanita* spp.
Horse chestnut	*Aesculus* spp.	Monkshood	*Aconitum* spp.
Horse nettle	*Solanum* spp.	Moonseed	*Menispermum canadense*
Horsebrush	*Tetradymia* spp.	Morning glory (seeds)	*Ipomoea tricolor*
Horsetail	*Equisetum arvense* & other spp.	Mountain fetterbrush	*Pieris* spp.
Hyacinth (bulbs)	*Hyacinthus orientalis*	Mountain laurel	*Kalmia latifolia*
Hydrangea	*Hydrangea* spp.	Narcissus (bulbs)	*Narcissus* spp.
Indian tobacco	*Lobelia* spp.	Nightshades	*Solanum* spp.
Irises (leaves, rhizomes)	*Iris* spp.	Oak trees	*Quercus* spp.
Ivy (leaves, berries)	*Hedera helix*	Oleander	*Nerium oleander*
Jack in the pulpit	*Arisaema* spp.	Panther	*Amanita pantherina*
Japanese pieris	*Pieris japonica*	Panther cap mushroom	*Amanita* spp.
Jessamine	*Gelsemium sempervirens*		

(continued)

Arrow Poisons

Toxic plant and animal products have been used for thousands of years in hunting, executions, and warfare. Usually the poisonous extracts were smeared on arrows or spears. The earliest reliable written evidence for these uses comes from the *Rigveda* from ancient India. Arrow poisons come in many different varieties, and most rain-forest hunters have their own secret blend. South American arrow poisons are generically called *curare*. There are more than seventy different plant species used in making arrow poisons. Two of the main arrow poison plants are woody vines from the Amazon: *Strychnos toxifera* and *Chondodendron tomentosum*. Some types of curare have proven medically useful. They are used as muscle relaxants in surgery, which lessens the amount of general anesthetic needed. A plant called *Strychnos nux-vomica* from Asia yields the poison strychnine, a stimulant of the central nervous system.

In ancient times, toxic plant products were also commonly used in executions. Many people were expert, professional poisoners in the ancient world. They could select a poison that would take days or even months to take effect, thus ensuring, for example, that an unfaithful spouse or lover would not suspect the reason for his or her lingering illness. On occasions when a more rapid result was required, a strong dose or more powerful poison could be prescribed.

Common Poisonous Plants and Fungi *(continued)*

Common Name	*Scientific Name*
Peach (leaves, seeds)	*Prunus persica*
Philodendron	*Philodendron* spp.
Pigweed	*Amaranthus* spp.
Pin cherry	*Prunus* spp.
Poinsettia	*Euphorbia* spp.
Poison hemlock	*Conium maculatum*
Poison ivy	*Toxicodendron radicans*
Poison oak	*Toxicodendron diversiloba*
Poison sumac	*Toxicodendron vernix*
Pokeweed	*Phytolacca americana*
Ponderosa pine	*Pinus ponderosa*
Poppies (inc. opium)	*Papaver* spp.
Potato	*Solanum* spp.
Prickly (Mexican) poppy	*Argemone mexicana*
Privet (leaves, berries)	*Ligustrum japonicum*
Ragworts	*Senecio* spp.
Red sage	*Lantana camara*
Rhodendron	*Rhodendron* spp.
Rhubarb (leaves)	*Rheum rhaponticum*
Rosary pea	*Abrus precatorius*
Senecio	*Senecio* spp.
Sensitive fern	*Onoclea sensibilis*
Sierra laurel	*Leucothoe davisiae*
Singletary pea	*Lathyrus* spp.
Snakeberry	*Actaea* spp.
Snow on the mountain	*Euphorbia* spp.
Sorghum or milo	*Sorghum* spp.
Spurges	*Euphorbia* spp.
Squirrel corn	*Dicentra* spp.
St. John's wort	*Hypericum perforatum*
Star of Bethlehem	*Ornithogalum umbellatum*
Stinging nettle	*Urtica* spp.
Sudan grass	*Sorghum* spp.
Sweet pea	*Lathyrus* spp.
Tall fescue	*Festuca arundinacea*
Tangier pea	*Lathyrus* spp.
Tobacco	*Nicotiana* spp.
Tomato (leaves, stems)	*Lycopersicon lycopersicum*
Tree tobacco	*Nicotiana* spp.
Tung oil tree	*Aleurites fordii*
Vetches	*Vicia* spp.
Virginia creeper (berries)	*Parthenocissus quinquefolia*
Water hemlock/ cowbane	*Cicuta* spp.
West Indian lantana	*Lantana camara*
White cohosh	*Actaea* spp.
White snakeroot	*Eupatorium rugosum*
White sweetclover	*Metilotus alba*
Wild cherries	*Prunus* spp.
Wisteria (pods, seeds)	*Wisteria* spp.
Wolfbane	*Aconitum* spp.
Yellow sage	*Lantana camara*
Yellow star thistle	*Centaurea solstitialis*
Yellow sweetclover	*Metilotus officinalis*
Yew	*Taxus cuspidata*

Note: spp. = species; This is a partial listing only: Parts of some plants that are particularly poisonous are identified in parentheses. However, it should *not* be assumed that other parts are necessarily benign or that plants not listed here are edible.

Source: Data are from Cornell University Poisonous Plants Informational Database, compiled by Dan Brown and staff, http://www.ansci.cornell.edu/plants/comlist.html (accessed April 11, 2002), and from Brian Capon, *Botany for Gardeners: An Introduction and Guide* (Portland, Oreg.: Timber Press, 1990), p. 96.

MICHAEL SHAW/PHOTO AGORA

*Poison ivy (*Toxicodendron radicans*) is well known for causing contact dermatitis. It is a widespread weed in the United States and southern Canada, growing in a variety of habitats: wetlands, disturbed areas, and the edges of forests. It has many forms, appearing as either a shrub or a woody vine which will grow up trees, houses, and fences.*

Poison Ivy

Toxicodendron radicans, commonly known as poison ivy, is well known for causing contact dermatitis. Poison ivy is a member of the *Anacardiaceae*, or cashew family, and is a widespread weed in the United States and southern Canada. It grows in a variety of habitats: wetlands, disturbed areas, and the edges of forests. It has many forms, appearing as either a shrub or a woody vine which will grow up trees, houses, fences, and fence posts. It has alternate leaves with three leaflets, forming the basis of the old saying "Leaves of three, let it be." After poison ivy flowers, it develops clusters of white or yellowish-white berries. Related species are poison oak, western poison oak, and poison sumac, which some scientists consider to be different types of poison ivy.

Roughly half the world's population is allergic to poison ivy. Very sensitive people develop a severe skin rash; about 10 percent of the people who are allergic require medical attention after exposure. The chemical compound causing the allergic reaction is called *urushiol*, a resin found in all parts of the plant. Urushiol is so potent that in some individuals, just one drop produces a reaction. Inhaling smoke from burning poison ivy can result in eye and lung damage. For some people, mere contact with the smoke from burning poison ivy can trigger a reaction. Urushiol lasts forever; in herbaria, dried plants one hundred years old have given unlucky botanists contact dermatitis.

Carol S. Radford

See also: Allelopathy; Animal-plant interactions; Ascomycetes; Bacterial resistance and super bacteria; *Basidiomycetes*; Basidiosporic fungi; Biochemical evolution in angiosperms; Biological invasions; Biological weapons; Carnivorous plants; Coevolution; Culturally significant plants; Deuteromycetes; Dinoflagellates; Horsetails; Legumes; Medicinal plants; Metabolites: primary vs. secondary; Mushrooms; Parasitic plants.

Sources for Further Study

Burrows, George E., and Ronald J. Tyrl. *Toxic Plants of North America*. Ames: Iowa State University Press, 2001. Comprehensive reference of toxic plants found in North America. Covers plant morphology and distribution, toxicants, treatments, and associate diseases. Includes illustrations and maps.

Levetin, Estelle, and Karen McMahon. *Plants and Society*. 2d ed. Boston: WCB/McGraw-Hill, 1999. An introductory text on the science of ethnobotany and uses of plants. Includes some basic botany.

Lewis, Walter H., and Memory P. F. Elvins-Lewis. *Medical Botany: Plants Affecting Man's Health*. New York: John Wiley and Sons, 1997. A thorough discussion and listing of medicinal and toxic plants.

Simpson, Beryl B., and Molly Conner Ogarzaly. *Economic Botany: Plants in Our World*. 3d ed. Boston: McGraw-Hill, 2000. An introductory text in basic botany and ethnobotany.

POLLINATION

Categories: Physiology; reproduction

Pollination involves the transfer of pollen from anther to stigma in flowering plants, or from male cone to ovules in gymnosperms. There are two different types of pollination: self-pollination and cross-pollination.

Pollination is the process, in sexually reproducing plants (both angiosperms and gymnosperms), whereby the male sperm and female egg are joined via transfer of *pollen* (male microspore). If the anthers and stigmas of the plants involved have the same genetic makeup or they are produced on the same plant, the type of pollination is called *self-pollination*. If anthers and stigmas are from plants with different genetic makeups, the type of pollination is called *cross-pollination*.

Self-pollination is efficient because pollen from the anther of a flower can be transferred easily onto the stigma of the same flower, owing to the proximity of the two parts. On the other hand, cross-pollination is risky because the transfer of pollen involves long distances and precise destinations, both of which depend on animal pollinators. In areas with few animal pollinators, the opportunities for cross-pollination may be greatly reduced (one of the many reasons that preserving biological diversity is an important ecological issue).

In spite of the risk associated with cross-pollination, most flowers have mechanisms that promote this kind of pollination. Cross-pollination increases the likelihood that offspring are vigorous, healthy, fertile, and able to survive even if the environment changes. Self-pollination leads to offspring that are less vigorous, less productive, and more subject to *inbreeding depression* (weakening of the offspring as a result of inbreeding).

When certain consumers forage among plants for food, they often come in contact with flowers. Many insects and other animals become dusted with pollen, and in the course of their travel they unintentionally but effectively bring about pollination. Throughout the evolutionary history of flowering plants, many pollinators have coevolved with plants. *Coevolution* occurs when the floral parts of a plant and the body parts and behavior of the pollinators become mutually adapted to each other, thereby increasing the effectiveness of their interaction. In many instances, the relationship between the plant and pollinator has become highly specialized, resulting in *mutualism*, which is interaction where both organisms benefit from each other.

In the case of pollination by animals, the pollinator receives a reward from the flower in the form of food. When the pollinator moves on, the plant's pollen is transferred to another plant. The adaptations between the flower and its pollinators can be intricate and precise and may even involve force, drugs, deception, or sexual enticement. In flowering plants, pollination is mostly due to insects or wind, but birds, bats, and rodents also act as pollinators for a number of plants.

Insects

Insect pollination occurs in the majority of flowering plants. There is no single set of characteristics for insect-pollinated flowers, because insects are a large and diverse group of animals. Rather, each plant may have a set of reproductive features that attracts mostly a specific species of insect. The principal pollinating insects are bees, although many other kinds of insects act as pollinators, including wasps, flies, moths, butterflies, ants, and beetles.

Bees have body parts suitable for collecting and carrying *nectar* and *pollen*. Their chief source of nourishment is nectar, but they also collect pollen for their larvae. The flowers that bees visit are generally brightly colored and predominantly blue or yellow—rarely pure red, because red appears black to bees. The flowers they visit often have distinctive markings that function as guides that lead them to the nectar. Bees can perceive ultraviolet (UV) light (a part of the spectrum not visible to humans), and some flower markings are visible only in UV light, making patterns perceived by bees sometimes different from those seen by humans. Many bee-pollinated flowers are delicately sweet and fragrant.

Moth- and butterfly-pollinated flowers are similar to bee-pollinated flowers in that they frequently have sweet fragrances. Some butterflies can detect red colors, and so red flowers are sometimes pollinated by them. Many moths forage only at night; the flowers they visit are usually white or cream-colored because these colors stand out against dark backgrounds in starlight or moonlight. With their long mouthparts, moths and butterflies are well adapted for securing nectar from flowers with long, tube-shaped *corollas* (the petals collectively), such as larkspur, nasturtium, tobacco, evening primrose, and amaryllis.

The flowers pollinated by beetles tend to have strong, yeasty, spicy, or fruity odors. They are typically white or dull in color, in keeping with the diminished visual sense of their pollinators. Although some beetle-pollinated flowers do not secrete nectar, they furnish pollen or other foods which are available on the petals in special storage cells.

Birds

Birds and the flowers that they pollinate are also adapted to each other. Birds do not have a highly developed sense of smell, but they have a keen sense of vision. Their flowers are thus frequently bright red or yellow and usually have little, if any, odor. The flowers are typically large or are part of a large inflorescence. Birds are highly active pollinators and tend to use up their energy very rapidly. Therefore, they must feed frequently to sustain themselves. Many of the flowers they visit produce copious quantities of nectar, assuring the birds' continued visitation. The nectar is frequently produced in long floral tubes, which prevent most

PhotoDisc

In a classic example of cross-pollination, this honeybee gathers nectar while inadvertently providing pollen from another plant to the ovule of this sunflower.

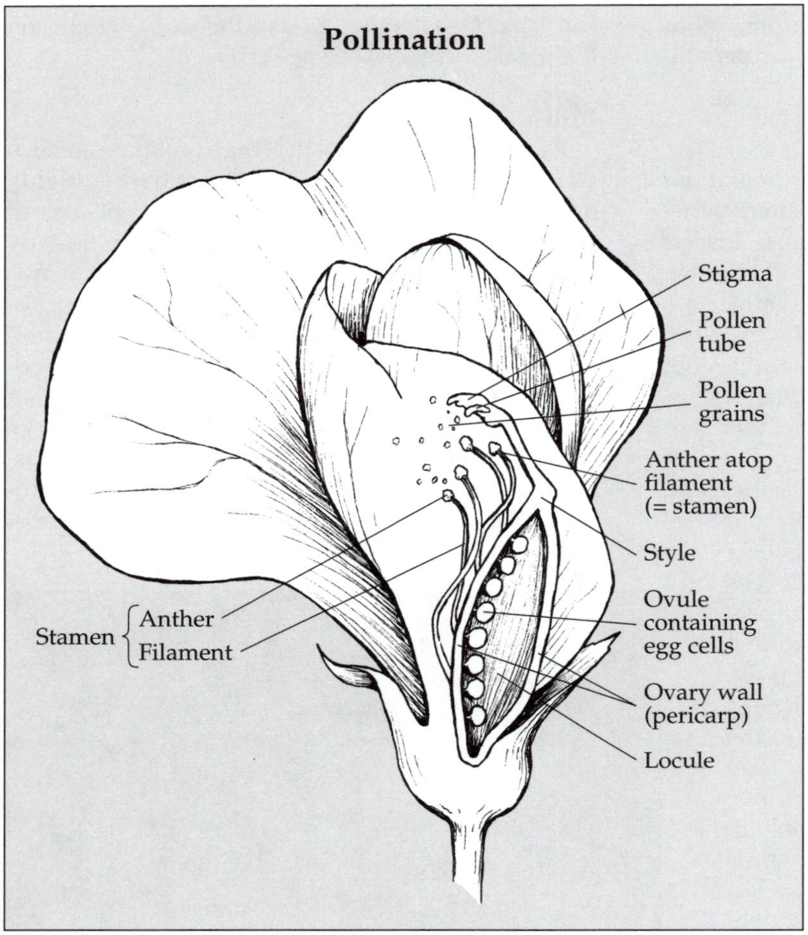

Kimberly L. Dawson Kurnizki

Pollination occurs as either self-pollination (shown here) or cross-pollination, in which the pollen from one plant is dispersed via wind, water, or animals to pollinate another plant. In both cases, pollen lands on the stigma and grows a pollen tube down through the style toward the ovule in the ovary, where it deposits the sperm into an egg cell inside the ovule.

insects from gaining access to it. Examples of bird-pollinated flowers are red columbine, fuchsia, scarlet passion flower, eucalyptus, hibiscus, and poinsettia.

Bats and Rodents

Bat-pollinated flowers are found primarily in the tropics, and they open only at night, when the bats are foraging. These flowers are dull in color, and like bird-pollinated flowers, they are large enough for the pollinator to insert part of its head inside. The plants may also consist of ball-like inflorescences containing large numbers of small flowers whose stamens readily dust the visitor with pollen.

Bat-pollinated flowers include bananas, mangoes, kapok, and sisal. Like moth-pollinated flowers, flowers that attract bats and small rodents open at night. Mammal-pollinated flowers are usually white and strongly scented, often with a fruity odor. Such flowers are large, to provide the pollinators enough pollen and nectar to fulfill their energy requirements. The flowers are also sturdy, to bear the frequent and vigorous visits of these small mammals.

Orchid Pollinators

The orchid family has pollinators among bees, moths and butterflies, and beetles. Some of the adaptations between orchid flowers and their pollinators are extraordinary. Many orchids produce their pollen in little sacs called *pollinia*, which typically have sticky pads at the bases. When a bee visits such a flower, the pollinia are usually deposited on its head. In some orchids, the pollinia are forcibly "slapped" on the pollinator through a trigger mechanism within the flower. In some orchids, a petal is modified so that it resembles a female wasp or bee. Male wasps or bees emerge from their pupal stage before the females and can mistake the orchids for potential mates. They try to copulate with these flowers, and while they are doing so, pollinia are deposited on their heads. When the wasps or bees visit other flowers, the pollinia are caught in sticky stigma cavities.

When moths and butterflies pollinate orchids, the pollinia become attached to their long tongues by means of sticky clamps instead of pads. The pollinia of certain bog orchids become attached to the eyes of the female mosquitoes that pollinate them. After a few visits, the mosquitoes are blinded and unable to continue their normal activities (a good example of a biological control within an ecosystem).

Among the most bizarre of the orchid pollination mechanisms are those whose effects are to dunk the pollinator in a pool of watery fluid secreted by the orchid itself and then permit the pollinator to escape underwater through a trap door. The route of the insect ensures contact between the pollinia and stigma surfaces. In other orchids with powerful narcotic fragrances, pollinia are slowly attached to the drugged pollinator. When the transfer of pollinia has been completed, the fragrance abruptly fades away, and the insect recovers and flies away.

Wind and Water

Wind pollination is common in those plants with inconspicuous flowers, such as grasses, poplars, walnuts, alders, birches, oaks, and ragweeds. These plants lack odor and nectar and are, hence, unattractive to insects. Furthermore, the petals are either small or absent, and the sex organs are often separate on the same plant. In grasses, the stigmas are feathery and expose a large surface to catch pollen, which is lightweight, dry, and easily blown by the wind. Because wind-pollinated flowers do not depend on animals to transport their pollen, they do not invest in the production of rewards for their visitors. However, they have to produce enormous quantities of pollen. Wind pollination is not efficient because most of the pollen does not end up on the stigmas of appropriate plants but on the ground, bodies of water, and in people's noses (a major cause of allergic reactions). Wind pollination is successful in cases where a large number of individuals of the same species grow fairly close together, as in grasslands and coniferous forests.

Water pollination is rare, simply because fewer plants have flowers that are submerged in water. Such plants include the sea grasses, which release pollen that is carried passively by water currents. In some plants, such as the sea-nymph, pollen is threadlike, thus increasing its chances of coming in contact with stigmas. In eelgrass, the entire male flower floats.

Danilo D. Fernando

See also: Animal-plant interactions; Angiosperm evolution; Angiosperm life cycle; Aquatic plants; Biochemical coevolution in angiosperms; Coevolution; Flower structure; Flower types; Metabolites: primary vs. secondary; Orchids; Population genetics; Reproduction in plants; Reproductive isolating mechanisms.

Sources for Further Study

Barth, Friedrich G. *Insects and Flowers: The Biology of a Partnership*. Princeton, N.J.: Princeton University Press, 1991. Describes the relationships of plants and insects with interesting emphasis on the pollinators. Includes photographs.

Proctor, Michael, Peter Yeo, and Andrew Lack. *The Natural History of Pollination*. Portland, Oreg.: Timber Press, 1996. Describes in detail mechanisms of pollination. Includes photographs.

Raven, Peter H., Ray F. Evert, and Susan E. Eichhorn. *Biology of Plants*. 6th ed. New York: W. H. Freeman/Worth, 1999. Contains a chapter that integrates pollination with the evolution of angiosperms. Includes photographs.

POLYPLOIDY AND ANEUPLOIDY

Categories: Genetics; reproduction and life cycles

In aneuploidy, one or more whole chromosomes has been lost or gained from the diploid state. In polyploidy, one or more complete sets of chromosomes has been gained from the usual state (generally diploid), as in triploidy and tetraploidy.

For each species of higher plants and animals, the base number of nuclear chromosomes is called the *haploid* number, denoted as *n*. Individuals of most species are *diploid*, having double the haploid number of chromosomes ($2n$) in each somatic cell. Aneuploid and polyploid organisms have abnormal numbers of whole chromosomes.

Aneuploidy

Strictly speaking, aneuploidy refers to any number of chromosomes in a cell or organism that is not an exact multiple of the haploid number. However, in common practice the term is used to refer specifically to situations in which an organism or cell has only one chromosome or a few chromosomes added or missing. In animals, aneuploidy is usually lethal and so is rarely encountered. In the plant kingdom, on the other hand, the addition or elimination of a small number of individual chromosomes may be better tolerated.

Nullisomy is the aneuploid condition in which two homologous chromosomes are missing, so that the organism has $2n - 2$ chromosomes. *Monosomy* refers to the absence of a single chromosome, giving $2n - 1$ total chromosomes. In *trisomy*, one extra chromosome is present ($2n + 1$). An example of aneuploidy in humans is the case of Down syndrome, trisomy-21, in which the individual has one extra copy of the twenty-first chromosome (thus, three total copies).

Aneuploidy is caused by *nondisjunction*, which occurs when a pair of homologous chromosomes fail to separate during cell division. If nondisjunction occurs in the first stage of meiosis, all four resulting gametes will be abnormal. Two of them will have no copy of the given chromosome, and two, correspondingly, will have one extra copy each. If nondisjunction occurs in the second stage of meiosis, one of the four resulting gametes will have no copy of the given chromosome, another will have an extra copy, and two will be normal.

Polyploidy

Polyploidy is caused by the addition of one or more complete chromosome sets to the normal diploid complement. In the animal kingdom polyploidy is lethal in nearly every case, but it is relatively common in plants. It is estimated that between 30 percent and 70 percent of extant angiosperms are polyploid. The process of sex determination is more sensitive to polyploidy in animals than in plants, and because many plants undergo self-fertilization, those with an even number of chromosome sets (such as those that are tetraploid) may still produce fertile gametes. This fact points to the crucial factor that determines whether a polyploid plant may be fertile: whether it has an even or an odd number of chromosome sets. Plants with an odd number of chromosome sets are almost always sterile. Because they always have an unpaired chromosome of each type, it is extremely unlikely for them to produce viable balanced gametes. On the other hand, there is potential for a polyploid plant with an even number of chromosome sets to produce a balanced gamete if multiple sets of conspecific chromosomes pair during meiosis.

Autopolyploidy and Allopolyploidy

Polyploid plants exist in two categories. *Autopolyploids* have a genome comprising multiple sets of chromosomes that are all from one species. In *allopolyploids*, the multiple sets of chromosomes come from multiple (usually related) species. Autopolyploidy can arise from situations in which a defect in meiosis creates a diploid or triploid gamete. If such a gamete is fused with a typical haploid gamete from the same species, the union leads to a polyploid zygote.

The most common such pairing, of a diploid and a haploid gamete, produces an autotriploid. As the previous section suggests, they are usually sterile. However, some sterile autotriploids that can be cultivated through vegetative propagation (by planting cuttings) are attractive food crops because they lack robust fertile seeds. For example, the cultivated banana is an effectively seedless (and therefore sterile) autotriploid; it cannot reproduce without human intervention.

Allopolyploidy arises through the interbreeding of different species. An allodiploid formed by the union of two haploid gametes from separate species will be sterile because there are no matching chromosomes to pair at meiosis. However, if the two sets of chromosomes become doubled within a cell, the result will be a potentially fertile allotetraploid. An organism in this condition is basically a double-diploid, in which homologous pairs of conspecific chromosomes can join at meiosis to produce a viable gamete. This type of polyploidy has played an important role in the natural history of many plants, including wheat, the most widely cultivated cereal in the world.

Allopolyploidy and Speciation

Contemporary domesticated wheat species can be identified as belonging to one of three groups, based on their number of chromosomes. One group has fourteen chromosomes, another twenty-eight, and a third has forty-two chromosomes. These groups form a series of polyploids based on a haploid number, *n*, equal to seven chromosomes. It is believed that these groups of domesticated wheat evolved in two major steps. First, members of a diploid genus *Triticum* ($2n$ = fourteen chromosomes) may have hybridized with one of the diploid goat grasses *Aegilops* ($2n$ = fourteen chromosomes) to form allotetraploid species of emmer and durum wheats ($4n$ = twenty-eight chromosomes). Then, it is believed, these species underwent a second round of hybridization with separate goat grass species to form the allohexaploid ($6n$ = forty-two chromosomes) species that is now known as bread wheat, or *Triticum aestivum*. Bread wheat, which probably appeared around eight thousand years ago, combines desirable qualities of all three of its diploid relatives, including nonshattering grains, a high protein content in the endosperm, and good tolerance to various environmental conditions. Allohexaploid wheat can reproduce, thanks to normal meiosis, in which homologous chromosomes pair to form a triploid gamete with twenty-one chromosomes.

The possibility of speciation by allopolyplidy is one danger of introducing plants to new regions. The American species of salt marsh grass *Spartina alterniflora* was introduced to the south coast of England around 1870, probably in ships' ballast water. It crossed with the native salt marsh grass *Spartina maritima* to form an allotetraploid species, *Spartina anglica*, which overran the native grass in the twentieth century and colonized coastal flats so aggressively as to create a floral monoculture that has proved inadequate for wintering populations of wading birds and wildfowl.

Agricultural Applications

Like wheat, many of today's other crops are polyploid. By creating polyploid lines, plant breeders can introduce desirable traits cumulatively. Among major polyploid crops are dietary staples, such as the white potato ($4n$ = forty-eight chromosomes), the domestic oat ($6n$ = forty-two chromosomes), the peanut ($4n$ = forty chromosomes), textile-producing plants such as cotton ($4n$ = fifty-two chromosomes), and the cash crops tobacco ($4n$ = forty-eight chromosomes) and coffee, of which existing species range from diploid to octoploid ($8n$ = eighty-eight chromosomes). As well as the aforementioned domesticated banana, which, as an autotriploid, is sterile and correspondingly seedless, some varieties of cultivated apple are triploid species.

Beyond engineering particular traits, another reason it is desirable to cultivate polyploid species is that polyploid plant cells are usually larger than the corresponding diploid cells. Consequently, the polyploid plants themselves are usually larger. Species of particularly large watermelons, marigolds, and snapdragons have been created through cultivation of polyploid lines.

Plant polyploidy is induced in the laboratory by treating dividing cells with the drug colchicine. It prevents the formation of a spindle during mitosis by disrupting the microtubules, causing the duplicated chromosomes to fail to separate. The most common method of application is to place the roots of a plant in a colchicine solution. Because colchicine inhibits the actual division of cells without affecting the duplication of chromosomes, when full rounds of mitosis commence upon removing the roots from the colchicine, the resulting cells contain an extra set of chromosomes.

Alistair Sponsel

See also: Agriculture: world food supplies; Biotechnology; Chromosomes; DNA in plants; Genetically modified foods; Genetics: Mendelian; Genetics: post-Mendelian; Grains; Green Revolution; High-yield crops; Hybrid zones; Hybridization; Mitosis and meiosis; Plant biotechnology; Population genetics; Species and speciation.

Sources for Further Study

Klug, William, and Michael Cummings. *Concepts of Genetics*. 5th ed. Upper Saddle River, N.J.: Prentice Hall, 1997. College-level textbook contains excellent diagrams relating to aneuploidy, engineering of polyploid lines, and the use of colchicine. Includes good photographs of the karyotypes and phenotypes of various polyploid and aneuploid organisms.

Levetin, Estelle, and Karen McMahon. *Plants and Society*. 2d ed. New York: WCB/McGraw-

Hill, 1999. Has a useful section on the evolution and breeding of domesticated wheat as an example of allopolyploidy.

Russell, Peter J. *Fundamentals of Genetics*. New York: HarperCollins College, 1994. College-level textbook contains a particularly thorough section on aneuploidy and the various forms of polyploidy.

POPULATION GENETICS

Categories: Evolution; genetics; reproduction and life cycles

Population genetics is concerned with analyzing the frequencies of the alleles, or forms, of genes in a population of individuals within a species. It examines how gene frequencies change across generations in response to external forces, such as mutation, natural selection, migration, or genetic drift.

The natural process that eliminates individuals of low fitness and advances those of high fitness is termed *natural selection*. Natural selection causes changes in allele frequency in a population, which is a process called *evolution*. Thus, population genetics combines Charles Darwin's ideas of natural selection and evolution with the basic principles of genetics set forth by Gregor Mendel.

The study of population genetics has many practical uses: It can help reveal how new diseases arise or why diseases persist in living organisms by examining genetic variation within populations. Population genetics principles also can be useful as guidelines for devising strategies for improving crop plants. Hardy-Weinberg equilibrium provides the framework for population genetics.

Hardy-Weinberg Theorem

The starting point for population genetics is the *Hardy-Weinberg theorem*. Derived from Mendelian principles, it states that in the absence of mutation, selection, migration, or genetic drift, the allele frequencies and genotype frequencies remain constant from generation to generation. This law applies to populations in a state of Hardy-Weinberg equilibrium, meaning large populations that are random-mating. If a population is not in this state, one generation of random mating restores equilibrium.

George H. Hardy, an English mathematician, and Wilhelm Weinberg, a German biologist and physician, independently developed this concept in 1908. In 1903 Harvard University professor William E. Castle also had shown that in the absence of selection, the composition of the randomly bred descendants of a population would remain constant thereafter. Castle did not, however, generalize the concept.

More specifically, Hardy-Weinberg equilibrium is maintained in a population as long as the following assumptions are true. First, there are a large (virtually infinite) number of individuals in the population. Second, there is random mating among individuals. Third, there are no new mutations. Fourth, there is no migration (in or out of the population). Fifth, there are no genotype-dependent differences for survival to reproductive age and transmission of genes to the next generation. This fifth assumption is often stated more succinctly as: There is no natural selection.

The Hardy-Weinberg theorem can also be expressed mathematically in the form of an equation for calculating genotype and phenotype frequencies. If two alleles at a locus, *A* and *a*, occur in a population with frequencies p and q, respectively, then $(p + q) = 1$. The proportion of individuals resulting from random matings will occur with the following frequencies:

$$AA = p^2,\ Aa = 2pq, \text{ and } aa = q^2$$

Putting these terms together results in the Hardy-Weinberg equation:

$$p^2 + 2pq + q^2 = 1$$

This simple equation can be modified in a variety of ways to mathematically model what happens when one or more of the Hardy Weinberg assumptions are violated. Violation of one or more of these assumptions will result in changes in allele frequencies, and thus, evolution. Changes in allele frequencies across a limited number of generations is often referred to as *microevolution*. Extending such changes across thousands of generations or more results in more extensive change and is often called *macroevolution*.

The Gene Pool

The Hardy-Weinberg equation and its modifications are used by population geneticists to describe changes in allele frequencies in *gene pools*. A gene pool represents all the genes carried and shared by individuals of an interbreeding population, with the assumption that each parent contributes equally to a large pool of gametes (eggs and pollen or sperm). Another assumption is that the parent and offspring generations are distinct, or nonoverlapping.

Species and Speciation

The gene pool of a population, when divided or isolated for a prolonged period of time, may form distinct subgroups. If, through *isolating mechanisms* such as genetic changes or geographical separation, the subpopulations are kept from interbreeding, new species may eventually arise after many generations when the isolated subpopulations evolve in different directions. The entire process of the splitting of a population into two or more reproductively isolated populations is termed *speciation*. Thus, a *species*, according to the commonly accepted *biological species concept*, is a group of interbreeding individuals that are capable of producing fertile offspring but are unable to do so with other such populations. Sometimes hybridization between two species can result in a fertile hybrid, thus exposing one of the weaknesses of the biological species concept. Speciation can be the result of either geographic separation or reproductive isolation at the cellular and molecular levels.

Manjit S. Kang

See also: Genetic drift; Genetics: Mendelian; Genetics: mutations; Genetics: post-Mendelian; Hardy-Weinberg theorem; Hybrid zones; Hybridization; Polyploidy and aneuploidy; Reproductive isolating mechanisms; Species and speciation.

Sources for Further Study

Baradat, P., W. T. Adams, and G. Müller-Starck, eds. *Population Genetics and Genetic Conservation of Forest Trees*. Amsterdam: SPB Academic, 1995. Papers from the International Symposium on Population Genetics and Gene Conservation in Carcans-Maubuisson, France, August 24-28, 1992, organized by the International Union of Forestry Research Organizations. Illustrations, maps.

Crow, J. F. *Basic Concepts in Population, Quantitative, and Evolutionary Genetics*. New York: W. H. Freeman, 1986. Chapters 1-4 and 7 contain excellent accounts of population genetics.

Gillespie, John H. *Population Genetics: A Concise Guide*. Baltimore, Md.: Johns Hopkins University Press, 1998. Includes bibliographical references and index.

Hadrick, Philip W. *Genetics of Populations*. 2d ed. Boston: Jones and Bartlett, 2000. Contains chapters on the diversity of genetic variation, measures of genetic variation, selection, alternative models to selection, inbreeding, genetic drift and effective population size, gene flow and population structure, mutation, molecular population genetics and evolution, multiple-gene models, and quantitative traits and evolution. Illustrated.

Hartl, Daniel L. *A Primer of Population Genetics*. 3d ed. Sunderland, Mass.: Sinauer Associates, 2000. Includes bibliographical references and index.

Rédei, G. P. *Genetics Manual*. River Edge, N.J.: World Scientific, 1998. Provides detailed definitions of the various forms of speciation and the Hardy-Weinberg theorem.

Roughgarden, Jonathan. *Theory of Population Genetics and Evolutionary Ecology: An Introduction*. Upper Saddle River, N.J.: Prentice Hall, 1996. A thorough, detailed introduction at more than 600 pages. Illustrated.

PROKARYOTES

Categories: Bacteria; cellular biology; microorganisms

Prokaryotes are one of two types of cell that form living organisms. Prokaryotic cells lack a nucleus and other organelles found in eukaryotic cells. Prokaryotes include the unicellular life-forms found in two of the three domains of life, Archaea *and* Bacteria, *whereas all protists, algae, fungi, plants, and animals are eukaryotic organisms, together forming the domain* Eukarya.

There are architecturally two distinct types of cells of living organisms: prokaryotic cells and eukaryotic cells. The defining difference between these two types of cells is that prokaryotic cells lack any of the internal membrane-bound structures (organelles) found in eukaryotic cells, such as a nucleus, mitochondria, chloroplasts, endoplasmic reticulum, and Golgi apparatus. Bacterial and archaeal cells are prokaryotes, while plants, animals, fungi, algae, and protozoa (protists) are composed of eukaryotic cells.

Structure

Although prokaryotic cells do not contain membrane-bound organelles, they do have a highly complex organization and structure. Like all cells, prokaryotes are surrounded by a *cytoplasmic membrane*. This membrane is composed of proteins and lipids and is semipermeable. This semipermeable layer regulates the flow of material into and out of the cell.

For most prokaryotes, the cell membrane is surrounded by a *cell wall*. The cell wall of almost every bacterial cell contains peptidoglycan, a cross-linked structure consisting of chains of sugar molecules, with the chains attached to one another through bridges composed of amino acids. This cell wall protects the bacterial cell from osmotic shock. Some bacterial cells also have an *outer membrane* linked to the peptidoglycan layer by lipoproteins. The outer membrane is a lipid bilayer that contains sugars and lipids and is known as lipopolysaccharide (LPS). LPS is often called endotoxin because this molecule can induce fever, shock, and death in animals. Archaeal cells may have cell walls composed of pseudopeptidoglycan, which is very similar to the peptidoglycan layer found in bacterial cells, or they may have cell walls composed of protein, polysaccharides, or other chemicals. Some bacteria and archaea lack cell walls entirely.

Some prokaryotic cells have structures external to the cell wall. These structures include capsules, slime layers, and S layers. *Capsules* are usually composed of polysaccharides, although some cells have proteinaceous capsules. Capsules are protective layers that are particularly important in allowing disease-causing bacteria to evade attack by mammalian immune systems. *Slime layers* are composed of polysaccharides and resemble less organized capsules. Slime layers help bacteria attach to surfaces, prevent dessication, and assist in trapping nutrients near the cell. *S layers* are crystalline protein layers of unknown function.

Many prokaryotic cells are motile due to the presence of *flagella*. Some bacteria have flagella attached only at one or both ends of the cell. These flagella are known as polar flagella. Other bacteria have flagella all around the cell, an arrangement known as peritrichous. Each flagellum is an inflexible, helical structure composed of molecules of the protein flagellin. Flagella rotate like propellers, causing bacteria to move in a corkscrew fashion.

Some prokaryotes produce *spores*, specialized structures that are extremely resistant to heat, cold, and desiccation. Spores are metabolically inert and can survive for extended periods, possibly for thousands of years. Spores form within prokaryotic cells when environmental conditions become unfavorable for survival. Once the spore is formed, the cell that produced the spore breaks open, releasing the spore. When the spore finds itself in favorable growth conditions, it germinates by swelling, breaking out of the spore coat, and resuming metabolic function.

Reproduction

Prokaryotic cells reproduce by *binary fission*. The cell cycle of prokaryotes has three parts: elongation, DNA synthesis, and cell division. During elongation, the cell synthesizes and secretes cytoplasmic membrane and cell wall material. Prokaryotes usually possess a single, double-stranded, circular DNA chromosome attached to the cytoplasmic membrane at one point. DNA synthesis, which occurs continuously in actively growing cells, results in two complete copies of the chromosome, each attached to the cytoplasmic membrane. As new membrane material is inserted into the cytoplasmic membrane during elongation, the two chromosomes are swept away from one another. During cell division, a septum forms in the center of the cell which eventually divides the cell into two daughter cells.

Binary fission is a type of asexual reproduction. Each of the daughter cells is identical to the parent cell, and there is no exchange of genetic material. Some prokaryotes, however, do engage in genetic recombination through a process called *conjugation*. Conjugation requires the presence of extrachromosomal pieces of DNA called *plasmids*. These plasmids are small, circular DNA molecules found in the cytoplasm of many prokaryotic cells. Some of these plasmids contain genes that encode a special structure, the *F-pilus*. The F-pilus is a proteinaceous rod that extends from the surface of cells. Cells that have an F-pilus are donor cells and can attach, via the F-pilus, to recipient cells which lack an F-pilus. Following attachment, the F-pilus contracts, drawing the donor and recipient close together. The donor then transfers DNA to the recipient. Although conjugation results in transfer of genes from one cell to another, it is not itself a method of reproduction.

Metabolism

Prokaryotes are metabolically diverse. Two basic nutritional pathways are found: autotrophy and heterotrophy. Autotrophic prokaryotes are capable of synthesizing their own energy-yielding compounds from simple inorganic compounds such as carbon dioxide and water. Some prokaryotic autotrophs, the cyanobacteria and the green and purple bacteria, utilize the energy from sunlight, in a process known as photosynthesis, to construct food molecules. It has been hypothesized that chloroplasts in plant cells evolved from cyanobacteria that were engulfed by a eukaryotic cell more than one billion years ago. Other autotrophs extract energy from metabolizing inorganic compounds such as hydrogen sulfide, iron sulfide, and ammonia.

Heterotrophic prokaryotes obtain energy from the metabolism of organic compounds. Various prokaryotes are capable of metabolizing a wide variety of organic molecules, including sugars, lipids, proteins, petroleum products, antibiotics, and methanol. Heterotrophs can metabolize food molecules using one of three methods: fermentation, aerobic respiration, and anaerobic respiration. Fermentation and anaerobic respiration do not require

A Prokaryotic Cell

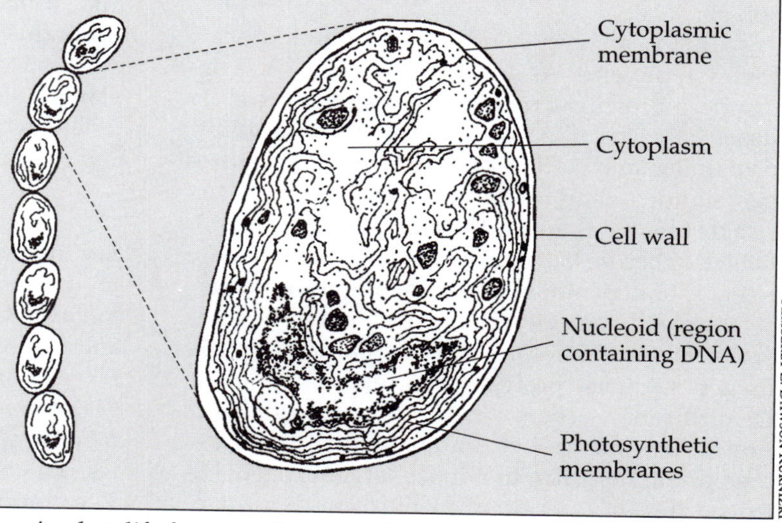

The simplest life-forms, archaea and bacteria, comprise prokaryotic cells. Based on an electron microscope image of one cell in a string forming a cyanobacterium, this depiction shows the basic features of a prokaryote: cell wall, cytoplasmic membrane, cytoplasm, and in this organism photosynthetic membranes and a series of internal membranes housing photosynthetic pigments such as chlorophyll. Prokaryotic cells lack the defined nucleus and organelles found in eukaryotic cells, and instead the genetic material is located in an unbound region called the nucleoid.

the presence of oxygen, while aerobic respiration does require oxygen. Fermentation often results in metabolic end products that include acids, carbon dioxide, alcohol, or a combination of these. The anaerobic respiration process is similar to aerobic respiration, except that molecules such as nitrate, sulfate, and iron are used instead of oxygen. The end products of aerobic respiration are carbon dioxide and water; for anaerobic respiration they are nitrite, hydrogen sulfide, or other reduced compounds.

Roles in the Global Ecosystem

Prokaryotes play important roles in the decay of organic matter as well as in three vital cycles of nature: the carbon, sulfur, and nitrogen cycles. The major categories of biological macromolecules (carbohydrates, lipids, proteins, and nucleic acids) are all carbon-containing compounds. Photosynthetic organisms, including photosynthetic prokaryotes, take carbon dioxide and convert it into carbohydrates. Those carbohydrates can be used for energy and biosynthesis by the photosynthetic organisms as well as by heterotrophs, which consume the photosynthetic organisms. Both heterotrophs and autotrophs also metabolize carbon-containing molecules, releasing carbon dioxide back into the atmosphere.

Sulfur is a component of certain amino acids found in proteins. As *decomposers*, prokaryotes decompose proteins deposited in water and soil by dead organisms and release the sulfur from sulfur-containing amino acids, often in the form of hydrogen sulfide. Some prokaryotes convert hydrogen sulfide to sulfates during their metabolism. The sulfates can then be taken up by plants, where they are reincorporated into sulfur-containing amino acids.

Nitrogen is an essential element in nucleic acids and proteins. Some prokaryotes, particularly soil microbes, digest proteins and release ammonia. Denitrification occurs when groups of symbiotic prokaryotes metabolize ammonia, first to nitrites, then to nitrates, then to atmospheric nitrogen. Nitrogen fixation occurs when nitrogen-fixing prokaryotes in the soil trap atmospheric nitrogen and convert it to ammonia that can be used by plants to synthesize new proteins and amino acids.

Disease

Infectious disease is a disturbance in normal organismal function caused by an infecting agent. Although most prokaryotes do not cause disease, some bacteria are capable of parasitizing a host and disrupting normal function. Prokaryotes capable of producing disease in plants are widely distributed and cause a number of diseases, including wilts, rots, blights, and galls. Some of these diseases are caused by soil-dwelling prokaryotes, while others are seedborne or are caused by obligate parasites, unable to survive outside plant tissue.

Commercial Uses

Prokaryotes are easily manipulated and therefore are useful for many commercial applications. Prokaryotes have been used for centuries in the production of food. Yogurt, sauerkraut, poi, kimchee, dry and semidry sausages, and vinegar are all examples of bacterially produced foods. Genetic engineering is more easily accomplished in prokaryotes than in eukaryotes. Prokaryotes now produce human insulin, antibiotics, plant hormones, and industrial solvents. Prokaryotes have been engineered to protect plants from frost damage, while plants have been genetically engineered, using bacterial vectors, to develop resistance to herbicides and to produce toxins that destroy insect pests.

Lisa M. Sardinia

See also: Anaerobes and heterotrophs; Anaerobic photosynthesis; *Archaea*; Bacteria; Cell wall; Chemotaxis; Chloroplasts and other plastics; Diseases and disorders; Eukaryotic cells; Evolution of cells; Flagella and cilia; Gene regulation; Genetic code; Membrane structure; Microbial nutrition and metabolism; Mitosis and meiosis; Molecular systematics; Paleoecology; Plasma membranes; *Protista*; Respiration; Systematics and taxonomy.

Sources for Further Study

Alcamo, I. Edward. *Fundamentals of Microbiology*. 6th ed. Boston: Jones and Bartlett, 2001. This college textbook is written for introductory microbiology courses with an emphasis on human disease, but it also contains sections on food and environmental microbiology. Illustrations, photographs, and "MicroFocus" boxes containing interesting facts.

Atlas, Ronald M. *Principles of Microbiology*. 2d ed. Boston: WCB/McGraw-Hill, 1997. Detailed college textbook focusing on prokaryotic biology with little reference to human diseases. Illustrations, photographs, bibliographic references, and an appendix containing systematic descriptions of groups of microorganisms.

Madigan, Michael T., John M. Martinko, and Jack Parker. *Brock Biology of Microorganisms*. 9th ed. Upper Saddle River, N.J.: Prentice Hall, 2000. Comprehensive college textbook covering all aspects of prokaryote biology, from metabolism to genetics, ecology, and diversity. Illustrations, micrographs, glossary, index.

Silverstein, Alvin, Robert Silverstein, and Virginia Silverstein. *Monerans and Protists*. Breckenridge, Colo.: Twenty-First Century Books, 1996. Describes single-celled organisms and explains their importance in maintaining the ecosystem. Aimed at middle schoolers.

PROTEINS AND AMINO ACIDS

Category: Cellular biology

Proteins are found in all cells and carry out a variety of important cellular functions. Within any one cell there may be thousands of different proteins having a variety of sizes, structures, and functions. Proteins are also important structural components of the cell wall.

Proteins are the most complex and abundant of the macromolecules. Within cells, many proteins function as enzymes in the catalysis of metabolic reactions, while others serve as transport molecules, storage proteins, electron carriers, and structural components of the cell. They are especially important in seeds, where they make up as much as 40 percent of the seed's weight and serve to store amino acids for the developing embryo. Proteins are also important structural components of the cell wall. Because proteins and their building blocks, amino acids, form such a large component of plant life, plants serve as an important dietary source of the eight to ten essential amino acids for humans and other animals.

Amino Acids

Amino acids are the molecular building blocks of proteins. Amino acids all share a structure, with a central carbon atom, the alpha carbon, covalently bonded to a hydrogen atom, an amino group, a carboxylic acid group, and a group designated as an R group, which varies in structure from amino acid to amino acid. It is the diverse nature of the R group that provides the protein with many of its structural and functional characteristics. Some R groups are either polar or electrically charged at physiological pH, making the R groups hydrophilic (water-loving). Other R groups are nonpolar and hydrophobic (water-avoiding). The twenty standard amino acids the cell uses to synthesize its proteins are alanine, arginine, aspartate (aspartic acid), asparagine, cysteine, glutamate (glutamic acid), glutamine, glycine, histidine, isoleucine, leucine, lysine, methionine, phenylalanine, proline, serine, threonine, tryptophan, tyrosine, and valine.

Each of the twenty amino acids differs from the other nineteen in the structure of its R group. Once incorporated into a protein, a standard amino acid may undergo modification to create nonstandard amino acids and an even greater diversity of protein structures. One of the more common nonstandard amino acids found in proteins is hydroxyproline, which is commonly found in plant cell-wall proteins. In addition to the twenty amino acids that build proteins, many nonstandard amino acids occur free in the cell and are not found in proteins. Canavanine, for example, occurs in the seeds of many legumes.

Based on the information in cellular deoxyribonucleic acid (DNA), the cell joins the twenty standard amino acids by peptide bonds in specific sequences, resulting in chains ranging from as few as two amino acids to many thousands. Shorter chains of amino acids are referred to as *peptides* or *oligopeptides*, while longer chains are referred to as *polypeptides*. The term "protein" is usually reserved for those oligopeptides and polypeptides that have biological functions, because single polypeptides often do not have biological functions unless associated with other polypeptides.

Primary Structure

Proteins differ from one another in the sequences of their amino acids. The sequence of amino acids of a protein is called its *primary structure*. Mutations have been shown to result in the change of as few as one amino acid in a protein. Because DNA specifies a protein's primary structure, protein sequence information is often used to study the evolutionary relationships among organisms.

Proteins are often complexed with other compounds in their biologically active state. These proteins are called *conjugated proteins*. Proteins complexed with metals, lipids, sugars, and riboflavin are called *metalloproteins*, *lipoproteins*, *glycoproteins*, and *flavoproteins*, respectively. Glycoproteins (literally, "sugar proteins") are important constituents of the plasma membrane. These sugar molecules can occur singly or in short, simple branched chains.

A protein chain may be folded into a variety of three-dimensional shapes. The three-dimensional shape a protein assumes is called its *conformation* and is determined by its amino acid sequence. In order for a protein to be active, it must assume a certain conformation. Any alteration in its conformation may result in reduced activity. Denaturing agents alter the structure of a protein so that it loses its conformation, biological function, and activity.

Secondary Structure

The *secondary structure* refers to the local folding or conformation of the polypeptide chain over relatively short (fifty amino acids or so) stretches. Two common secondary structures, the alpha helix and the beta sheet, occur regularly in proteins. On average, only about half of the polypeptide chain assumes the alpha or beta conformation, while the remainder exists in turns and random structures. Some proteins show only alpha structure, others only the beta structure, while still others show either a mixture of the two structures or neither secondary structure. Both the alpha and the beta structures increase the structural stability of the protein. The amino acid sequence determines whether a particular sequence of amino acids in a protein will assume the alpha or beta structure.

Tertiary Structure

The overall spatial orientation of the entire polypeptide chain in space is referred to as its *tertiary structure*. Generally, two tertiary structures are recognized. Fibrous or filamentous proteins are arranged as fibers or sheets, while globular proteins are arranged roughly as spherical or globular structures. The amino acid sequence determines the overall folding of the protein tertiary structure. Fibrous proteins are primarily involved with structural functions, whereas globular proteins function as enzymes, transport molecules, electron carriers, and regulatory proteins.

Quaternary Structure

Some proteins are composed of more than one polypeptide chain. A protein composed of only one polypeptide is called a *monomer*, while proteins composed of two, three, four, and so on are referred to as *dimers*, *trimers*, and *tetramers*, and so on, respectively.

Charles L. Vigue

See also: Carbohydrates; Cell wall; Chromosomes; Cytoskeleton; Cytoplasm; Cytosol; DNA in plants; DNA replication; Endomembrane system and Golgi complex; Endoplasmic reticulum; Energy flow in plant cells; Evolution of cells; Lipids; Membrane structure; Nuclear envelope; Nucleic acids; Nucleolus; Nucleoplasm; Nucleus; Oil bodies; Plasma membranes; RNA.

Sources for Further Study

Dey, P. M., and J. B. Harborne, eds. *Plant Biochemistry*. San Diego: Academic Press, 1997. Describes various topics in plant biochemistry, each covered by a different specialist. Illustrated and referenced, the text emphasizes plant metabolism but also includes enzymes, functions, regulation, and molecular biology.

Mathews, Christopher K., K. E. Van Holde, and Kevin G. Ahern. *Biochemistry*. 3d ed. New York: Addison Wesley Longman, 2000. Covers all aspects of biochemistry for the advanced student. Profusely illustrated and referenced.

Nelson, David L., and Michael M. Cox. *Lehninger Principles of Biochemistry*. 3d ed. New York: Worth, 2000. Covers general biochemistry, from structures to metabolism. Excellently il-

lustrated and referenced. Comes with a CD of art and animations.
Zubay, Geoffrey L. *Biochemistry*. 4th ed. Boston: Wm. C. Brown, 1998. General, advanced biochemistry text has especially good coverage of proteins. Includes illustrations and references and comes with a CD of art and animations.

PROTISTA

Categories: Microorganisms; *Protista*; taxonomic groups; water-related life

The Protista *form one of the four kingdoms of eukaryotic organisms and include the algae, protozoans, slime molds, and oomycota.*

Included among the diverse organisms called protistans (kingdom *Protista*) are algae, protozoans, slime molds, and the oomycota. Algae were long considered to be simple plants and were assigned to kingdom *Plantae* but lack the more highly differentiated tissues and organs characteristic of "higher plants" such as mosses, ferns, and seed plants. Protozoa, all of which are unicellular, were assigned to the kingdom *Animalia*; considered more animal-like than plant-like, they are not considered here in detail.

The two remaining groups of protists, the slime molds and oomycota, were traditionally considered fungi, as indicated by their names (*mycoto* means "fungus"), and accordingly were assigned to the kingdom *Fungi*, largely due to their being heterotrophs (nonphotosynthetic). As information about the true nature of these life-forms has accumulated, however, they were grouped together with the protists. Nevertheless, *Protista* is a heterogeneous kingdom, composed of both the heterotrophic slime molds and oomycetes and the autotrophic (photosynthetic) algae. Thus, the kingdom *Protista* is a "catch-all" group, containing various organisms that seem not to fit into any of the other kingdoms.

Biology of the *Protista*

Most protistans are unicellular, but some, notably the large species of marine algae called kelps and seaweed, are multicellular. All are eukaryotic (with cells that have nuclei, making them members of the domain *Eukarya*, along with fungi, plants, and animals), in contrast to prokaryotic bacteria (which are divided between the domains *Archaea* and *Bacteria*).

In addition to having a distinct nucleus surrounded by a nuclear membrane (envelope), the cytoplasm of a eukaryotic cell contains various types of organelles, each specialized for performing a particular task (or related ones). Examples are mitochondria (cellular respiration), chloroplasts (photosynthesis), Golgi bodies (packaging of molecules), and the endoplasmic reticulum (to lend rigidity). As a result of this specialization, eukaryotic cells are able to function more efficiently than do prokaryotic cells, which must perform the same functions but within the cell as a whole. In this way, protistans are similar to plants, animals, and fungi. Furthermore, the evolution of eukaryotic cells from prokaryotic ones was a pivotal event in the evolution of life, leading not only to protistans but also from them to the more highly evolved kingdoms.

As previously stated, protists exist in great variety. Included are some that, due to the absence of a rigid cell wall, are able to change shape rapidly. Other protists have cell walls surrounding their cell membranes, resulting in a more permanent shape. Some possess fringelike cilia or whiplike flagella which they use to swim; others move by other means; many are nonmotile. Some protists live solitary lives, while others aggregate to form colonies. Some are parasitic, whereas most are free-living (nonparasitic).

In size there is also great variation: Many are unicellular and microscopic, while some algae, such as the kelps and seaweeds, grow to many meters in length. Furthermore, many protists have complex life cycles, with the various stages possessing different combinations of these characteristics.

Biologists generally agree that fungi, plants, and

Phyla of *Protista*

Phylum	*Common Name*	*Approx. Species*
Slime molds		
Dictyosteliomycota	Cellular slime molds	50
Myxomycota	Plasmodial slime molds	700
Water molds		
Oomycota	Oomycetes	700
Algae		
Bacillariophyta	Diatoms	100,000+
Chlorophyta	Green algae	17,000
Chrysophyta	Chrysophytes	1,000
Cryptophyta	Cryptomonads	200
Dinophyta	Dinoflagellates	4,000
Euglenophyta	Euglenoids	1,000
Haptophyta	Haptophytes	300
Phaeophyta	Brown algae	1,500
Rhodophyta	Red algae	6,000

Source: Data on numbers of species adapted from Peter H. Raven et al., *Biology of Plants*, 6th ed. (New York: W. H. Freeman/Worth, 1999).

animals are derived from ancient protists. Thus, the study of protists, which continue to inhabit the earth, sheds light on the origin of these groups of more highly evolved organisms. Also, each of these modern protists plays an ecological role together with the other organisms that occupy its ecosystem. Some protists affect humans more directly as agents of disease; some serve as a source (or potential source) of medicines that can be used to combat various diseases.

Algae: Classification

The term "algae" (singular "alga"), when unqualified, refers generally to an organism, usually inhabiting water or a wet habitat, that is somewhat plantlike (photosynthetic) but that lacks the more specialized tissues characteristic of plants. Furthermore, nearly all algae produce reproductive cells, spores or gametes, which lack surrounding specialized enclosures such as are typical of plants.

The blue-green "algae," historically considered to be algae, are now recognized as a separate group of organisms. As they are prokaryotic, they are now considered to be a special type of bacteria, the *cyanobacteria*. They differ from other bacteria primarily because they possess chlorophyll and therefore have the ability to photosynthesize.

Within the kingdom *Protista*, algae are currently divided by phycologists (scientists who study algae) among anywhere from four to thirteen phyla. Here, a system is used in which nine phyla of the kingdom *Protista* include algae. The nine phyla include the following.

Euglenophyta (euglenoids). This small group of protists (about nine hundred species) is a good starting point for studying algae, as euglenoids combine traits characteristic of plants, fungi, and animals. All are unicellular, and nearly all are mobile by means of two flagella that emerge from a groove at the end of the cell. By means of an eyespot near the base of the flagella, euglenoids can detect light and swim toward it. Most members of the phylum lack chlorophyll and therefore must absorb food from external sources. Others, though, such as *Euglena*, have chloroplasts with chlorophylls *a* and *b*, along with carotenoids and accessory pigments. They reproduce by fission (cell division). Most are freshwater organisms, but a few occur in brackish or marine environments.

Cryptophyta (cryptomonads). These algae are single-celled flagellates that are commonly brownish, blue-green, or red. Their name (Greek *kryptos* means "hidden") refers to their small size (3-50 micrometers), which makes them inconspicuous. Like euglenoids, they include both colorless (unpigmented) and pigmented photosynthetic members. Pigments includes chlorophylls *a* and *c* and carotenoids. Evidence exists to indicate that cryptomonads arose from the fusion of two different kinds of eukaryotic cells. Of the two hundred known species, some are marine; others live in fresh water.

Rhodophyta (red algae). Red algae are primarily multicellular marine organisms and are commonly referred to as *seaweeds*. However, about one hundred of the five hundred species are unicellular and live in fresh water. The chloroplasts of red algae contain chlorophyll *a*, but the presence of phycobilins (red pigments) usually masks the chlorophyll, giving them a red or reddish appearance. The red pigment aids in light absorption in deep water, where many red algae are found. As chloroplasts of red algae resemble those of cyanobacteria, there is reason to believe that they probably evolved from these prokaryotic organisms.

One group of red algae, the *coralline algae*, deposit calcium carbonate in their cell walls, resulting in stony formations in the ocean. They are often associated with coral reefs, which they help to stabilize.

Many red algae have complicated life histories. The simplest type is that in which a haploid gametophyte alternates with a diploid sporophyte. This pattern, known also throughout the plant kingdom, is known as alternation of generations. In most red algae, there are three generations: a haploid gametophyte, a carposporophyte, and a tetrasporophyte (both diploid). It has now been recognized that some red algae previously considered to be different species are actually different stages of the same species.

Dinophyta (dinoflagellates). Most dinoflagellates are unicellular marine algae, each with two flagella. Of the two, one is within an equatorial groove around the cell; the other passes down a longitudinal groove before extending outward. In addition to chlorophyll, they have accessory pigments that give them a golden-brown color. Many dinoflagellates are associated symbiotically with various invertebrates; for example, many corals benefit from the food they derive from these algae. Other dinoflagellates are nonphotosynthetic and live as parasites within other marine organisms.

Many dinoflagellates are bioluminescent; they emit a faint light that can be seen in darkness from a passing ship. Others are responsible for fish kills when they become superabundant in warm stagnant water. The death of the fish within these "red tides" is often due to toxins produced by the dinoflagellates.

Haptophyta (haptophytes). This group, consisting of only about three hundred known species, includes a diversity of primarily marine species, but some freshwater and even terrestial species are known. Included are both unicellular and colonial flagellates, along with others that are nonmotile. The distinctive feature of haptophytes is the haptonema, a threadlike structure that bends and coils as it apparently helps the cell to catch food particles. It differs from a flagellum by lacking the 9 + 2 arrangement of microtubules, which is characteristic of flagella of eukaryotic cells. Most haptophytes are photosynthetic and possess chlorophylls *a* and *c* together with an accessory pigment such as fucoxanthin. Although the phenomenon is not as well documented as for dinoflagellates, haptophytes also cause marine fish kills by releasing toxins.

Chrysophyta (chrysophytes). Some one thousand species of chrysophytes exist, including both unicellular and colonial organisms that are often abundant in both fresh and marine habitats. Some lack chlorophylls *a* and *c*. However, the golden color of fucoxanthin, which they possess, usually masks the green pigments, giving them their characteristic golden hue and accounting for their name (*chrysos* means "gold"). Some chrysophytes feed on bacteria. Some are responsible for "brown tides" that causes damage to shellfish and salmon fisheries.

Bacillariophyta (diatoms). The name "diatom" comes from the two overlapping shells that fit together like the two parts of a candy box. Composed of silica (silicon dioxide), the shells persist long after the living cell inside has died. Most species are photosynthetic, possessing chlorophylls *a* and *c* as well as fucoxanthin. The life cycles of diatoms include both asexual (cell division) and sexual phases.

It would be difficult to overestimate the importance of diatoms. There are more than 100,000 known species, and they are abundant in practically all aquatic and marine habitats. Due to the persistence of their shells, it is known that there aremany extinct species also. Diatoms are responsible for as much as 25 percent of total world food production (photosynthesis). Especially in polar waters, they are the primary food source for aquatic animals. Large accumulations of diatom shells, known as "diatomaceous earth," are mined, cleaned, and used in filters, in gas masks, in toothpaste, and for a variety of other purposes.

Phaeophyta (brown algae). Found only in salt water, this group includes large, conspicuous, multicellular forms generally called *kelps* or seaweeds. Often seen on rocky shores, various of the fifteen hundred species inhabit the ocean, especially in temperate and cooler waters. *Laminaria* are called kelps and often form "kelp forests" along the gently sloping shores off the coast of California.

Although lacking true roots, stems, and leaves, kelps have the most highly differentiated bodies of any of the algae. Some of their cells resemble the phloem (food-conducting cells) of vascular plants. The brown pigment fucoxanthin is present in addition to chlorophylls *a* and *c*.

Chlorophyta (green algae). The seventeen thousand or more species of green algae are perhaps the most diverse group of the algae. Most live in fresh water, but some live in the ocean, and others live in soil or on tree trunks. Many form symbiotic associations with sponges, protozoa, and other invertebrates; others are associated with fungi within lichens. As green algae resemble plants more than do any other group of algae, plants are believed to have evolved from green algae. They, like plants, possess chlorophyll *a* and *b*; food is stored in specialized cytoplasmic organelles called plastids.

The phylum is divided into three classes. In the class *Chlorophyceae* are included a diversity of forms, nearly all of which are freshwater species. *Chlamydomonas* is a motile unicellular species which, nevertheless, exhibits a complex life cycle. *Volvox* and several other large spherical colonial forms are composed of cells, each of which strongly resembles *Chlamydomonas*. Other members of this class are filamentous.

The class *Ulvophyceae* includes primarily marine species. A common example is the *Ulva* species (sea lettuce), composed of flat sheets of cells; it is found in shallow seas around the world.

Included in the class *Charophyceae* is the familiar *Spirogyra*, a freshwater filamentous species with spiral chloroplasts.

Human Uses of Algae

References have already been made to ways that algae are involved in the overall "economy of nature." Because algae are autotrophs (producers) and photosynthesize, they generate food that is made available to the heterotrophic animals (consumers). At the same time, oxygen, which results as a by-product, is made available to these same animals. As algae perform functions in marine and freshwater environments that are similar to those performed by grasses (and other plants) on land, algae have been called "the grasses of many waters."

Algae are often involved also in human affairs in more direct ways. In Asian countries especially, kelps and other multicellular algae have been used for food for centuries. Nori, a red alga of the genus *Porphyra* has been collected and eaten as a vegetable in Japan and China. It is now cultivated on a large scale, thus increasing its availability and popularity. Unfortunately, most seaweeds are not high in food value, although they do provide some needed minerals and vitamins. Another problem is their taste, which is not acceptable to many Westerners.

Of more commercial value in Europe and North America are a number of products derived from kelps and various seaweeds: alginates, carrageenans, and agar. *Alginates* are hydrophobic (water-attracting) compounds derived from various brown algae such as *Laminaria*. After they are harvested mechanically from the ocean, the algae are processed. The resulting products are salts of sodium and potassium alginate. These alginates are used in the paper industry as a sizing and polishing agent and in the manufacture of paints, cosmetics, and a wide variety of foods. In each case, the role of the alginate is to improve the consistency of the product and to prevent the separation of its ingredients.

Carrageenans are obtained primarily from Irish moss (*Chondrus crispus*), a red alga found off the coast of New England. After processing, the resulting carrageenans are used for some of the same purposes as alginates. However, because of their higher melting point, they have been found superior for uses in many kinds of foods, especially desserts.

The pioneer German bacteriologist Robert Koch popularized the use of *agar* for the culture of bacteria. Agar is obtained from certain red algae. After processing and cleaning, agar is added in small quantities (1-2 percent) to water along with nutrients required by the bacteria. The result is a solid medium on which bacteria can be isolated from a mixed culture.

Although Asians have long used certain algae for folk medicinal purposes, their use in Western medicine has until recently been limited largely to serving as a binder in medicinal tablets or as a laxative. Their potential as a source of therapeutic drugs is being pursued by an increasing number of researchers. Included are those from which antibiotics and anticancer drugs may be extracted.

Slime Molds

As the name indicates, these organisms resemble molds (kingdom *Fungi*) and thus have historically been considered to be fungi. Like fungi, they are heterotrophic and are found growing on decaying organic matter. However, the accumulation of more data, including molecular information, indicates that they are a group distinct from fungi. Slime molds are typically divided between two phyla.

Plasmodial slime molds (phylum *Myxomycota*) often exist as a conspicuous fan-shaped mass of protoplasm that creeps along a surface somewhat as do amoebas. From this stage, known as a plasmodium, are formed sporangia with spores inside. From the spores are formed the single-celled amoebalike stage; they converge to form the plasmodium.

Cellular slime molds (phylum *Dictyosteliomycota*) are amoeba-like organisms that combine at one stage to form "slugs," or pseudoplasmodia. Like that in plasmodial slime molds, reproduction in cellular slime molds is both sexual and asexual, but, unlike plasmodial slime molds, no flagellated cells are known. In both types of slime molds, particulate food such as bacteria can be ingested (fungi absorb only digested food).

Oomycetes

Also previously considered to be fungi, oomycetes (phylum *Oomycota*) are probably more nearly related to certain algae than to fungi. Like algae, their cell walls are of cellulose. Some species are unicellular, whereas others are filamentous or highly branched. Some of the filamentous forms are coenocytic (no cell walls separate adjacent cells). The name of the group reflects the large female gamete or egg; this type of sexual reproduction is called oogamy.

Some terrestial oomycetes are plant pathogens of considerable importance. Downy mildew of grapes, caused by *Plasmopara viticola*, has often threatened the wine industry of France. Species of the genus *Phytophthora* cause diseases of many fruit crops and other plants of economic importance. Among these is *P. infestans*, which causes the potato late blight. One particular outbreak of this parasite caused the infamous Irish Potato Famine of the mid-1840's, during which more than a million people (some estimates say four million) were affected.

Saprolegnia is a prominent member of a group of aquatic oomycetes called "water molds." Most are saprophytic on dead plants and animals, but a few are parasitic.

Thomas E. Hemmerly

See also: Algae; Brown algae; Cellular slime molds; *Charophyceae*; *Chlorophyceae*; Chrysophytes; Cryptomonads; Diatoms; Dinoflagellates; Euglenoids; Eukaryotic cells; Evolution of cells; Green algae; Haptophytes; Oomycetes; Phytoplankton; Plasmodial slime molds; Red algae; *Ulvophyceae*.

Sources for Further Study

Bold, H. C., and Michael J. Wynne. *Introduction to the Algae: Structure and Reproduction*. 2d ed. Englewood Cliffs, N.J. 1985. This classic by renowned phycologists provides detailed coverage of this important group of organisms.

Purves, William K., et al. *Life: The Science of Biology*. 6th ed. New York: W. H. Freeman, 2001. This university-level textbook includes an excellent overview of the *Protista* and the origin of eukaryotic cells (chapters 26 and 27).

Raven, Peter H., Ray F. Evert, and Susan E. Eichhorn. *Biology of Plants*. 6th ed. New York: W. H. Freeman, 1999. Chapters 16 and 17 include a detailed look at the various groups of organisms classified as protists.

PSILOTOPHYTES

Categories: Evolution; paleobotany; *Plantae*; seedless vascular plants; taxonomic groups

Psilotophyte is the common name for members of the phylum Psilotophyta *(from the Greek word* psilos*, meaning "bare"). Molecular evidence points to the likelihood of psilotophytes as being highly reduced (and therefore derived) ferns. If psilotophytes are indeed reduced ferns, they probably diverged from the fern lineage early, after ferns arose some 400 million years ago during the Devonian period.*

The family *Psilotaceae* is the only family of psilotophytes. There are two living genera: *Psilotum* and *Tmesipteris*. *Psilotum*, the whiskfern, is widespread throughout tropical and subtropical regions. Lacking leaves and roots, *Psilotum* species grow in a variety of soil conditions, including very warm soils near active volcanoes, or they may be epiphytic, growing on the trunks of host trees. *Psilotum nudum*, the best-known species of psilotophyte, has an upright growth habit and is easily maintained in greenhouse culture. One epiphytic species of *Psilotum* has a pendulous, or hanging, growth habit.

Tmesipteris is found only in the South Pacific region, including Australia, New Zealand, and some South Pacific islands. Terrestrial species of *Tmesipteris* may be upright in growth habit, while some species are epiphytes growing on the stems of ferns and trees or in mounds of moist humus. Epiphytic *Tmesipteris* species express a pendulous growth habit.

Although psilotophytes have a rather restricted range and do not appear to be a highly diverse group, they are important members of the ecosystems in which they are found. *Psilotum* has been cultivated as a horticultural specimen for many years in Asia. One particular variety, called *Bunryu-zan*, has been selected to express no prophylls and to produce synangia (fused sporangia, the sites of spore production) at the tips of the aerial branches, which makes this *Psilotum* variety appear even more similar to plants known only from the fossil record.

Life Cycle

Psilotophytes exhibit a life cycle pattern called alternation of generations. Haploid spores produced in the sporangia of the diploid sporophytes (the diploid, spore-producing, generation) are released when the sporangium splits open. The spores fall to the ground or into rock or bark crevices, where they germinate in complete darkness.

Haploid gametophytes (the haploid, gamete-producing generation) develop from the spores. The nonphotosynthetic underground gametophytes are completely dependent on their endophytic fungal partners for energy. Male and female sex organs develop on a single gametophyte. Usually, the male antheridia and the female archegonia develop at slightly different times to reduce the likelihood of self-fertilization. When gametes (egg and sperm) are mature and when liquid water is present, multiflagellated sperm are released from the antheridia and swim to a nearby archegonium. The sperm swim through the neck of the archegonium, where fertilization takes place in the venter. The resulting zygote develops into an embryo, which in turn develops into a young sporophyte. The young sporophyte stage is dependent on the gametophyte for support until it grows through the soil and begins photosynthesis.

Rob and Ann Simpson

Members of the genus Psilotum, *one of two living genera of psilotophytes, are known as whiskferns and are widespread throughout tropical and subtropical regions. Lacking leaves and roots,* Psilotum *species grow in a variety of soil conditions. Shown is* Psilotum nudum, *the best-known species of psilotophyte.*

Sporophyte Anatomy

As is the pattern in vascular plants, the diploid sporophyte generation is the dominant phase of the life cycle. Psilotophytes have many features that are similar to primitive vascular plants. Like fossil vascular plants, psilotophytes have a horizontal stem called a rhizome. Aerial branches grow from the rhizome. The lower portions of the aerial stems in *Psilotum* are often five-sided, while the upper aerial stems are usually three-sided. Both the rhizome and the aerial stem branch dichotomously, which means they tend to produce two equal branches at each node. *Psilotum* usually produces several dichotomies, or branching series, while *Tmesipteris* may have only one dichotomy on a particular aerial stem. Dichotomous branching is considered a primitive characteristic.

The outermost layer of the stem is composed of epidermal tissue. A waxy cuticle covers the epidermis of the aerial branches. Stomata are present in spaces between the ribs, which run lengthwise along the branch. Interior to the epidermis is a wide cortex region. The cortex is composed of parenchyma cells. The cortex parenchyma is important for storage, as is evident from the presence of the many starch granules in each cell. Lying inward from the cortex in *Psilotum* is a layer of cells called the endodermis. Endodermal cells have a layer of suberin (fatty material) called a Casparian strip embedded in part of their cell walls to restrict water movement into and out of the vascular cylinder. The endodermis is well developed throughout the aerial part of the *Psilotum* plant body as well as the underground parts. In *Tmesipteris*, the endodermis is present only in the rhizome. A layer of cells containing tannins and phenolic compounds lies between the phloem cylinder and the cortex in Tmesipteris stems. This layer may be the physiologically active equivalent of an endodermis.

Although psilotophytes lack true roots, they do possess rootlike epidermal extensions called rhizoids. The rhizoids aid in absorption of water and mineral nutrients and also act as the points of entry for symbiotic fungi, whose presence may be essential for the survival of the organism. The association of mutualistic fungi with plant roots or rhizoids is called a mycorrhiza. Because the fungi invade the body of the plant, they are sometimes referred to as endophytic fungi.

The xylem tissue of psilotophytes is composed of tracheids. Phloem tissue is made up of sieve cells, along with some parenchyma helper cells called albuminous cells. The stele, or vascular cylinder, of the psilotophyte rhizome is usually interpreted as a type of protostele. Protosteles have no pith; they have a solid core of vascular tissue. In the protostelic actinostele of *Psilotum*, the solid core of xylem has armlike extensions that reach into the surrounding tissues. Phloem surrounds the xylem. In the lower portions of the aerial branches of *Psilotum*, the stele is considered to be a type of siphonostele. In this case, the center of the stele is interpreted as a pith made up of sclerotic (hard, thick-walled) cells. This is similar to the aerial stem pattern in *Tmesipteris*. The tips of the aerial branches in *Psilotum* are strictly actinosteles.

Psilotum lacks true leaves. The leaflike appendages of the *Psilotum* shoot are called enations, or prophylls. The prophylls are composed of small flaps of photosynthetic tissue. Small traces of vascular tissue end at the base of the prophyll and do not actually enter the structure. In contrast, the foliar appendages of *Tmesipteris* are larger and are considered to be a type of true leaf called a microphyll. Microphylls have a single vein extending into the blade of the leaf.

Spores are produced in sporangia located on very short shoots on the sides of aerial branches. The lateral placement of sporangia in psilotophytes contrasts with the terminal placement of sporangia in primitive vascular plants such as *Rhynia*. Because each sporangium appears to represent a fusion of two sporangia (as in *Tmesipteris*) or three sporangia (as in *Psilotum*), they are generally called synangia. Cells within the synangium undergo meiosis to produce haploid spores. The spores are described as monolete, meaning that they have a single ridge. The monolete character is considered a derived trait, which is different from the *trilete*, or three-ridged, spores of primitive vascular plants.

Gametophyte Anatomy

The haploid psilotophyte gametophytes are very small and grow underground. In general, they are cylindrical, brown in color, covered with rhizoids, and are often branched. The branching may be dichotomous, as in the sporophyte, or irregular in response to the wounding of the gametophyte's apical meristems.

Gametophytes rely on endophytic fungi to obtain energy. Fungi invade nearly all of the cells of the gametophyte body except the apical meristems and the gametangia, or gamete-producing organs.

Both male and female sex organs are found scattered throughout a single plant. The antheridia (male organs) are small, multicellular, hemispherical structures that protrude from the surface of the gametophyte. Cells inside the antheridia undergo mitosis to produce multiflagellated sperm cells. The archegonia (female organ) has a swollen base called the venter. The venter is sunken below the surface of the gametophyte.

A cell within the venter undergoes mitotic cell division to produce an egg. Four rows of cells form the neck of the archegonium, which surrounds a neck canal. Two cells initially fill the neck canal. These neck canal cells break down when the archegonium is ready for fertilization.

Darrell L. Ray

See also: Evolution of plants; Ferns; Horsetails; Lycophytes; Seedless vascular plants.

Sources for Further Study

Bell, Peter R., and Alan R. Hemsley. *Green Plants: Their Origin and Diversity*. 2d ed. New York: Cambridge University Press, 2000. Contains an introduction to psilotophytes. Includes photos, diagrams.

Bold, Harold C., Constantine J. Alexopoulos, and Theodore Delevoryas. *Morphology of Plants and Fungi*. 5th ed. New York: HarperCollins, 1987. A classic text on plant morphology, including a chapter on psilotophytes. Includes photos, diagrams.

Gifford, Ernest M., and Adriance S. Forster. *Morphology and Evolution of Vascular Plants*. 3d ed. New York: W. H. Freeman, 1989. A standard plant morphology text. Devotes chapter to psilotophytes. Includes photos, diagrams.

Raven, Peter H., Ray F. Evert, and Susan E. Eichhorn. *Biology of Plants*. 6th ed. New York: W. H. Freeman/Worth, 1999. A standard text for introductory botany courses at the university level. Discusses structure and significance of major groups of seedless vascular plants, including psilotophytes. Includes photos, diagrams.

RAIN-FOREST BIOMES

Categories: Biomes; ecosystems; forests and forestry

Tropical forests exist in frost-free areas between the Tropic of Cancer and the Tropic of Capricorn. Temperatures range from warm to hot year-round. These forests are found in northern Australia, the East Indies, Southeast Asia, equatorial Africa, and parts of Central America and northern South America.

Rain forests are complicated tropical ecosystems with extremely high levels of *biodiversity*. Occupying only 6 percent of the earth's surface, they contain at least 50 percent of the world's known plant and animal species. The rain forests are being destroyed at such an unprecedented rate, however, that if the trend continues, no sustainable tropical rain forests will remain by the middle of the twenty-first century.

Tropical rain forests are the most complex ecosystems on earth, consisting of interacting systems of vegetation and animal species, all interdependent and coexisting in a concentrated cacophony of life. Because species found in one area may differ radically from those in another region only miles away, rain forests are individualistic and highly diversified.

The rain forests surrounding the Amazon basin in South America represent the last great contiguous expanse of tropical rain forest remaining in the world. Containing at least 20 percent of the earth's higher plant species and an equal percentage of the world's birds, Amazonia is being systematically devastated by human actions. About 20 percent has already been destroyed, and the rate has accelerated. In the mid-1990's about 14,000 acres were being cleared every day.

The environmental, economic, and social consequences of large-scale tropical deforestation are numerous and severe for the entire earth. The implications of losing the rain forests' biodiversity are dire. Moreover, rain forests act as giant solar-powered engines that pump water, nutrients, and carbon dioxide through the *biosphere*. Water captured by vegetation is returned to the atmosphere by evaporation and then to the forest as rain. The forest also stores large quantities of water in the vines and roots of the plants, water that is slowly released to streams and rivers. When large tracts of rain forest are cleared, the system's ability to store water decreases, disrupting the local climate and causing alternating periods of drought and floods.

Rain forests consist of lush, abundant growth, but the soil is deficient in nutrients and only marginally fertile. Although organic materials decompose rapidly in warm, humid climates, heavy rains leach nutrients from the soil. Plant life has adapted by rapidly ingesting nutrients as they become available; thus most nutrients are stored in the vegetation itself, not in the soil. When the land is cleared for agriculture or livestock grazing, the necessary crop-sustaining nutrients are depleted within a few growing seasons. Rain-forest destruction also adversely impacts the environment and human life by reducing the conversion of atmospheric carbon dioxide into oxygen.

Food Resources

Tropical rain forests are important sources of genetic material for improving crop plants or breeding new varieties. As the population size of a species shrinks, genetic diversity shrinks in direct proportion because as a species diminishes in number, genes disappear even if the species survives. A reduction of the gene pool renders a species less adaptable to changing environments and more susceptible to extinction. Future generations will be unable to benefit from currently unidentified but potentially useful properties of these species if their genetic diversity is extinguished.

Without periodic infusions of new germ plasm, crops bred specifically for humans, such as coffee, bananas, and cocoa, cannot continue to produce high yields at low cost. As the products of generations of selective breeding, these crops continually require the amalgamation of new genetic material to maintain productivity and flavor, to counteract new diseases and insect strains, and to endure environmental stresses such as unusual cold or drought.

In recent decades, a number of important crops, including cocoa, coffee, bananas, and sugarcane, have been saved from viruses and other pathogens by being crossbred with wild species having naturally acquired resistance. Although crop diseases can be contained or eliminated by applying pesticides, the cost is more than many farmers can afford. It is much less expensive, and more environmentally benign, to find a resistant strain of the same species in the wild.

Future contributions from wild germ plasm, in addition to breeding new disease-resistant varieties of crops, might include the creation of hybrid perennial varieties of annual crops, eliminating the need for yearly plowing and sowing. Another possibility may be new varieties of conventional crops which could survive in conditions or environments that are currently unsuitable, extending the plants' cultivation range.

Rain forests also provide opportunities for humans to develop and cultivate entirely new crops. Many staples, such as rice, peanuts, yams, and pineapples, originated in ancient tropical forests. These staples are not necessarily the best possible sources of nutrition and protein; they were merely the crops most easily cultivated by Neolithic humans.

While temperate forests produce fewer than two dozen edible fruits, more than twenty-five hundred palatable fruits grow in the tropics. Only about 10 percent of these are typically consumed, and only about one dozen are exported. The variety of fruits that Americans purchase from grocery stores is but a small fraction of those potentially available.

Because Americans consume an enormous quantity of sugar annually, there is a need for a sweetening agent lacking the potentially undesirable side effects of synthetic sweeteners. Natural sweeteners found in common fruits are problematic because Americans already consume more of these than is healthy. However, a new class of nonfattening natural sweeteners made of protein compounds has been identified. At least one thousand times sweeter than sucrose, and with no known detrimental side effects, these tropical sweeteners may prove a viable replacement for the sucrose commonly added to food items ranging from salad dressing to fish products.

African Rain Forests

Although the rain forests of central Africa are at lower risk of deforestation than are tropical forests in some other parts of the world, they are still in great danger. First, they contain many valuable timber trees that are being exploited actively. Second, these forested areas are in demand for other uses. In some areas, excessive timber harvesting and clearing to make way for oil palm and rubber plantations leaves little untouched forest. Some of the worst damage has occurred in Liberia and the Ivory Coast, where populations are on the rise. Much research is being done to develop sustainable and regenerative ways to harvest trees. In many areas, certain parts of the forest are considered sacred by local tribes and are jealously guarded from all encroachment. Medicinal plants are obtained from these areas, and they also are used as burial grounds that shelter the tribes' ancestral spirits.

Medicinal Plants

Rain-forest plants contain a plethora of yet uninvestigated biodynamic compounds with undiscovered potential for use in modern medicine. Future generations may benefit from these substances only if the species containing them are preserved and studied. Although the medicinal properties of plants have been known for at least thirty-five thousand years, modern science has reduced human dependence on medicinal plants. Nevertheless, approximately 50 percent of all U.S. prescription drugs contain substances of natural origin, and at least half of these contain active ingredients derived from plants.

Medicinal plants from the tropics aid modern pharmacologists in four ways. Plants may be used as a direct source of therapeutic drugs that cannot effectively be synthesized in the laboratory. Plant extracts may become the starting point for more complex compounds. Derived substances can serve as "blueprints" for new synthetic compounds. Finally, plants may assist as taxonomic markers for uncovering new healing compounds. Of the hundreds of thousands of plant species inhabiting the rain forests, only a small fraction have been identified and studied, and the potentially beneficial pharmacological properties of many of these are yet to be ascertained.

Endangered Biome

Although extinction is a natural process, tropical rain forests are disappearing at a rate four hundred times faster than during the recent geological past. One prominent and direct cause of the destruction of the rain forest in Amazonia is the production and transportation of oil. The fragile rain forest environment is easily contaminated by leaks, spills, and the ejection of effluents during pumping operations. Of even greater environmental impact is the oil companies' practice of building roads from inhabited areas to the well sites. Pipelines are built along the roads to carry the oil out of the jungle, but unemployed urban residents often follow the roads into the jungle and become squatters on adjoining land. They clear a small section of rain forest, using the *slash-and-burn* method, to eke out a living as subsistence farmers.

Unfortunately, when the rain forest is gone, so are most of the nutrients needed for local agriculture; the land cannot sustain crops for long. Then the farmers and their families must move on and claim more of the forest, regardless of the effects on the jungle or its native species. The tens of thousands of squatters engaging in this practice contribute to the rapid rate of rain-forest destruction.

Although aware of the problem, the governments of most South American oil-exporting countries have a strong incentive to underplay or ignore the negative impact of oil production in their rain forests: Their economies depend on oil, which is one of South America's largest exports. Ultimately the demand for oil, driven by high consumption rates in the industrialized nations of the Northern Hemisphere, particularly the United States, is one of the primary factors causing rain-forest destruction.

George R. Plitnik

See also: African flora; African agriculture; Asian flora; Asian agriculture; Australian flora; Australian agriculture; Biomes: types; Endangered species; Plant domestication and breeding; Rain forests and the atmosphere; Slash-and-burn agriculture; South American flora; South American agriculture.

Sources for Further Study

Branford, Susan, and Oriel Glock. *The Last Frontier: Fighting Over Land in the Amazon*. Totowa, N.J.: Biblio Distribution Center, 1985. Explains the battles being fought and lost in Amazonia and presents cogent arguments on why it is happening.

Farnsworth, N. "Screening Plants for New Medicines." In *Biodiversity*, edited by E. O. Wilson. Washington, D.C.: National Academy Press, 1988. Presents arguments for preserving the vast pharmacological potential of rain forests.

Myers, Norman. *A Wealth of Wild Species: Storehouse for Human Life*. Boulder, Colo.: Westview Press, 1983. Discusses the importance of tropical rain forests as a viable and necessary source of new genetic material important to the future of humanity.

Place, Susan, ed. *Tropical Rainforests: Latin American Nature and Society in Transition*. Rev. ed. Wilmington, Del.: Scholarly Resources, 2001. Offers comprehensive coverage of complex rain-forest issues.

RAIN FORESTS AND THE ATMOSPHERE

Categories: Biomes; ecosystems; environmental issues; forests and forestry; pollution

Because photosynthesis releases large amounts of oxygen into the air, a curtailment of the process by rain-forest deforestation may have negative effects on the global atmosphere.

Rain forests are ecosystems noted for their high *biodiversity* and high rate of *photosynthesis*. The rapid *deforestation* of such areas is of great concern to environmentalists both because it may lead to the extinction of numerous species and because it may reduce the amount of photosynthesis occurring on the earth.

All living things on the earth—plants, animals, and microorganisms—depend on the "sea" of air surrounding them. The atmosphere includes abundant, permanent gases such as nitrogen (78 percent) and oxygen (21 percent) as well as smaller, variable amounts of other gases such as water vapor and carbon dioxide. Organisms absorb and use this air as a source of raw materials and release into it by-products of their life activities.

Cellular Respiration

Cellular respiration is the most universal of the life processes. A series of chemical reactions beginning with glucose and occurring in cytoplasmic organelles called mitochondria, cellular respiration produces a chemical compound called adenosine triphosphate (ATP). This essential substance furnishes the energy cells need to move, to divide, and to synthesize chemical compounds—in essence, to perform all the activities necessary to sustain life. Cellular respiration occurs in plants as well as animals, and it occurs during both the day and the night. In order for the last of the series of chemical reactions in the process to be completed, oxygen from the surrounding air (or water, in the case of aquatic plants) must be absorbed. The carbon dioxide that forms is released into the air.

For cellular respiration to occur, a supply of glucose (a simple carbohydrate compound) is required. Photosynthesis, an elaborate series of chemical reactions occurring in chloroplasts, produces glucose, an organic carbon compound with six carbon atoms. Energy present in light must be trapped by the chlorophyll within the chloroplasts to drive photosynthesis. Therefore, photosynthesis occurs only in plants and related organisms, such as algae, and only during the daytime. Carbon dioxide, required as a raw material, is absorbed from the air, while the resulting oxygen is released into the atmosphere. The exchange of gases typically involves tiny openings in leaves, called stomata.

AP/Wide World Photos

A rain cloud hangs over the rain forest in Lope Reserve, Gabon. In such tropical forests, the many layers of forest vegetation result in energy from sunlight being efficiently used as it passes downward. The large amounts of oxygen released are available for use not only by the forests but also, because of global air movement, by other ecosystems throughout the world.

Oxygen Cycle

Oxygen is required for the survival of the majority of microorganisms and all plants and animals. From the surrounding air, organisms obtain the oxygen used in cell respiration. Plants absorb oxygen through the epidermal coverings of their roots and stems and through the stomatal openings of their leaves.

The huge amounts of oxygen removed from the air during respiration must be replaced in order to maintain a constant reservoir of oxygen in the atmosphere. There are two significant sources of oxygen. One involves water molecules of the atmosphere that undergo a process called *photodissociation*: Oxygen remains after the lighter hydrogen atoms are released from the molecule and escape into outer space.

The other source is *photosynthesis*. Chlorophyll-containing organisms release oxygen as they use light as the energy source to split water molecules in a process called photolysis. The hydrogen is transported to the terminal phase of photosynthesis called the *Calvin cycle*, where it is used as the hydrogen source necessary to produce and release molecules of the carbohydrate glucose. In the meantime, the oxygen from the split water is released into the surrounding air.

Early in the history of the earth, before certain organisms evolved the cellular machinery necessary for photosynthesis, the amount of atmospheric oxygen was very low. As the number and sizes of photosynthetic organisms gradually increased, so did the levels of oxygen in the air. A plateau was reached several million years ago as the rate of oxygen release and absorption reached an equilibrium.

Ozone

Another form of oxygen is ozone. Unlike ordinary atmospheric oxygen, in which each molecule contains two atoms, ozone molecules have three oxygen atoms each. Most ozone is found in the stratosphere at elevations between 10 and 50 kilometers (6 and 31 miles). This layer of ozone helps to protect life on earth from the harmful effects of ultraviolet radiation. Scientists, especially ecologists, are concerned as the amount of ozone has been reduced drastically over the last few decades. Already, an increase in the incidence of skin cancer in humans and a decrease in the efficiency of photosynthesis has been documented. Another concern related to ozone is that of an increase in ozone levels nearer to the ground, where living things are harmed as a result. The formation of ozone from ordinary oxygen within the atmosphere is greatly accelerated by the presence of gaseous pollutants released from industrial processes.

Carbon Cycle

All forms of life are composed of organic (carbon-containing) molecules. Carbohydrates include glucose as well as lipids (fats, oils, steroids, and waxes), proteins, and nucleic acids. The ability of carbon to serve as the backbone of these molecules results from the ability of carbon atoms to form chemical bonds with other carbon atoms and also with oxygen, hydrogen, and nitrogen atoms.

Like oxygen, carbon cycles in a predictable manner between living things and the atmosphere. In photosynthesis, carbon is "fixed" as carbon dioxide in the air (or dissolved in water) is absorbed and converted into carbohydrates. Carbon cycles to animals as they feed on plants and algae. As both green and nongreen organisms respire, some of their carbohydrates are oxidized, releasing carbon dioxide into the air. Each organism must eventually die, after which decay processes return the remainder of the carbon to the atmosphere.

Greenhouse Effect

Levels of atmospheric carbon dioxide have fluctuated gradually during past millennia, as revealed by the analysis of the gas trapped in air bubbles of ice from deep within the earth. In general, levels were lower during glacial periods and higher during warmer ones. After the nineteenth century, levels rose slowly until about 1950 and then much more rapidly afterward. The apparent cause has been the burning of increased amounts of fossil fuels associated with the Industrial Revolution and growing energy demands in its wake. The global warming that is now being experienced is believed by most scientists to be the cause of increased carbon dioxide levels. The *greenhouse effect* is the term given to the insulating effects of the atmosphere with increased amounts of carbon dioxide. The earth's heat is lost to outer space less rapidly, thus increasing the earth's average temperature.

Forest Ecosystems

The *biotic* (living) portions of all ecosystems include three ecological or functional categories: *pro-*

ducers (plants and algae), *consumers* (animals), and *decomposers* (bacteria and fungi). The everyday activities of all organisms involve the constant exchange of oxygen and carbon dioxide between the organisms of all categories and the surrounding atmosphere.

Because they release huge quantities of oxygen during the day, producers deserve special attention. In both fresh and salt water, algae are the principal producers. On land, this role is played by a variety of grasses, other small plants, and trees. Forest ecosystems, dominated by trees but also harboring many other plants, are major systems that produce a disproportionate amount of the oxygen released into the atmosphere by terrestrial ecosystems.

Forests occupy all continents except for Antarctica. A common classification of forests recognizes these principal categories: coniferous (northern evergreen), temperate deciduous, and tropical evergreen, with many subcategories for each. The designation "rain forest" refers to the subcategories of these types that receive an amount of rainfall well above the average. Included are tropical rain forests (the more widespread type) and temperate rain forests. Because of the ample moisture they receive, both types contain lush vegetation that produces and releases oxygen into the atmosphere on a larger scale than do other forests.

Tropical and Temperate Rain Forests

Tropical rain forests exist at relatively low elevations in a band about the equator. The Amazon basin of South America contains the largest continuous tropical rain forest. Other large expanses are located in western and central Africa and the region from Southeast Asia to Australia. Smaller areas of tropical rain forests occur in Central America and on certain islands of the Caribbean Sea, the Pacific Ocean, and the Indian Ocean. Seasonal changes within tropical rain forests are minimal. Temperatures, with a mean near 25 degrees Celsius, seldom vary more than 4 degrees Celsius. Rainfall each year measures at least 400 centimeters.

Tropical rain forests have the highest biodiversity of any terrestrial ecosystem. Included is a large number of species of flowering plants, insects, and animals. The plants are arranged into layers, or strata. In fact, all forests are stratified but not to the same degree as tropical rain forests. A mature tropical rain forest typically has five layers. Beginning with the uppermost, they are an *emergent* layer (the tallest trees that project above the next layer); a *canopy* of tall trees; *understory* trees; shrubs, tall herbs, and ferns; and low plants on the forest floor.

Several special life-forms are characteristic of the plants of tropical rain forests. *Epiphytes* are plants such as orchids that are perched high in the branches of trees. Vines called *lianas* wrap themselves around trees. Most tall trees have trunks that are flared at their bases to form buttresses that help support them in the thin soil.

This brief description of tropical rain forests helps to explain their role in world photosynthesis and the related release of oxygen into the atmosphere. As a result of the many layers of forest vegetation, the energy from sunlight as it passes downward is efficiently utilized. Furthermore, the huge amounts of oxygen released are available for use not only by the forests themselves but also, because of global air movement, by other ecosystems throughout the world. Because of this, tropical rain forests are often referred to as "the earth's lungs."

Temperate rain forests are much less extensive than tropical rain forests. Temperate rain forests occur primarily along the Pacific Coast in a narrow band from southern Alaska to central California. Growing in this region is a coniferous forest but one with higher temperatures and greater rainfall than those to the north and inland. This rainfall of 65 to 400 centimeters per year is much less than that of a tropical rain forest but is supplemented in the summer by frequent heavy fogs. As a result, evaporation rates are greatly reduced. Because of generally favorable climatic conditions, temperate rain forests, like tropical ones, support a lush vegetation. The rate of photosynthesis and release of oxygen is higher than in most other world ecosystems.

Ecologists and conservationists are greatly concerned about the massive destruction of rain forests. Rain forests are being cut and burned at a rapid rate to plant crops, to graze animals, and to provide timber. The ultimate effect of deforestation of these special ecosystems is yet to be seen.

Thomas E. Hemmerly

See also: Acid rain; Air pollution; Biosphere concept; Calvin cycle; Carbon cycle; Deforestation; Ecosystems: overview; Ecosystems: studies; Greenhouse effect; Hydrologic cycle; Ozone layer and ozone hole debate; Photorespiration; Photosynthesis; Rain-forest biomes; Respiration; Savannas and deciduous tropical forests.

Sources for Further Study

Laurance, William F., and Richard O. Bierregaard, eds. *Tropical Forest Remnants: Ecology, Management, and Conservation of Fragmented Communities*. Chicago: University of Chicago Press, 1997. A compilation of articles on tropical forest habitat fragmentation and its effects on global biology. Each section of the book deals with a different component of the problems.

Shipp, Steve. *Rainforest Organizations: A Worldwide Directory of Private and Governmental Entities*. Jefferson, N.C.: McFarland, 1997. Contains contact and mission information for 120 environmental, governmental, nongovernmental, religious, and educational agencies dealing with rain-forest issues.

Townsend, Janet G. *Women's Voices from the Rainforest*. New York: Routledge, 1995. Discusses the many issues of rain-forest management which are obfuscated by the language and methods of international development policy.

Vandermeer, John. *Breakfast of Biodiversity: The Truth About Rain Forest Destruction*. Oakland, Calif.: Institute for Food and Development Policy, 1995. Covers human involvement in tropical forests in general and ramifications of U.S. policy in the tropics.

RANGELAND

Categories: Animal-plant interactions; economic botany and plant uses; ecosystems.

Open land of a wide variety of types, including grasslands, shrublands, marshes, and meadows as well as some desert and alpine land, is known as rangeland.

Rangeland is a valuable and resilient *ecosystem* resource that supports considerable plant and animal life. Rangeland generally refers to a kind of land rather than a use of that land. The Society for Range Management defines rangelands as "land on which the native vegetation (climax or natural potential) is predominantly grasses, grass-like plants, forbs, or shrubs." Rangeland "includes lands revegetated naturally or artificially" as well as "natural grasslands, savannas, shrublands, most deserts, tundra, alpine communities, coastal marshes and wet meadows."

Rangelands usually have some limitation for intensive agriculture, such as low and erratic precipitation, lack of soil fertility, shallow or rocky soil, or steep slopes. In addition to livestock *grazing*, rangelands serve multiple-use functions such as providing recreational opportunities, *watersheds*, mining locations, and habitat for many animal species. Renewable natural resources associated with rangelands are plants and animals (and, in some senses, water). Nonrenewable resources include minerals and other extractable materials.

Location and Characteristics

Rangelands are extensive and extremely variable. As defined by the Society for Range Management, they occupy more than 50 percent of the world's total land surface and about 1 billion acres in the United States alone. Rangelands are home to nomadic herders on nearly every continent. They vary from high-elevation alpine tundra and high-latitude Arctic tundra to tropical grasslands. The *tall-grass prairies* in the United States (now mostly plowed for intensive agriculture) and the rich grasslands of eastern Africa are among the most productive.

Rangelands grade into woodlands and forests as woody species and trees become more abundant. Some forests are grazed by wild and domestic animals, and the distinction between rangeland and forest is often not clear. The other difficult distinction is between rangeland and pastureland. Pastureland is generally improved by seeding, fertilization, or irrigation, whereas rangelands support native plants and have little intensive improvement.

AP/Wide World Photos

Horses graze amid the sagebrush in Pinedale, Wyoming. In addition to livestock grazing, rangelands serve multiple-use functions, providing recreational opportunities, watersheds, mining locations, and habitat for many animal species.

In the United States, rangeland improvements during the twenty years following World War II often included brush control, grazing management, and seeding, but rangelands were not irrigated. After the 1970's, when fuel costs increased and environmental concerns about pesticide use increased, brush control practices were reduced considerably. Today environmental concerns include rangeland degradation from *overgrazing*, especially on riparian vegetation along streams, and concern for endangered animal and plant species. These issues have become controversial in the United States.

Rangelands as Ecosystems

Rangelands constitute natural ecosystems with nonliving environmental factors such as soil and climatic factors. Life-forms are *primary producers* (grasses, forbs, and shrubs), *herbivores* (livestock; big game animals such as deer and bison; and many rodents and insects), *carnivores* (such as coyotes, bears, and eagles), and *decomposers* (fungi and bacteria) that break down organic matter into elements that can be utilized by plants. Plants convert carbon dioxide and water into complex carbohydrates, fats, and proteins that nourish animals feeding on the plants.

Individual chemical elements are circulated throughout the various components. Many of these elements are present in the soil, including phosphorus, magnesium, potassium, and sulfur. Nitrogen, on the other hand, is present in large amounts in the atmosphere but must be converted (fixed) into forms that can be utilized by plants before it can be cycled. Energy is fixed through the process of photosynthesis and transformed to forms useful for the plants, then the animals that feed on plants.

When chemicals are taken up by plant roots from the soil, they become available to a wide group of herbivores, from small microbes to large ungulates. Eventually nutrients are passed on to organisms at higher *trophic levels* (omnivores and carnivores). Both plant and animal litter is eventually broken down by decomposers—bacteria, fungi, and other soil organisms—and returned to the soil or, in the case of carbon or nitrogen, given off to the atmosphere.

However, energy is degraded at each step along the way; energy is transferred but not cycled. Grazing animals on rangelands influence plants by removing living tissue, by trampling, and by altering competitive relations with other plants. Large grazing animals tend to compact the soil, reducing infiltration and increasing surface runoff.

Rangeland Dynamics

Rangelands vary considerably with time. Scientists are gaining a better understanding of some factors related to rangeland change. Pollen records and, in the southwestern United States, packrat middens have been used to reconstruct past climate and vegetational conditions. Some areas have become drier and others more mesic. The formation and retreat of glaciers influenced climatic patterns and soil development. A recent general trend in many rangelands is an increase in woody plants at the expense of grasses. Many factors are probably responsible for these shifts, but fire control, overgrazing, climatic shifts, introduction of exotic species, and influence of native animals are likely causal agents.

Rangelands are being threatened by encroachment from crop agriculture as worldwide development increases. Nomadic herders traditionally met periodic drought conditions by having the flexibility to move to areas not impacted by drought. Now, with area lost to livestock grazing and other political restrictions, herders are often forced to maintain higher livestock numbers to support those directly dependent on livestock. Despite various kinds of disturbances and stresses on rangelands, these areas have supported many large grazing animals and people for centuries.

Rex D. Pieper

See also: Agriculture: history and overview; Agriculture: modern problems; Forest and range policy; Forests; Grasses and bamboos; Grasslands; Grazing and overgrazing; Trophic levels and ecological niches; Tundra and high-altitude biomes.

Sources for Further Study

Heady, Harold F., and R. Dennis Child. *Rangeland Ecology and Management*. 2d ed. Boulder, Colo.: Westview, 2000. General text on rangeland grazing, animal management, and vegetation manipulation.

Holechek, Jerry L., Rex D. Pieper, and Carlton H. Herbel. *Range Management: Principles and Practices*. 4th ed. Upper Saddle River, N.J.: Prentice Hall, 2001. General text on rangelands and their management.

Jacobs, Lynn. *Waste of the West: Public Lands Ranching*. Tucson, Ariz.: L. Jacobs, 1991. This critical account of public lands ranching deals with historical and contemporary situations and issues of environment, economics, politics, animal welfare, livestock production, and more. Includes ideas for activism, bibliography, and index.

Longworth, John W., and Gregory J. Williamson. *China's Pastoral Region: Sheep and Wool, Minority Nationalities, Rangeland Degradation, and Sustainable Development*. Wallingford, England: CAB International, 1993. Comprehensive picture of this increasingly important part of the world focuses on improving and preserving the livelihood of pastoral communities in northern and northwestern China.

Sayre, Nathan F. *The New Ranch Handbook: A Guide to Restoring Western Rangelands*. Santa Fe, N.Mex.: Quivera Coalition, 2001. Describes practices of New Ranches in New Mexico, Arizona, and Colorado, where vegetation and soil have responded positively to innovative ways of managing livestock.

RED ALGAE

Categories: Algae; economic botany and plant use; food; taxonomic groups; water-related life

Red algae, from the phylum Rhodophyta *(from the Greek* rhodos, *meaning "red"), are named for their reddish color because of their pigment phycoerythrin. Despite their name, not all rhodophytes are red. Some of them have very little phycoerythrin and appear green or bluish. Most are multicellular and vary greatly in shape; platelike, coral-like, crustlike, leathery, and featherlike forms are known.*

The bodies of some rhodophytes are relatively complex, characterized by a great deal of branching of their leaflike structures as well as the presence of a *holdfast* that resembles plant roots. These rhodophytes are commonly known as seaweeds. Red algae are distinctive from other eukaryotic algae in that they lack flagella (or motile cells of any kind) in their vegetative cells, spores, and gametes.

There are four thousand to six thousand species of red algae, and although some rhodophytes do inhabit fresh water (about fifty species), red algae are most common in tropical marine environments. Many are found at great depths, living 210-260 meters below the surface of the ocean. The pigment *phycoerythrin* allows red algae to live and photosynthesize at these depths. Phycoerythrin absorbs blue light, which penetrates water to a greater depth than light of longer wavelengths normally used in photosynthesis. Reds are the deepest-growing photosynthetic eukaryotes.

Habitats

Rhodophytes are important members of many *periphyton* communities (which typically grow attached to substrata) from tropical to polar seas. Epiphytism is common among red algae. Epiphytic rhodophytes grow on the surface of larger, typically brown, algae. Although their holdfasts penetrate the tissue of their hosts, they do not obtain any nutrients from host algae.

Nearly 15 percent of red algae are parasites, often living on closely related species. Parasitic reds transfer nuclei into host cells and transform them. Host reproductive cells may then carry the parasite's genes. Other *Rhodophyta* are extremophiles. For example, the red alga *Cyanidium* lives in acidic (pH 2-4) hot springs at 55 degrees Celsius. Some calcified rhodophytes are major contributors to the formation of tropical reefs. Reef-building red algae are called coralline algae.

Structure and Properties

A red algal cell is surrounded by a cell wall. In many species, the main wall component is cellulose (similar to cell walls of various other algae and plants), but other reds have mannans (polymers of mannose) and xylans (polymers of xylose). Other polymers associated with the cell walls include *agar* and *carrageenan*. The majority of coralline red algae contain calcium carbonate, which forms limestone in the cell walls. Because of their ability to secrete calcium carbonate, red algae do not decay and have a better preserved fossil record than many other algae. Fossils of red algae have been found in rocks 500 million years old. Production of calcium carbonate is linked to photosynthetic carbon fixation. Apparently, carbon dioxide fixation results in a pH increase (an increase in alkalinity), which facilitates calcium carbonate precipitation.

An unusual feature of red algae compared to other algae is the occurrence of protein plugs (pit connections) in cell walls between the cells, although some red algae lack them. All plugs consist of protein, and in some species protein polysaccharides are an additional component. Cells of the *Rhodophyta* may contain several nuclei as a result of either the fusion of nongamete cells or mitosis without cytokinesis. Cell fusion is a very important feature of parasitic reds. Red algae lack centrioles, but the mitotic spindle radiates from the "nuclear-associated organelle," which often appears as a pair of short, hollow cylinders. Some red algae have large vacuoles in the centers of their cells. Cells of the *Rhodophyta* may produce mucilage, which plays an important role in the attachment of their reproductive cells. Mucilages are polymers of D-xylose, D-glucose, D-glucuronic acid, and galactose and are produced within Golgi apparatuses.

Pigments of red algae include chlorophyll *a* and two classes of accessory pigments: phycobilins and carotenoids. Phycoerythrin, phycocyanin, and allophycocyanin are phycobilins. They attach to proteins known as phycobiliproteins, which occur in highly organized structures called *phycobilisomes*. Phycoerythrin occurs in at least five forms in the *Rhodophyta* (B-phycoerythrin I and II, R-phycoerythrin I, II, and II). Carotenoids are also found in plants, and those in red algae are similar in structure and function.

Some parasitic forms of red algae lack photosynthetic pigments. In red algae that have pigments, all pigments are located in the chloroplasts. Red algae chloroplasts have a highly distinctive ultrastructure. Two membranes surround each chloroplast. Chloroplasts of red algae probably originated from cyanobacteria that formed an ancient symbiotic relationship with the reds. Both red algal chloroplasts and cyanobacteria share same phycobilin pigments. Inside the chloroplast are thylakoids, which are not stacked. This is the same arrangement found in cyanobacteria, but it is different from that of other algae and from plants. On the thylakoid surface there are many phycobilisomes. Some red algae have chloroplasts that contain pyrenoids, which have no known function.

Photoautotrophy is the principal mode of nutrition in red algae; in other words, they are "self-feeders," using light energy and photosynthetic apparatuses to produce their own food (organic carbon) from carbon dioxide and water. A few *Rhodophyta* are heterotrophic, and these organisms are generally obligate parasites (parasites that must live off a host) of other algae. Carbon and nitrogen metabolism in red algae is similar to that in other algae. Various rhodophytes produce unusual carbohydrates, such as digeneaside, which is used to regulate osmotic status of cells in response to drought stress in shoreline environments. Some red algae are covered by surface-protein *cuticle*, which is different from that found in higher plants. The food storage of red algae is a unique polysaccharide floridean starch. This starch differs from that synthesized by green algae and plants. Floridean starch grains are formed in the cytoplasm. Red algae store inorganic nitrogen in the form of phycobilin pigments.

Reproduction

Red algae reproduce both asexually and sexually. Methods of asexual reproduction include discharging spores and fragmentation of the algal bodies. Sexual reproduction, as well as alteration of generations, is widespread among the *Rhodophyta*, but two classes of red algae (floridean and bangean) have particular variations.

In contrast to the two phases in an alteration of generations of other algae and plants (gametophyte and sporophyte, haploid and diploid stages, correspondingly), most species of red floridean algae have three phases: free-living, haploid gametophytes, diploid carposporophytes, and diploid tetrasporophytes. Male and female gametophytes are often separate. The male gametophytes produce male nonflagellated gametes called spermatia. Female gamethophytes produce a special branch, the carpogonial branch, that produces a terminal carpogonium (oogonium, an egg-bearing structure).

Contact between spermatia and carpogonia is facilitated by water movements. The carposporophyte is a diploid stage that develops from the zygote (fertilized carpogonium) and produces carpospores. Diploid tetrasporophytes develop from carpospores. Tetrasporophytes form tetrasporangia, which produce four haploid tetraspores. When released, tetraspores develop into new gametophytes. The gametophyte and tetrasporophyte may appear nearly identical, and therefore can be said to be isomorphic, as in the *Polysiphonia*. Alternatively, the tetrasporophyte and gametophyte may be very different in size and appearance (heteromorphic), as in *Phyllophora*.

Diversity of Red Algae

Red algae are divided into two subclasses or classes: *Florideophyceae* (florideophyceans or floridean) and *Bangiophyceae* (bangiophyceans or bangean). Floridean algae have numerous small chloroplasts and a complex life cycle. Bangean algae have life cycles without carpogonia and carposporophyte development and have a single central chloroplast. Representative species of *Florideophyceae* are *Batrachospermum*, *Chondrus*, *Corallina*, *Gelidium*, and *Polysiphonia*. Representatives of *Bangiophyceae* include *Porphyra*, *Bangia*, and *Cyanidium*.

Uses

People have used red algae for thousands of years. Most are collected along seashores for use in human food or for the extraction of gelling compounds. A few red algae, such as *Porphyra*, *Eucheuma*, and *Gracilaria*, are cultivated. More than 60,000 hectares of sea along Japanese coasts are occupied by "red algal culture." Thousands of people worldwide are engaged in cultivating red seaweeds.

The most valuable of all algae is *Porphyra*. The annual *Porphyra* harvest worldwide has been estimated to be worth 2.5 billion dollars. *Porphyra* (in Japanese, *Nori*; in Chinese, *Zicai*) is used as a wrapper for sushi or may be eaten mixed with rice and fish and in salads. It is very rich in vitamins B and C as well as minerals, including iodine. There are about seventy species of *Porphyra*, but the most widely used species is *Porphyra yezoensis*. Two important compounds derived from red algae are agar and carrageenan, both of which are polymers of galactose. Agar is used as a medium for culturing microorganisms, including algae; as a food gel (for jams and jelly); and in pharmaceutical capsules. In the United States, agar is used in the canning industry as a protective agent against the unwanted effects of metals. In addition, agar is the source of agarose, which is widely used in recombinant DNA (deoxyribonucleic acid) technology for gel electrophoresis. The first agar was produced in 1670 in Japan, and Japan is still the largest producer of agar.

The red algae *Gelidium*, *Gracelaria*, and *Pterocladia* are harvested for extraction of agar. Carrageenan is used in toothpaste, cosmetics, and food, such as ice cream and chocolate milk. *Eucheuma*, *Kappaphycus*, and *Chondus* (the so-called Irish moss) are the sources of carrageenan. The most important producer of carrageenan is Europe, followed by the Philippines and Indonesia.

Sergei A. Markov

See also: Agriculture: marine; Algae; Brown algae; Cell wall; *Charophyceae*; *Chlorophyceae*; Chrysophytes; Cryptomonads; Diatoms; Dinoflagellates; Electrophoresis; Euglenoids; Eutrophication; Green algae; Haptophytes; Marine plants; Photosynthesis; Phytoplankton; Pigments in plants; *Protista*; *Ulvophyceae*.

Sources for Further Study

Graham, Linda E., and Lee W. Wilcox. *Algae*. Upper Saddle River, N.J.: Prentice Hall, 2000. This textbook focuses on the major algal types and gives a comprehensive overview of the phylum *Rhodophyta*.

Van Den Hoek, C., et al. *Algae: An Introduction to Phycology*. Cambridge, England: Cambridge University Press, 1995. A comprehensive text on major algal groups, including red algae.

REFORESTATION

Categories: Forests and forestry; environmental issues

Reforestation is the growth of new trees in an area that has been cleared for human activities. It can occur naturally or be initiated by people.

Many areas of the eastern United States, such as the New England region, reforested naturally in the nineteenth and early twentieth centuries after farmland that had been abandoned was allowed to lie fallow for decades. After an area has been logged, environmentalists, as well as the commercial logging industry, advocate planting trees rather than waiting for natural regrowth because the process of natural regeneration can be both slow and unpredictable. In natural regeneration, the mixture of trees in an area may differ significantly from the forest that preceded it. For example, when nineteenth century loggers *clear-cut* the white pine forests of the Great Lakes region, many logged-over tracts grew back primarily in mixed hardwoods.

Nursery workers harvest ponderosa pine seedlings. A decline in timber harvests in the western United States means less demand by national forests for fresh stocks of trees.

Land that has been damaged by industrial pollution or inefficient agricultural practices sometimes loses the ability to reforest naturally. In some regions of Africa, soils exposed by *slash-and-burn agriculture* contain high levels of iron or aluminum oxide. Without a protective cover of vegetation, even under cultivation, soil may undergo a process known as *laterization*. Laterite is a residual product of rock decay that makes soil rock-hard. Such abandoned farmland is likely to remain barren of plant life for many years. In polluted areas such as former mining districts, native trees may not be able to tolerate the toxins in the soil; in these cases, more tolerant species must be introduced.

Safeguarding Timber Resources

Reforestation differs from *tree farming* in that the goal of reforestation is not always to provide woodlands for future harvest. Although tree farming is a type of reforestation (trees are planted to replace those that have been removed), generally only one species of tree is planted, with explicit plans for its future harvest. The trees are seen first as a crop and only incidentally as wildlife habitat or a means of erosion control.

As foresters have become knowledgeable about the complex interactions within forest ecosystems, however, tree farming methods have begun to change. Rather than *monocropping* (planting only one variety of tree), the commercial forest industry has begun planting mixed stands. Trees that possessed no commercial value, once considered undesirable weed trees, are now recognized as nitrogen fixers necessary for the healthy growth of other species. In addition to providing woodlands for possible use in commercial forestry, goals of reforestation include wildlife habitat restoration and the reversal of environmental degradation.

Early Efforts

Reforestation to replace trees removed for commercial purposes has been practiced in Western

Europe since the late Middle Ages. English monarchs, including Queen Elizabeth I, realized that forests were a vanishing resource and established plantations of oaks and other hardwoods to ensure a supply of ship timbers. Similarly, Sweden created a corps of royal foresters to plant trees and watch over existing woodlands. These early efforts at reforestation were inspired by the reduction of a valuable natural resource. By the mid-nineteenth century it was widely understood that the removal of forest cover contributes to soil erosion, water pollution, and the disappearance of many species of wildlife.

Ecological and Environmental Aspects

Water falling on hillsides made barren by clear-cutting timber washes away *topsoil* and causes rivers to choke with sediment, killing aquatic life. Without trees to slow the flow of water, rain can also run off slopes too quickly, causing rivers to flood. For many years, soil conservationists advocated reforestation as a way to counteract the ecological damage caused by erosion.

In the mid-twentieth century, scientists established the vital role that trees, particularly those in tropical rain forests, play in removing carbon dioxide from the earth's atmosphere through the process of photosynthesis. Carbon dioxide is a greenhouse gas: It helps trap heat in the atmosphere. As forests disappear, the risk of global warming—caused in part by an increase in the amount of carbon dioxide in the atmosphere—becomes greater. Since the 1980's, scientists and environmental activists concerned about global warming have joined foresters and soil conservationists in urging that for every tree removed anywhere, whether to clear land for development or to harvest timber, replacement trees be planted. As the area covered by tropical rain forests shrinks in size, the threat of irreversible damage to the global environment becomes greater.

Reforestation Programs

In 1988 American Forests, an industry group, established the Global ReLeaf program to encourage reforestation efforts in an attempt to combat global warming. In addition to supporting reforestation efforts by government agencies, corporations, and environmental organizations, Global ReLeaf and similar programs encourage people to practice reforestation in their own neighborhoods. Trees serve as a natural climate control, helping to moderate extremes in temperature and wind. Trees in a well-landscaped yard can reduce a homeowner's energy costs by providing shade in the summer and serving as a windbreak during the winter. Global ReLeaf is one of many programs that support reforestation efforts.

Arbor Day, an annual day devoted to planting trees for the beautification of towns or the forestation of empty tracts of land, was established in the United States in 1872. The holiday originated in Nebraska, a prairie state that seemed unnaturally barren to homesteaders used to eastern woodlands. Initially emphasizing planting trees where none had existed before, Arbor Day is observed in U.S. public schools to educate young people about the importance of forest preservation. Organizations such as the National Arbor Day Foundation provide *saplings* (young trees) to schools and other organizations for planting in their own neighborhoods.

Nancy Farm Männikkö

See also: Deforestation; Erosion and erosion control; Forest management; Logging and clear-cutting; Rain-forest biomes; Rain forests and the atmosphere; Slash-and-burn agriculture; Soil management; Sustainable forestry; Timber industry.

Sources for Further Study

Cherrington, Mark. *Degradation of the Land*. New York: Chelsea House, 1992. Addresses large-scale ramifications of deforestation and reforestation.

TreePeople, with Andy Lipkis and Katie Lipkis. *The Simple Act of Planting a Tree: A Citizen Forester's Guide to Healing Your Neighborhood, Your City, and Your World*. Los Angeles: St. Martin's Press, 1990. Specific and local approaches to the ramifications of deforestation and reforestation.

Weiner, Michael. *Plant a Tree: Choosing, Planting, and Maintaining This Precious Resource*. New York: Wiley, 1992. Specific and local approaches to the ramifications of deforestation and reforestation.

REPRODUCTION IN PLANTS

Categories: Genetics; physiology; reproduction and life cycles

Plants have evolved a remarkable number of ways to increase their numbers. These include not only a variety of forms of sexual reproduction, in which two individuals produce specialized cells that fuse to become a new offspring, but also many ways of achieving asexual reproduction, in which a single plant produces offspring.

In unicellular organisms three steps result in *cellular reproduction*. Among prokaryotic organisms (made of cells that have no nuclei), the single loop of DNA (deoxyribonucleic acid, the molecule that carries genetic information) replicates; then one copy is carried to each daughter cell as the original cell elongates and then "pinches" in two, a process called *fission*.

In eukaryotic organisms (whose cells have nuclei), the DNA is located within a nucleus in discrete *chromosomes* that must be precisely divided between the two daughter cells. Nuclear division to produce two identical daughter cells (asexual reproduction) is called *mitosis*. A chromosome during cell division consists of two halves, *sister chromatids*, each of which is identical to the other. During mitosis, every chromosome in the nucleus splits in half so that one chromatid will migrate to the first daughter cell, and the second chromatid migrates to the other. When cell division is complete, the result is two genetically identical daughter cells.

In *sexual reproduction*, a nucleus must divide by *meiosis*. In sexually reproducing organisms, at least some cells will have pairs of every type of chromosome. Such cells are *diploid*, or $2n$, where n is the number of different types of chromosomes. During meiosis, the pairs separate so that each daughter cell has only one of each type of chromosomes and is *haploid*, or n. If two different haploid cells fuse, the resulting cell will have pairs of every type of chromosome but with one of each pair contributed by each of the two parents. The offspring will thus be different from either parent.

Sexual Reproduction: Alternation of Generations

Sexual reproduction provides an opportunity for an organism to have different kinds of cells at different stages of its life cycle. The most familiar example is what is known from animals, including humans. The body cells of the adult are diploid, but in the reproductive organs, meiosis occurs to form haploid cells, either *eggs* or *sperm*. If these haploid reproductive cells (*gametes*) fuse, a new diploid cell is formed, the *zygote*. The zygote divides mitotically to form an *embryo* and eventually a new adult consisting of diploid body cells similar to those of both parents. This is a *gametic life cycle*, in which the gametes are the only haploid cells, and they are formed directly by *meiosis*. Fertilization occurs immediately after meiosis.

Some algae, particularly diatoms and some of the green algae, also have a gametic life cycle. In many plant species the gametes are large, immotile eggs and small, motile sperm, just as in animals. This condition is *oogamy*. However, many other plants are not oogamous. In some cases the two gametes appear to be identical (*isogamous*), while in others there may be two distinctive sizes of gametes, but their shape and motility are the same (*anisogamous*).

Some of the green algae have a life cycle exactly the opposite of the gametic cycle described above. In these plants, the diploid zygote divides by meiosis to produce haploid daughter cells, which multiply to form either a population of haploid unicellular plants or a multicellular plant with a haploid body. Eventually some of these haploid cells will differentiate into gametes, and two gametes will fuse to form a new zygote. In this type of *zygotic life cycle*, the zygote is the only diploid cell in the plant's life cycle, and fertilization is delayed after meiosis occurs.

Even more interesting is the *sporic life cycle*, in which a plant will have both a haploid *gametophyte* and diploid *sporophyte* body at different stages of its life. This type of life cycle is often called *alternation of generations*. These bodies may look the same (*isomorphic*), or they may look completely different (*heteromorphic*). In some cases two different species have been described and later discovered to be simply the haploid and diploid forms of the same species. Most plants, including most algae and fungi, have some form of a sporic life cycle. In a sporic life cycle, the diploid adult sporophyte plant forms reproductive organs in which meiosis occurs to form *spores*. These spores germinate and undergo mitosis to form a multicellular haploid gametophyte body. The gametophyte forms reproductive organs in which gametes are produced. Following fertilization, the resulting zygote undergoes mitosis to form the new sporophyte.

Many red algae and some of the green and brown algae are isomorphic; that is, the gametophyte and sporophyte bodies look identical. Bryophytes, some green algae, some brown algae, and most fungi are heteromorphic, with the gametophyte being the dominant, more conspicuous body. For instance, in mosses the green leafy plant is the gametophyte. When it is mature, the gametophyte will produce two types of *gametangia*, archegonia and antherida. *Archegonia* are flask-shaped structures that produce an egg, and *antheridia* are saclike structures that produce multiple sperm. After a sperm cell fertilizes the egg, the zygote grows into a sporophyte plant, still attached to and growing out of the gametophyte. The tip of the sporophyte swells to become a *capsule*, or *sporangium*, where meiosis occurs to form haploid spores. Each spore that falls onto a suitable place will germinate and grow into a new, leafy gametophyte plant.

Vascular plants and many brown algae are heteromorphic, with the sporophyte dominant. For instance, leafy ferns are the leaves of a sporophyte plant with an underground stem and roots. Sporangia typically develop on the underside of the fern, and meiosis occurs inside the sporangia to produce haploid spores. If a spore falls onto a suitable habitat it will germinate to form a small, inconspicuous gametophyte plant. The gametophyte typically produces archegonia and antheridia. Following fertilization, a new sporophyte plant grows out of the archegonium, but it is usually not seen until the first leaves enlarge enough to be visible above the soil surface.

Asexual Reproduction

In theory, any cell from a plant body should be capable of generating an entire new plant, because the nucleus of each cell has identical genetic information to every other cell in that body. In fact, since the mid-1950's it has been possible to *clone* many plants—produce entire new plants from single cells of a parent plant. The techniques of plant cell and tissue culture are used to propagate many commercially important species of ornamental plants. These techniques are also valuable tools in plant research. The basis for these tools is found in nature—the variety of methods of asexual reproduction found in plants.

In nonvascular plants, particularly fungi and filamentous algae, fragmentation can be an effective way of increasing the number of individuals of a plant. If the plant body is physically broken into pieces, each piece may continue to grow and develop as an independent plant. Most algae also form *sporangia*, asexual reproductive organs that produce motile *zoospores*. When the unicellular zoospores are released, they will swim for a period and, if they settle in a suitable environment, will germinate and grow to form a new plant. In some cases specialized multicellular asexual propagules are formed. For instance, lichens may form *isidia* or *soridia*, and liverworts may form *gemmae*. If the propagule breaks off and lands in a suitable environment, it will grow into a new plant.

Asexual reproduction may also occur in vascular plants. If a stem is laid horizontally on the ground, it may produce *adventitious roots*, which will allow that portion of the stem to grow as a separate plant. Similarly, some plants have roots that form *buds* and produce new stems at some distance from the original stem. The most dramatic examples of this are aspen groves in which all the trees are *clones* of one another. These natural phenomena are the basis of horticultural plant propagation by means of stem, leaf, or root cuttings.

Other plants form specialized structures for asexual propagation. For instance, strawberries produce *stolons*, specialized stems that grow out from a plant, then root and form new plantlets. *Bulb*-forming plants, such as gladiolus, frequently form new bulblets, and if a *rhizome* is split, such as on an iris, each half will continue to grow as a new plant. Some plants, such as kalenchoe, produce complete plantlets on the edges of leaves, which fall off and disperse to propagate the plant.

Marshall D. Sundberg

See also: Algae; Angiosperm life cycle; Angiosperm plant formation; Bacteria; Bryophytes; Chromosomes; Cloning of plants; DNA in plants; DNA replication; Fungi; Germination and seedling development; Hybridization; Lichens; Mitosis and meiosis; Nonrandom mating; Nucleus; Plant biotechnology; Plant life spans; Pollination; Seeds; Shoots.

Sources for Further Study

Bell, Peter Robert. *Green Plants: Their Origin and Diversity.* 2d ed. New York: Cambridge University Press, 2001. Includes detailed life cycles of green plants.

Hartmann, Hudson T., et al. *Plant Propagation: Principles and Practices.* 5th ed. Englewood Cliffs, N.J.: Prentice Hall, 1990. A handbook of techniques for reproducing plants with details for most commercially important species.

Raven, Peter H., Ray F. Evert, and Susan E. Eichhorn. *Biology of Plants.* 6th ed. New York: W. H. Freeman/Worth, 1999. Chapters on plant diversity cover basic life cycles.

REPRODUCTIVE ISOLATING MECHANISMS

Categories: Evolution; genetics; reproduction and life cycles

Reproductive isolating mechanisms are genetically influenced, as opposed to factors such as geographic isolation. These built-in isolating mechanisms prevent interbreeding between species and thereby promote reproductive efficiency.

Reproductive isolating mechanisms prevent interbreeding between species. The term, which was first used by Theodosius Dobzhansky in 1937 in his landmark book *Genetics and the Origin of Species*, refers to mechanisms that are genetically influenced and built-in. Geographic isolation can prevent interbreeding among populations, but it is an external factor. The standard model for speciation requires that populations be geographically isolated long enough to diverge genetically. Later, if the geographic barriers break down, built-in isolating mechanisms maintain reproductive isolation between the divergent populations. As these mechanisms continue to prevent hybridization, continued divergence leads to new species.

Reproductive isolating mechanisms function only between sexually reproducing species. They have no applicability to forms that reproduce only by asexual means. *Hermaphrodites*, organisms with both male and female reproductive organs that reproduce only by self-fertilization (rare in animals, more common in plants), represent a distortion of the sexual process that produces essentially the same results as asexual reproduction. Many lower animals, many plants, and protists regularly employ both asexual and sexual means of reproduction, and the significance of isolating mechanisms in such forms is essentially the same as in normal sexual species.

Prezygotic Mechanisms

Reproductive isolating mechanisms are usually classified into two main groups. *Premating*, or *prezygotic*, mechanisms operate prior to mating, or the release of *gametes*, and therefore do not result in a waste of the reproductive potential of the individual. By contrast, *postmating*, or *postzygotic*, mechanisms come into play after mating, or the release of gametes, and could result in a loss of the genetic contribution of the individual to the next generation. This distinction is important in the theoretical sense, in that *natural selection* should favor genes that promote premating isolation. Genes that do not promote premating isolation presumably would be lost more often through mating with an individual from another species, which often leads to no offspring or infertile offspring, in turn leading to a reinforcement of premating isolation.

Ecological isolation (habitat isolation) often plays an important role in both animals and plants. Different forms may be adapted to different habitats in the same general area and may meet only infrequently at the time of reproduction. Different plant species may occur on different soils, on different drainage profiles or exposures, or at different altitudes. This type of isolation, although frequent and widespread, is often incomplete, as the different forms may come together in transitional habitats. The importance of ecological isolation, however, is attested by the fact that when *hybrid swarms* (groups of organisms that show signs of extensive hybridization) are produced between forms that normally remain distinct, they have often been found to result from disruption of the environment, usually by humans.

Mechanical isolation is a less important type of premating isolation in plants, though it does occur in some. Complex floral structures in certain plants (such as orchids) may favor one species of animal pollinator over others. Finally, temporal differences often contribute to premating isolation. The most common type of *temporal isolation* is seasonal isolation: Species may reproduce at different times of the year. One type of western pine normally sheds its pollen in February, while another does not shed its pollen until April. Differences can also involve the time of day. In one species of desert plant, the flowers open in the early morning, while in another species of the same genus the flowers open in the late afternoon. Such differences, as in the case of ecological isolation, are often incomplete but may be an important component of premating isolation.

Postzygotic Mechanisms

If premating mechanisms fail, postmating mechanisms can come into play. If gametes are released, there still may be a failure of fertilization (*intersterility*). In plants, the pollen may not germinate on the foreign stigma or the pollen tube may fail to develop. Fertilization failure is almost universal between remotely related species and occasionally occurs even between closely related species.

If fertilization does take place, other postmating mechanisms may operate. The hybrid may be inviable (*zygotic inviability*). In other cases, development may be essentially normal, but the hybrid may be ill-adapted to survive in any available habitat (*hybrid adaptive inferiority*). Even if hybrids are produced, they may be partially or totally sterile (*hybrid sterility*). Hybrids between closely related forms are more likely to be fertile than those between more distantly related species, but the correlation is an inexact one. The causes for hybrid sterility are complex and can involve genetic factors, differences in gene arrangements on the chromosomes that disrupt normal chromosomal pairing and segregation at meiosis, and incompatibilities between cytoplasmic factors and the chromosomes.

If the hybrids are fertile and interbreed or backcross to one of the parental forms, a subtler phenomenon known as *hybrid breakdown* sometimes occurs. It takes the form of reduced fertility or reduced viability in the offspring. The basis for hybrid breakdown is poorly understood but may result from an imbalance of gene complexes contributed by the two species.

A Fail-Safe System

In most cases of reproductive isolation that have been carefully studied, more than one kind of isolating mechanism has been found. Even though one type is clearly of paramount importance, it is usually supplemented by others. Should the predominant type fail, others may come into play. In this sense, reproductive isolation can be viewed as a fail-safe system.

A striking difference in the overall pattern of reproductive isolation between animals and plants is the much greater importance of premating isolation in animals and the emphasis on postmating mechanisms in plants. *Behavioral isolation*, together with other premating mechanisms, is highly effective in animals, and postmating factors usually function only as a last resort. In contrast, behavioral factors contribute little to premating isolation in plants. Pollen in many forms is widely distributed either by the wind or by unselective animal pollinators, and postmating factors consequently are much more likely to come into play. This difference is reflected in a much higher incidence of natural hybridization in plants as compared with animals.

John S. Mecham

See also: Gene flow; Genetic drift; Genetics: Mendelian; Genetics: mutations; Genetics: post-Mendelian; Hybrid zones; Hybridization; Nonrandom mating; Pollination; Population genetics; Species and speciation.

Sources for Further Study

Baker, Jeffrey J. W., and Garland E. Allen. *The Study of Biology*. 4th ed. Reading, Mass.: Addison-Wesley, 1982. Offers coverage of isolating mechanisms and speciation. Includes list of suggested readings.

Dobzhansky, Theodosius. *Genetics of the Evolutionary Process*. New York: Columbia University Press, 1970. Dobzhansky's contributions over the years have had a major impact on genetics and evolutionary biology. This book is a revision and expansion of his *Genetics and the Origin of Species*, which first appeared in 1937. The book is directed at the advanced reader and presupposes a foundation in genetics. Much of chapter 10, however, which is devoted to an excellent review of reproductive isolation, can be understood by the general reader. Includes extensive technical bibliography.

Futuyma, Douglas J. *Evolutionary Biology*. 3d ed. Sunderland, Mass.: Sinauer Associates, 1998. This process-oriented textbook assumes some undergraduate training in biology, especially genetics, but most topics are prefaced with background information that should help the less prepared. Treatment of isolating mechanisms is brief, but the book provides one of the best introductions to current evolution theory available. Major references are given at the end of each chapter; includes extensive list of references.

Grant, Verne. *Plant Speciation*. 2d ed. New York: Columbia University Press, 1981. A comprehensive review of speciation and related topics in plants that should be required reading for the advanced student with a special interest in this area. The approach is essentially classical, however, and little attention is given to more recent contributions of molecular biology. Two short chapters are devoted to isolating mechanisms. Includes technical bibliography.

Stebbins, G. Ledyard, Jr. *Variation and Evolution in Plants*. New York: Columbia University Press, 1950. When it appeared, this book represented a landmark synthesis of evolution as applied to the plant kingdom. Although the book is now out of date, most of the material still rings true. Chapter 6 is largely devoted to a very complete treatment of isolating mechanisms in plants, including numerous examples. Includes numerous technical references.

RESISTANCE TO PLANT DISEASES

Categories: Anatomy; diseases and conditions; physiology

Plants have a number of defense mechanisms by which they can resist biotic or abiotic stresses that might cause their death or inhibit their growth. Such mechanisms are referred to as resistance.

Plants, the primary producers in all food chains, are besieged by a host of biological agents throughout their life cycles. Each species of plant is attacked by at least one hundred different kinds of mycoplasmas, viruses, bacteria, fungi, nematodes, and insects. In addition, regional weather patterns change, and plants are often faced with unfavorable environmental conditions such as heat, drought, or cold. Plants must contend with the chemicals used by people to deal with unwanted plants. These *herbicides* are designed to kill the plant or to restrict its growth.

When plants are exposed to *biotic* or *abiotic* stress, varying degrees of damage may occur, but many plants manage to protect themselves. Collectively, their *defense mechanisms* make up what is referred to as plant resistance. Although all plants possess some means of defense against stress, the term *resis-*

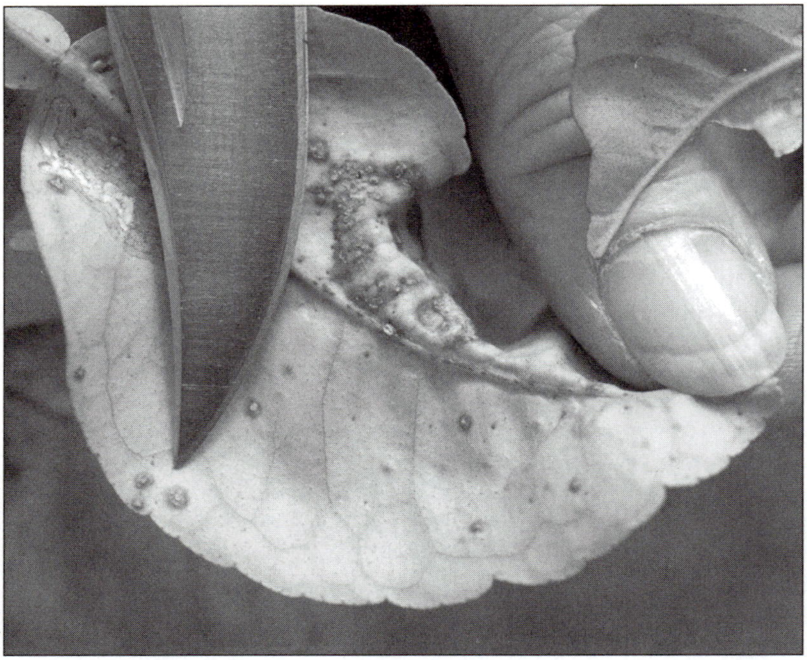

AP/Wide World Photos

The bacterial disease citrus canker attacks all parts of citrus trees. Symptoms include these brown, raised lesions. When faced with this and other diseases, various plants' defense structures include the waxy surfaces of many leaves, thickness of the cuticle that covers the epidermal cells, characteristics of openings into the plant, and thickness and toughness of the cell walls of the plant cells.

tant plant is usually reserved for varieties that have the ability to produce a larger crop of good quality fruit than would other varieties placed under the same stress conditions.

Types of Resistance

Resistance is often categorized according to its intensity. An *immune cultivar* is one that will never be consumed or injured by a particular stress under any condition. Few, if any, cultivars are considered immune to biotic or abiotic factors known to induce stress in other cultivars of the same plant species. A *high-resistance cultivar* is one that exhibits only slight damage from a specific biotic or abiotic stressing agent under a given set of conditions. A *low-resistance cultivar* is one that demonstrates less damage from a stressing agent than the average for the species. *Susceptible* and *highly susceptible* cultivars show increasing damage from a biotic or abiotic agent greater than the average for the species.

Resistance also varies according to environmental conditions, genetic control, number of pests, and plant age. *Multiple resistance* refers to cultivars that are resistant to multiple factors. Resistance under field conditions (called *field resistance*) may be considerably different from the resistance observed in the laboratory or greenhouse.

Resistance may be controlled by a single gene (monogenic), by a few genes (oligogenic), or by many genes (polygenic). The terms "horizontal resistance" or "general resistance" are used when describing resistance that is expressed equally against all biotypes of a pest species, and "vertical resistance," or "specific resistance," refers to resistance expressed against only some of the biotypes of a pest species. Resistance may be expressed at any stage of the life cycle from seedling through maturity. On occasion, host plants may pass through a particularly susceptible stage of growth quickly, thereby avoiding infestation by a large number of pests.

Mechanisms of Resistance

The mechanisms of plant resistance are generally grouped into three main categories: *tolerance*, *nonpreference*, and *antibiosis*. Tolerance is the sum of all plant responses that give the species the ability to withstand a particular degree of stress. The term "tolerance" is particularly applicable to mechanisms of resistance associated with environmental stresses. Plants commonly develop tolerances to stresses such as heat, cold, drought, or salt. Nonpreference is a phenomenon in which the plant is merely ignored by a particular pest. The plant has no food or shelter to offer the pest and is not suitable for egg-laying; therefore, the plant is not a potential host. Antibiosis refers to a mechanism in which the plant exerts some deleterious action on the pest. For example, the plant may produce a substance that inhibits some essential function of the pest's biology, such as reproduction or development, usually leading to death of the pest.

Structural Defenses

Both structural and biochemical defenses can be preexisting or induced by stress. Preexisting defense structures include the waxy surfaces of many leaves, thickness of the cuticle that covers the epidermal cells, characteristics of openings into the plant, and thickness and toughness of the cell walls of the plant cells. After a pest invades a plant, inducible changes in structure can provide some degree of defense. After invasion by pests such as fungi, bacteria, viruses, and nematodes, some plants will form layers of cork tissue that seal off the invading organisms and prevent them from reaching the remainder of the plant.

Other structural defense strategies include the formation of structures called *tyloses* to seal off the infected vascular tissue or the deposition of gums around lesions. Both tyloses and gum deposits prevent the spread of the agent. In some instances, plants will form abscission layers that seal off a section of leaf and cause it to die along with the pest.

Biochemical Defenses

Although structural barriers provide some degree of defense against invading organisms, chemicals produced by the plant during or after the induction of stress appear to be much more important in conferring resistance. There are several preexisting biochemical defense systems. Although plants do not produce antibodies to specific invading pests, some type of immunological response appears to be operating. Plants that are resistant to specific pathogens do not contain the *antigens*, chemicals that induce the resistance response, that are found in the susceptible plants. Some cultivars maintain resistance by limiting the production of certain chemicals that are essential nutrients for invading pathogens. Other preexisting defense mechanisms include the presence in the plants' cells of chemicals that inhibit the growth of an invading pest or the release into the environment of chemicals that either inhibit or kill potential pathogens.

When injured by a biotic agent, chemicals, or environmental factors, plants respond with a series of biochemical reactions aimed at limiting the injury and healing the wound. This response is much more pronounced in resistant plants than in susceptible plants. The biochemical response to stress shows tremendous variation. Many resistant plants respond to a pest invasion by releasing phenolics or other toxic compounds. Fungi produce a group of toxic substances called *phytoalexins* in response to an invasion.

Many plants respond to stress by the induced synthesis of proteins and other enzymes that form an immune layer around the infected site. When resistant plants are confronted with the oxidative stress that usually accompanies environmentally induced stress, they increase the production of antioxidant enzymes. Enzymes produced by invading organisms are often responsible for the damage suffered by the host plant, but some resistant plants produce substances that either resist or inactivate these enzymes. Some invading organisms produce toxins that damage the host plant, and plants resistant to these organisms generally produce chemicals that detoxify the toxins. Other plants develop resistance by altering certain biochemical pathways or initiating a hypersensitive response.

Genetically Engineered Resistance

With the advent of recombinant DNA (deoxyribonucleic acid) technology in the 1970's, the development of new traits such as resistance was no longer limited to mutation or natural selection from a limited pool of genes. Scientists first developed transgenic animals and plants in the early 1980's, and industry has made widespread use of genetically modified organisms. Despite the beneficial applications, potential risks and ethical issues associated with the technology have led to controversy.

Genetic engineering has been used extensively in agriculture. Products of modified organisms are used to protect plants from frost and insects. In 1986 the U.S. Environmental Protection Agency approved the release of the first genetically modified crop plant; by the end of the 1990's, more than one thousand others had been field-tested. Plants have been designed to resist disease, drought, frost, insects, and herbicides as well as to improve the nutritional value or flavor of foods.

D. R. Gossett, updated by Bryan Ness

See also: Biopesticides; Biotechnology; Cloning of plants; Diseases and disorders; DNA: recombinant technology; Genetically modified foods; Herbicides; Integrated pest management; Pesticides; Plant biotechnology; RNA.

Sources for Further Study

Campbell, Neil A., and Jane B. Reece. *Biology*. 6th ed. San Francisco: Benjamin Cummings, 2002. Introductory textbook provides an excellent discussion of the biology associated with plant reproduction. Includes superb graphics, lists of suggested readings, and glossary.

Chrispeels, Maarten J., and David E. Sadava. *Plants, Genes, and Agriculture*. Boston: Jones and Bartlett, 1994. Outstanding treatise on the use of biotechnology in crop production. Contains sections related to the use of biotechnology to transfer resistance to susceptible plants. Although written at an advanced level, it will interest the general reader. The well-illustrated text has a supplemental reading list.

Hale, M. G., and D. M. Orcutt. *The Physiology of Plants Under Stress*. New York: John Wiley & Sons, 1987. An excellent treatise on the way in which plants respond and develop resistance to environmental stress. Written for the advanced student, it contains much the general reader will understand. Sparsely illustrated; includes an excellent glossary.

Maxwell, Fowden G., and Peter R. Jennings, eds. *Breeding Plants Resistant to Insects*. New York: Wiley, 1980. Written for more advanced students, this book contains material of interest to the general reader. One section explains how plants respond and defend themselves against an insect invasion. Sparsely illustrated. Has a list of general references and a glossary.

Smith, Michael C. *Plant Resistance to Insects: A Fundamental Approach*. New York: Wiley, 1989. An excellent treatise on the fundamental principles associated with insect resistance in plants. One section is devoted to the methods of studying resistance. The general reader will not find the book overly technical. Well illustrated; includes supplemental reading list.

RESPIRATION

Categories: Cellular biology; photosynthesis and respiration; physiology

All cells must have a source of energy in order to survive. Almost all cells utilize ATP as their energy currency. In other words, ATP is produced and stored up until it is needed to supply energy for metabolic activity. Respiration is the process by which cells oxidize a fuel, usually the simple sugar glucose, and use the energy released during this oxidation to produce ATP.

The term "metabolism" refers to the sum total of all the chemical activity that occurs within an organism. Metabolism can be further divided into two large categories, *anabolism* and *catabolism*. Anabolism refers to those metabolic reactions associated with the synthesis of molecules, such as proteins or carbohydrates, while catabolism includes those reactions involved in the degradation of molecules. Respiration is a catabolic process.

Carbohydrates, especially the simple sugars glucose and fructose, serve as the initial substrates for the respiratory process. In plants, these substrates are produced by *photosynthesis* in the chloroplasts. Solar energy is used to convert carbon dioxide and water into small sugar-phosphate molecules which are then combined to form fructose. Energy is required to form each carbon-to-carbon bond (called a *covalent bond*), and a specified amount of energy is stored in each covalent bond. In other words, solar energy is converted to chemical energy.

The fructose may directly enter *glycolysis*, the first series of reactions in cellular respiration, or it may be converted to glucose. The glucose may also enter glycolysis directly, or it may be combined with a fructose molecule to form sucrose (table sugar), which can be transported to other parts of the plant. Once sucrose reaches the target cells, it can be converted back to glucose and fructose, which can then be metabolized via glycolysis. Glucose can also be polymerized into large starch molecules, which can be stored. When the plant experiences an energy deficit, starches can be broken down to glucose molecules.

During respiration fuel molecules, such as glucose, undergo a series of reactions in which the molecule is oxidized into smaller molecules, and in aerobic respiration, the glucose will ultimately be degraded to carbon dioxide and water. As the molecules are degraded, the energy stored in the chemical bonds that held the glucose molecule together is released, and the cells trap this energy in the form of *adenosine triphosphate* (ATP) molecules. In a sense, the oxidation of glucose as a fuel is similar to the oxidation (burning) of any other organic fuel, such as gas, fuel oil, or coal, except the biological oxidation of glucose is a stepwise process and results in the formation of ATP. Each step requires an *enzyme*, an organic catalyst. There are many enzymes associated with respiration, but the three most common types are kinases, decarboxylases, and oxioreductases. *Kinases* catalyze reactions associated with ATP formation or utilization; *decarboxylases* catalyze decarboxylation reactions which remove chemical groups called carboxyls as carbon dioxide; and *oxioreductases* catalyze oxidation/reduction reactions. *Oxidation* is the removal of electrons from a molecule, and *reduction* is the addition of electrons. In general, oxidation involves the removal of two electrons and two hydrogen ions (H^+) from the substrate molecule. The release of H^+ into the cytosol would result in acidification of the cell; therefore, oxioreductases require the presence of *coenzymes* which will accept the electrons and H^+. In other words, these coenzymes are reduced as glucose is oxidized. The two most important coenzymes in respiration are *nicotine adenine dinucleotide* (*NAD*) and *flavin adenine dinucleotide* (FAD). The reduced forms of these coenzymes are NADH + H^+ and $FADH_2$, respectively.

Glycolysis and Fermentation

Glycolysis, the first series of reactions in the respiratory pathway, consists of nine or ten separate steps, depending on whether the initial substrate is fructose or glucose. Because sugar molecules are not very reactive, the first two or three steps in the process use two ATP molecules to convert the six-carbon glucose or fructose to a very reactive molecule called *fructose-1,6-bisphosphate*. This six-carbon compound is then broken down into the equivalent of two molecules of a three-carbon compound called *glyceraldehyde-3-phosphate* (*PGAL*). Each molecule of PGAL undergoes a series of reactions in which it is converted to *pyruvic acid*. During this conversion, one NADH + H^+ is formed, and enough energy is released to produce two ATP molecules per PGAL. In summary, glycolysis is the conversion of glucose to two molecules of pyruvic acid, two molecules of NADH + H^+, and a net gain of two ATP molecules (four were produced, but two were used to initiate the process). Glycolysis occurs in the cytosol and is entirely *anaerobic*, meaning that it occurs in the absence of oxygen.

In anaerobic organisms, such as yeast, each molecule of pyruvic acid is decarboxylated to produce carbon dioxide and a molecule called acetaldehyde. The NADH + H^+ produced during glycolysis is then used to convert the acetaldehyde to ethanol. This anaerobic process is called *fermentation*. Overall, the process of fermentation results in the conversion of glucose to two molecules of carbon dioxide, two molecules of ethanol, and a net gain of two ATP molecules.

The Krebs Cycle and Oxidative Phosphorylation

In aerobic plants, each molecule of pyruvic acid is transported to the matrix of the mitochondria where it is oxidatively decarboxylated to produce a carbon dioxide molecule, a molecule of NADH + H^+, and a compound called acetyl coenzyme A (acetyl CoA), which contains two carbon atoms from the initial glucose molecule. Acetyl CoA then enters a second series of reactions called the *Krebs cycle*, also known as the tricarboxylic acid cycle or the citric acid cycle. The two carbons from the glucose molecule combine with a four-carbon compound called *oxaloactic acid* to form a six-carbon compound called *citric acid*. The citric acid is decarboxylated twice, and the remaining four-carbon fragment is ultimately converted back to oxaloacetic acid. During this process, two molecules of carbon dioxide, three molecules of NADH + H^+, one $FADH_2$ molecule, and one ATP molecule are produced. During glycolysis, two molecules or pyruvic acid are produced per glucose; therefore, after two turns of the Krebs cycle, glucose is converted to six molecules of carbon dioxide, six (four net) ATP molecules, ten molecules of $NADH + H^+$, and two molecules of $FADH_2$.

A third series of reactions, referred to as *electron transport*, takes place within the mitochondrial membranes. The electrons and H^+ ions bound to the $NADH + H^+$ and $FADH_2$ are transferred to initial electron receptors and then passed through a series electron transporters, each with a lower *reduction potential* (the tendency to accept electrons). Several of these electron transporters are iron-sulfur containing proteins called cytochromes. Hence, this electron transport system is sometimes referred to as the cytochrome system. The final electron acceptor in this system is oxygen. When two electrons and H^+ ions are transferred to oxygen, a molecule of water is formed. This provides the cells with a safe means of removing excess H^+ ions, but more important, additional ATP is produced. As electrons are transported from $NADH + H^+$ and $FADH_2$ through the electron transport chain and ultimately to oxygen, energy is released. This energy can be used to synthesize ATP from ADP (adenosine diphosphate). Since the energy is stored in the phosphate bond (ADP to ATP) and occurs only in the presence of oxygen, the production of ATP during aerobic respiration is referred to as *oxidative phosphorylation*. Each mole of $NADH + H^+$ produced within the mitochondria results in the production of three moles of ATP via the electron transport system. The $NADH + H^+$ from glycolysis and $FADH_2$ enter the electron transport system downstream from the $NADH + H^+$ derived from inside the mitochondria. As a result, each mole of these molecules produces only two moles of ATP. The total ATP production per mole of glucose from electron transport and oxidative phosphorylation is thirty-two moles.

3×8 $NADH + H^+$ from within the mitochondria = 24; 2×2 $NADH + H^+$ from glycolysis = 4; 2×2 $FADH_2$ from the Krebs cycle = 4: 24 + 4 + 4 = 32.

Overall, aerobic respiration uses glycolysis, the Krebs cycle, and electron transport and results in the conversion of one mole of glucose to six moles of carbon dioxide, twelve moles of water and a net of thirty-six moles of ATP.

2 net ATPs from glycolysis + 2 ATPs from the Krebs cycle + 32 ATPs from electron transport = 36 ATPs.

Each mole of ATP represents about 8,000 calories of energy; therefore the oxidation of one mole of glucose can produce a total of 288,000 calories of energy available for cellular work.

D. R. Gossett

See also: ATP and other energetic molecules; Carbohydrates; Energy flow in plants; Ethanol; Glycolysis and fermentation; Krebs cycle; Microbial nutrition and metabolism; Mitochondria; Oxidative phosphorylation; Photosynthesis.

Sources for Further Study

Hopkins, William G. *Introduction to Plant Physiology*. 2d ed. New York: John Wiley and Sons, 1999. This general text for the beginning plant physiology student contains a very clear discussion of respiration in plants. Includes diagrams, illustrations, and index.

Nicholls, D. G., and S. J. Ferguson. *Bioenergetics 2*. London: Academic Press, 1992. An excellent source of information dealing with all aspects of respiration, for the more advanced student. Includes bibliographical references.

Raven, Peter H., Ray F. Evert, and Susan E. Eichhorn. *Biology of Plants*. 6th ed. New York: W. H. Freeman/Worth, 1999. A very good general botany text with a simplified discussion of respiration in plants. Includes diagrams, illustrations, and index.

Salisbury, Frank B., and Cleon W. Ross. *Plant Physiology*. 4th ed. Belmont, Calif.: Wadsworth, 1992. One of the standard texts for the more advanced plant physiology student, with an excellent discussion on respiration in plants. Includes biographical references, diagrams, illustrations, and index.

Taiz, Lincoln, and Eduardo Zeiger. *Plant Physiology*. 2d ed. Sunderland, Mass.: Sinauer Associates, 1998. A text written for the advanced plant physiology student, with an outstanding discussion of respiration in plants. Includes biographical references, diagrams, illustrations, and index.

RHYNIOPHYTA

Categories: Evolution; paleobotany; *Plantae;* seedless vascular plants; taxonomic groups

When first proposed by Harlan Banks in 1968, the Rhyniophyta *were the first and oldest vascular land plants. The trimerophytes subsequently evolved from them.*

In 1908, Octave Lignier developed a model of what the sporophyte of the earliest vascular land plants might look like. The sporophyte is a diploid (2*n*) plant that produces spores in a sporangium, while its counterpart, the gametophyte, is a haploid (*n*) plant that bears the male (antheridia) and female (archegonia) sex organs. Lignier proposed that the first vascular land plants would consist of a forked, photosynthetic stem lacking both roots and leaves. At the point of branching, the fork looked like a capital Y. Running along the ground or just below the soil surface was a horizontal stem (rhizome) bearing hairlike filaments (rhizoids) on its lower surface that functioned in anchorage and absorption. Spores were produced either within the unmodified stem tips of some of the aerial branches or in a specialized reproductive structure, a thick-walled, elongate sporangium, that terminated some branch tips.

In 1912, some unique plant fossils were found in the Rhynie Chert (406 million to 401 million years old) in Scotland. The Rhynie Chert is a hot-spring chert formed when silica-saturated water from a hot spring flooded a low-lying marsh, seeped into the plants, and hardened within and about them. Because they are encased in chert, the Rhynie plants were the first fossil plants to yield both structural and anatomical data. Two of the plants, *Rhynia major* (now *Aglaophyton*) and *R. gwynne-vaughanii*, were reconstructed as near-perfect matches to Lignier's model of the earliest vascular land plant. With their discovery, the science of paleobotany came of age.

As knowledge of the rhyniophytes increased, a revision of the group seemed in order. As originally defined, the *Rhyniophyta* included both vascular (tracheophytes *Rhynia* and *Cooksonia pertoni*) and nonvascular plants that are intermediate between bryophytes and vascular plants (*Aglaophyton* and *Horneophyton lignieri*). These plants are placed in two groups, the *Rhyniophyta* and *Horneophyta*, respectively.

Horneophyta or Protracheophytes

Aglaophyton's rhizome was in contact with the ground at intervals, and rhizoids were concentrated at those positions rather than spread evenly along the axis. The tissue of the rhizome contained a fungus, representing the earliest known example of a plant-fungus symbiosis. The vast majority of land plants require this symbiotic relationship to live. The land plant provides the fungus with shelter within its cells and access to the products of photosynthesis. The fungus absorbs water and minerals from the soil, which it passes on to the land plant.

This partnership is so important to living plants that scientists were not surprised to find it very early in the fossil history of the land plants. Dense clusters of buds (possibly dormant stem tips) were found on the rhizome at the base of the aerial stems. The aerial stem was naked and branched by forking into two daughter axes of equal size. Elongate sporangia were borne at the tips of some aerial branches. The water-conducting cells of *Aglaophyton* lack the internal wall thickenings characteristic of tracheids (thick-walled, dead cells found in the xylem of vascular land plants) and resemble the thin-walled, water-conducting cells (hydroids) of the mosses. Because of the absence of tracheids, *Aglaophyton* is considered a nonvascular plant and must be removed from the *Rhyniophyta*. A male reproductive structure (*Lyonophyton rhyniensis*) represents the gametophyte of *Aglaophyton*.

Another nonvascularized sporophyte found in the Rhynie Chert was *Horneophyton*. Instead of having obvious sporangia, the spores of *Horneophyton* were produced within the stem tip and released by means of a terminal pore. Multicellular projections were found on the stem, which terminated in stomates. The aerial stems did not grow from a rhizome. The aerial stems each terminated in a bulbous, nonvascularized structure called a corm. Several corms were found attached to one another, but no vascular tissue connected them. As in *Aglaophyton*, the water-conducting cells were thin-walled and lacked internal wall thickenings. *Langiophyton mackiei* represents the female gametophyte of *Horneophyton*.

When gametophytes are known in the protracheophytes, they have an anatomical and structural complexity similar to that of their corresponding sporophytes. In contrast, the gametophytes of vascular land plants are small and inconspicuous (often subterranean). Scientists know more about the gametophytes of the plants of the Rhynie Chert than they do about almost all other groups of fossil plants.

Rhynia

Rhynia gwynne-vaughanii was reconstructed as a smaller version of *Aglaophyton*. The aerial stem of *Rhynia* was naked and branched by forking. The rhizome bore rhizoids. Thick-walled water-conducting cells (tracheids) were present. Although *Rhynia* was reconstructed bearing sporangia, none were originally found attached to the aerial stems. In addition to lacking sporangia, *Rhynia* had bumps along the aerial stem that were interpreted as female reproductive structures (archegonia). As a result, *Rhynia* was reinterpreted as the gametophyte of *Aglaophyton*. Subsequently, sporangia were found attached to *Rhynia*, making the plant a sporophyte once more. The sporangia were borne laterally along the stem on short side branches, and they were shed after releasing their spores. The gametophyte of *Rhynia* is unknown.

The small, circular vascular strands and fleshy stems associated with the rhyniophytes and horneophytes indicate that these plants did not grow very tall. Water pressing against the inside of the cell walls in their stems (turgor pressure) held these plants upright.

Cooksonia

The oldest vascular land plant (412 million to 406 million years old) and only cooksonioid with identifiable vascular tissue is *Cooksonia pertoni*. The aerial stems of *Cooksonia* branched by forking and bore at their branch tips sporangia shaped like kidney beans. Plants whose sporangia resemble those of *C. pertoni* are known from older sediments (414 million to 412 million years in age), but no vascular tissue has been found in their stems. The oldest fossil plant from the Western Hemisphere that resembles *Cooksonia* comes from Bathurst Island in the Canadian Arctic (420 million to 414 million years in age). No roots or rhizomes are known for any *Cooksonia*. If the older specimens of Cooksonia lacked vascular tissue, *Cooksonia* may not be a valid genus. These plants are best referred to collectively as cooksonioids, and plants showing these traits could be bryophytes, protracheophytes, or rhyniophytes.

Fossil Spores

Fossils spores are known from sediments that are about 40 million years older than those from which cooksonioids are recovered. The most abundant spores were not produced by the vascular land plants and are called cryptospores. Banded tubules and sheets bearing cell outlines (collectively called nematoclasts) are recovered with the cryptospores. Similar structures are produced by the sporangial epidermis of modern mosses (cellular sheets) and liverworts (banded tubes). The cryptospores had a worldwide distribution and showed limited diversity. These spores were probably produced by early bryophytes.

To identify the type of plant that produced a specific spore, researchers must find the spore within a sporangium. Although free spores are common, no spores are known from sporangia for the first 65 million years of land plant evolution. The presence of cryptospores and the occasional spore that might have been produced by a vascular land plant indicates that land plants first appeared about 500 million years ago. Most of the 65-million-year period from which sporangia are unknown was characterized by high sea levels, and little continental deposition occurred. The missing continental deposits are where the necessary sporangia would have been found. The oldest spores from continental deposits are about 435 million years old. Cryptospores are found in sporangia from England that are between 414 million and 412 million years old. Although they predominated early, cryptospores form a very minor component of the spore flora about by 414 million years ago. The decrease in the number of cryptospores reflects the growing importance (diversification) of the vascular land plants.

Ancestors of the Rhyniophytes

The first land plants, bryophytes, protracheophytes, and tracheophytes, evolved from green algal (charophycean) ancestors that lived in freshwater habitats whose appearance was unpredictable (not seasonal) and of short duration. Their algal ancestors were preadapted to life on land. They were resistant to microbial attack due to the structure of their cell walls. They had a genetic (heritable) basis for lowering or suspending metabolic activity during drought (dry conditions that could last for days or months) and desiccation (dry conditions that could last for decades). Their reproductive bodies did not dry out during air transport from one water body to another. The first to colonize the land successfully were the bryophytes. The protracheophytes and tracheophytes appeared after the bryophytes were well established on land.

Gary E. Dolph

See also: Adaptive radiation; Algae; *Charophyceae*; Evolution of plants; Fossil plants; Hornworts; Liverworts; Mosses; Paleobotany; Plant tissues; Seedless vascular plants; Shoots; Species and speciation; Stems; *Trimerophytophyta*; *Zosterophyllophyta*.

Sources for Further Study

Gensel, Patricia G., and Henry N. Andrews. *Plant Life in the Devonian*. New York: Praeger, 1984. Good descriptions of the rhyniophytes with valuable illustrations. Illustrations, bibliographical references, index.

Gensel, Patricia G., and Dianne Edwards. *Plants Invade the Land: Evolutionary and Environmental Perspectives*. New York: Columbia University Press, 2001. Provides the current information on the rhyniophytes in the context of the first land plants. Illustrations, bibliographical references, index.

Kenrick, Paul, and Peter R. Crane. *The Origin and Early Diversification of Land Plants: A Cladistic Study*. Washington, D.C.: Smithsonian Institution Press, 1997. Detailed cladistical treatment of the rhyniophytes. Excellent line drawings, bibliographical references, index.

RIBOSOMES

Categories: Cellular biology; genetics

Ribosomes are complex cellular structures found in all cells and are responsible for making proteins. They are composed of ribosomal RNA (rRNA) and protein and are most abundant in the cytoplasm, although some functional ribosomes can be found in the nuclei of eukaryotic cells. Chloroplasts and mitochondria also have their own ribosomes.

Ribosomes are responsible for synthesizing the proteins in all cells by a process called *translation*. It is called translation because ribosomes use messenger ribonucleic acids (mRNAs) as their guide and must "translate" the message contained in the nucleotides of mRNAs. The general structure of ribosomes is the same in all cells, but ribosomes of prokaryotes are smaller than ribosomes in the cytoplasm of eukaryotes. The ribosomes in chloroplasts and mitochondria are more similar to the smaller ribosomes of prokaryotes but are often smaller yet.

Ribosome Structure

Because ribosomes, and rRNAs, are so large, they are not described on the basis of their molecular weight but rather in *Svedbergs*, or Svedberg units. Svedbergs (denoted S) are a measure of how quickly a large molecule or aggregation of molecules sediment, or sink to the bottom of a centrifuge tube while being spun around. Higher S values mean faster sedimentation and thus greater mass. For example, prokaryotic ribosomes sediment at 70's, whereas eukaryotic ribosomes, which are larger, sediment at 80's. Svedbergs cannot be added together, because they are not directly related to mass but to a combination of mass and overall molecular size and shape. Consequently, if two molecules that are both 30's are joined together, their combined sedimentation rate would be about 50's or a little more but not 60's.

Prokaryotic ribosomes constitute three rRNAs and fifty-two different proteins. Like all ribosomes, they are composed of two subunits, referred to simply as large and small subunits. Large subunits have a sedimentation rate of 50's and are composed of a 23S rRNA, a 5S rRNA, and thirty-one proteins. Small subunits have a sedimentation rate of 30's and are composed of a single 16S rRNA and twenty-one different proteins. To function, a small and large subunit must come together.

Eukaryotic ribosomes are more complex, with four rRNAs and more than eighty proteins. Large subunits have a sedimentation rate of 60's and are composed of a 28S rRNA, a 5.8S rRNA, a 5S rRNA, and about forty-nine proteins. Small subunits have a sedimentation rate of 40's and are composed of a single 18S rRNA and about thirty-three proteins. Assembly of the subunits takes place in a region of the nucleus called the *nucleolus*. When completed, the subunits are transported through nuclear pores to the cytoplasm. Although translation of mRNAs takes place primarily in the cytoplasm, a small amount occurs in the nucleus.

Translation in Prokaryotes

Before translation can take place, *transcription* of a gene must occur. Transcription converts the deoxyribonucleic acid (DNA) code of a gene into a complimentary RNA code, in the form of an mRNA. Even while RNA polymerase catalyzes the joining of nucleotides for the mRNA, translation can begin. Translation includes three steps: initiation, elongation, and termination.

In *initiation*, special proteins called initiation factors (IFs) enable the small subunit of a ribosome to bind to an mRNA and form the initiation complex. Next, a special tRNA called fMet-tRNA (N-formylmethionyl-tRNA), which carries a specially modified amino acid derivative of methionine, binds to the start codon of the mRNA (AUG). This modified methionine is the first amino acid in all prokaryotic proteins, but many proteins are modified after transcription, often by removing the first few amino acids. Lastly, the large ribosomal subunit binds.

Elongation is a complex process in which the ribosome adds amino acids, one at a time, to the growing protein. Ribosomes have three sites where events in translation takes place. The A site (aminoacyl-tRNA site) is where new tRNAs, with their attached amino acids, bind to the ribosome and mRNA. The P site (peptidyl-tRNA site) is where the tRNA with the growing protein (polypeptide) is attached to the ribosome and the mRNA. The E site (exit site), only recently recognized, is where tRNAs leave the ribosome after they have completed their work.

The sequence of events in elongation can be difficult to visualize, but it is a cyclical process that repeats until elongation is done. The steps are as follows. First, while the tRNA with the growing protein (called a peptidyl-tRNA) rests in the P site, a tRNA with an amino acid (called an aminoacyl-tRNA) enters the A site and binds to the codon on the mRNA. Second, the peptidyl-tRNA attaches the growing protein to the amino acid on the aminoacyl-tRNA. Now the tRNA in the P site has no amino acids attached to it. Third, the tRNA in the P site now leaves the ribosome by passing through the E site. Fourth, as the tRNA leaves the P site, the tRNA at the A site (now a peptidyl-tRNA) is translocated (moved) to the P site by moving the mRNA over one codon, leaving the A site open for the next aminoacyl-tRNA to attach. These steps repeat until the stop codon, near the end of the mRNA, is reached. Many of the events of elongation were long believed to be catalyzed by proteins in the ribosome. It is now known that some of the ribosome's catalytic properties are due to the rRNAs, which, as a result, are sometimes called *ribozymes*.

The stop codon is not recognized by any of the tRNAs. *Termination* occurs when some special proteins called *releasing factors* bind to the ribosome. When they bind, the protein that is attached to the tRNA in the P site is released from the tRNA, the tRNA leaves through the E site, and the two subunits come apart.

Translation in Eukaryotes

The process of translation in eukaryotes varies only in minor details from translation in prokaryotes, and these differences are due to the greater complexity of eukaryotic mRNA and ribosomes. Eukaryotic mRNA must go through extensive processing after being transcribed, including intron excision/exon splicing and addition of a special 5′ cap and a 3′ poly-adenosine tail. Eukaryotic mRNA cannot be translated until these modifications are completed. Although some mRNA is translated in the nucleus, most must be transported to the cytoplasm first. Consequently, transcription and translation are completely separate processes, unlike their coupling in prokaryotes.

Initiation in eukaryotes involves a larger number of initiation factors. Once the small ribosomal subunit binds, it then starts at the 5′ end of the mRNA and searches for the start codon (AUG). When the start codon is found, a methionyl-tRNA binds to the ribosome and the start codon of the mRNA. In eukaryotes, like in prokaryotes, the first amino acid in all proteins is methionine, but in eukaryotes it is not a N-formyl-methionine. Elongation and termination are essentially the same in both eukaryotes and prokaryotes. As in prokaryotes, completed proteins are typically modified in various ways before being used by the cell.

Many ribosomes will typically translate an mRNA simultaneously. An mRNA with several ribosomes lined up along it, all of them translating at once, is often called a *polyribosome*. Polyribosomes are observed in both eukaryotes and prokaryotes.

Distribution in Eukaryotic Cells

After being synthesized in the nucleolus, a few ribosomes will function for an unknown length of time in the nucleus. Most, though, are transported to the cytoplasm as separate subunits. Once in the cytoplasm, ribosomes will either bind to endoplasmic reticulum (ER), in regions called rough ER because of the fuzzy appearance of these areas in electron micrographs, or they will remain free in the cytoplasm. Ribosomes bound to the rough ER specialize in making proteins that will either be embedded in membranes or transported in vesicles to the Golgi complex. The different fates of proteins made at the rough ER are determined by special *signal sequences*, sequences of amino acids at the beginning of proteins which are typically removed later.

Ribosomes that are free in the cytoplasm make proteins of various kinds that are needed for processes occurring in the cytoplasm. Regardless of location, ribosomes are usually found disassembled, and the two subunits come together only when needed for translation. Cells have anywhere from several thousand ribosomes to as many a few million in cells that have particularly high rates of protein synthesis.

Bryan Ness

See also: Angiosperm cells and tissues; Cell cycle; Cell theory; Chloroplasts and other plastids; Cytoplasm; Cytoskeleton; Cytosol; DNA: historical overview; DNA in plants; DNA replication; Endomembrane system and Golgi complex; Endoplasmic reticulum; Eukaryotic cells; Membrane structure; Microbodies; Nucleus; Oil bodies; Peroxisomes; Photosynthesis; Plant cells: molecular level; Plasma membranes; Respiration; RNA.

Sources for Further Study

Garrett, Roger A., Stephen R. Douthwaite, and Anders Liljas, eds. *The Ribosome: Structure, Function, Antibiotics, and Cellular Interaction*. Washington, D.C.: ASM Press, 2000. A collection of papers based on presentations given on a ribosome conference. Detailed and up to date information on ribosomes, best for college-level and more advanced readers.

Gunning, Brian E. S., and Martin W. Steer. *Plant Cell Biology, Structure, and Function*. Sudbury, Mass.: Jones & Bartlett, 1996. An integrated textbook copiously illustrated with electron micrographs and containing detailed information on plant cell and molecular biology.

Smith, H., ed. *The Molecular Biology of Plant Cells: Botanical Monographs*. Berkeley: University of California Press, 1978. Somewhat out of date but provides a good overview for all aspects of plant cell biology, including ribosomes and translation.

Spedding, Gary. *Ribosomes and Protein Synthesis: A Practical Approach*. Oxford, England: Oxford University Press, 1990. A practical guide to the methods of studying the structure and function of ribosomes in light of the ability or rRNA to act as a catalyst.

Spirin, Alexander S. *Ribosomes*. New York: Plenum, 2000. The author is considered the world authority on ribosomes, and this book covers every aspect of ribosome structure and function. Written for college-level and more advanced readers.

RICE

Categories: Agriculture; economic botany and plant uses; food

Rice, the starchy seeds of an annual cereal grass, is the most commonly consumed food grain for a majority of the world's population.

The rice plant, *Oryza sativa*, is a member of the grass family, classified into *indica* and *japonica* varieties. World production of rice exceeds 500 million tons. Most countries—particularly in Asia—cultivate rice for domestic consumption, so less than 5 percent enters the export market. The United States generates only about 2 percent of world rice production, but almost half of U.S. production is exported. Rice cultivation almost certainly began in India, where it dates back to about 3000 B.C.E. During medieval times it spread westward to southern Europe.

Digital Stock

Rice paddies in Indonesia.

Cultivation

Monsoon tropics are ideal for indica rice, which is commonly cultivated in China and Southeast Asia. The plants can adapt to uncertain conditions. The japonica type of rice requires precise water control as well as weed and insect control. It is cultivated in temperate zones such as the United States, Australia, Japan, Korea, and certain parts of China.

Rice is self-pollinated, and the grain is enclosed in the *palea*, or hull. Harvested but unmilled rice is called *paddy* or rough rice. Milling of rough rice by any of several processes yields the polished grain that is ready for consumption. Rough rice contains approximately 10 percent protein, 65 percent starch, 2 percent lipids, 5 percent minerals, and 18 percent hull/bran. The unhulled whole rice kernel also contains thiamine, niacin, and riboflavin. Parboiled rice can be stored for long periods.

The International Rice Research Institute in the Philippines has contributed significantly to the development

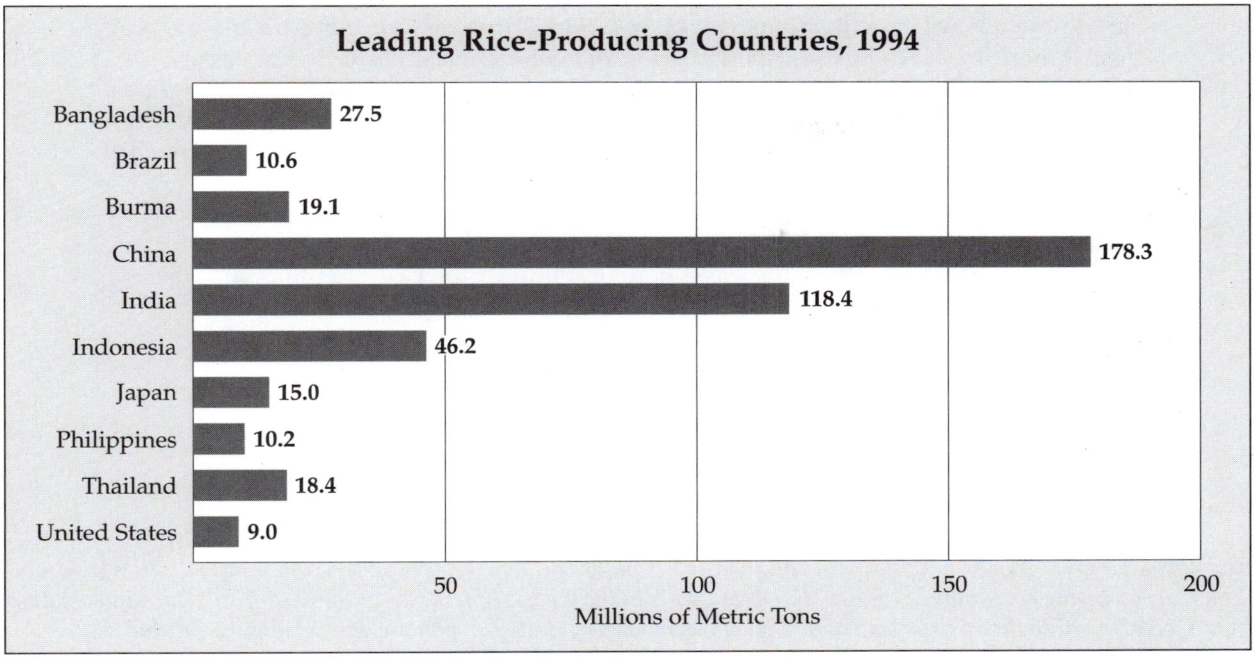

Note: World total for 1994 was approximately 535 million metric tons.
Source: U.S. Department of Commerce, *Statistical Abstract of the United States, 1996*, 1996.

of high-yielding types of rice, beginning in the mid-1960's. The development of these plants is considered a significant part of the 1960's *Green Revolution* in agriculture. Some of these varieties demand complete irrigation systems year-round that help keep the soil submerged under about 6 inches (15 centimeters) of water. Next to corn, rice provides the farmer with the greatest yield when plants are cultivated with the necessary care. The crop grows well in irrigated and flooded areas.

Cooked rice is mostly consumed in its whole grain form. Puffed rice and flaked rice are common breakfast cereals, and rice flour is used in bakery products. Laundry starch is made from rice starch. Rice hull is used in cattle feed as well as fertilizers. The rice plant produces oil for food and industry and thatching material for roofs and mats. The Japanese alcoholic beverage sake is made from a process that involves the fermentation of rice.

Wild Rice

The plant commonly known as wild rice, *Zizania aquatica*, is actually a separate genus found in North America. Like rice, wild rice is an annual grass, and it grows mostly in lakes and streams. Lakes in Minnesota, Wisconsin, and southern Canada provide a good harvest of wild rice. Wild rice, once a staple of the diet of American Indians in those regions, has become a popular side dish.

Mysore Narayanan

See also: African agriculture; Agriculture: world food supplies; Alternative grains; Aquatic plants; Asian agriculture; Australian agriculture; Caribbean agriculture; Central American agriculture; Green Revolution; Grains; High-yield crops; Hybridization; North American agriculture; Pacific Island agriculture; Plant biotechnology; Plant domestication and breeding; South American agriculture.

Sources for Further Study

Carney, Judith Ann. *Black Rice: The African Origins of Rice Cultivation in the Americas*. Cambridge, Mass.: Harvard University Press, 2001. Establishes the independent domestication of rice in West Africa and its vital significance there well before Europeans arrived, discrediting the notion that Europeans introduced rice to West Africa and then brought the knowledge of its cultivation to the New World. Carney asserts that Africans and African American slaves transferred rice seeds and cultivation skills to the Americas.

Piper, Jacqueline M. *Rice in South-East Asia: Cultures and Landscapes*. New York: Oxford University Press, 1993. Shows how rice, as a staple food and as a source of livelihood, determines the yearly rhythm as its major ceremonials mark the changing seasons for much of the population of Southeast Asia.

Shimamoto, Ko. *Molecular Biology of Rice*. New York: Springer, 1999. Describes how advances in genome analysis and transformation have made rice a model monocot for molecular biologists. Discusses these technological impacts on rice breeding.

RNA

Categories: Cellular biology; genetics

Ribonucleic acid (RNA), a molecule that plays many roles in the effective usage of genetic information, exists in several forms, each with its own unique function. RNA functions in the process of protein synthesis, during which information from DNA is used to direct the construction of a protein, and possesses enzymatic and regulatory capabilities.

Ribonucleic acid (RNA) is a complex biological molecule that is classified along with deoxyribonucleic acid (DNA) as a *nucleic acid*. Chemically, RNA is a polymer (long chain) consisting of subunits called *ribonucleotides* linked together by phosphodiester bonds. Each ribonucleotide consists of three parts: the sugar ribose (a five-carbon simple sugar), a negatively charged phosphate group, and a nitrogen-containing base. There are four types of ribonucleotides, and the differences between them lie solely in which of four possible bases they contain. The four bases are adenine (A), guanine (G), cytosine (C), and uracil (U).

The structures of DNA and RNA are very similar, but there are three important differences. The sugar found in the nucleotide subunits of DNA is *deoxyribose*, which is related to but differs slightly from the ribose found in the ribonucleotides of RNA. In addition, while DNA nucleotides also contain four possible bases, there is no uracil in DNA; instead, DNA nucleotides may contain a different base, called thymine (T). Finally, while DNA exists as a double-stranded helix in nature, RNA is almost always single-stranded.

Folding of RNA Molecules

The significance of many types of RNA lies in the order of their nucleotides, which represents information transcribed from DNA. This nucleotide order is called the *primary structure* of the molecule. An important aspect of many biological molecules, however, is the way their primary structures fold to create a three-dimensional shape. A single strand of DNA, for example, associates with another strand in a particular way to form the famous double helix, which represents its actual three-dimensional shape in nature. Similarly, protein molecules, especially enzymes, must be folded into a very specific three-dimensional shape if they are to perform their functions; loss of this shape will cause their inactivation.

Since RNA is single-stranded, it was recognized shortly after the discovery of some of its major roles that its capacity for folding is great and that this folding might play an important part in the functioning of the molecule. The nucleotides in an RNA molecule can form hydrogen-bonded *base pairs*, according to the same rules that govern DNA base pairing. Cytosine binds to guanine, and uracil binds to adenine. What this means is that in a particular single-stranded RNA molecule, complementary portions of the molecule are able to fold back and form base pairs with one another. These are often local interactions, and a common structural element that is formed is called a "hairpin loop" or "stem loop." A hairpin loop is formed when two complementary regions are separated by a short stretch of bases so that when they fold back and pair, some bases are left unpaired, forming the loop. The net sum of these local interactions is referred to as the RNA's secondary structure and is usually important to an understanding of how the RNA works. All *transfer RNAs* (tRNAs), for example, are folded into a secondary structure that contains three stem loops and a fourth stem arranged onto a "cloverleaf" shape.

Finally, local structural elements may interact with other elements in long-range interactions, causing more complicated folding of the molecule in space. The cloverleaf arrangement of a tRNA undergoes further folding so that the entire molecule takes on a roughly L-shaped appearance. An understanding of the three-dimensional shape of an RNA molecule is crucial to understanding its function. By the late 1990's, the three-dimensional structures of many tRNAs had been worked out, but it had proven difficult to do X-ray diffraction analyses on most other RNAs because of technical problems. More advanced computer programs and alternate structure-determining techniques have now enabled research in this field to proceed.

Three Classes of RNA

While all RNAs are produced by transcription, several classes of RNA are created, and each has a particular function. By the late 1960's, three major classes of RNAs had been identified, and their respective roles in the process of protein synthesis had been elucidated. In general, protein synthesis refers to the assembly of a protein using information encoded in DNA, with RNA acting as an intermediary to carry information and assist in protein building. In 1956, Francis Crick, one of the scientists who had discovered the double-helical structure of DNA, referred to this information flow as the "central dogma," a term that continues to be used.

Messenger RNA (mRNA) is the molecule that carries a copy of the DNA instructions for building a particular protein. It usually represents the information provided by a single gene and carries this information to the ribosome, the site of protein synthesis. This information must be decoded so that it will specify the order of amino acids in a protein. Nucleotides are read in groups of three (*codons*). In addition to the information required to order amino acids, the mRNA contains signals that tell the protein-building machinery where to start and stop reading the genetic information.

Ribosomal RNA (rRNA) exists in three distinct sizes and is part of the structure of the ribosome. The three ribosomal RNAs interact with many proteins to complete the ribosome, the cell structure that directs the events of protein synthesis. One of the functions of the rRNA is to interact with mRNA at a particular location and orient it properly so that reading of its genetic code can begin at the correct location. Another rRNA acts to facilitate the transfer of the growing polypeptide chain from one tRNA to another (peptidyl transferase activity).

Transfer RNA (tRNA) serves the vital role of decoding the genetic information. There are at least twenty and usually fifty to sixty different tRNAs in a given cell. On one side, they contain an "anticodon" loop, which can base-pair to the mRNA codon according to its sequence and the base-pairing rules. On the other side, they contain an amino acid binding site, to which is attached the appropriate amino acid for its anticodon. In this way, tRNA allows the recognition of any particular mRNA codon and matches it with the appropriate amino acid. The process continues until an entire new protein molecule has been constructed.

Split Genes and mRNA Processing

In bacterial genes, there is a colinearity between the segment of a DNA molecule that is transcribed and the resulting mRNA. In other words, the mRNA sequence is complementary to its template and is the same length, as would be expected. In the late 1970's, several groups of scientists made a seemingly bizarre discovery regarding mRNAs in eukaryotes (organisms whose cells contain a nucleus, including all living things that are not bacteria or archaebacteria): The sequences of mRNAs isolated from eukaryotes were not collinear with the DNA from which they were transcribed. The coding regions of the corresponding DNA were interrupted by seemingly random sequences that served no immediately obvious function. These *introns*, as they came to be known, were apparently transcribed along with the coding regions (*exons*) but were somehow removed before the mRNA was translated. This completely unexpected observation led to further investigations that revealed that mRNA is extensively processed, or modified, after its transcription in eukaryotes.

After a eukaryotic mRNA is transcribed, it contains all of the intervening sequences and is referred to as immature, or a "pre-mRNA." Before it can become mature and functional, three major processing events must occur: splicing, the addition of a "cap," and the addition of a "tail."

The process of splicing is a complex one that occurs in the nucleus with the aid of the "spliceosome," a large complex of RNAs and proteins that identify intervening sequences and cut them out of the pre-mRNA. In addition, the spliceosome must rejoin the sequences from which the intron-encoded nucleotides were removed so that a complete, functional mRNA results. The process must be extremely specific, since a mistake that caused the removal of only one extra nucleotide could change the protein product of translation so radically that it might fail to function.

While splicing is occurring, two other vital events are being performed to make the immature mRNA ready for action. A so-called cap, which consists of a modified guanine, is added to one end of the pre-mRNA by an unconventional linkage. The cap appears to function by interacting with the ribosome, helping to orient the mature mRNA so that translation begins at the proper location. A tail, which consists of many adenines (often two hundred or more), is also attached to the other end of the pre-mRNA. This so-called poly-A tail, which virtually all eukaryotic mRNAs contain, seems to be involved in determining the relative stability of an mRNA. These important steps must be performed after transcription in eukaryotes to enable the creation of a mature, functional messenger RNA molecule that is now ready to be translated. Most mRNAs must be transported out of the nucleus before they are used to make proteins, but about 10 to 15 percent of the mRNAs produced remain in the nucleus, where they are used to make proteins.

Other Specialized Functions

The traditional roles of RNA in protein synthesis were originally considered the only roles RNA was capable of performing. RNA in general, while considered an important molecule, was thought of as a "helper" in translation. This all began to change in 1982, when the molecular biologists Thomas Cech and Sidney Altman, working independently and with different systems, reported the existence of RNA molecules that had catalytic or enzymelike activity, meaning that RNA molecules can function as enzymes. Until this time, it was believed that all enzymes were protein molecules. The importance of these findings cannot be overstated, and Cech and Altman ultimately shared the 1989 Nobel Prize in Chemistry for the discovery of these RNA enzymes, or *ribozymes*. Both of these initial ribozymes catalyzed reactions that involved the cleavage of other RNA molecules—that is, they acted as nucleases. Subsequently, many ribozymes have been found in various organisms, from bacteria to humans. Some of them are able to catalyze different types of reactions, and there are new ones reported every year. Thus ribozymes are not a mere curiosity but play an integral role in the molecular machinery of many organisms. Their discovery also gave rise to the idea that at one point in evolutionary history, molecular systems composed solely of RNA performing many roles existed in an "RNA world."

At around the same time as these momentous discoveries, still other classes of RNAs were being discovered, each with its own specialized functions. In 1981, Jun-Ichi Tomizawa discovered *RNAI*, the first example of what would become another major class of RNAs, the *antisense RNAs*. The RNAs in this group are complementary to a target molecule (usually an mRNA) and exert their function by binding to that target via complementary base pairing. These antisense RNAs usually play a regulatory role, often acting to prevent translation of the relevant mRNA to modulate the expression of the protein for which it codes.

Another major class of RNAs, the *small nuclear RNAs* (snRNAs), was also discovered in the early 1980's. Molecular biologist Joan Steitz was working on the autoimmune disease systemic lupus when she began to characterize the snRNAs. There are six different snRNAs, now called U1-U6 RNAs. These RNAs exist in the nuclei of eukaryotic cells and play a vital role in mRNA splicing. They associate with proteins in the spliceosome, forming so-called *ribonucleoprotein complexes* (snRNPs), and play a prominent role in detecting proper splice sites and directing the protein enzymes to cut and paste at the proper locations.

It has been known since the late 1950's that many viruses contain RNA, and not DNA, as their genetic material. This is another fascinating role for RNA in the world of biology. The viruses that cause influenza, polio, and a host of other diseases are RNA viruses. Of particular note are a class of RNA viruses known as *retroviruses* because they have a particularly interesting life cycle. Retroviruses, which include human immunodeficiency human immunodeficiency virus (HIV), the virus that causes acquired immunodeficiency syndrome (AIDS) in humans, use a special enzyme called reverse transcriptase to make a DNA copy of their RNA instructions when they enter a cell. That DNA copy is inserted into the DNA of the host cell, where it is referred to as a *provirus*, and never leaves. Clearly, understanding the structures and functions of the RNAs associated with these viruses will be important in attempting to create effective treatments for the diseases associated with them.

An additional role of RNA was noted during the elucidation of the mechanism of DNA replication. It was found that a small piece of RNA, called a primer, must be laid down by the enzyme primase before DNA polymerase adds DNA nucleotides to this initial RNA sequence, which is subsequently removed.

Matthew M. Schmidt

See also: Chloroplast DNA; DNA: historical overview; DNA in plants; DNA replication; Gene regulation; Genetic code; Genetics: mutations; Mitochondrial DNA; Nucleic acids; Proteins and amino acids.

Sources for Further Study

Eckstein, Fritz, and David M. J. Lilly. *Catalytic RNA* New York: Spring, 1996. Offers a comprehensive overview of ribozyme diversity and function. Illustrated. Bibliography, indexes.

Erickson, Robert P., and Jonathan G. Izant, eds. *Gene Regulation: Biology of Antisense RNA and DNA*. New York: Raven Press, 1992. Provides a comprehensive overview of natural antisense RNA function and prospects for its uses in gene therapy. Illustrations, bibliographical references, index.

Murray, James A. H., ed. *Antisense RNA and DNA*. New York: Wiley-Liss, 1992. Covers antisense RNA. Illustrations, bibliography, index.

Simons, Robert W., and Marianne Grunberg-Manago, eds. *RNA Structure and Function*. Cold Spring Harbor, N.Y.: Cold Spring Harbor Laboratory Press, 1998. An advanced and hefty text (more than seven hundred pages) that, like Murray's book, takes a detailed look at the various structures of RNA, their relationships to function, and the techniques for determining RNA structure. Illustrations, bibliography, index.

Watson, James D., et al. *Molecular Biology of the Gene*. 4th ed. 2 vols. Menlo Park, Calif.: Benjamin/Cummings, 1987. The codiscoverer of DNA's double-helix structure provides an insightful discussion of the various structures of RNA and their relationships to function. Bibliographies, indexes, illustrations.

ROOT UPTAKE SYSTEMS

Categories: Anatomy; physiology; transport mechanisms; soil

Root uptake systems are processes by which root cells transport water and nutrients from the soil, across the root surface, and to the tissues that will move the water and nutrients throughout the plant.

Fertile soil is a complex mixture of a variety of minerals, many different types of organic matter in different stages of decay, and a host of living microorganisms. This complex medium holds a large quantity of water, which it supplies to plants. In addition to water, the soil supplies the plants with the thirteen mineral nutrients required for normal growth and development. These nutrients (and the ionic forms taken up by the root) are nitrogen, phosphorus, potassium, sulfur, calcium, magnesium, iron, manganese, boron, chlorine, zinc, copper, and molybdenum. Varying amounts of these mineral nutrients exist both as constituents of the soil particles and as dissolved ions in the soil

Most plant roots extend below the soil surface.

water. Root uptake systems are responsible for taking these nutrient ions and water from the soil and moving them into the root tissues.

Root Structure

Most plant roots extend below the soil surface as either a *fibrous root system* or *taproot system*. In a fibrous root system, the major root branches numerous times, and each branch divides again and again, until a meshlike network is formed within the soil. This system does not penetrate very deep into the soil, but it does cover considerable area close to the surface.

In a taproot system, only small secondary lateral roots branch off the main root (the taproot) as it grows downward into the soil. The taproot may extend several meters in depth. Along the outer periphery of either of these root systems, large numbers of small *filamentous root fibers* and *root hairs* can be found. The root hairs are filamentlike projections of the epidermal (outer layer) cells. These root hairs, in conjunction with the cells of the filamentous root fibers, are responsible for the vast majority of water and nutrient uptake.

Each root cell, as is the case in all plant cells, is surrounded by a porous cell wall. Immediately inside the cell wall is a semipermeable membrane, which regulates the movement of ions and molecules into the *cytosol* of the cell. This semipermeable membrane is a fluid mosaic structure composed primarily of lipid and protein. A double layer of lipid provides the basic stable structure of the membrane. The protein is then interspersed periodically throughout the lipid bilayer. Some of these proteins, called *peripheral proteins*, penetrate only one of the layers of lipid, while *integral proteins* extend through both lipid layers to interface with the environment both inside and outside the cell. The rate and extent of water and ion movement through membranes are largely determined by this structural configuration.

Ion Uptake Mechanisms

Uptake of ions across membranes is accomplished by three mechanisms: *simple diffusion*, *facilitated diffusion*, and *active transport*. The first two are passive processes, with no direct input of cellular energy required. The latter, as the name indicates, is an active process requiring the cell to expend energy.

In its simplest analysis, diffusion is the net movement of suspended particles down a *concentration gradient*. Thus, certain ions dissolved in the soil solution will move into the root cell cytosol as long as the external concentration is higher than the concentration inside the cell.

Because of the lipid nature of the membrane, the rate of this movement will be determined by the *lipid solubility* of the particle. Those particles with high lipid solubility will diffuse across the membrane much faster than those with low lipid solubility.

Many nutrients exhibit low lipid solubility, yet still diffuse across the membrane. This is accomplished by facilitated diffusion. The ion with low lipid solubility combines with a membrane protein, which then facilitates its uptake across the membrane.

In both free and facilitated diffusion, the particles move down a concentration gradient. In numerous instances, however, the ion concentration of the root cell cytosol is greater than the concentration of the ion in the soil solution. The ion will nevertheless continue to accumulate within the root tissues.

Diffusion of any sort cannot account for this "uphill" movement against a concentration gradient. In this instance, the nutrient ion will combine with a membrane protein referred to as a *carrier*. This protein carrier will transport the particle across the membrane. Uptake mediated by this protein carrier system is called active transport and requires the input of energy supplied by the hydrolysis of adenosine triphosphate (ATP), the cell's primary energy currency.

Osmotic Potential

Regardless of the mechanism utilized to transport ions into the root cytosol, the final result is the establishment of an *osmotic potential* across the membrane. Osmotic potential is a measure of the tendency of water to move across a semipermeable membrane in response to a difference in solute concentration. The addition of a solute to water lowers its osmotic potential, and water will always move from the side of the membrane containing the solution with the lower solute concentration (higher osmotic potential) to the side of the membrane containing the solution with the higher solute concentration (lower osmotic potential). Hence, the uptake of the mineral ions by the plant root cells establishes a lower osmotic potential within the cytosol than exists in the soil solution, and water flows across the membrane into the cell.

In order for the water and ions to move across the root from the epidermal layer to the internal transport vessels called *xylem*, several layers of *cortex cells* and the *endodermis* must be crossed. There are two pathways water and ions can take. One is through the cytosol of the epidermal cells, cortical cells, and endodermal cells by means of the *plasmodesmata*, which are cytoplasmic strands that pass between plant cells, thereby connecting the cells as microscopic "bridges." These plasmodesmata provide a means by which the cytosol from all cells can exist as a continuous mass referred to as the *symplast*, and transport through this system is called *symplastic transport*.

A second pathway, referred to as *apoplastic transport*, occurs through the apoplast (the region of continuous cell walls among cells). Because this apoplast is freely permeable to water and ions, the cells of the epidermis, cortex, and endodermis are in intimate contact with the soil water. Apoplastic transport, however, cannot occur across the endodermis because of the presence of an impermeable waxy layer in the cell walls called the *Casparian strip*. Thus, water and ions can travel through the apoplast until reaching the endodermis. There, movement into the endodermal cytosol by one of the three mechanisms mentioned above is required.

After passing the Casparian strip, the water and ions can move back into the apoplast. Although there is some disagreement among plant scientists as to which pathway is most important, it is highly probable that both pathways are involved. Regardless of whether the transport across the root is apoplastic or symplastic, the water and ions reach the xylem tubes within the interior of the root, where subsequent transport throughout the plant can take place.

D. R. Gossett

See also: Active transport; Angiosperm cells and tissues; Cells and diffusion; Liquid transport systems; Osmosis, simple diffusion, and facilitated diffusion; Plant tissues; Roots; Vesicle-mediated transport; Water and solute movement in plants.

Sources for Further Study

Begon, Michael, and Alastair Fitter, eds. *Advances in Ecological Research*. Vol. 27. San Diego: Academic Press, 1997. Includes chapters on ecology of root life span and age-related decline in forest productivity.

Campbell, Neil A., and Jane B. Reece. *Biology*. 6th ed. San Francisco: Benjamin Cummings, 2002. An introductory college-level textbook for science students. The chapter "Traffic Across Membranes" provides a clear, concise, detailed description of the membrane transport process. The well-written text, combined with superb graphics, furnishes the reader with a very clear understanding of membrane structure and transport. List of suggested readings at the end of the chapter. Includes glossary.

Mader, Sylvia S. *Inquiry into Life*. 10th ed. Dubuque, Iowa: McGraw-Hill, 2002. An introductory-level textbook for the nonscience student. The chapter "Plant Form and Function" provides a simple overview of the root uptake and transport process. Clear, concise diagrams with a very readable text establish a general picture of the process. Key terms and suggested reading list at the end of each chapter. Includes glossary.

Nissen, P. "Uptake Mechanisms: Inorganic and Organic." *Annual Review of Plant Physiology* 25 (1974): 53-79. Detailed review article outlining much of the pertinent information related to the control of root uptake. A historical perspective as well as a discussion of the recent advances is presented by the reviewer. A very informative article with a detailed list of complex literature citations.

Raven, Peter H., Ray F. Evert, and Susan E. Eichhorn. *Biology of Plants*. 6th ed. New York: W. H. Freeman/Worth, 1999. An introductory college-level textbook for science students. The chapters "The Root: Primary Structure and Development" and "The Movement of Water and Solutes in Plants" provide a discussion of root anatomy and transport across the root. This profusely illustrated text furnishes a good pictorial study of the root uptake process. Includes glossary.

Steward, F. C. *Plant Physiology: A Treatise*. 15 vols. New York: Academic Press, 1959-1991. A technical treatise for the more advanced student. An excellent review of plants in their relation to water and solutes. A historical perspective, an overview, and in-depth treatment are provided for each topic. Includes extensive bibliography.

ROOTS

Categories: Anatomy; physiology

Roots are those underground portions of a plant that store food, absorb water and minerals from the soil, and anchor the plant in the earth.

Roots account for more than 80 percent of plants' biomass in ecosystems such as tundra and shortgrass prairies. In many plants, roots are longer and spread wider than the shoots. The extensive root systems of plants are effective collectors of water and minerals necessary for the life of the plant.

Root Cap and Quiescent Center

The root's structure facilitates each of its functions. Tips of roots are covered by a thimble-shaped *root cap*. At the base of the root cap is a *meristem* that produces cells that form the cap. The meristem pushes cells forward into the cap, which protects the tip of the growing root as it forces its way

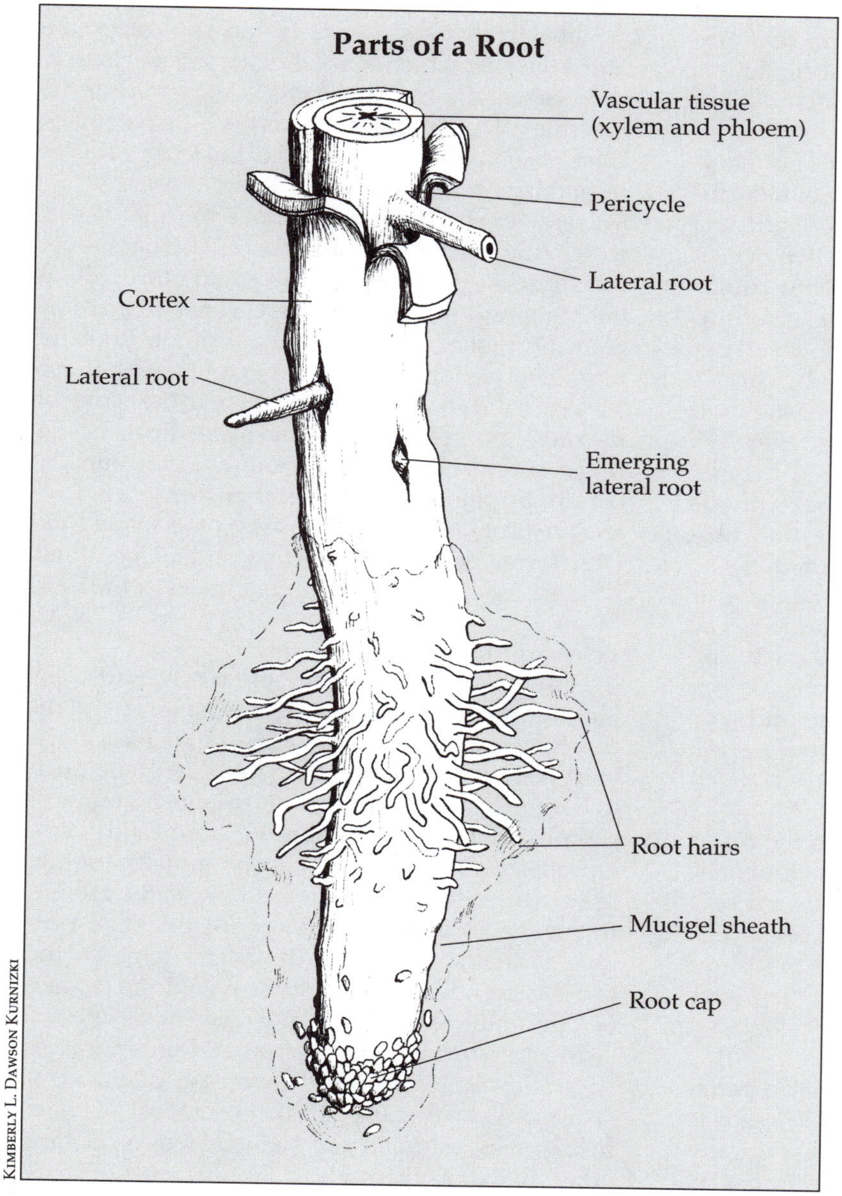

through the soil. As these cells move through the cap, they differentiate into rows of *columella cells*. Columella cells contain *plastids* that sediment, in response to gravity, to the lower side of the cell. This sedimentation is how roots perceive gravity.

Surrounding the outside of the root cap are *peripheral cells*. Peripheral cells produce and secrete large amounts of a slimy, water-soluble substance called *mucigel*. Mucigel has several important functions. It protects roots from desiccation and contains compounds that diffuse into the soil and inhibit growth of other roots. Mucigel also lubricates roots as they force their way between soil particles. Soil particles cling to mucigel, thus increasing the root's contact with the soil, which helps roots absorb water.

Just behind the root cap is the *quiescent center*, which is made up of five hundred to one thousand seemingly inactive cells. Cells of the quiescent center divide only about once every twenty days, while those of the adjacent meristem divide as often as twice per day. Cells of the quiescent center become active when the tip of the root is damaged. When this occurs, the quiescent cells divide rapidly to form cells to repair the damaged root tip. The quiescent center also organizes the patterns of primary growth in roots.

Subapical Region

The *subapical region* of roots consists of three zones: the *zone of cellular division*, the *zone of cellular elongation*, and the *zone of cellular maturation*. These regions of the root intergrade and are not sharply defined. The zone of cellular division surrounds the quiescent center and is a dome-shaped meristem 0.5 to 1.5 millimeters behind the root tip. Thus, the meristem of a root is subterminal and is made of small, multisided cells. Meristematic cells divide between one and two times per day. In some plants, the meristem produces almost twenty thousand new cells per day. The rate of these divisions is influenced by hormones such as *ethylene*.

The zone of cellular elongation is 4 to 15 millimeters behind the root tip. Cells in this zone elongate rapidly by filling their vacuoles with water. As a result, the elongating zone is easily distinguished from the root cap and zone of cellular division by its long, vacuolate cells. Cellular elongation in the elongating zone pushes the root cap and apical meristem through the soil as fast as 2 to 4 centimeters per day. Cellular elongation is typically inhibited by the hormone *auxin* and stimulated by low concentrations of ethylene.

Cells behind the elongating zone do not elongate. Elongation begins the process of cellular differentiation, or specialization. Differentiation is completed in the zone of cellular maturation, which is 1 to 5 centimeters behind the root tip. The maturation zone is distinguished by the presence of numerous root hairs—as many as forty thousand per square centimeter. Root hairs increase the surface area of the root by a factor of several thousands and are usually less than 1 millimeter long. They live only a few days and form only in the mature, nonelongating region of the root. Because root hairs are fragile extensions of epidermal cells, they usually break off when plants are transplanted.

All mature tissues of roots form behind the zone of cellular maturation. The root is surrounded by an *epidermis*, which is usually only one cell thick. Epidermal cells usually lack a cuticle. The epidermis covers all of the root except the root cap and typically has no openings.

Cortex

Immediately inside the epidermis is the *cortex*. The cortex occupies most of a root's volume and consists of three concentric layers: the *hypodermis*, storage *parenchyma cells*, and the *endodermis*. The hypodermis is a waxy, protective layer that slows outward movement of water. Thus, the hypodermis helps roots retain water and nutrients that have been absorbed.

Most of the cortex consists of thin-walled parenchyma cells that store carbohydrates. These cells are separated by large intercellular spaces occupying as much as 30 percent of the root's volume. The innermost layer of the cortex is the endodermis. Unlike other cortical cells, endodermal cells are packed tightly together and lack intercellular spaces. Their radial and transverse walls, furthermore, are impregnated with a *Casparian strip* of lignin and suberin. If endodermal cells are compared to bricks in a brick wall, then the Casparian strip is analogous to the mortar surrounding each brick. The Casparian strip prevents inward movement of water and nutrients through the cell wall and intercellular spaces. The endodermis functions somewhat like a valve that regulates movement of nutrients.

Collectively, the tissues inside the cortex are called the *stele*, which consists of the *pericycle*, *vascular tissues*, and sometimes a *pith*. The *pericycle* is the outermost layer of the stele and is a meristematic layer of cells one to several cells thick; it produces secondary, or lateral, roots.

Inside the pericycle is the root's vascular tissue, which consists of *xylem* and *phloem*. Vascular tissues transport water, minerals, and sugars throughout the plant and differentiate in response to auxin, a plant hormone, coming from the shoot. Roots of most dicotyledons (dicots) and gymnosperms have a lobed, solid core of primary xylem in the center of the root. Roots of monocots and a few dicots have a ring of vascular tissue that surrounds a pith. Bundles of primary phloem differentiate between lobes of xylem. In dicots and gymnosperms, a *vascular cambium* later forms between the xylem and phloem and produces secondary growth that thickens the root.

Types of Root Systems

Different kinds of plants often have different kinds of root systems. Most gymnosperms and dicots have a *taproot system* consisting of a large primary root and smaller branch roots. In plants such as the carrot, fleshy taproots store large amounts of carbohydrates. Not all taproots store food. Long taproots of plants such as poison ivy and mesquite are modified for reaching water deep in the ground rather than storing food. Many plants have very long taproots. Engineers digging a mine in the southeastern United States uncovered the taproot of a mesquite tree more than 50 meters down.

Most monocots such as corn and other grasses have a *fibrous root system* that consists of an extensive mass of similarly sized roots. Most of these roots are *adventitious roots*, which form on organs other than roots themselves. Fibrous roots of some plants are edible—for example, sweet potatoes are fleshy parts of fibrous root systems of ipomoea (morning glory) plants. Plants with fibrous root systems reduce erosion because their root systems are extensive and cling tightly to soil particles.

Adventitious roots are common in ferns, club mosses, and horsetails. In plants such as tree ferns, adventitious roots form in stems, grow down through the cortex, and finally emerge at the base of the stem. Adventitious roots are a primary means of *asexual reproduction* in many plants. For example, prairie grasses and forests of quaking aspen trees are often derived from a single individual propagated by adventitious roots. Humans use adventitious roots to propagate plants such as raspberries, apples, and brussels sprouts. Formation of adventitious roots is controlled by hormones such as auxin, which is often an ingredient in "rooting" compounds sold commercially.

Functions

The structure of a root relates directly to its four primary functions: *absorption*, *anchorage*, *conduction*, and *storage*. For example, most water and nutrients are absorbed by root hairs in the zone of maturation of the root. The water thereafter moves through the root either inside cells or in spaces between cells. Water seeping between cells finally encounters the endodermis, which is the primary barrier to absorption. The Casparian strip in the endodermis ensures that water and nutrients enter the stele via the plasmodesmata (narrow strands of cytoplasm that connect the cytoplasms of adjacent cells). Most nutrients are absorbed and accumulate in the apical 0.3 to 0.5 meter of the root.

Few nutrients are absorbed past a few centimeters beyond the root tip because these parts of the

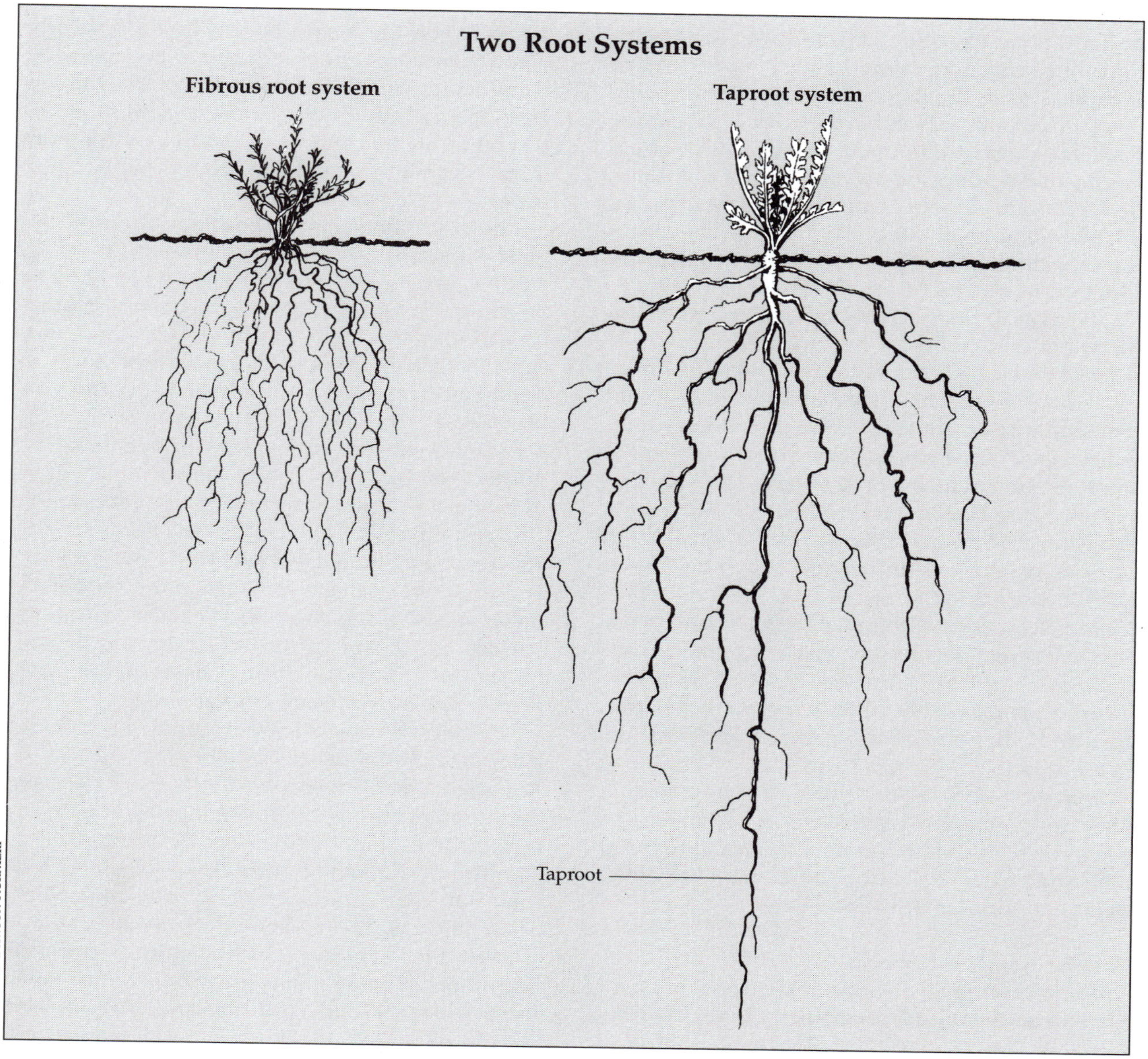

Kimberly L. Dawson Kurnizki

root lack root hairs and have a waxy endodermis. These nonabsorptive regions of roots anchor plants and may later produce branch roots. Water and dissolved nutrients absorbed by roots move to the shoot in xylem. Roots receive nutrients from the shoot via the phloem. These nutrients either are used for growth or are stored in cortical cells for future use.

Roots of many plants are modified for special functions. For example, roots of plants such as beets, radishes, dandelions, and cassava store large amounts of starch. Sweet potato roots store carbohydrates as sugars. Roots of other plants are used for asexual reproduction.

Roots of cherry, apple, and teak possess adventitious buds that form shoots called *suckers*. When separated from the parent plant, suckers become new individuals. Adventitious buds are a common means of propagating many other plants. For example, most groups of creosote bushes are clones derived from a single plant. Some of these clones are more than twelve thousand years old—meaning that the first seed germinated approximately four thousand years before humans began writing.

The roots of many plants minimize competition for water and nutrients by growing in different parts of the soil. For example, mesquite trees growing in deserts often have taproots more than 20 meters long that obtain water from the underground water table. Nearby cacti, however, survive in the same environment by producing a shallow root system that spreads as far as 30 meters. Cactus roots do not reach the water table; rather, they quickly absorb water after the infrequent and often heavy rains that occur in the desert.

Roots can protect the plant from other organisms. Most root defenses against soil pathogens are chemical rather than structural. Roots often secrete noxious chemicals that inhibit the growth of pathogens and other organisms, including other plants, in some cases.

Prop roots are aerial roots that grow into the soil. They are common in plants such as corn and banyan trees. Banyan trees produce thousands of prop roots that grow down from horizontally oriented stems and form pillarlike supports.

Environmental Influences

The growth and distribution of roots are controlled by several environmental factors. For example, the short days and cooler temperatures of winter typically cause roots to become inactive. Microbes living in the soil also affect how roots grow. Most microbes in soil live within 10 micrometers of the root and secrete compounds that significantly affect growth and distribution of roots. These compounds can affect the anatomy, morphology, number of root hairs, and branching patterns of root systems.

Microbes also affect how plants absorb and transport minerals from the soil. Plants growing in sterile soil absorb fewer minerals than those growing in soil containing microbes. Beneficial fungi called *mycorrhizae* live in and on roots of almost all plants in a form of mutualism, meaning that both the plant and the fungus benefit from the association. The mycorrhizae absorb nutrients from the environment, while the host plant provides the fungus with carbohydrates, amino acids, vitamins, and other organic substances. Plants with mycorrhizae tolerate drought and other types of stress better than uninfected plants.

Roots of legumes such as beans are often infected with *Rhizobium* (from *rhiza*, the Greek word for "root"), a genus of the nitrogen-fixing bacteria. Swellings in response to these infections are called *nodules*. Bacteria receive carbohydrates and other substances from the host, while the host plants receive large amounts of usable nitrogen from the bacteria.

Many organisms compete with roots to collect nutrients and water in the soil. For example, 1 gram of fertile soil contains approximately 10^9 bacteria, 10^6 actinomycetes (a group of fungilike bacteria), 10^5 fungi, and 10^3 algae. Plants have evolved several different strategies for competing with these organisms. One is to produce an extensive root system consisting of many roots that permeate the soil. Most roots grow in the upper 3 meters of soil, however, where nutrients are most abundant.

The narrow zone of soil surrounding a root and subject to its influence is called the *rhizosphere*. Roots modify the rhizosphere by secreting organic matter, compressing the soil, and absorbing nutrients. As a result of these effects, the rhizosphere is significantly different from bulk soil. It usually contains large amounts of energy-rich molecules. These molecules are eaten by fungi and bacteria.

Roots tend to grow best in moist, loosely packed soil. Roots of many plants grow two to four times faster in loose, sandy soil than in tightly packed clay. The slow growth of roots in poorly aerated soil is probably caused by the accumulation of ethylene, a plant hormone that slows root growth.

Roots of plants that grow in wet areas are usually small and modified for gas exchange. They possess small amounts of xylem, lack root hairs, and contain large intercellular spaces, thereby improving ventilation. Roots of aquatic plants are also modified for gas exchange. For example, the black mangrove produces specialized roots called *pneumatophores*, which grow up into the air, where they function like snorkels through which oxygen diffuses to submerged roots.

Epiphytes, including some bromeliads and orchids, are plants that grow on other plants but are not parasites. Adaptations among these "air plants" include a thickened root epidermis which protects the cortex and retards water loss. The "flower pot plant" (*Dischidia rafflesiana*) grows a "pot" that collects water and debris; the plant's aerial roots grow into the pot, where they can absorb the minerals collected there.

Roots of plants that grow in dry areas are often extensive and modified for rapid transport of water. They contain large amounts of xylem, which allows them to move water rapidly to the shoot after rainfall. Plants such as witchweed and broomrape use their roots to parasitize other plants. Witchweed is a red-flowered plant that infects grains such as corn and sorghum; it is the second leading cause of cereal famine in Africa.

Randy Moore, updated by Elizabeth Slocum

See also: Angiosperm cells and tissues; Bulbs and rhizomes; Gas exchange in plants; Germination and seedling development; Growth and growth control; Liquid transport systems; Nitrogen fixation; Plant tissues; Root uptake systems; Water and solute movement in plants.

Sources for Further Study

Bohm, W. *Methods of Studying Root Systems*. New York: Springer-Verlag, 1979. Excellent account of the various ways of studying root systems. Includes numerous diagrams and photographs.

Epstein, Emanuel. "Roots." *Scientific American* 228 (May, 1973): 48-58. This article describes the structure and function of roots, including roots modified for unusual functions. Diagrams of root structure are included to show the integration of structure and function, especially as related to ion uptake.

Raven, Peter H., Ray F. Evert, and Susan E. Eichhorn. *Biology of Plants*. 6th ed. New York: W. H. Freeman/Worth, 1999. The chapter "The Root: Primary Structure and Function" is a well-illustrated description of the primary structure of roots. Numerous micrographs and diagrams help explain the relationship of root structure to root function.

Waisel, Yoav, Amram Eshel, and U. Kafkafi, eds. *Plant Roots: The Hidden Half*. 3d rev. ed. New York: Marcel Dekker, 2002. This overview of root systems addressees structure and development, methods of study, growth and metabolism, ion and water relations, the rhizosphere, root pests, and more.

RUBBER

Category: Economic botany and plant uses

Rubber is a product of rubber tree bark or a macromolecule or polymer of repeated chains of carbon and hydrogen atoms. Rubber's properties of extensibility, stretchability, toughness, and resilience make it a useful commodity in applications ranging from tires to clothing.

The name "rubber" originates from the material's ability to erase pencil marks; its chemical designation is *polyisoprene* with several isomers. About 60 to 65 percent of the rubber produced today is synthetic. When explorer Christopher Columbus arrived in Haiti in 1492 he found Indians playing a game with a ball made from the *latex* of rubber. American Indians were also known to have used latex for making footwear, bottles, and cloaks. By 1735 latex had been described as *caoutchouc* by a French geographical expedition in South America.

The role that rubber could play in clothing and footwear attracted the attention of chemists and inventors throughout the world in the late eighteenth and mid-nineteenth centuries. Charles Macintosh and Thomas Hancock, working as colleagues, discovered two separate means of using rubber in fabrics and footwear. Macintosh found that placing rubber between layers of fabric resulted in a fabric with no sticky and brittle surfaces. Hancock developed the rubber masticator, which welded rubber scraps to be used for further manufacturing.

The dramatic increase in the use of rubber that occurred in the twentieth century is attributable largely to the development of the automobile industry and advances in industrial technology. Although rubber's percentage of use compared with other elastomers decreased from the end of World War II to the late 1970's, the development of radial automobile tires in Europe in the late 1940's and their popularization in the United States in the late 1960's resulted in increased use of natural rubber.

Origin of Rubber

The early use of rubber involved all-natural rubber formed from a number of different plant species belonging to the *Euphobiaceae* family, of which the rubber tree (*Hevea brasiliensis*), native to Brazil, has become the exclusive commercial source. As a coagulated milk substance, rubber is obtained from a fluid in latex vessels located in the bark of the tree. A number of other tropical and subtropical plant species also contain such latex vessels, including *Manihot*, *Castilla*, the Russian dandelion, guayule

AP/Wide World Photos

Rubber plantation workers in the Philippines test the sap of a rubber tree.

(*Parthenium argentatum*), and *Funtumia elastica*.

Both the Russian dandelion and guayule were widely used during World War II. Research has continued on guayule, a plant native to the southwestern United States and northern Mexico. Similarly, *Funtumia elastica*, native to West and Central Africa, has received some research attention. Guayule, used by American Indians, is still considered a possible alternative rubber source to synthetic rubber in North America, particularly the southwestern United States.

Today the production of natural rubber is based on *Hevea brasiliensis*, which is grown mostly in tropical and subtropical environments. While production is concentrated in developing countries, consumption occurs mostly in industrialized countries. Between 1955 and 1988 production of rubber more than doubled, with Malaysia the leading world producer and the United States the world's largest consumer.

Growing Rubber Plants

Trees for commercial rubber plantations are vegetatively propagated by means of bud grafting. The bud from a high-yielding tree is cut and inserted under the bark of a rootstock. Upon a successful take, the bud grows, and the rootstock is topped or removed at the point of growth. It is then transplanted from the nursery to the field. The tree is ready for tapping in five to seven years, when tree girth reaches 50 centimeters at 1.60 meters from ground level. Crown budding may also be done before budded stumps are transferred to the field. This type of budding is used to provide a crown that is tolerant of or resistant to disease or wind damage. Stand density in rubber plantations ranges from 250 trees to 400 trees per hectare at an average spacing of about 6 meters by 6 meters.

Rubber grows best in deep, well-drained soil but can be grown on a wide range of soils. Rainfall should exceed 2,000 millimeters yearly, evenly distributed without any marked dry season. Temperature should average 25 to 28 degrees Celsius, with high (80 percent) humidity and bright sunshine of about six hours per day year-round. These conditions exist in the major rubber-producing countries of the world.

In *Hevea*, latex is obtained from latex vessels called secondary *laticifers*. The quantity of laticiferous tissue in the tree is determined by a number of anatomical factors, such as vessel rings, size of laticifers, girth of trees, and the distribution of latex and latex vessel rows. The flow of latex and, subsequently, the yield of a rubber tree is dependent on these anatomical features.

Latex Processing

Until about 1913 Brazil was the major producer of natural rubber, obtained mostly from wild rubber trees growing in the jungles of the Amazon basin. In the early twentieth century, plantation production of rubber began, based on the work of Henry N. Riley in Singapore around 1890. Riley developed the "tapping" method for extracting latex from *Hevea*. Later improvements to this method included the mechanization of the tapping knife.

During tapping, a slice of bark is systematically removed from one side (panel) of the tree, starting from an upper left corner and shaving to a lower right corner, with care being taken not to damage the cambium. The cut usually has an angle of 25 to 30 degrees. Once the cut is made, latex flows into a collecting cup through a spout inserted on the tree. Generally, tapping is done from just before sunrise to about 10:00 A.M., to take advantage of maximum turgor pressure within the tree in the early morning hours. Stoppage of latex flow is attributable to a coagulum that plugs latex vessels.

About four to five hours after tapping, the latex is collected from the trees. Field latex or cuplumps and "tree-lace" latex (strips or sheets of latex coagulated on a tapping cut) are collected and taken to a factory, laboratory, or small-holder processing center. At the processing center, latex is sieved to remove foreign objects, such as stones, branches, and leaves, and is then blended by the addition of water or dilute acetic or formic acid. About 10 percent of the latex is shipped as latex concentrate, following blending. Concentrates of natural rubber latex are obtained by the process of centrifugation and creaming. Meanwhile, the remainder of the latex and field coagulum are processed, either into conventional types of rubber or into technically specified rubber (TSR).

A number of fundamental weaknesses associated with manufactured rubber were resolved in 1839 with the development of vulcanization by Charles Goodyear, an American inventor. Vulcanization is the process of treating natural rubber with sulfur and lead and subjecting the compounds to intense heat, resulting in what Goodyear first called "fire proof gum" but later called vulcanized rubber.

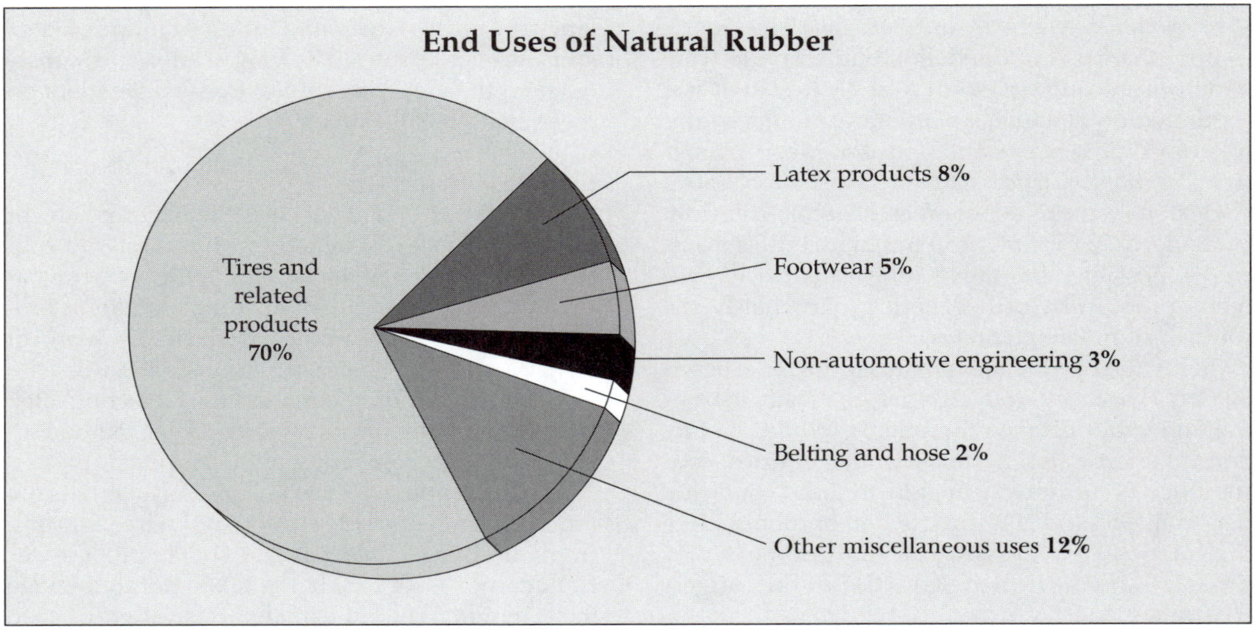

Source: Adapted from R. E. Brice and K. P. Jones, in *Natural Rubber Science and Technology*, edited by A. D. Roberts, 1988.

Current vulcanization technology is a modification of Goodyear's invention. New forms of vulcanization are available based on diurethanes, which are stable at processing temperatures as high as 200 degrees Celsius or more. Vulcanized rubber can then be processed into a wide range of applications, including tires, fabrics, bridge constructions, and other latex products such as adhesives and footwear.

Future Uses of Natural Rubber

There is continuing interest and effort on the part of research scientists and natural rubber producers to find new uses for natural rubber. Thus, projections for new uses range from snowplow blades to earthquake-resistant building construction materials. Although many proposed uses are engineering applications, there are other applications in the area of wood products that may eventually make the large acreages of rubber plantations important sources for environmental restoration, given the increasing deforestation that is taking place in the natural rubber-producing areas of the world. In rubber plantations that are more than forty years old, the regeneration of secondary forests with associated wildlife species occurs frequently. Thus, natural rubber is both an important industrial crop species and a major renewable resource.

Synthetic Rubber

Much of what people typically consider rubber today is actually synthetic rubber. Synthetic rubber is a polymer of several hydrocarbons; its basis is monomers such as butadiene, isoprene, and styrene. Almost all monomers for synthetic rubber are derived from petroleum and petrochemicals. The emulsion polymerization process occurs at very high temperatures. There are different types of synthetic rubbers, three of which are dominant in the rubber industry. These are styrene-butadiene rubber (SBR), polyisoprene rubber (IR), and polybutadiene rubber (BR). Unlike natural rubber, synthetic rubber is produced mainly in industrialized countries. The United States is the world's leading producer.

Theoretically, synthetic rubber production dates back to 1826, when scientist Michael Faraday indicated that the empirical formula for synthetic rubber was $(C_5H_8)_x$. The technology for synthetic rubber production was not developed until 1860, however, when Charles Williams found that natural rubber was made of isoprene monomers. Great interest in using synthetic rubber as a substitute for natural rubber began during World War II, when the Germans were looking for alternatives to natural rubber. The severe shortages of natural rubber during and immediately after the war stimulated significant research in synthetic rubber and its technology. Today, synthetic rubber is used in a wide range of applications, and it constitutes 60 to 65 percent of the total rubber produced and consumed.

Oghenekome U. Onokpise

See also: Asian agriculture; Asian flora; Central American flora; Forest management; Metabolites: primary vs. secondary; Plants with potential; South American flora; Vacuoles.

Sources for Further Study

Cosner, Sharon. *Rubber.* New York: Walker, 1986. Concise, factual history describes the earliest use of rubber and traces it to the present.

Johnson, Peter S. *Rubber Technology: An Introduction*. Cincinnati, Ohio: Hanser Gardner, 2001. Overview deals with all aspects of rubber processing. For newcomers to the industry, suppliers, processors, or end-product users.

Korman, Richard. *The Goodyear Story: An Inventor's Obsession with Rubber at the Dawn of the Industrial Revolution*. San Francisco: Encounter Books, 2002. Biography describes how Charles Goodyear, in his quest to find the recipe for rubber, mired his family in poverty and provoked powerful enemies, ultimately becoming an American industrial legend.

Stokes, Charles E., Jr. *The Amazon Bubble: World Rubber Monopoly.* Fort McKavett, Tex.: C. E. Stokes, Jr., 2000. Constructs the background and history of the Amazon rubber trade, from the perspective of all the South American countries involved.

RUSTS

Categories: Diseases and conditions; fungi; pests and pest control; poisonous, toxic, and invasive plants; taxonomic groups

Rust fungi belong to the taxonomic class Teliomycetes *within the basidiosporic phylum and include some seven thousand known species, all plant parasites, with many being extremely important plant pathogens.*

Rust diseases have caused serious problems for centuries and continue to cost billions of dollars annually worldwide. Important examples include: leaf, stem, and crown rusts of wheat, barley, and oats; the blister, fusiform, and gall rusts of pines; coffee rust; many field crop rusts, such as soybean, peanut, sunflower, and flax rusts; and rusts of many horticultural crops and ornamental plants, such as cedar-apple rust, bean rust, and rose rust.

Symptoms and Signs

Symptoms, the visible effects on a host caused by rust diseases, are often less noticeable than are the signs of infection, that is, the visible presence of a pathogen, such as reproductive spores. Symptoms such as bushy overgrowths, *witches' brooms* (dense clusters of shoots), and large galls are more common in some tree hosts than in herbaceous hosts. In most herbaceous rusts and the foliar rusts of some trees, such as coffee and poplar, little, if any, distortion of the host occurs. Most noticeable are the rust's reproductive spores borne in various types of pustules (or sori) that rupture leaf and stem epidermis. Loss of water vapor through ruptured epidermis can cause wilt symptoms during hot, dry periods. Several kinds of spores and pustules may develop during different stages of the rust's life cycle. Combined results of parasitism, water loss, cell death, and other effects on the host can result in significant reductions in yield.

Life Cycle and Related Spore Stages

Rust fungi have a complex life cycle that may contain up to five different spore types, or stages, each performing specific functions in the cycle. Some rust species lack one or more spore stage or stages. *Autoecious* forms of rusts, such as bean rust, complete their life cycle on one host species. *Heteroecious* rusts require two unrelated hosts on which to complete their life cycle. *Puccinia graminis*, the wheat stem rust pathogen, which produces all five spore stages on two unrelated hosts, is a good example of a rust with the heteroecious life cycle.

Distribution and Dissemination

Initial distribution of rusts between continents, especially if separated by oceans, is often the result of human activity, such as the inadvertent introduction of the pathogen on seeds, cuttings, or other plant parts. Once established in a region, rusts are most often spread locally and long distance by wind-blown urediniospores. In North America, south-to-north dispersal of urediniospores of wheat stem and leaf rust fungi occurs annually from Mexico and south Texas through the Great Plains into Canada. Coffee rust, long absent in the Western Hemisphere, rapidly spread across South and Central America by wind dispersal following its 1970 introduction into Brazil.

Obligate Parasitism

Although a few rusts can be grown on chemically defined, nonliving nutrients, most rusts require a living host on which to grow and reproduce. Consequently, rusts are termed *obligate parasites* or *biotrophs*, in contrast to the majority of plant pathogens, which are *facultative parasites* or *heterotrophs*. Those can grow and reproduce on many nutritional media in culture, in dead plant tissue, or as parasites of living plants. Within infected host tissue, rust fungi grow primarily as filamentous hyphae in intercellular spaces between host cells. Rusts obtain nutrition through specialized branches, haustoria, that extend into and absorb nutrients from living host cells. Those absorbed nutrients are then transported through intercellular hyphae to expanding or spore-producing portions of the rust colony.

RICHARD MARTIN/PHOTO AGORA

All rusts, such as this cedar apple rust, are parasites. Most rusts require a living host on which to grow and reproduce.

Disease Control

Typically, total control of rust diseases is impossible, both biologically and economically; hence, the usual goal is management of rust diseases, not control. Several different management methods are commonly pursued, depending on the host, the type of rust involved, its major means of dissemination, and economics or aesthetic value of the host plant or crop. The ideal and most cost-effective management tool is host resistance, either naturally present in the host population or resulting from plant breeding.

Wheat Stem Rust: Five Stages

Stage	*Spore-Bearing Structure*	*Spores*	*Spores' Nuclear Condition*
0	Pycnia, bearing[1]	Pycniosporesand receptive hyphae	Haploid (n)
I	Aecia, bearing[1]	Aeciospores	Dikaryotic ($n + n$)
II	Uredinia, bearing[2]	Urediniospores	Dikaryotic ($n + n$)
III	Telia, bearing[2]	Teliospores	Dikaryotic, later diploid ($2n$)
IV	Basidia, bearing	Basidiospores[3]	Haploid (n)

1. Produced on the barberry host.
2. Produced on the wheat or barley host.
3. Produced upon germination of teliospores, on crop debris or on soil.

Breeding for rust resistance has been especially important in rusts of cereals and other major agronomic plants. Unfortunately, many resistant varieties, or cultivars, may eventually succumb to new, more virulent or aggressive strains or races of the pathogen that arise by natural selection from populations of the pathogen in nature. Foliar fungicide sprays can be cost-effective management tools for rusts of some cereal crops, beans, coffee, and roses. In some instances, removing or eradicating the alternate host, such as common barberry, in heteroecious rusts has controlled such diseases but has usually proven ineffective economically.

Larry J. Littlefield

See also: Basidiosporic fungi; Diseases and disorders; Fungi; Parasitic plants; Resistance to plant diseases.

Sources for Further Study

Agrios, G. *Plant Pathology*. 4th ed. San Diego: Academic Press, 1997. University-level plant pathology text; in-depth discussion of all types of plant pathogens and plant diseases.

Alexopoulos, C. J., C. W. Mims, and M. Blackwell. *Introductory Mycology*. 4th ed. New York: John Wiley & Sons, 1996. Comprehensive university-level mycology text; covers all groups of fungi, including pathogens, nonpathogens, beneficial fungi, and taxonomy, morphology, and ecology.

Littlefield, L. J. *Biology of the Plant Rusts: An Introduction*. Ames: Iowa State University Press, 1981. A broad, introductory overview of rust fungi and rust diseases, including taxonomy, morphology, aerobiology, life cycles, and historical and economic importance.

________. "Smut and Rust Diseases." In *Plant Pathology: Concepts and Laboratory Exercises*, edited by R. N. Trigiano et al. Boca Raton, Fla.: CRC Press, 2002. College-level text provides a broad, introductory presentation of all aspects of plant pathology. The smut and rust diseases chapter covers those pathogens and diseases. Includes CD of illustrations.

Raven, Peter H., Ray F. Evert, and Susan E. Eichhorn. *Biology of Plants*. 6th ed. New York: W. H. Freeman/Worth, 1999. Outstanding basic botany text includes a brief, well illustrated introduction to rust fungi. With glossary, index.

Schumann, G. L. *Plant Diseases: Their Biology and Social Impact*. St. Paul, Minn.: APS Press, 1991. An overview of a wide range of plant diseases and pathogens, including rusts, with emphasis on both basic biology and economic, social, and historical impacts of plant diseases.

SAVANNAS AND DECIDUOUS TROPICAL FORESTS

Categories: Biomes; ecosystems; forests and forestry

Savannas are areas of continuous grass or sedge cover beneath trees that range from scattered, twisted, and gnarled individuals to open woodlands. Deciduous tropical forests have continuous to open forest cover and undergo a leafless period during a seasonally lengthy dry season.

Where the annual rainfall in tropical regions is less than 2,000 millimeters and three to six months out of the year are dry, savannas and deciduous forests are common. Both savannas and deciduous tropic forests often occur where the annual rainfall is less than that of savannas. Together, the two biomes are referred to here as the *dry tropical biome*.

A pronounced pattern of seasonally wet and dry periods is the most important factor affecting the distribution of these types of plant cover. Higher soil fertility favors forest over grasses and savanna such as in the cerrado of Brazil, which occurs only on certain geological formations and low-nutrient soils. Fire has been a dominant feature of these biomes, and human influences—fires, agriculture, and grazing of animals—have interacted with climate to produce a varied landscape.

The dry tropical biome is most geographically widespread on the continents of Africa, South America, and Australia, with smaller enclaves in Asia. The world's largest expanses of dry forest—the Brachystegia woodland across Central Africa, the *cerrado* (savanna) and *caatinga* (dry forest) of the Amazon basin, and much of interior Australia—are notable examples. "Elephant grass savanna," with tall grasses up to 4 meters tall and scattered trees, occurs exclusively in Africa. In the West Indies, dry forest occurs in rain-shadow zones on the leeward sides of islands affected by the tradewinds.

Plant Adaptations and Diversity of Life-Forms

As the rainfall decreases below 2,000 millimeters, and especially below 1,000 millimeters, the height of the forest decreases and the proportion of trees that are deciduous increases. In the dry tropics, leaf fall occurs in response to drought, and therefore the lengthy dry season becomes a selective pressure to which plants have adapted. Tree leaves tend to be compound, with small leaflets that help plants exchange heat with their surroundings better than large, simple leaves; rates of leaf respiration and transpiration are thereby reduced. Sclerophyllous leaves are common, aiding in moisture retention, and the drier, more open woodlands may have cacti or other succulents.

The dry forest is far less species-rich than the rain forest, but the diversity of life-forms and the proportion of endemics are greater. For example, dry forests may contain xerophytic (dry-adapted) evergreens, either obligatively or facultatively deciduous trees, trees with photosynthetic bark, plants that use the crassulacean acid metabolism (CAM) photosynthesis as well as C_3 and C_4 dicots, grasses, bromeliads, lianas, epiphytes, and hemiparasites. Trees from *Fabaceae* (the legume family) are the most well-represented family among trees.

Dry forests contain a higher proportion of wind-dispersed species than wetter forests, and many trees will have their flowering and fruiting controlled by the duration and intensity of the dry season. Synchronous flowering within and among species is common, and many produce seed during the dry season. Flowers are often conspicuous and visited by specialized pollinators such as hawkmoths, bats, and bees.

It is incorrect to generalize about savannas and dry tropical forests because, although they both occur in the drier tropics, the two vegetation types occur in different habitats and are adapted differently to their respective environments.

Trees of the cerrado in northeast Brazil are deeply rooted, tap groundwater, and have high rates of transpiration. Drought here is atmospheric, as water is always available below 2 meters of soil depth. The deciduous caatinga of central Brazil, however, receives only 500 millimeters of rain yearly, and transpiration of trees is low. Here, trees suffer significant water deficits during the long, dry season, are truly xerophytic, and exhibit classic adaptations to drought.

Trees of the cerrado have a number of adaptations that confer resistance to fire. These include a thick, corky bark, the ability to form adventitious roots from buds on roots following the burning of the stem, and the cryptophyte or hemicryptophyte life-form (cryptophytes produce buds underground). Many herbaceous species are induced to flower by fire.

Human Impacts and Conservation

Fires have occurred in the Brazilian cerrado for thousands of years based on carbon 14 dating of charcoal fragments. Fire is thus an environmental factor to which the vegetation has become adapted. Yet, the human influence has been to increase the incidence of fire. The cerrado has changed as a result to a more open form of plant cover with fewer trees and shrubs. In addition, timber extraction, charcoal production, and ranching have altered the savanna landscape. The ability of belowground organs to survive such types of disturbance has increased the ability of the cerrado to persist. Yet it is estimated that 50 percent of the cerrado has been destroyed, much of this the result of clearing for agriculture since the 1960's.

Because of better soils and fewer pests, humans in tropical areas of Central America have mostly chosen the dry and moist life zones as places to live rather than the wetter rain-forest zones. As a result, dry forest ecosystems have been subject to massive disturbance. Today, only a small fraction of the original dry forest remains. Fire has been used as a means of clearing the forest for farming, but, unlike the savanna, the dry forest is not adapted to fire. At Guanacaste, Costa Rica, a well-known tropical conservation project, restoration of dry forest, is dependent on controlling annual fires set by farmers and

CORBIS

Savannas are landscapes of dense grass and scattered trees, such as these yellow fever trees growing in Nakura National Park in Kenya. Common on the continent of Africa, savannas are also found in India, Australia, and the northern part of South America.

ranchers and supporting the return of forest vegetation to dry areas. In Africa, large areas of dry forest are burned annually by farmers, and areas of dense, dry forest have been converted to more open forest or even savanna. Sustainable land-use systems are urgently needed for dry tropical regions.

Allan P. Drew

See also: African flora; Australian flora; Biomes: types; Biomes: definitions and determinants; Biomes: types; Central American flora; Community-ecosystem interactions; Drought; Slash-and-burn agriculture; South American flora; Sustainable agriculture.

Sources for Further Study

Allen, W. *Green Phoenix: Restoring the Tropical Forests of Guanacaste, Costa Rica*. New York: Oxford University Press, 2001. Discusses Daniel Janzen's work with ecological restoration of dry forest to form the Guanacaste Conservation Area. Photographs, bibliographical references, index.

Bullock, S. H., H. A. Mooney, and E. Medina. *Seasonally Dry Tropical Forests*. New York: Cambridge University Press, 1995. Geographical distribution of dry forests, plant and life-form diversity, drought responses, plant reproduction, plant-herbivore interactions, nutrient cycling, conversion to pasture and agriculture, and ethnobotany. Tables, figures, photographs, maps, references, index.

Rizzini, C. T., A. F. C. Filho, and A. Houaiss. *Brazilian Ecosystems*. Rio de Janeiro: ENGE-RIO, Index Editora, 1988. Major ecosytems are presented through excellent photographs and knowledgeable commentary. Map, references, side notes providing scientific names.

SEEDLESS VASCULAR PLANTS

Categories: Evolution; paleobotany; *Plantae*; seedless vascular plants; taxonomic groups

Seedless vascular plants possess vascular tissues (xylem and phloem) for transport of materials through the body but do not produce seeds bearing dormant embryos as part of the reproductive process. They are among the oldest of land plants.

Modern seedless vascular plants include species from several different phyla, including the club mosses, spike mosses, and quillworts of the phylum *Lycophyta*, the horsetails of the phylum *Sphenophyta*, the whiskferns of the phylum *Psilotophyta*, and the great diversity of ferns in the phylum *Pterophyta*. Lycophytes, sphenophytes, and psilotophytes are generally referred to as *fern allies*. Carolus Linnaeus used the word *Cryptogamia* (from the Greek *kryptos*, meaning "hidden," and *gamos*, meaning "marriage," or reproduction) as an inclusive taxonomic category for a wide range of organisms such as bryophytes, ferns, fern allies, algae, and fungi whose sexual reproductive parts were concealed from observation. The term *cryptogam* in modern usage refers to plants that do not produce seeds.

Origin and Relationships

Theories on the origin of vascular plants suggest that they probably arose from an ancestor in the group of nonvascular plants known as the bryophytes. Fossil spores from the Ordovician period (about 500 million years ago) are linked to the earliest land plants, which were likely an early liverwort (phylum *Hepatophyta*) or hornwort (phylum *Anthocerophyta*). The most derived bryophytes, the mosses (phylum *Bryophyta*), share an ancestor with the line that gave rise to vascular plants. A line that gave rise to the phylum *Lycophyta* appears to have diverged early in vascular plant development, soon after the advent of phloem and the rise of the dominance of the branching sporophyte generation. The remaining line diverged into a branch that gave rise to modern sphenophytes (horsetails) and ptero-

Seedless Vascular Plants: Phyla and Characteristics

Phylum	*Status*	*Anatomical Characteristics*	*Reproduction Exhibits*
Lycophyta	Living	Stems, roots, leaves, some branching	Homospory, heterospory
Psilotophyta	Living	Stems, no leaves or roots, branching	Homospory
Pterophyta	Living	Stems, roots, leaves, no branching	Homospory, minimal heterospory
Rhyniophyta	Extinct	Stems, no leaves or roots, often branching	Homospory
Sphenophyta	Living	Stems, roots, leaves, no branching	Homospory, some (extinct) heterospory
Trimerophytophyta	Extinct	Stems, no leaves or roots, branching rare	Homospory
Zosterophyllophyta	Extinct	Stems, no leaves or roots, often branching	Homospory, some heterospory

Source: Some data adapted from Peter H. Raven et al., *Biology of Plants*, 6th ed. (New York: W. H. Freeman/Worth, 1999).

phytes (ferns) and a branch that gave rise to modern seed plants.

Seedless vascular plants are well represented in the fossil record. *Cooksonia*, a representative of the extinct phylum *Rhyniophyta*, is thought to be one of the earliest vascular plants, dating to the mid-Silurian period (around 430 million years ago). Numerous other examples dating to the Devonian period (around 400 million years ago) are well documented.

Life Cycle and Gametophyte Anatomy

Seedless vascular plants express the typical life cycle pattern called *alternation of generations* found in many algae and members of the kingdom *Plantae*. As in all vascular plants, the diploid sporophyte generation, which produces haploid spores for the asexual reproductive phase, is dominant. Haploid spores are produced by meiosis in special structures called sporangia (singular, sporangium). In psilotophytes, some lycophytes, the sphenophytes, and the pterophytes, only one kind of spore is produced in a phenomenon called *homospory*. Sexually reproducing haploid gametophytes that have both male and female sex organs arise from these spores. In some lycophytes, two kinds of spores are produced in a phenomenon called *heterospory*. Large spores called megaspores are produced in megasporangia. Megaspores develop into female gametophytes. Small spores called microspores are produced in microsporangia. Microspores develop into male gametophytes.

The haploid gametophytes of seedless vascular plants, which produce gametes (sex cells) for the sexual reproductive phase, are independent and smaller than the sporophytes. Gametophytes range in size from a few millimeters to a few centimeters in length or diameter, while sporophytes can range in size from a few millimeters to several meters in height. Since there is a distinct difference between the appearance of the sporophyte and the gametophyte, these organisms are said to express *heteromorphy*. In psilotophytes and some lycophytes, gametophytes are nonphotosynthetic and underground, while in most of the other groups gametophytes grow on the soil surface and are photosynthetic. Underground gametophytes must rely on mutualistic fungi in mycorrhizal relationships to supply them with energy and carbon sources for growth and development.

Gametes are produced by mitotic cell division. Male gametes are produced in organs called *antheridia* (singular, *antheridium*). Sperm cells are multiflagellated and require liquid water for transmission to the female organs of a nearby plant. Eggs are produced in organs called *archegonia* (singular, *archegonium*). Sperm cells swim through the neck of the archegonium and into a swollen region at the base called the venter, which contains the egg. Fertilization results in the production of a diploid zygote, which develops into an embryo that then grows into a sporophyte.

Sporophyte Anatomy

According to the fossil record, the earliest vascular plants were quite similar in structure to the modern genus *Psilotum*, of the phylum *Psilotophyta*: the whiskferns. The *Psilotum* sporophyte lacks true leaves and roots (that is, the organs lack vascular tissue). Small, nonvascularized flaps of tissue called prophylls are found on the stem of *Psilotum*, which some theorize either are reduced leaves or may be suggestive of the evolutionary origin of leaves—that is, that true leaves are the result of the evolutionary vascularization of prophylls. The other living psilotophyte genus, *Tmesipteris*, possesses true leaves, each with a single vein, referred to as microphylls. Other groups produce either the one-veined microphylls or megaphylls that have numerous, often branched veins. The most dramatic leaf expression in seedless vascular plants is found in the delicate, pinnately compound, megaphyllous fronds of many of the ferns in phylum *Pterophyta*. In lycophytes, sporangia are produced on specialized leaves called sporophylls. Some interpret the spore-bearing sporangiophores that make up the cones of sphenophytes to be specialized leaves, as well.

The aerial stem of *Psilotum* exhibits a primitive pattern of isotomous dichotomous branching, with two equal branches arising from one branch point, or node. Other seedless vascular plants may show different patterns: For example, lycophytes often show an unequal dichotomous branching pattern, or anisotomy.

In most cryptogams, aerial stems arise from underground stems called rhizomes. Rhizomes of psilotophytes have many rhizoids, rootlike epidermal extensions, to aid in absorption of water and mineral nutrients. The sporophytes of all other groups of seedless vascular plants possess vascularized roots for absorption and anchorage.

The vascular tissues of all members of this broad group include the primitive type of xylem called tracheids and the primitive type of phloem called sieve cells. The vascular cylinder, or stele, of *Psilotum* is composed of a solid core of xylem surrounded by a layer of phloem. This arrangement is called a *protostele*. Variations on the protostele pattern are common in the roots, rhizomes, and shoots of members of the *Lycophyta*. Protosteles are the most common stele types found in vascular plant roots, in general.

The shoot steles of sphenophytes and pterophytes are composed of a ring of xylem surrounding a pith made of parenchyma tissue. A ring of phloem surrounds the xylem ring. This arrangement is called a *siphonostele*. Variations on the siphonostele pattern are common in the shoots of all vascular plants from the sphenophytes through the seed plants.

The endodermis is a layer of cells that surrounds the vascular cylinder of roots and rhizomes and some shoots in certain psilotophytes; it is is usually localized in the roots and rhizomes of lycophytes, sphenophytes, and pterophytes. The waterproofing suberin deposited in the Casparian strip of the endodermis regulates water flow to the vascular tissue.

Lycophyta

Fossil evidence suggests that the phylum *Lycophyta*, the lycophytes, arose during the Devonian period. There are between ten and fifteen genera, consisting of about one thousand species, divided among three broad groups: the club mosses of the order *Lycopodiales*, the spike mosses of the order *Selaginellales*, and the quillworts of the order *Isoetales*. All possess microphylls; however, the homosporous club mosses differ significantly from the heterosporous spike mosses and quillworts. Some extinct lycophytes were woody and grew to tree size. Lycophytes were among the dominant species during the Carboniferous period (around 350 million years ago), during which the plants that eventually became coal deposits were produced.

Sphenophyta

Phylum *Sphenophyta* arose during the Devonian period and reached its peak during the late Devonian and Carboniferous periods. Among the largest sphenophytes were members of the genus *Calamites*, treelike plants that exceeded 18 meters (60 feet) in height and nearly 0.5 meter (18 inches) in basal stem diameter. There are fifteen living herbaceous species in the only remaining genus, *Equisetum*, the horsetails.

Psilotophyta

The fossil record is incomplete regarding the ancestry of modern psilotophytes, so it is difficult to assign a date of origin. Although once considered to possess characteristics that allied them to the most primitive vascular plants, molecular studies have suggested that psilotophytes are possibly closely related to pterophytes (ferns), perhaps representing a reduction from an early fern ancestor. The phylum includes only two living genera, consisting of a total of six to eight species.

Pterophyta

Pterophytes, or ferns, arose during the Devonian period and reached the height of abundance and diversity during the Carboniferous. Today, there are about eleven thousand species of ferns ranging in size from a few millimeters across (such as *Azolla*) to several meters in height (such as *Dicksonia*). Excluding flowering plants, ferns are the most abundant and diverse plants on earth, exhibiting a variety of growth habits, from floating aquatics to tropical epiphytes to temperate terrestrial plants. They are important both ecologically and economically, serving as habitat and food sources for many organisms. Humans use ferns as food, fuel, and decoration.

Darrell L. Ray

See also: Bryophytes; Evolution of plants; Ferns; Hornworts; Horsetails; Liverworts; Lycophytes; Mosses; Mycorrhizae; Plant tissues; Psilotophytes; Reproduction in plants; *Rhyniophyta*; *Tracheobionta*; *Trimerophytophyta*.

Sources for Further Study

Bell, Peter R., and Alan R. Hemsley. *Green Plants: Their Origin and Diversity*. 2d ed. New York: Cambridge University Press, 2000. Reviews relationships among major groups of seedless vascular plants. Photographs, diagrams.

Ennos, Roland, and Elizabeth Sheffield. *Plant Life*. Malden, Mass.: Blackwell Science, 2000. Introduces major groups of seedless vascular plants. Photographs, diagrams.

Gifford, Ernest M., and Adriance S. Forster. *Morphology and Evolution of Vascular Plants*. 3d ed. New York: W. H. Freeman, 1989. A standard among plant morphology texts. Extensive discussions of origins and relationships of seedless vascular plants. Photographs, diagrams.

Raven, Peter H., Ray F. Evert, and Susan E. Eichhorn. *Biology of Plants*. 6th ed. New York: W. H. Freeman, 1999. A standard text for college botany. Introduces history and relationships among major groups of seedless vascular plants. Photographs, drawings, diagrams.

SEEDS

Categories: Anatomy; reproduction and life cycles

A seed is a mature, fertilized plant ovum containing an embryo, a food supply, and a protective covering called the testa, or seed coat.

A mature seed typically consists of a mature plant *ovum* containing a minute, partially developed young plant, the *embryo*, surrounded by an abundant supply of food and enclosed by a protective *seed coat*. Seed plants are divided into two main groups: the *gymnosperms*, primarily cone-bearing plants such as pine, spruce, and fir trees, and the *angiosperms*, the flowering plants. The gymnosperms have naked *ovules* which, at the time of *pollination*, are exposed directly to the pollen grains. Their food supply in the seed is composed of a female gametophyte, rather than the endosperm found in angiosperms.

In angiosperms, seeds develop from ovules that are enclosed in a protective *ovary*. The ovary is the basal portion of the *carpel*, typically a vase-shaped structure located at the center of a flower. The top of the carpel, the *stigma*, is sticky, and when a pollen grain lands upon it, the grain is firmly held. The germinating pollen grain produces a *pollen tube* that grows down through the stigma and *style* into the ovary and pierces the ovule.

Two male sperm nuclei are released from the pollen grain and travel down the pollen tube into the ovule. One of the sperm nuclei fuses with an egg cell inside the ovule. This fertilized egg divides many times and develops into the embryo. The second male nucleus unites with other parts of the ovule and develops into the *endosperm*, a starchy or fatty tissue that is used by the embryo as a source of

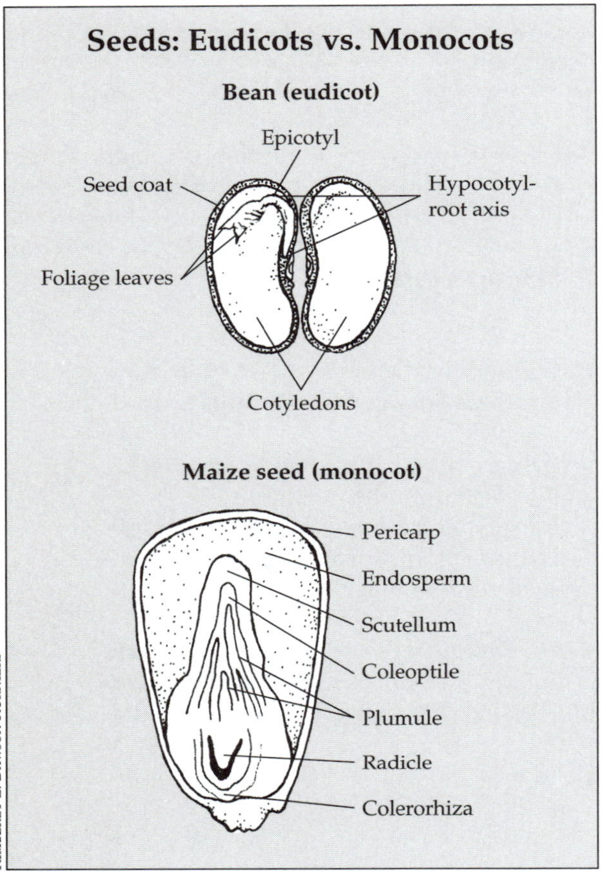

In angiosperm seeds, the embryo may have either one or two cotyledons. Angiosperms with one cotyledon are plants called monocots; those with two cotyledons are called eudicots (formerly dicots). A typical example of a eudicot is the bean (Phaseolus vulgaris), *whereas a typical monocot is corn, or maize* (Zea mays).

food during germination. Angiosperm seeds remain protected at maturity. While the seed develops, the enclosing ovary also develops into a hard shell, called a seed coat or *testa*, often enclosed in a fibrous or fleshy fruit.

Structure

Although the characteristics of different plant seeds vary greatly, some structural features are common to all seeds. Each seed contains an embryo with one, two, or several *cotyledons*. In angiosperm seeds, the embryo may have either one or two cotyledons. Angiosperms with one cotyledon are plants are called *monocots*; those with two cotyledons are called *eudicots* (formerly *dicots*). A typical example of a monocot is corn, or maize (*Zea mays*), whereas the bean (*Phaseolus vulgaris*) is a typical eudicot. In gymnosperms, the embryo may have between two and sixteen cotyledons; for example, the embryo of Scots pine (*Pinus sylvestris*) possesses eight cotyledons.

Immediately below the cotyledons is the *hypocotyl*, at the tip of which lies the growing point of the root. Above the cotyledons lies the *epicotyl*, which consists of a miniature shoot tip and leaves. Upon germination of the seed, the epicotyl develops into the stem and leaves of the new young plant. Almost all seeds carry with them a supply of food, which in angiosperms is the endosperm. Although the embryo is usually surrounded by the endosperm, in some seeds (such as maize) embryos and endosperm lie side by side. In the seeds of the pea family (*Leguminosae*), the food reserves of the endosperm are absorbed by the embryo, resulting in enlarged cotyledons. Gymnosperm seeds differ from those of angiosperms in the origin of their stored food. In gymnosperms the stored food is provided by a female gametophyte housed with the embryo inside the seed, whereas in angiosperms the food reserve is the endosperm.

All seeds are surrounded by a seed coat, the testa. Variability in the appearance of the testa is considerable, and these variations are used by taxonomists as an aid in distinguishing among different genera and species. The testa is of great importance to the seed; it is often the only barrier protecting the embryo from the external environment. The seed coats of some plants swell and produce a jellylike layer in response to contact with water. The gel retains water needed by the seed for germination. Cotton fibers are formed as extensions from some of the outermost cells of the seed coat in cotton plants (*Gossypium*). The seed coats of nutmeg contain aromatic substances.

Size and Chemistry

The range of seed size is extreme—more than nine orders of magnitude. The largest known seed is that of the double coconut (*Lodoicea maldivica*); the seed and fruit together weigh as much as 27 kilograms. At the other end of the scale, the dustlike seeds of some orchids, begonias, and rushes weigh only about 5 milligrams per seed. It is thought that the size of seed displayed by each species represents a compromise between the requirements for dispersal (which would favor smaller seeds that can be borne on wind or picked up by animals) and the requirements for establishment of the seedling (which would favor larger seeds that can adhere to a growth medium).

The chemical composition of seeds varies widely among species. In addition to the normal compounds found in all plant tissues, seeds contain unique food reserves that are used to support early seedling growth. About 90 percent of plant species use *lipids* (fats and oils) as their main seed reserves. The cotyledons of soybeans and peanuts are rich in oil, whereas in other legumes such as peas and beans, *starch* is the reserve material. Sixty-four percent of the weight of a castor bean is derived from the oil stored in the endosperm. In seeds of cereal crops, the endosperm stores much starch; in corn it can be up to 80 percent of the weight of the seed. All seeds, particularly legumes, also store protein as a reserve substance.

Dispersal

A seed can be regarded as a vessel in which lies a partially developed young plant in a condition of arrested growth, waiting for the correct conditions for growth to resume. Successful reproduction depends on seed dispersal to places appropriate for germination to occur. During the evolutionary history of plants, seeds and fruits have developed a great variety of specialized structures that enhance seed dispersal.

Wind is one major means of seed distribution. Very small seeds, such as the dustlike seeds of orchids, heathers, and some rushes and grasses, are dispersed by wind. Such seeds have been recovered from the atmosphere by airplanes at elevations up to 1,000 meters. Heavier seeds have evolved a variety of structures to ensure wind dispersal. For example, some members of the daisy family, such as dandelions, bear numerous one-seeded fruits to which are attached a feathery, tuftlike structure that acts as a parachute. Similar structures aid in the dispersal of the seeds of many other plants, such as cattails and milkweeds. Heavier seeds, such as those of ash, maple, and pine trees, have developed large, flat wings that allow the seeds to fly in a propeller-like manner for a considerable distance from the parent plant.

Often adaptations for wind dispersal of seeds can be seen not only in the seed's structure but also in structures of the parent plant. Many plants offer their seeds to the wind by bearing them on long flower stalks that tower above the surrounding vegetation. Tumbleweed bushes are small and almost spherical. When mature, they develop a weakness of the main stem at soil level. Wind can break the stem of the bush from which, as it rolls over the ground, the seeds are shaken loose and are scattered.

The seeds or fruits of many plants are dispersed by sticking to the outsides of birds or land animals.

Digital Stock

The size of seed displayed by each species is thought to represent a compromise between the requirements for dispersal (which would favor smaller seeds that can be borne on wind, as are these dandelion seeds, or picked up by animals) and the requirements for establishment of the seedling (which would favor larger seeds that can adhere to a growth medium).

Seeds are transported in mud that sticks to the feet of animals. The large number of species whose seeds show no obvious special dispersal adaptation are probably spread in this manner. Many seeds, however, can attach themselves to a passerby by means of adhesive substances, hooks (such as the fruit of bedstraw), or burrs (such as the seeds of the burdock plant).

Some plant species have seeds that are adapted for dispersal by animals and birds that transport them internally. The attractive and tasty fleshy fruits and berries of plants can be considered an adaptation to aid in seed dispersal. The seeds of most fruits eaten by animals and birds have a digestion-resistant coat. The animal deposits excrement containing seeds at a location at some distance from the parent plant, where the seeds grow into new plants. In some species, germination will not even occur unless the seed has passed through an animal's digestive tract. The presence of seeds in bird droppings is responsible for the appearance of some plants on remote, barren, volcanic islands. Various animal behaviors, including the collecting behavior of ants and the seed-burying activities of mice, squirrels, and jays, also aid in seed dispersal.

Several plants have evolved mechanisms that expel seeds explosively away from the parent plant. The pods of *Impatiens*, for example, develop strongly unbalanced tension forces as they ripen. When the pod is fully mature, the tension is so intense that the slightest disturbance causes the pod to split open, violently expelling the seeds.

Dormancy

Seeds are adapted for conditions other than geographic dispersal. When a seed arrives at its destination after dispersal, conditions may not be suitable for establishment of the plant. The delaying mechanism which prevents germination under adverse conditions is called *dormancy*. Seeds can remain in a condition of dormancy for varying lengths of time, depending on the species, until the correct balance of oxygen, moisture, and temperature triggers germination.

Viability varies greatly from species to species and may last only a few weeks or many years. Seeds of the cocoa plant are viable for only ten weeks. Some seeds, however, remain viable for decades or even hundreds of years. Seeds of the Indian lotus have been shown to remain viable for almost one thousand years. No claims for long-term viability have surpassed those made for the Arctic lupine, however: Seeds of this species have been successfully germinated after having been buried in the Arctic tundra for ten thousand years.

Iain Miller and Randy Moore

See also: Angiosperm life cycle; Dormancy; Flower structure; Fruit: structure and types; Germination and seedling development; Gymnosperms; Plant life spans; Pollination; Reproduction in plants.

Sources for Further Study

Bewley, J. Derek, and Michael Black. *Seeds: Physiology of Development and Germination*. 2d ed. New York: Plenum Press, 1994. Intermediate-level college text presenting a general yet rigorous treatment of the essential aspects of seed physiology and biochemistry. Assumes some knowledge of plant physiology, although fundamentals are explained where necessary. Well illustrated.

Campbell, Neil A., and Jane B. Reece. *Biology*. 6th ed. San Francisco: Benjamin Cummings, 2002. A well-written textbook suitable for high school and college students as well as the layperson. The chapter "Plant Reproduction" covers the reproductive process in angiosperms from pollen and ovule development through fruit development and seed dispersal. The chapter "Plants and the Colonization of the Land" compares and contrasts sexual reproduction in gymnosperms and angiosperms. Easy to read; good illustrations.

Fenner, Michael. *Seed Ecology*. New York: Chapman and Hall, 1993. Lower-level college text that stresses an ecological approach to the study of seeds. Highlights the simplicity of many ecology experiments and field observations and thus is useful to both the student and interested amateurs wishing to make some experimental investigations of their own.

Raven, Peter H., Ray F. Evert, and Susan E. Eichhorn. *Biology of Plants*. 6th ed. New York: W. H. Freeman/Worth, 1999. Lower-level college text presenting a detailed and comprehensive look at plant biology. The chapter "Seed Plants" describes the life cycles of both gymnosperms and angiosperms in some detail. The chapter "Early Development of the Plant Body" addresses embryo development and seed germination. Well-drawn diagrams and sharp color photographs enhance understanding of the complex processes involved in plant reproduction. Glossary, index, references.

Stidworthy, John, ed. *Plants and Seeds: Through the Microscope*. New York: Gloucester Press, 1990. The main strength of this book is its striking collection of photographs of enlarged microscopic images.

SELECTION

Categories: Evolution; genetics

Selection refers to any process by which some individuals are allowed to reproduce at the expense of others, leading to a shift in the composition of a population over generational time.

In the context of animal and plant breeding by humans, the process of selection is referred to as *artificial selection*. In the wild, it is referred to as *natural selection* and is viewed by scientists as the principal means by which *adaptations* (traits that promote the survival and reproduction of organisms) arise and new species evolve over geological time frames.

Natural Selection

In *On the Origin of Species by Means of Natural Selection*, published in 1859, Charles Darwin presented arguments for two related theories. The first argument documents evidence for evolution on the basis of an extensive study of biogeographical, embryological, and fossil data. Within ten years of its publication, nearly all individuals who today would be considered scientists were convinced that evolution had occurred.

Darwin's second argument develops the case for what he identified as the chief mechanism of evolution: his theory of natural selection. This argument elegantly draws attention to the probabilistic consequences of three conditions in nature. First, organisms vary from one another in ways that affect their ability to survive and reproduce. Second, at least some of this variation is heritable. Third, there is *competition*, owing to the fact that organisms produce more offspring than can possibly survive. Darwin's genius was to recognize that as a consequence of these conditions, members of a population possessing favored variations would be more likely to leave offspring than those that did not, and as such, the composition of the population as a whole would change over generational time in the direction of the favored form.

Evidence

In addition to evidence for the existence of the above conditions in nature, Darwin also gave indirect evidence in support of the long-term consequences of these conditions he had deduced. In addition to thought experiments, he developed an analogy between artificial and natural selection. He pointed out that during the relatively short time that pigeon and dog species had been domesticated, breeders had been able, by picking which individuals were allowed to reproduce, to create new varieties as distinct in appearance as true species in nature. Darwin then drew attention to the fact that conditions in nature (coupled with the occasional introduction of new variants by mutation) operating over geological time periods could have a similarly dramatic effect on populations in the wild.

Although the vast time and spatial frames involved make it difficult to observe directly the origin of new adaptations and species, many field experiments in the years since Darwin first wrote have confirmed both the power and ubiquity of natural selection in nature. A particularly important example in the context of botany was provided in the late 1960's by Janis Antonovics and others, who demonstrated that the evolution of heavy metal tolerance in many plant species was the result of natural selection. Other contemporary areas of research on selection include the study of *developmental plasticity*, or the ability of an organism to respond to environmental conditions during its development, as occurs when a growing sapling forms leaves to maximize its light exposure in the presence of partial shading by other trees in a forest. This ability is itself the object of natural selection.

Genotypes vs. Phenotypes

Although evolution is often discussed in terms of a change in the frequency of genes in a population, in fact, natural selection acts directly on the *phenotypes* (observable characteristics of organisms) and only indirectly on the *genotypes* (the specific forms of the gene, or *alleles*, an individual has inherited from its parents) responsible for them. While often portrayed as a process that removes genetic variability from populations, natural selection can promote variability, as occurs when the population exists in an environment where one form of a trait is advantageous in some areas but another form is advantageous in other areas. Such processes may lead to *clines*, or gradations, in the frequencies of genes over the range of the population.

Ecosystems

The foregoing has discussed selection with reference to isolated populations. In ecosystems composed of multiple populations of distinct species, natural selection can promote the evolution of *ecotypes*, or populations having a distinct set of characteristics unique to the region they inhabit and the mode of life they pursue. Natural selection can favor the *coevolution* of populations of distinct species with one another. This occurs in the evolution of predator and prey species, in which, for example, the origin of an adaptation that allows a predator to consume a grass species that was previously toxic to it changes the selective environment of the prey species, leading to the evolution of entirely new plant defenses.

Convergent vs. Divergent Evolution

The presence of common selective conditions in distinct locations may lead to the independent evolution of similar characteristics in separate species; that is, *convergent evolution*. A good example is provided by cacti in North and South American deserts and euphorbs in African deserts: Both have thornless leaves with a similar structure that has evolved to maximize water retention. Natural selection can also lead to a partitioning of resource space, a phenomenon known as *divergent evolution*, in which distinct populations of a species evolve divergent modes of life. This may reflect the imposition of a barrier that prevents interbreeding or competition among cohabiting populations leading to a partitioning of niche space. A good example is the common mistletoe, *Viscum album*, a parasitic higher plant having three distinct races that specialize on deciduous trees, firs, and pines.

David W. Rudge

See also: Adaptations; Clines; Coevolution; Competition; Evolution: convergent and divergent; Evolution: historical perspective; Evolution of plants; Genetic drift; Genetics: Mendelian; Genetics: mutations; Genetics: post-Mendelian; Plant domestication and breeding; Population genetics; Species and speciation.

Sources for Further Study

Bell, Graham. *The Basics of Selection*. New York: Chapman & Hall, 1997. A college-level introduction to different types of selection and how they are studied.

Bowler, Peter J. *Evolution: The History of an Idea*. Berkeley, Calif.: University of California Press, 1984. An overview of the history of the concept of natural selection in connection with debates on evolution.

Cragg, J. B. *Advances in Ecological Research*. London: Academic Press, 1971. Contains Janis Antonovics et al.'s classic paper reviewing research on the evolution of heavy metal tolerance in plants.

Endler, John. *Natural Selection in the Wild*. Princeton, N.J.: Princeton University Press, 1986. A survey of many studies documenting the presence and ubiquity of natural selection.

Jones, Steve. *Darwin's Ghost: The Origin of Species Updated*. New York: Random House, 2000. An assessable retelling of Charles Darwin's argument, augmented in light of contemporary research.

SERPENTINE ENDEMISM

Category: Soil

Serpentine endemics are plants that grow only in serpentine soils. These uncommon soils present serious challenges to plants and are often sparsely covered with dwarfed vegetation.

Serpentine rock is one form of *ultramafic rock*, an uncommon rock found in mountain-building zones. Soils derived from ultramafic rock are called serpentine soils. The most important chemical characteristics of ultramafic rock and of the serpentine soils formed from it are high magnesium concentrations; low calcium concentrations; low calcium/magnesium ratios; low concentrations of other macronutrients (such as nitrogen, phosphorus and potassium); high concentrations of toxic heavy metals (such as nickel, chromium, and cobalt); and low micronutrient (molybdenum, boron) concentrations. All these factors are detrimental to plant growth.

Physical factors also tend to be harsh in serpentine soils. They are rocky and low in humus (the organic matter formed from decomposing plants), and temperatures tend to fluctuate widely. In part, these physical factors are caused by the chemical factors discussed above and their restriction of plant growth. Sparse plant cover results in less plant material to decompose, thus less humus; and a greater plant cover would mitigate temperature changes. Whatever their causes, these physical factors further restrict plant growth. Not all serpentine soils have all these characteristics, but those present combine in various ways to severely restrict plant growth on serpentine soils.

Plants of Serpentine Soils

Many plant species cannot grow in serpentine soils. Others can grow in serpentine soils or in other soils, but the serpentine varieties are often dwarfed and require special adaptations. Still others, the serpentine endemics, grow only in serpentine soils. Two examples are *Quercus durata*, a shrubby oak that grows in some western serpentines, and *Aster depauperatus*, a small aster of some Appalachian serpentine soils.

One common adaptation of serpentine plants is a high tolerance for nickel, the most troublesome of the serpentine heavy metals. A second is the ability to use calcium efficiently in the presence of excess magnesium. Magnesium is an essential nutrient, but in high concentrations it interferes with the plant's use of calcium, another essential nutrient. These two adaptations are widespread among serpentine endemics, suggesting that they are two of the adaptations important to serpentine endemics. Additional adaptations are probably necessary because a plant growing on serpentine must overcome all of the soil's troublesome characteristics.

Competition may explain why these plants are unable to grow in more favorable soils. Many serpentine endemics cannot compete with other plants outside the serpentine environment and are thus excluded from nonserpentine soils. Because most plants cannot grow in serpentine soil, serpentine endemics are freed from competition and can grow successfully in those soils.

Because they only occur on these uncommon soils, serpentine endemics are especially susceptible to extinction. Preservation efforts have focused on some of these areas and their plant life.

Carl W. Hoagstrom

See also: Adaptations; Agronomy; Halophytes; Nutrient cycling; Selection; Soil.

Sources for Further Study

Brooks, Robert R. *Serpentine and Its Vegetation: A Multidisciplinary Approach*. Portland, Oreg.: Dioscorides Press, 1987. This world survey offers an understandable explanation of serpentine endemics.

Dann, Kevin T. *Traces on the Appalachians: A Natural History of Serpentine in Eastern North America*. New Brunswick, N.J.: Rutgers University Press, 1988. Nontechnical explanation of the history of serpentine ecosystems.

Roberts, Bruce A., and John Proctor, eds. *The Ecology of Areas with Serpentinized Rocks: A World View*. Boston: Kluwer Academic, 1992. This understandable coverage of serpentine ecology includes extensive examples.

SHOOTS

Category: Anatomy

"Shoot" is a term used to refer collectively to the stem and associated leaves of a vascular plant.

The *stem* of a vascular plant is typically a cylindrical, aboveground axis that grows upright, away from the pull of gravity. Young stems are green and photosynthetic, and in some plants, such as horsetails and most cacti, the stems are the primary site of photosynthesis. In most plants, however, the primary function of the stem is to support leaves in a position where they are well exposed to light. The part of the stem where a leaf attaches is a *node*. The section of stem between nodes is an *internode*. Internode elongation spaces the leaves and is responsible for most of the increase in length of the stem every growing season.

The cylindrical shape of the stem is a structural adaptation that provides maximum strength per volume of tissue. The *vascular tissues*, which contain tough, thick-walled fiber cells, contribute to the strength of the stem. In non-seed plants, such as ferns, these tough cells form a core in the center of the stem with softer tissues around them. In seed plants, such as conifers and flowering plants, the vascular tissues tend to lie in a subperipheral ring below the surface, surrounding a softer core of tissue. This is analogous to the arrangement of steel reinforcing bars in a column of reinforced concrete. In gymnosperms and dicots, the vascular tissues are formed in a single ring. In monocots, strands of vascular tissue form throughout the stem but even so tend to be more numerous and more tightly packed around the periphery of the stem.

Leaves are lateral appendages of the stem that are typically flattened to expose a maximum amount of surface area to sunlight. They are the primary site of photosynthesis in most vascular plants. The flattened portion of the leaf is the *blade*, or *lamina*. In some plants, such as grasses, the lamina appears to attach directly to the stem. In most plants, however, there is a stemlike stalk, the *petiole*, that extends the lamina away from the stem. In many cases the cells of the petiole enable the blade to be repositioned to maximize exposure to light. The *leaf base* is the attachment zone where the leaf connects to the stem. In some plants the cells of the leaf base enlarge to form *stipules* on each side of the petiole. The leaf base merges with the node. One or more *axillary buds* arise in the angle formed by this merger.

Shoot Development

At the tip of every shoot is a *terminal bud*. If the leaves of this bud are carefully peeled away, a tiny, rounded dome of tissue is exposed—the *shoot apical meristem*. A meristem is a region of the plant where cell division is concentrated. Cell division and differentiation in the shoot apical meristem produce the tissues that form the stem and leaves of that shoot.

The structure of the shoot apical meristem varies, depending on the plant being examined. Seedless vascular plants, such as ferns, have a large, pyramidally shaped apical cell that ultimately gives rise to all the cells and tissues of the shoot. The seed plants have a multicellular apical meristem differentiated into several zones. Each zone is associated with one of three primary meristems that form the tissues of the shoot system. A surface layer of cells covers the dome and differentiates into *protoderm*, the primary meristem that forms *epidermal cells*. The epidermis is a continuous protective layer that covers the stem and leaves of the shoot system. Within the apical meristem, certain cells begin to elongate and form the *procambium*, the primary meristem that forms vascular tissues. The remainder of the internal cells of the apical meristem become ground meristem, the primary meristem that forms ground tissues, such as cortex and pith.

As the apical meristem grows and new cells are formed, the dome of tissue enlarges. This growth is not symmetric; rather, surface or subsurface cells of a localized region will proliferate more rapidly than the rest to form a small bulge of tissue. This bulge is the first sign of initiation of a new leaf, a *leaf primordium*. As the leaf primordium enlarges, the protoderm also proliferates to maintain a continuous covering over the developing leaf that is contiguous with the developing stem. This continuity is maintained as cells differentiate into mature epidermal cells.

After initiation of a leaf primordium, the shoot apical meristem continues to grow, and procambium begins to differentiate within the developing node. These procambial cells connect to already formed procambium in older portions of the stem to form the template for the vascular system of the stem. Simultaneously, as the leaf primordium enlarges, procambial strands differentiate in the leaf base region. Further development is bidirectional. Cells from the basal end of each strand differentiate into the node and connect with the already-formed

Parts of the Shoot

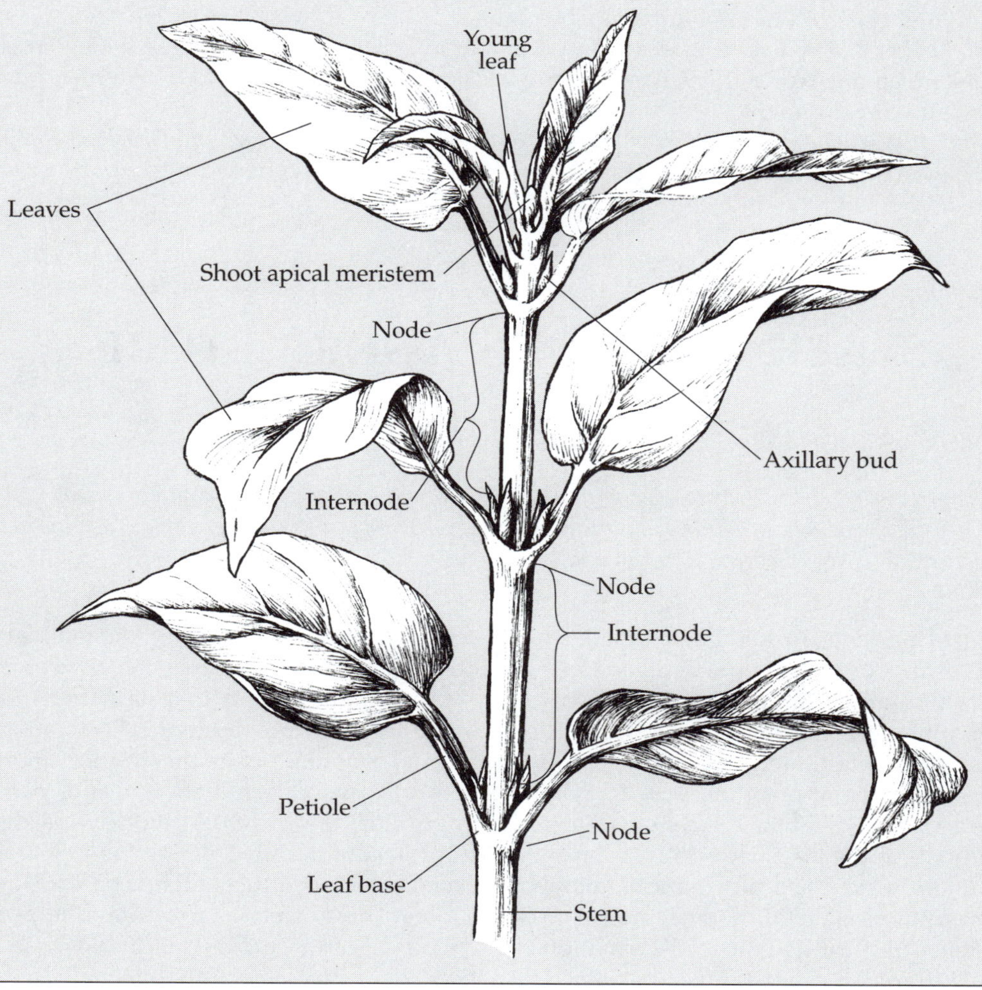

Kimberly L. Dawson Kurnizki

procambium there. Cells from the upper end of each strand continue to develop into the enlarging leaf primordium. In this way the vascular tissues of the leaf and stem are integrated into a continuous system. Similarly, the ground tissues of leaf and stem are continuous as the leaf primordia are initiated and develop.

Specialized Shoots

Specialized shoots are usually associated with storage or vegetative reproduction. *Bulbs*, such as onions, have a shortened stem with enlarged, overlapping, fleshy leaves. *Tubers*, such as potatoes, are subterranean storage shoots with thick, starch-filled stem cells and undeveloped leaves subtending axillary buds (the "eyes"). *Rhizomes* are horizontal shoots running on, or below, the soil surface. Their leaves are typically much reduced.

Marshall D. Sundberg

See also: Angiosperm cells and tissues; Bulbs and rhizomes; Cloning of plants; Growth habits; Leaf anatomy; Leaf arrangements; Stems; Plant tissues; Thigmomorphogenesis; Tropisms; *Tracheobionta*.

Sources for Further Study

Bell, Adrian D. *Plant Form: An Illustrated Guide to Flowering Plant Morphology*. New York: Oxford University Press, 1991. Photos and line art illustrate the common and specialized structure of flowering plants. Includes bibliographical references, index.

Harris, James G., and Melinda Woolf Harris. *Plant Identification Terminology: An Illustrated Glossary*. Spring Lake, Utah: Spring Lake, 1994. Clear line drawings illustrate specific terminology used to identify and describe plants.

Raven, Peter H., Ray F. Evert, and Susan E. Eichhorn. *Biology of Plants*. 6th ed. New York: W. H. Freeman/Worth, 1999. Basic botany text covers all aspects of shoot structure, function, and development.

Rudall, Paula. *Anatomy of Flowering Plants: An Introduction to Structure and Development*. 2d ed. New York: Cambridge University Press, 1993. Basic introduction to angiosperm structure and development.

SLASH-AND-BURN AGRICULTURE

Categories: Agriculture; economic botany and plant uses; environmental issues

Slash-and-burn agriculture, also called swidden agriculture, is a practice in which forestland is cleared and burned for use in crop and livestock production. While yields are high during the first few years, they rapidly decline in subsequent years, leading to further clearing of nearby forestland.

Slash-and-burn agriculture has been practiced for many centuries among people living in tropical *rain forests*. Initially, this farming system involved small populations. Therefore, land could be allowed to lie *fallow* (unplanted) for many years, leading to the full regeneration of the *secondary forests* and hence a restoration of the ecosystems. During the second half of the twentieth century, however, several factors led to drastically reduced fallow periods. In some places such fallow systems are no longer in existence, resulting in the transformation of forests into *shrub* and *grasslands*, negative effects on agricultural productivity for small farmers, and disastrous consequences to the environment.

Among the factors that have been responsible for reduced or nonexistent fallow periods are increased population in the tropics, increased demand for wood-based energy, and, perhaps most important, the increased worldwide demand for tropical commodities during the 1980's and 1990's, especially for products such as palm oil and natural rubber. These last two factors have helped industrialize slash-and-burn agriculture, which was practiced for centuries mainly by small farmers. Ordi-

A bean farmer in Nicaragua walks away from cut vegetation, or slash, he has set on fire. The land is burned in preparation for planting, and ashes from the burned trees fertilize the soil.

narily, small farmers are able to control their fires so that they are similar to a small forest fire triggered by lightning in the northwestern or southeastern United States. However, the continued reduction in fallow periods, coupled with increased burning by *subsistence farmers* and large agribusiness, especially in Asia and Latin America, is resulting in increased environmental concern.

While slash-and-burn agriculture seldom takes place in temperate regions, some agricultural burning occurs in the Pacific Northwest of the United States, where it is estimated that three thousand to five thousand agricultural fires are set each year in Washington State alone. These fires also create problems for human health and the environment.

Habitat Fragmentation

One of the most easily recognized results of slash-and-burn agriculture is *habitat fragmentation*, which leads to a significant loss of the vegetation needed for the maintenance of effective *gaseous exchange* in tropical regions and throughout the world. For every acre of land lost to slash-and-burn agriculture, 10 to 15 acres (4 to 6 hectares) of land are fragmented, resulting in the loss of habitat for wildlife, plant species, and innumerable macro- and microorganisms yet to be identified. This also creates problems for management and wildlife conservation efforts in parts of the world with little or no resources to feed their large populations. Fragmentation has also led to intensive discussions on *global warming*. While slash-and-burn agriculture by itself is not completely responsible for global warming, the industrialization of the process could make it a significant component of the problem, as more and more vegetation is fragmented.

Human Health

The impact of slash-and-burn agriculture on human health and the environment is best exemplified by the 1997 Asian fires that resulted from such practices. Monsoon rains normally extinguish the fires set by farmers, but a strong El Niño weather phenomenon delayed the expected rains, and the fires burned out of control for months. Thick smoke caused severe health problems. It is estimated that more than 20 million people in Indonesia alone were treated for asthma, bronchitis, emphysema, and eye, skin, and cardiovascular problems as a result of the fires. Similar problems have been reported for smaller agricultural fires.

Three major problems are associated with air pollution: *particulate matter*, *pollutant gases*, and *volatile organic compounds*. Particulate compounds of 10 microns or smaller that are inhaled become attached to the alveoli and other blood cells, resulting in severe illness. Studies by the U.S. Environmental Protection Agency (EPA) and the University of Washington indicate that death rates associated with respiratory illnesses increase when fine particulate air pollution increases. Meanwhile, pollutant gases such as carbon monoxide, nitric oxide, nitrogen dioxide, and sulfur dioxide become respiratory irritants when they combine with vapor to form *acid rain* or fog. Until the Asian fires, air pollutants stemming from the small fires of slash-and-burn agriculture that occur every planting season often went unnoticed. Thus, millions of people in the tropics experience environmental health problems because of slash-and-burn agriculture that are never reported.

Soil and Water Quality

The loss of vegetation that follows slash-and-burn agriculture causes an increased level of soil *erosion*. The soils of the humid tropics create a hard pan underneath a thick layer of organic matter. Therefore, upon the removal of vegetation cover, huge areas of land become exposed to the torrential rainfalls that occur in these regions. The result is severe soil erosion. As evidenced by the impact of Hurricane Mitch on Honduras during 1998, these exposed lands can give rise to large mudslides that can lead to significant loss of life. While slash-and-burn agriculture may not be the ultimate cause for sudden mud slides, it does predispose these lands to erosional problems.

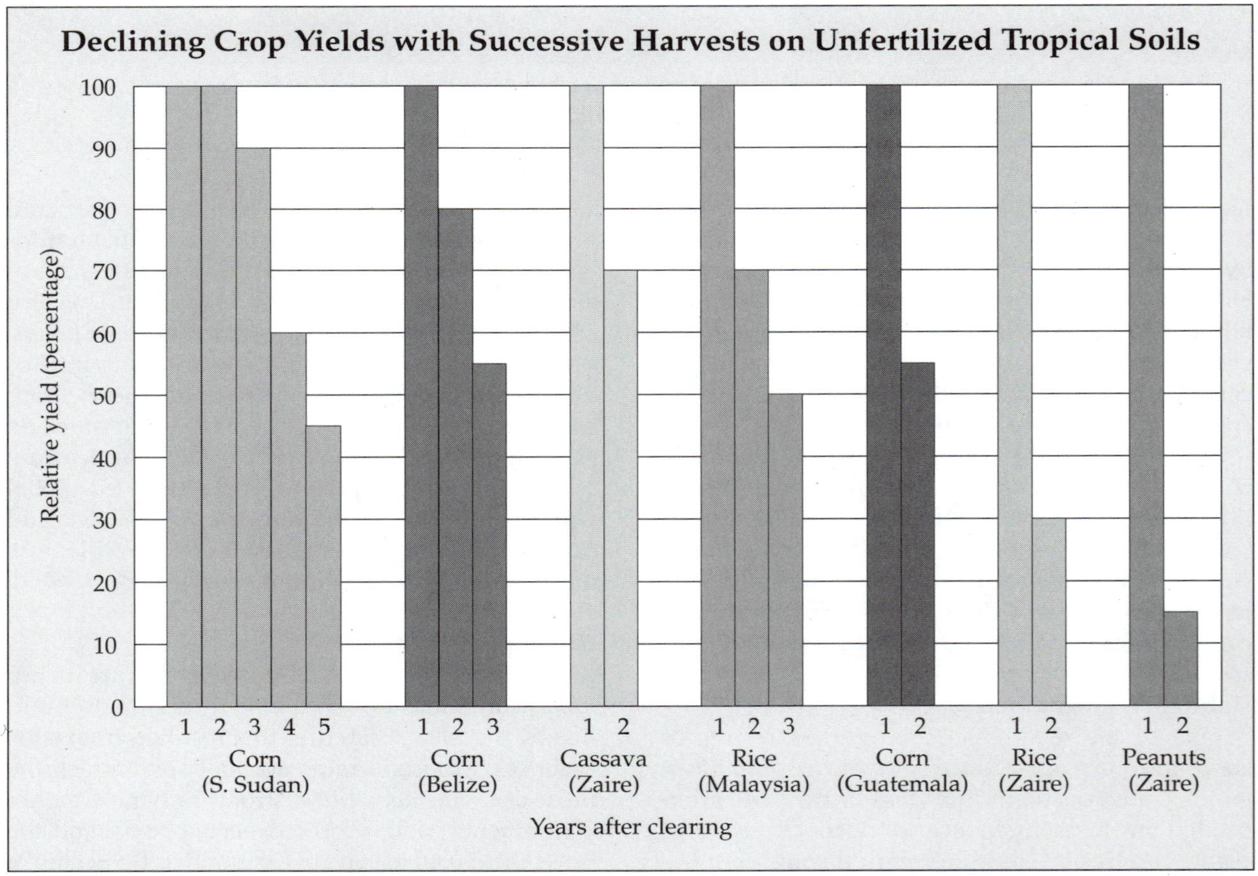

Source: Data adapted from John Terborgh, *Diversity and the Tropical Rain Forest*. New York: W. H. Freeman, 1992.

Associated with erosion is the impact of slash-and-burn agriculture on water quality. As erosion continues, *sedimentation* of streams increases. This sedimentation affects stream flow and freshwater discharge for catchment-area populations. Mixed with the sediment are minerals such as phosphorus and nitrogen-related compounds that enhance algal growth in streams and estuaries, which depletes the supply of oxygen that aquatic organisms require to survive. Although fertility is initially increased on noneroded soils, nutrient deposition and migration into drinking water supplies continues to increase.

Controlling Slash-and-Burn Agriculture

Given the fact that slash-and-burn agriculture has significant effects on the environment not only in regions where it is the mainstay of the agricultural systems but also in other regions of the world, it has become necessary to explore different approaches to controlling this form of agriculture. However, slash-and-burn agriculture has evolved into a sociocultural livelihood; therefore, recommendations must be consistent with the way of life of a people who have minimal resources for extensive agricultural systems.

Among the alternatives are new *agroecosystems* such as *agroforestry systems* and *sustainable agricultural systems* that do not rely so much on the slashing and burning of forestlands. These systems allow for the cultivation of agronomic crops and livestock within forest ecosystems. This protects soils from being eroded. Another possibility is the education of small rural farmers, absentee landlords, and big agribusiness concerns in developing countries to understand the environmental impact of slash-and-burn agriculture. While small rural farmers may not have the resources for renovating utilized forestlands, big business can organize ecosystems restoration, as has been done in many developed nations of the world.

Oghenekome U. Onokpise

See also: Agriculture: modern problems; Air pollution; Deforestation; Erosion and erosion control; Forest fires; Rain-forest biomes; Rain forests and the atmosphere; Soil degradation; Sustainable agriculture.

Sources for Further Study

Jordan, C. F. *An Amazonian Rain Forest: The Structure and Function of a Nutrient Stressed Ecosystem and the Impact of Slash and Burn Agriculture*. Boca Raton, Fla.: CRC Press, 1990. A collection of studies on the effects of slash-and-burn agriculture in the Amazon.

Simons, L. M., and M. Yamashita. "Indonesia's Plague of Fire." *National Geographic* 194 (August, 1998): 100-120. Details the extensive effects of the 1997 Asian fires.

Terborgh, J. *Diversity and the Tropical Rain Forest*. New York: Scientific American Library, 1992. Contains a good overview of tropical rain forests and the problems associated with slash-and-burn agriculture.

Youth, Howard. "Green Awakening in a Poor Country." *World Watch* 11, no. 5 (September/October, 1998): 28-37. Describes attempts to counter the effects of slash-and-burn agriculture in Honduras.

SOIL

Categories: Nutrients and nutrition; soil

Soil is a product of the physical and chemical breakdown of the earth's surface into small fragments, including sand, silt, and clay. Soil contains the products of organic matter decomposition—the composting of dead plant and animal debris.

Soils are classified on the basis of *soil profile* and *soil formation*. They can be grouped according to a number of characteristics, including agronomic use, color, organic matter content, texture, and

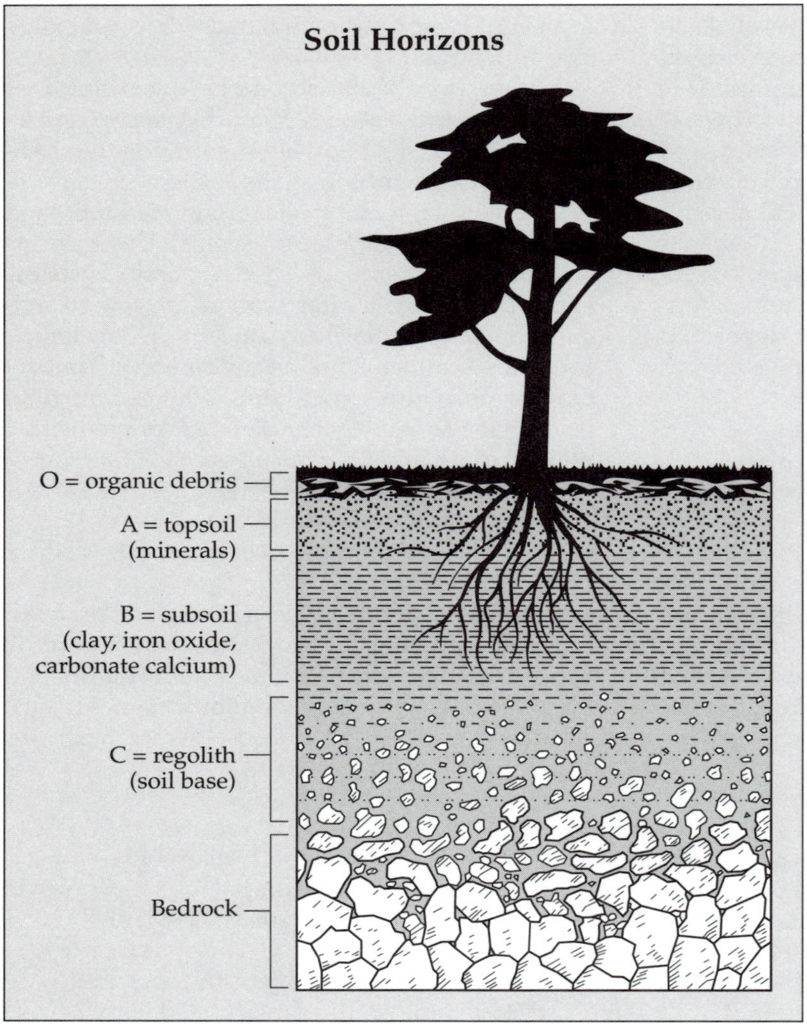

moisture condition. Typical soil is about 45 percent minerals and about 5 percent organic matter. The other 50 percent of soil consists of pores that hold either water or air. The liquid portion of soil contains dissolved minerals and organic compounds, produced by plants and microorganisms. The gases found in soil often are the same as those found in the air above it. Soil can support plant life if climate and moisture are suitable. It is a changing and dynamic body, adjusting to conditions of climate, topography, and vegetation. In turn, soil influences plant and root growth, available moisture, and the nutrients available to plants. While "the soil" is a collective term for all soils, "a soil" means one individual soil body with a particular length, depth, and breadth.

Soil Profile

In a typical soil, the top layer is usually dark with decomposing organic matter; the layers below are sand, silt, clay, or some combination of the three. Soil scientists classify soils on the basis of soil profile and soil formation.

Typically the top soil layer is called the *O horizon*, or organic matter horizon. It has rotten logs, leaf litter, and other recognizable bits of plants and animals. Underneath the O horizon is the *A horizon*. It is characterized by thoroughly decomposed organic matter. Water passing through the A horizon carries clay particles and organic acids through it into the *B horizon*. Clay or organic substances passing into the B horizon glue soil particles together, forming soil aggregates. Soil aggregates—granular, columnar, and so on—are indicators of a mature, healthy soil. The lowest level of the soil profile is the *C horizon*. It contains bedrock or soil parent material that shows little or no evidence of plant growth or soil formation.

Soil Formation

Soil formation takes hundreds, even thousands, of years. Parent material, climate, organisms, topography, and time all contribute. Sources of parent material include igneous, sedimentary, and metamorphic rocks (fragments of which may be deposited by water, wind, and ice), and plant and animal deposits.

Soil formation is the result of the physical, chemical, and biochemical breakdown of parent material. It also reflects the processes of weathering and change within the soil mass. Many substances are added to soil—rain, water from irrigation, nitrogen from bacteria, sediment, salts, organic residues, and a variety of substances created by humans. However, many substances are also removed from the soil—water-soluble minerals, clay, plants, carbon dioxide, and nitrogen. Other transformations also are occurring: Organic matter is decomposing, and minerals are solubilizing and changing chemical form. Clays and soluble salts that move along with the soil water cause color and chemical changes in the soil.

Parent material is a primary determinant of soil type or soil classification. All soils at the lowest category of soil classification are distinct if the parent material differs. The differences in parent materials—weathering rates, the plant nutrient content, and soil texture resulting from parent material breakdown—contribute to the formation of distinctive soils. For example, sandstone yields sandy soil with low fertility.

Chemical Weathering

Soils slowly change color and density as a result of wetting and drying, warming and cooling, and freezing and thawing. During *weathering*—the rubbing, grinding, and moving of rocks by water, wind, and gravity—rocks are split into smaller and smaller fragments. Soil is composed of fragments 2 millimeters or less in diameter.

The expansion force of water as it freezes is sufficient to split minerals. However, water also is involved in chemical weathering—solution, hydrolysis, carbonation, reduction, oxidation, and hydration. A simple example of solution, the dissolving of minerals in liquid, is the dissolving of salt in water. The salts then move along with the liquid. In hot arid climates, salts can move to the surface as water evaporates, creating salt flats. In wetter climates, salts can move through the soil, depleting it of necessary plant nutrients and contaminating groundwater.

Hydrolysis is the splitting of a water molecule to form hydroxides and soluble hydroxide compounds, such as sodium hydroxide. *Hydration* is the addition of water to minerals in rock. When a mineral such as hematite (an oxide of iron) hydrates, it expands, softens, and changes color. *Carbonation* is the reaction of a compound with carbonic acid, a weak acid produced when carbon dioxide dissolves in water. Water often contains carbonic acid and other organic acids produced by organic matter decomposition. These acids increase the power of the water to disintegrate rock. *Oxidation* is the addition of oxygen to a mineral, and *reduction* is the removal of oxygen from a mineral.

Biological Weathering

Biological weathering is a combination of physical and chemical disintegration of rocks to produce soil. The roots of plants can crack rocks and break them apart. Plant roots also produce carbon dioxide, which combines with water to produce carbonic acid. Carbonic acid dissolves certain minerals, speeding the breakdown of parent material and chemically changing the soil.

Plants and animals also add humus (organic matter) to soil, increasing its fertility and water-holding capacity and speeding rock weathering. Animals such as earthworms, ants, prairie dogs, gophers, and moles also contribute to soil aeration and fertility by mixing the soil. In areas where animal populations are large, they can influence both the formation and destruction of soil.

Climate and Topography

Climate also influences soil formation indirectly through its action on vegetation. Soils in arid climates have sparse vegetation, less organic matter, and little soil profile development. Wet soil, however, usually has thick vegetation and high organic matter, and therefore a deep soil profile.

The shape of the land is referred to as its topography. Each landform—valleys, plains, hills, and mountains—is covered with a crazy quilt of different soil types. For example, the steep sides of the Sandia Mountains near Albuquerque, New Mexico, which are severely eroded by wind and summer rains, contain a variety of soil types—forest soils, sandy soils, and rocky soils. Sand, silt, and clay eroded from the mountains and nearby extinct volcanoes combine in the moist and fertile Rio Grande Valley. The valley has deep sandy soils, layered sand and clay soils, and soils eroded by flash floods.

Soils located in similar climates that develop from similar parent material on steep hillsides usually have thin A and B horizons because less water moves through the soil. Similar materials on shallower slopes allow more water to pass through them. Topography and climate work together either to allow or to prohibit plant growth and organic matter deposition. Without moisture, plants cannot grow to impede soil erosion, and soil development is slow. With moisture, plants can grow, hold the soil in place, add organic matter to the soil, and speed soil development.

The age of a soil may be reckoned in tens, hundreds, or thousands of years. Under ideal conditions, a soil profile may develop in two hundred years; however, under less favorable conditions soil development may take several thousand years.

Soil Classification

Scientists have identified and classified soils for hundreds of years. Soils can be grouped according to agronomic use, color, organic matter content, texture, moisture condition, and other characteristics. Each of these groupings serves a particular purpose. U.S. soil scientists adopted a system of soil classification on January 1, 1965, that was based on the knowledge they had about soil genesis, morphology, and classification. The U.S. system is divided into six categories: order, suborder, great group, subgroup, family, and series. *Soil taxonomy* is patterned after the worldwide system of plant and animal taxonomy, which contains phylum, class, order, family, genus, and species.

Changes to the system have proceeded through a number of major revisions or approximations.

The Twelve Soil Orders in the U.S. Classification System

Soil Order	*Features*
Alfisols	Soils in humid and subhumid climates with precipitation from 500 to 1,300 millimeters (20 to 50 inches), frequently under forest vegetation. Clay accumulation in the B horizon and available water most of the growing season. Slightly to moderately acid soils.
Andisols	Soils with greater than 60 percent volcanic ash, cinders, pumice, and basalt. They have a dark A horizon as well as high absorption and immobilization of phosphorus and very high cation exchange capacity.
Aridisols	Aridisols exist in dry climates. Some have horizons of lime or gypsum accumulations, salty layers, and A and slight B horizon development.
Entisols	Soils with no profile development except a shallow A horizon. Many recent river floodplains, volcanic ash deposits, severely eroded areas, and sand are entisols.
Gelisols	Soils that commonly have a dark organic surface layer and mineral layers underlain by permafrost, which forms a barrier to downward movement of soil solution. Common in tundra regions of Alaska. Alternate thawing and freezing of ice layers results in special features in the soil; slow decomposition of the organic matter due to cold temperatures results in a peat layer at the surface in many gelisols.
Histosols	Organic soils of variable depths of accumulated plant remains in bogs, marshes, and swamps.
Inceptisols	Soils found in humid climates that have weak to moderate horizon development. Horizon development may have been delayed because of cold climate or waterlogging.
Mollisols	Mostly grassland soils, but with some broadleaf forest-covered soils with relatively deep, dark A horizons, a possible B horizon, and lime accumulation.
Oxisols	Excessively weathered soils. Oxisols are over 3 meters (10 feet) deep, have low fertility, have dominantly iron and aluminum oxide clays, and are acid. Oxisols are found in tropical and subtropical climates.
Spodosols	Sandy leached soils of the cool coniferous forests, usually with an organic or O horizon and a strongly acidic profile. The distinguishing feature of spodosols is a B horizon with accumulated organic matter plus iron and aluminum oxides.
Ultisols	Strongly acid and severely weathered soils of tropical and subtropical climates. They have clay accumulation in the B horizon.
Vertisols	Soils with a high clay content that swell when wet and crack when dry. Vertisols exist in temperate and tropical climates with distinct dry and wet seasons. Usually vertisols have only a deep self-mixing A horizon. When the topsoil is dry, it falls into the cracks, mixing the soil to the depth of the cracks.

The system can be used to classify soils anywhere in the world, especially with the addition of a new soil order, the andisols. The new soil classification continues to be tested, and minor modifications may be anticipated. The approximation being used as of 1997 treated soil as a collection of three-dimensional entities that can be grouped based on similar physical, chemical, and mineralogical properties. The minimum volume of soil that scientists consider when they classify a soil is the pedon, which can range from 1 to 10 meters square and is as deep as roots extend into a soil.

By 1999, the U.S. soil classification system recognized twelve soil *orders*. The differences among orders reflect the dominant soil-forming processes and the degree of soil formation. Each order is identified by a word ending in "-sol." Each order is divided into suborders, primarily on the basis of properties that influence soil genesis, are important to plant growth, and reflect the most important variables within the orders. The last syllable in the name of a suborder indicates the order. An example is "aquent," meaning water, plus "-ent," from "entisol."

Suborders are distinctive to each order and are not interchangeable between orders. Each suborder is divided into great groups on the basis of additional soil properties and horizons resulting from differences in soil moisture and soil temperature.

Great groups are denoted by a prefix that indicates a property of the soil. An example is "psammaquents" ("psamm" referring to sandy texture and "aquent" being the suborder of the entisols that has an aquic moisture regime). Soil scientists have identified more than three hundred great groups in the United States. Great groups are distinguished on the basis of differing horizons and soil features. The differing soil horizons include those with accumulated clay, iron, or organic matter and those hardened or cemented by soil cultivation or other human activities. The differentiating soil features include self-mixing of soil due to clay content, soil temperature, and differences in content of calcium, magnesium, sodium, potassium, gypsum, and other salts.

There are more than twenty-four hundred *subgroups*, and each great group is divided into three kinds of *subgroups*: a *typic* subgroup, an *intergrade* subgroup, and an *extragrade* subgroup. The typic subgroup represents the central spectrum of a soil group. The intergrade subgroup represents soils with properties like those of other orders, suborders, or great groups. The extragrade subgroup represents soils with some properties that are not representative of the great group but do not indicate transitions to any other known kind of soil. Each subgroup is identified by one or more adjectives preceding the name of the great group.

Families are established within a subgroup on the basis of physical and chemical properties and other characteristics that are important to plant growth or that are related to the behavior of soils that are important for engineering concerns. Among the properties and characteristics considered are particle size, mineral content, temperature regime, depth of the root zone, moisture, slope, and permanent cracks. A family name consists of the name of a subgroup preceded by terms that indicate soil properties. Several thousand families have been identified in the United States.

Finally, the *series* is the lowest soil category, with more than nineteen thousand recognized in the United States as of 1999. A series might share one or more properties with those of an entire family, but for at least one property only a narrower range is permitted.

Texture, Structure, and Consistency

Soil texture is determined by the percent of sand, silt, and clay in a soil sample. Most fertile or productive soils have a loam texture, or about equal amounts of sand, silt, and clay, and a high organic matter content (about 5 to 10 percent). Soil texture determines the water-holding and nutrient-holding capacity of a soil. Thus, clay soils have a high nutrient-holding capacity, but they waterlog easily. Sandy soils have a lower nutrient-holding capacity but dry out easily. Farmers base their plans of how to fertilize and irrigate their crops partly on the texture of the soil.

Soil structure refers to how soil particles are glued together to form aggregates. During soil formation, soil particles are glued together with clay, dead microorganisms, earthworm slime, and plant roots, and they form air and water channels. Plants need these channels so they can absorb nutrients, water, and air. Soil structure may be destroyed when farmers cultivate wet or waterlogged soils with heavy farm machinery. Destroying soil structure makes a soil unsuitable for plant growth.

Soil consistency is the "feel" of a soil and the ease with which a lump can be crushed in one's fingers.

Common soil consistencies are loose, friable, firm, plastic, sticky, hard, and soft. Clay soils, for example, are sticky or plastic when they are wet, but they become hard or harsh when they are dry. The best time to work a clay soil is when it is soft or friable. Sandy soils, on the other hand, do not become plastic or sticky when they are wet or hard or harsh when they are dry. They have a tendency to stay loose, which makes them easier to work. Loam and silt loam soils are intermediate in behavior. When farmers are trying to determine whether to work the soil or wait for better soil moisture conditions, they usually check the soil consistency.

Aeration and Moisture

Soil aeration relates to the exchange of soil air with atmospheric air. Growing roots need oxygen and are constantly expiring carbon dioxide. Unless there is a continuous flow of oxygen into soil and carbon dioxide out of the soil, oxygen becomes depleted. When their oxygen supply is cut off, the roots will die.

Soil moisture refers to water held in soil pores. A plant draws water from soil the same way a child draws water from a cup with a straw. When the cup is full, it is easy for the child to draw up the water, but as the cup empties, the child must work harder to get water. Similarly, plants draw water from soil easily when the soil has plenty of water. As the soil dries, however, plants must work harder to pull water out of the soil until they reach wilting point.

Fertility

Plants absorb many of the nutrients they need from soil, including phosphorus, potassium, calcium, magnesium, sulfur, boron, chlorine, cobalt, copper, iron, manganese, molybdenum, and zinc. They may obtain carbon, hydrogen, and nitrogen from the air and water.

Soil testing services give farmers specific fertilizer and lime recommendations based on soil texture and chemical analysis. Farmers use soil tests to determine if their soil has enough essential nutrients for a crop to grow. The absence of one essential nutrient can limit overall crop growth. Nitrogen, phosphorus, and potassium are commonly applied to the soil as commercial fertilizer and manure. Calcium and magnesium are applied as lime, which is also used to reduce the acidity of soil and to increase the solubility of some minerals. Manure and other organic matter added to soils increase water-holding and nutrient-holding capacity and therefore boost crop yields.

Agricultural extension services offer guidelines for the maximum amounts of manure, sewage sludge, fertilizer, and other chemicals that farmers should apply to soils. Farmers are encouraged to apply nitrogen fertilizer in small applications at times when plants are growing rapidly. This soil management practice decreases deep percolation losses that could pollute groundwater.

With an understanding of soil characteristics, farmers and gardeners can learn to manage a wide variety of soils. Some soils are naturally fertile and need few amendments to promote high crop yields. Other soils, whether because of their parent material or climate, are naturally infertile and might best be used for purposes other than agriculture. Like the water and the air, the soil is a crucial natural resource. From an airplane, all soils look about the same, but from an ant's view, soils are all different. Differences in soil type make huge differences to plants, animals, humans, and the environment.

J. Bradshaw-Rouse, updated by Christina J. Moose

See also: Agronomy; Composting; Hydroponics; Nutrients; Soil conservation; Soil contamination; Soil degradation; Soil management; Soil salinization.

Sources for Further Study

Donahue, Roy L., Roy Hunter Follett, and Rodney W. Tulloch. *Our Soils and Their Management: Increasing Production Through Environmental Soil and Water Conservation and Fertility Management*. 6th ed. Danville, Ill.: Interstate Publishers, 1990. Covers soil management; emphasis is on agricultural applications. Bibliography, indexes.

Harpstead, Milo I., Thomas J. Sauer, and William F. Bennett. *Soil Science Simplified*. 3d ed. Ames: Iowa State University Press, 1997. A concise introduction. Illustrations, maps, index.

Miller, Raymond, and Duane T. Gardiner. *Soils in Our Environment*. 9th ed. Upper Saddle River, N.J.: Prentice Hall, 1997. This is the authoritative textbook on soil formation. This edition includes a CD-ROM as well as sixty full-color photographs and expanded and updated coverage touching on the latest technologies and environmental issues. Glossary, index.

SOIL CONSERVATION

Categories: Environmental issues; soil

Soil conservation is the effort by farmers and other landowners to prevent the buildup of salts and fertilizer acids in the soil, as well as the loss of topsoil from wind erosion, water erosion, desertification, and chemical deterioration. This can be achieved by implementing management practices to maintain soil quality and reduce pollution.

According to the United Nations Environment Programme, approximately 17 percent of the earth's vegetated land is degraded, which poses a threat to agricultural production around the world. The introduction of minerals, metals, nutrients, fertilizers, pesticides, bacteria, and pathogens suspended in *topsoil runoff* into waterways is a significant source of water pollution and is a threat to fisheries, wildlife habitat, and drinking water supplies. The introduction of soil particles into the air through wind *erosion* is a significant source of air pollution.

Threats and Responses

The Industrial Revolution of the nineteenth century and the population explosion of the twentieth century encouraged people to till new land, cut down forests, and disturb soil for the expansion of towns and cities. The newly exposed topsoil quickly succumbed to erosion from rainfall, floods, wind, ice, and snow. The Dust Bowl, which occurred in the Great Plains in the United States during the 1930's, is one example of the devastating effects of wind erosion.

Hugh Hammond Bennett, the so-called father of soil conservation, lobbied for congressional establishment of the United States Soil Conservation Service (approved in 1937) and the establishment of voluntary Soil Conservation Districts in each state. Bennett was named the first chief of the U.S. Soil Conservation Service in 1937. On August 4, 1937, the Brown Creek Conservation District in Bennett's home county, Anson County, North Carolina, became the first Soil Conservation District in the United States. Local landowners voted to establish the district by three hundred to one, proving that farmers were concerned about soil conservation. A reporter for the *Charlotte Observer* newspaper sought out the one negative voter; after having the program explained to him, he changed his opinion. By 1948, more than twenty-one hundred districts had been established nationwide. They were eventually renamed Soil and Water Conservation Districts. There are more than three thousand such districts in the United States.

The U.S. Food Security Act of 1985 authorized the Conservation Reserve Program to take land highly susceptible to erosion out of production and required farmers to develop soil conservation plans for the remaining susceptible land. The Natural Resource Conservation Service estimates that the loss of topsoil was nearly cut in half, reduced from 1.6 billion tons per year to 0.9 billion tons. The European Community and Australia also adopted soil conservation measures during the 1990's.

Practices

Soil conservation practices include covering the soil with *vegetative cover*, reducing *soil exposure* on tilled land, creating wind and *water barriers*, and installing *buffers*. Vegetative cover slows the wind at ground level, slows water runoff, protects soil particles from being detached, and traps blowing or floating soil particles, chemicals, and nutrients. Because the greatest wind and water erosion damage often occurs during seasons in which no crops are growing or natural vegetation is dormant, soil conservation often depends on permitting the dead residues and standing stubble of the previous crop to remain in place until the next planting time. An-

PhotoDisc

Soil conservation practices include protecting soil particles from becoming detached and trapping blowing or floating soil particles, chemicals, and nutrients.

nual tree-foliage loss serves as a natural ground mulch in forested areas. Planting grass or legume cover crops until the next planting season, or as part of a *crop rotation* cycle or *no-till planting* system, also reduces erosion.

Modern no-till and *mulch-till planting* systems reduce soil exposure to wind and rain, while plowing the land brings new soil to the surface and buries the ground cover. No-till systems leave the soil cover undisturbed before planting and insert crop seeds into the ground through a narrow slot in the soil. Mulch-till planting keeps a high percentage of the dead residues of previous crops on the surface when the new crop is planted. Row crops are planted at right angles to the prevailing winds and to the slope of the land in order to absorb wind and rainwater runoff energy and trap moving soil particles. Crops are planted in small fields to prevent avalanching caused by an increase in the amount of soil particles transported by wind or water as the distance across bare soil increases. As the amount of soil moved by wind or water increases, the erosive effects of the wind and water also increase. Smaller fields reduce the length and width of unprotected areas of soil.

Wind and water barriers include tree plantings and crosswind strips of perennial shrubs and 1-meter-high (3-foot-high) grasses, which act as wind breaks to slow wind speeds at the surface of the soil. The protected area extends for ten times the height of the barrier. In *alley cropping*, which is used in areas of sustained high wind, crops are planted between rows of larger mature trees. Contour strip farming on slopes, planting grass waterways in areas where rainwater runoff concentrates, and planting 3-meter-wide (10-foot-wide) grass field borders on all edges of cultivated or disturbed soil are additional methods for reducing wind speed and rainwater runoff and trapping soil particles, chemicals, and nutrients.

Buffers filter runoff to remove sediments and chemicals. Riparian buffers are waterside plantings of trees, shrubs, and grasses, usually 6 meters (20 feet) in width. Riparian buffers planted only in grass are called *filter strips*. Grassed waterways, field borders, water containment ponds, and contour grass strips are other types of soil conservation buffers.

Gordon Neal Diem

See also: Agriculture: modern problems; Agronomy; Composting; Erosion and erosion control; Hydroponics; Nutrients; Slash-and-burn agriculture; Soil; Soil contamination; Soil degradation; Soil management; Soil salinization; Sustainable agriculture.

Sources for Further Study

Gershuny, Grace, and Joseph Smillie. *The Soul of Soil: A Guide to Ecological Soil Management*. 3d ed. Davis, Calif.: agAccess, 1995. Soil conservation and related issues are discussed from an ecological perspective. Illustrations, bibliography, index.

Morgan, R. P. C. *Soil Erosion and Conservation*. New York: Wiley, 1995. This nearly two-hundred-page discussion is supplemented by illustrations, maps, bibliography, and index.

Richter, Daniel D., and Daniel Markowitz. *Understanding Soil Change: Soil Sustainability over Millennia, Centuries, and Decades*. New York: Cambridge University Press, 2001. Aimed at farmers and agricultural professionals, this book discusses the forces of erosion and procedures to sustain soil fertility.

Sanders, David W., Paul C. Huszar, and Samran Sombatpanit, eds. *Incentives in Soil Conservation: From Theory to Practice*. Enfield, N.H.: Science Publishers, 1999. A series of studies on the results of incentives to soil conservation in worldwide soil management programs.

Troeh, Frederick R., et al. *Soil and Water Conservation: Productivity and Environmental Protection*. Upper Saddle River, N.J.: Prentice Hall, 1998. Considers various hazards—erosion, sedimentation, pollution—with reference to various world soils. Takes an agronomic approach. Appendices present conversion factors and common and scientific plant names.

Yudelman, Montague, John C. Greenfield, and William B. Magrath. *New Vegetative Approaches to Soil and Moisture Conservation*. Washington, D.C.: World Wildlife Fund and the Conservation Foundation, 1990. Examines soil and water conservation in developing countries. Bibliographical references, illustrations.

SOIL CONTAMINATION

Categories: Environmental issues; pollution; soil

Soils contaminated with high concentrations of hazardous substances pose potential risks to human health and the earth's thin layer of productive soil.

Productive soil depends on bacteria, fungi, and other soil microbes to break down wastes and release and cycle nutrients that are essential to plants. Healthy soil is essential for growing enough food for the world's increasing population. Soil also serves as both a filter and a buffer between human activities and natural water resources, which ultimately serve as the primary source of drinking water. Soil that is contaminated may serve as a source of water pollution through *leaching* of contaminants into groundwater and through *runoff* into surface waters such as lakes, rivers, and streams.

The U.S. government has tried to address the problem of soil contamination by passing two landmark legislative acts. The Resource Conservation and Recovery Act (RCRA) of 1976 regulates hazardous and toxic wastes from the point of generation to disposal. The Comprehensive Environmental Response, Compensation, and Liability Act

(CERCLA) of 1980, also known as Superfund, identifies past contaminated sites and implements remedial action.

Sources of Contamination

Soils can become contaminated by many human activities, including fertilizer or pesticide application, direct discharge of pollutants at the soil surface, leaking of underground storage tanks or pipes, leaching from landfills, and atmospheric deposition. Additionally, soil contamination may be of natural origin. For example, soils with high concentrations of heavy metals can occur naturally because of their close proximity to metal ore deposits. Common contaminants include inorganic compounds such as nitrate and heavy metals (for example, lead, mercury, cadmium, arsenic, and chromium); volatile hydrocarbons found in fuels, such as benzene, toluene, ethylene, and xylene BTEX compounds; and chlorinated organic compounds such as polychlorinated biphenyls (PCBs) and pentachlorophenol (PCP).

Contaminants may also include substances that occur naturally but whose concentrations are elevated above normal levels. For example, nitrogen- and phosphorus-containing compounds are often added to agricultural lands as *fertilizers*. Since nitrogen and phosphorus are typically the limiting nutrients for plant and microbial growth, accumulation in the soil is usually not a concern. The real concern is the leaching and runoff of the nutrients into nearby water sources, which may lead to oxygen depletion of lakes as a result of the *eutrophication* encouraged by those nutrients. Furthermore, nitrate is a concern in drinking water because it poses a direct risk to human infants (it is associated with blue-baby syndrome).

Contaminants may reside in the solid, liquid, and gaseous phases of the soil. Most will occupy all three phases but will favor one phase over the others. The physical and chemical properties of the contaminant and the soil will determine which phase the contaminant favors. The substance may preferentially adsorb to the solid phase, either the inorganic minerals or the organic matter. The attraction to the solid phase may be weak or strong. The contaminant may also *volatize* into the gaseous phase of the soil. If the contaminant is soluble in water, it will dwell mainly in the liquid-filled pores of the soil.

Contaminants may remain in the soil for years or wind up in the atmosphere or nearby water sources. Additionally, the compounds may be broken down or taken up by the biological component of the soil. This may include plants, bacteria, fungi, and other soil-dwelling microbes. The volatile compounds may slowly move from the gaseous phase of the soil into the atmosphere. The contaminants that are bound to the solid phase may remain intact or be carried off in runoff attached to soil particles and flow into surface waters. Compounds that favor the liquid phase, such as nitrate, will either wind up in surface waters or leach down into the groundwater.

Metals display a range of behaviors. Some bind strongly to the solid phase of the soil, while others easily dissolve and wind up in surface or groundwater. PCBs and similar compounds bind strongly to the solid surface and remain in the soil for years. These compounds can still pose a threat to waterways because, over long periods of time, they slowly dissolve from the solid phase into the water at trace quantities. Fuel components favor the gaseous phase but will bind to the solid phase and dissolve at trace quantities into the water. However, even trace quantities of some compounds can pose a serious ecological or health risk. When a contaminant causes a harmful effect, it is classified as a pollutant.

Treatments

There are two general approaches to cleaning up a contaminated soil site: treatment of the soil in place (*in situ*) or removal of the contaminated soil followed by treatment (*non-in situ*). In situ methods, which have the advantage of minimizing exposure pathways, include biodegradation, volatilization, leaching, vitrification (glassification), and isolation or containment. Non-in situ methods generate additional concerns about exposure during the process of transporting contaminated soil. Non-in situ options include thermal treatment (incineration), land treatment, chemical extraction, solidification or stabilization, excavation, and asphalt incorporation. The choice of methodology will depend on the quantity and type of contaminants, and the nature of the soil. Some of these treatment technologies are still in the experimental phase.

John P. DiVincenzo

See also: Agronomy; Composting; Environmental biotechnology; Hydroponics; Nutrients; Soil; Soil conservation; Soil degradation; Soil management; Soil salinization.

Sources for Further Study

Pierzynski, Gary M., J. Thomas Sims, and George F. Vance. *Soils and Environmental Quality.* Boca Raton, Fla.: Lewis, 1994. A good overview of the environmental health of soils.

Sparks, Donald L. "Soil Decontamination." In *Handbook of Hazardous Materials*, edited by Morton Corn. San Diego, Calif.: Academic Press, 1993. Provides a good introduction to the sources of soil contamination and then gives a brief overview of the options for decontamination.

Testa, Stephen M. *The Reuse and Recycling of Contaminated Soil*. Boca Raton, Fla.: Lewis, 1997. Includes a good introduction that provides an overview of the problem and the regulatory aspects of soil contamination. The remainder of the book is geared for the more advanced reader and discusses technical aspects of decontaminating soil.

SOIL DEGRADATION

Categories: Environmental issues; soil

A decline in soil quality, productivity, and usefulness due to natural causes, human activities, or both, is known as soil degradation. It is often caused by unfavorable alterations in one or all of a soil's physical, chemical, and biological attributes.

In 1992, for the first global study of soil degradation, the World Resources Institute in Washington, D.C., reported that 3 billion acres of land worldwide had been seriously degraded since World War II. They also stated that 22 million acres of once usable land could no longer support crops.

Natural Processes and Human Activities

Of the total acreage lost to soil degradation, almost two-thirds is in Asia and Africa; most of the loss is attributable to water and wind *erosion* resulting from agricultural activities, *overgrazing*, *deforestation*, and firewood collection. There are also seriously degraded soils in Central America, where degradation is caused primarily by deforestation and overgrazing. In Europe, industrial and urban wastes, pesticides, and other substances have poisoned soils in much of Poland, Germany, Hungary, and southern Sweden. In the United States, the U.S. Department of Agriculture estimates that a quarter of the nation's croplands have been depleted through deep plowing, removal of crop residue, conversion to permanent pasture, and other conventional agricultural practices. Although unwise management practices contribute significantly to soil degradation, soil degradation also involves three natural soil processes: physical, chemical, and biological degradation.

Physical Degradation

Physical soil degradation involves deterioration in soil structure, leading to compaction, crusting, accelerated erosion, reduced water-holding capacity, and decreased aeration. Soil *compaction* is the compression of soil particles into a smaller volume. Excessively compacted soil suffers from poor aeration and reduced gas exchange, which can restrict the depth of root penetration. Soil compaction also causes accelerated runoff and erosion of soils.

Crusting is the formation of a hard layer a few millimeters or a few tens of millimeters thick at the soil surface. Crusts affect drainage, leading to waterlogging at the soil surface and to salinity or alkalinity problems. Once crusts called *duricrusts* form, soil moisture recharge declines, and vegetation cannot root. Sheet and gully erosion increases as the land fails to absorb precipitation. Hard layers can also form below the cultivation depth and are called *hard pans* (other names are plow soles, traffic pans, and plow pans). These compacted layers can restrict root growth, making crops and trees vulnerable to drought and lodging (falling over).

Chemical Degradation

Chemical degradation comprises changes in soil's chemical properties that regulate nutrient availability. *Nutrient depletion* is the major factor in chemical soil degradation. Soil nutrient depletion may be caused or exacerbated by many factors, including *monocropping*, leaching of nutrients, and salt buildup.

A historic example of nutrient depletion is the depletion of soils in the southeastern United States by the growing of cotton. As late as 1950, "King Cotton" was the most valuable farm commodity produced in Alabama, Arkansas, Georgia, Louisiana, Mississippi, South Carolina, Tennessee, and Texas. In the eighteenth and nineteenth centuries, the growing of cotton ruined soil fertility as it spread westward from the Atlantic to the Texas panhandle. Cotton growth without regard to topography in hilly regions contributed to soil erosion. Topsoil was eventually removed from many fields, which further depleted nutrients. One reason that peanuts became a major crop in the South is that they are *nitrogen-fixing* plants that can grow in soils depleted of nitrogen by cotton.

Nutrient *leaching* is another problem. Continuous irrigation can leach nutrients and cause salt buildup in soils where drainage is poor. Leaching can move essential but soluble nutrients past the root zone deeper into the soil and into groundwater. In addition, the water used to irrigate soil often contains salts that can accumulate to toxic levels and inhibit plant growth where evaporation occurs readily. Thick crusts of salt on farmland in Pakistan, Australia, Ethiopia, Sudan, and Egypt have made soil unfit for crops.

Laterization refers to the product and process of wetting and drying that leads to the irreversible consolidation and hardening of aluminum and iron-rich clays into hard pans, sometimes of great thickness, called *plinthitic materials* (Greek *plinthos* means "brick"). Laterization is particularly common in the humid and subhumid tropics.

PhotoDisc

Of the total acreage lost to soil degradation, most of the loss is attributable to water and wind erosion resulting from agricultural activities, overgrazing, deforestation, and firewood collection. Soil compaction also causes accelerated runoff and erosion of soils.

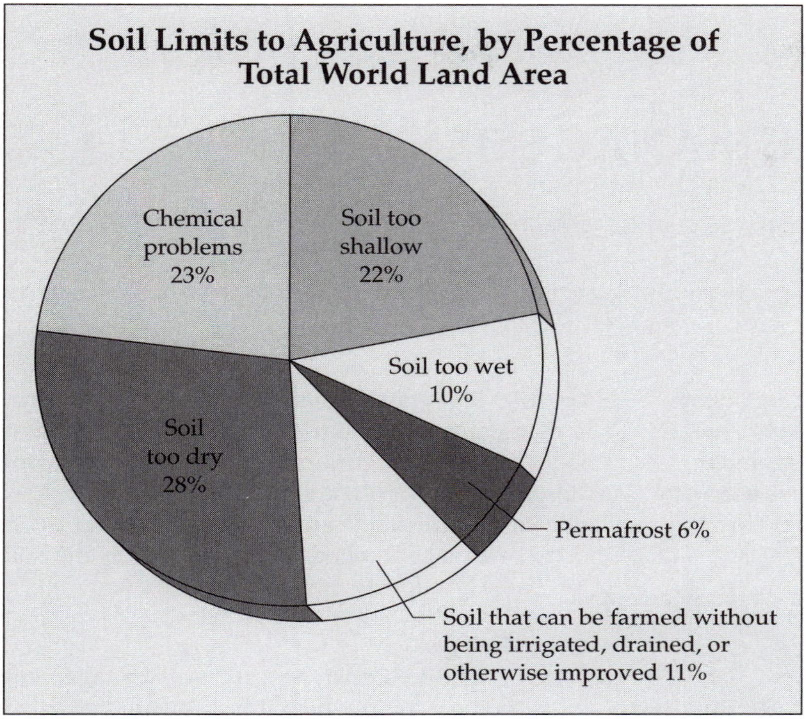

Source: United Nations Food and Agriculture Organization (FAOSTAT Database, 2000).

Biological Degradation

The loss of organic matter and soil nutrients needed by plants can occur in any environment, but it is most dramatic in hot, dry regions. Organic matter is important in maintaining soil structure, supporting microorganisms, and retaining plant nutrients. Because organic matter is near the soil surface, it is generally the first soil component to be lost. Organic matter may be lost through brush fires, stubble-burning, overgrazing, or the removal of crops, fodder, wood, and dung. Loss of organic matter can be accelerated when soil moisture is reduced, when soil aeration is increased, or both. For example, peat soils that are drained decompose rapidly and subside. In drier climates, the loss of organic matter reduces the soil's moisture-holding capacity and lowers soil fertility, which leads to lower crop yields and thus to less organic matter being returned to the soil.

Tropical rain forests such as those of the Amazon basin in South America seem lush, so people widely assume tropical soils to be fertile and high in organic matter. Although tropical forests do produce considerable organic matter, the amount that stays in the soil is surprisingly small, and the soils actually have low nutrient levels. Soil microorganisms in the rain forest break down the organic matter and release nutrients that are absorbed by growing plants. However, warm temperatures and high rainfall cause accelerated nutrient loss if plants are absent. Nutrients that would buffer the pH of the soil are lost. Consequently, the clearing of rain forests exposes the soil to erosion, leaching, acidification, and rapid nutrient depletion.

J. Bradshaw-Rouse

See also: Agronomy; Composting; Deforestation; Erosion and erosion control; Fertilizers; Grazing and overgrazing; Herbicides; Hydroponics; Monoculture; Nutrients; Pesticides; Slash-and-burn agriculture; Soil; Soil conservation; Soil contamination; Soil management; Soil salinization.

Sources for Further Study

Bridges, E. M., ed. *Responses to Land Degradation*. Enfield, N.H.: Science Publishers, 2001. Assesses legislative responses to soil degradation.

Lindert, Peter. *Shifting Ground*. Cambridge, Mass.: MIT University Press, 2000. Quantitative historical study of soil fertility in China and Indonesia suggests that modern agricultural practices actually remedy soil degradation.

Murnaghan, Niamh, and Michael A. Stocking, eds. *A Handbook for the Assessment of Land Degradation*. London: Earthscan Publications, 2001. A guidebook for farmers, intended to help assess soil degradation and choose appropriate options for ameliorating the problems.

SOIL MANAGEMENT

Categories: Agriculture; disciplines; environmental issues; soil

Tillage, conservation, and cropping practices are soil management techniques that are used to preserve soil resources while optimizing soil use.

Soils are managed differently depending on their intended use. Soil management groups are soil types with similar adaptations or management requirements for specific purposes, such as use with crops or cropping rotations, drainage, fertilization, forestry, highway engineering, and construction. In managing soil for agriculture, soil management includes all tillage and planting operations, cropping practices, fertilization, liming, irrigation, herbicide and insecticide application, and other treatments conducted on or applied to the soil surface for the production of plants.

PhotoDisc

Soil management extends to the way in which soils are manipulated. Terraces, for example, are raised horizontal strips of earth constructed along the contour of a hill to slow the movement of downward-flowing water.

Tillage

The most basic aspect of soil management is the way in which it is cultivated or tilled for crop growth. *Tillage* is the mechanical manipulation of the soil profile to modify soil conditions, manage crop residues or weeds, or incorporate chemicals for crop production. Tillage can be exhaustive or minimal. Conventional tillage uses multiple tillage operations to bury existing crop residue and prepare a uniform, weed-free seed bed for planting. This method breaks up soil aggregates in the process and destroys soil structure. Consequently, it can result in excessive wind and water erosion.

Conservation tillage, or minimum tillage, involves soil management practices that leave much more crop residue on the soil surface and cause much less soil disruption. As a result, the soil is less susceptible to erosion, and the plant residue acts as a mulch to protect the soil surface from the destructive impact of rainfall as well as to reduce evaporation.

No-tillage, or chemical tillage, is a soil management practice adapted to sloping soils in which herbicides rather than tillage are used to control weeds, while the disruption of soil structure is limited to a narrow slit in the soil surface in which the seeds are planted.

Hillsides and Wetlands

Soil management extends to the way in which soils are manipulated. *Terraces*, for example, are raised horizontal strips of earth constructed along the contour of a hill to slow the movement of downward-flowing water. Tile drains are perforated ceramic or plastic pipes buried in poorly drained soils that act as underground channels to carry water away, lower the water table, and allow a soil to drain faster after rainfall. The benefit of managing potentially erodible soils on hill slopes as permanent pastures is being recognized as another way of managing hillside soils.

Likewise, the value of retaining wet soils as *wetlands* has been acknowledged. Wetlands provide wildlife habitat, assist in flood control, and act as buffers to protect surface waterways from nutrient and soil runoff from cultivated fields.

Chemical Management

Soil management also involves the addition of chemicals to soil: lime to make acid soils more neutral, fertilizers to increase the nutrient level, herbicides and insecticides to control weed and insect pests, soil conditioners to improve soil aggregation, structure, and permeability. A growing technology is the use of mobile global positioning system (GPS) units attached to the equipment that applies these chemicals to soil. Called *site-specific management*, it uses computer technology to regulate chemical addition based on the exact position in a field and previous yield or fertility maps that indicate whether the soil needs to be amended. The goal of site-specific management is to optimize chemical use and profit while minimizing potential chemical loss to other environments by only applying the chemicals to areas where they are needed.

J. Bradshaw-Rouse

See also: Agronomy; Composting; Fertilizers; Hydroponics; Nutrients; Soil; Soil conservation; Soil contamination; Soil degradation; Soil salinization; Sustainable agriculture.

Sources for Further Study

Donahue, Roy L., Roy Hunter Follett, and Rodney W. Tulloch. *Our Soils and Their Management: Increasing Production Through Environmental Soil and Water Conservation and Fertility Management*. 6th ed. Danville, Ill.: Interstate Publishers, 1990. Covers soil management; emphasis is on agricultural applications. Bibliography, indexes.

Fisher, Richard, and Dan Brinkley. *Ecology and Management of Forest Soils*. 3d ed. New York: John Wiley & Sons, 2000. Provides basics of soil management for forestry professionals.

Prasad, Rajendra, and James F. Power. *Soil Fertility Management for Sustainable Agriculture*. Boca Raton, Fla.: Lewis, 1997. Provides a broad picture of soil management practices appropriate to different environmental conditions.

Richter, Daniel D., and Daniel Markowitz. *Understanding Soil Change: Soil Sustainability over Millennia, Centuries, and Decades*. New York: Cambridge University Press, 2001. Aimed at farmers and agricultural professionals, this book discusses the forces of erosion and procedures to sustain soil fertility.

SOIL SALINIZATION

Categories: Agriculture; environmental issues; pollution

Soil salinization is a process in which water-soluble salts build up in the root zones of plants, blocking the movement of water and nutrients into plant tissues.

Soil salinization rarely occurs naturally. It becomes an environmental problem when it occurs as a result of human activity, denuding once-vegetated areas of all plant life. Rainwater is virtually free of dissolved solids, but surface waters and underground waters (*groundwater*) contain significant quantities of dissolved solids, ultimately produced by the weathering of rocks. *Evaporation* of water at the land surface results in an increase in dissolved solids that may adversely affect the ability of plant roots to absorb water and nutrients.

Arid Climates

In arid regions, evaporation of soil water potentially exceeds rainfall. Shallow wetting of the soil followed by surface evaporation lifts the available dissolved solids to near the surface of the soil. The near-surface soil therefore becomes richer in soluble salts. In natural arid areas, soluble salts in the subsurface are limited in quantity because rock weathering is an extremely slow process. Degrees of soil salinization detrimental to plants are uncommon.

Irrigating arid climate soils with surface or groundwater provides a constant new supply of soluble salt. As the irrigation water evaporates and moves through plants to the atmosphere, the dissolved solid content of the soil water increases. Eventually, the increase in soil salt will inhibit or stop plant growth. It is therefore necessary to apply much more water to the fields than required for plant growth to flush salts away from the plant root zone. If the excess water drains easily to the groundwater zone, the groundwater becomes enriched in dissolved solids, which may be detrimental.

If the groundwater table is near the surface, or if there are impermeable soil zones close to the surface, *overirrigation* will not alleviate the problem of soil salinization. This condition requires the installation of subsurface drains that carry the excess soil water and salts to a surface outlet. The problem with this method is that disposing of the salty drain water is difficult. If the drain water is released into surface streams, it degrades the quality of the stream water, adversely affecting downstream users. If the water is discharged into evaporation ponds, it has the potential to seep into the groundwater zone or produce a dangerously contaminated body of surface water, as occurred at the Kesterson Wildlife Refuge in California, where concentrations of the trace element selenium rose to levels that interfered with the reproduction of resident birds.

Robert E. Carver

See also: Agronomy; Composting; Deserts; Halophytes; Hydroponics; Nutrients; Soil; Soil conservation; Soil contamination; Soil degradation; Soil management.

Sources for Further Study

Abrol, I. P., J. S. P. Yadev, and F. I. Massoud. *Salt-Affected Soils and Their Management*. Rome: Food and Agriculture Organization, 1988. A United Nations-sponsored study of soil salinization and its remedies.

Frenkel, H., and A. Mieri, eds. *Soil Salinity: Two Decades of Research in Irrigated Agriculture*. New York: Van Nostrand Reinhold, 1985. A collection of studies on soil salinization in Israel.

Ghassemi, F., A. J. Jakeman, and H. A. Nix. *Salinization of Land and Water Resources: Human Causes, Extent, Management, and Case Studies*. Wallingford, England: CAB International, 1995. Practical assessment of soil salinization.

McClug, Sue. "Salts: Old Dilemma, New Solutions." *Western Water* (September/October, 1995): 4-13. Approaches being taken to address soil salinization in the San Joaquin and Imperial Valleys in California.

SOUTH AMERICAN AGRICULTURE

Categories: Agriculture; economic botany and plant uses; food; world regions

Increasing urbanization in South America raises questions regarding use of both land that is farmed and land that is abandoned.

In South America as elsewhere, landforms, soil, water, climate, and culture interact to produce an arrangement of agricultural production specific to the region. South America can be divided into six general landform regions: the Andes Mountains, the plateaus of the interior of the continent, the river lowlands, the coastal lowlands, the *tierra templada*, and the *tierra fria*.

Andes Mountains

The Andes Cordillera reaches from Venezuela in the north to Tierra del Fuego at the south tip of the continent. Some Andean peaks exceed 20,000 feet (6,100 meters) in height. The mountain soils are rocky and steeply graded. This makes farming difficult but not impossible. Farmers use terrace systems, building up steplike fields carved into the side of a mountain. Because of the limitations of the soils and the difficulty in farming them, the region supports only subsistence agricultural settlements. People in the highlands grow small plots of corn, barley, and especially potatoes on the high-altitude soils.

Highland Growing Regions

In the central eastern region of Brazil are the Brazilian Highlands. To the north of the Amazon basin lies another plateau region called the Guiana Highlands. Both plateaus, which are not much higher than 9,000 feet (2,743 meters), are old geologic structures with relatively rough surfaces to farm. The Brazilian Highlands constitute the world's primary coffee-growing region. More than one-third of the world's coffee is grown there, along with soybeans and oranges. The Guiana Highlands, which stretch through southern Venezuela and Guyana, are covered in savannas (grasslands with trees and shrubs). People there use slash-and-burn agriculture to grow corn and rice.

River Lowlands

Some of the most remote regions in South America are its fastest-developing areas. The northern interior of Brazil in the Amazon basin is covered with the world's largest tropical rain forest, the size of the forty-eight contiguous United States. Within some areas of this rich habitat, gold and huge iron ore deposits have been discovered. Development programs have encouraged mining, hydroelectric projects, ranching, and farming there. This has produced a need for clearing the forests and caused a large migration of people into this otherwise remote, isolated region. It has been estimated that 30,000 square miles (78,000 square kilometers) of tropical rain forest were destroyed annually in the 1990's.

The rain forest is not a highly productive growing region for crops. Soils there are thin and have few nutrients. To grow crops and grasses for cattle, *slash-and-burn agriculture* is practiced. Trees and brush on the land are cut down and burned, and the ashes enrich the soils for a time. This allows for limited production of crops such as corn and rice, but after about three years the nutrients are washed out of the soil by the heavy tropical rains. Farmers and ranchers move on and find new forest area to cut, and the cycle goes on. Rain forests cover only about 5 percent of the total land area of the earth but contain more than half the different types of plants and animals on earth. These forests provide lumber products, medicines, and food.

The Orinoco River drains the Guiana Highlands and the *llanos* region of Venezuela, flowing into the Atlantic Ocean. The *llanos* form a large, expansive plain of grassland between the Andes in Venezuela and the Guiana Highlands. Soils are flooded in this area during the rainy seasons, providing for rich grass development and the support of large cattle ranches.

Selected Agricultural Products of South America

The Amazon River drains the north and western region of Brazil. Although more water moves through the Amazon River than any other river in the world, its location in dense tropical forests under a humid tropical climate makes agriculture there a challenge. Amerindians of the region practice subsistence agriculture, growing yams and bananas and raising some small animals. Cassava, or manioc (a root crop from which tapioca is made) and sugarcane are also grown there in slash-and-burn fashion.

The Paraná, Paraguay, and Uruguay Rivers, south of the Amazon lowlands, dissect a great grassland region known as the Pampas and a forested region known as the Gran Chaco, which extends to northern Argentina. The Pampas is not unlike the Midwest of North America. It is a huge grassy plain nearly 400 miles (640 kilometers) long in the central part of Argentina. Agriculture is the primary industry in Argentina and encompasses 60 percent of the country's land. Large *estancia* (cattle ranches) are common in the region. The soils are well suited to wheat and other grains as well as alfalfa, a grass grown for cattle and horse feed. The level character of the land makes it easy to work, but it is difficult to drain and is prone to flooding.

Paraguay's agricultural zones are divided by the Paraguay River. Tobacco, rice, and sugarcane grow to the east of the river in the more humid climates. To the west, in the drier climates and the Chaco region, is the unique growing area of the Quebracho Forest. Quebracho is a hardwood that grows only in the Chaco. The wood contains tannin, which is used to produce tannic acid, a chemical used for tanning leather. This area is suited for cattle, but because of its rocky and steep topography, the population of goats and sheep rises as one moves farther west toward the Andes. Irrigated with mountain streams from the eastern slopes of the Andes, grapes also grow well and are made into wine.

Coastal Lowlands

The coastal lowlands, up to an altitude of about 2,500 feet (750 meters), encompass an area known as the hot land (*tierra caliente*). Temperatures there average between 75 and 80 degrees Fahrenheit (22 and 24 degrees Celsius), and plantation agriculture abounds. Plantations are huge commercial farming operations that grow large quantities of crops that are usually sold for export. Because of their easy access to port facilities, coastal lowlands historically have been linked with the markets of Europe and North America.

The banana is one of the best-known examples of a plantation crop. It grows well in the wet, hot climate of this zone and has been cultivated there for U.S. and European markets since 1866. In the 1990's, more bananas were traded on the international fruit market than any other commodity. In South America, Ecuador and Brazil are the leading banana producers, with Colombia third. Cacao, the

PhotoDisc

Farmers sell local produce at a market in Ecuador.

bean pods from which cocoa and chocolate are made, is also grown on plantations in this zone. The largest producing area for cacao is Ghana in West Africa, but Brazil and Ecuador are fifth and sixth in world yearly production.

Although sugarcane is grown in almost every country in South America, it does well as a plantation crop in the lowlands of eastern Brazil, the world's largest exporter. There the crop is not used just to produce sugar but also to produce ethanol for gasohol, an alcohol-based gasoline that fuels more than half the automobiles in Brazil. This type of commercial agriculture has made agricultural business the fastest-growing part of the Brazilian economy. Yams, cassava, and other root crops used as staple foods also grow well in this humid, hot climate.

Tierra Templada

Just above the *calienta* zone lies a zone of cooler temperatures, the *tierra templada*, that extends to about 6,000 feet (1,850 meters). Temperatures there range from 65 to 75 degrees (17 to 22 degrees Celsius), and the commercial crop that dominates the landscape is coffee. It is grown on large plantations called *fazendas*. Brazil, the largest South American coffee producer, exports about one-quarter of the world's coffee, producing nearly forty million bags of about 132 pounds each (60 kilograms) annually. Colombia is South America's second-largest coffee producer. Coffee was once the leading export from Colombia, but as a result of a coffee-worm infestation and lower world prices, other products are taking the lead.

Other commercial crops from this zone include fresh fruit. In the central valley of Chile, grape vineyards and apple orchards have emerged. Produce from this region enters stores in the United States and elsewhere as the growing seasons of the domestic products are finishing up. Chilean grapes, apples, peaches, and plums are now sold worldwide. Corn and wheat are also grown in this zone. These staple foods are produced for local consumption and sold only at local markets.

Tierra Fria

A cooler climate, the *tierra fria*, extends from 6,000 feet (1,850 meters) to about 15,000 feet (4,570 meters). Average temperatures there range from 55 to 64 degrees Fahrenheit (12 to 17 degrees Celsius). This zone extends throughout the Andes and can maintain only that plant life that can withstand the limited soils and the cold conditions. In this region of subsistence-type farming, crops and animals are grown and raised mostly for family use.

In the lower reaches of this zone, barley grows well because it requires a short growing season. In the cooler portions of this zone, the potato began. A tuber, the potato can flourish in cold conditions with moderate moisture. The loose soils of this zone are perfect for its production, but it requires a considerable amount of cultivation. Although the potato is used throughout South America, potato production for the continent constitutes only about 5 percent of the world total production. Alpacas, a type of goat, and llamas are also raised there.

M. Mustoe

See also: Central American agriculture; Corn; Ethanol; Grains; Grasslands; Rain-forest biomes; Rain forests and the atmosphere; Rangeland; Rice; Savannas and deciduous tropical forests; South American flora; Wheat.

Sources for Further Study

Caufield, Catherine. *In the Rainforest: Report from a Strange, Beautiful, Imperiled World*. Chicago: University of Chicago Press, 1991. A comprehensive study of the world's rain forests looks at these threatened resources from historical, political, economical, and biological viewpoints. The afterword includes addresses of organizations working to save rain forests.

Grigg, David B. *An Introduction to Agricultural Geography*. 2d rev. ed. London: Routledge, 1995. Provides a comprehensive introduction to agriculture in developing and developed nations, describing both human and environmental issues. Covers the physical environment, economic behavior and demands, institutional, social and cultural influences, and the impact of farming upon the environment.

Stanfield, Michael Edward. *Red Rubber, Bleeding Trees: Violence, Slavery, and Empire in Northwest Amazonia, 1850-1933*. Albuquerque: University of New Mexico, 1998. Explores the complex transformation of northwestern Amazonia by the rubber boom from 1850 to 1933. Examines the historical myths and realities of northwest Amazonia before its incorporation and then shows how the Indians and environment were radically altered by the rubber boom and international trade.

SOUTH AMERICAN FLORA

Category: World regions

South America is the most diverse continent in terms of flora, primarily because of its location and geography.

South America's floristic diversity is increased by its high mountains, especially the Andes Mountains, which extend from north to south along the western part of the continent for much of its length. South America has such diverse biomes as tropical *rain forests*, tropical *savannas*, extremely dry deserts, temperate forests, and alpine *tundra*. The largest of these biomes are deserts, savanna, and tropical forest. With the rapid rate of deforestation in places like the Amazon basin, some plants may become extinct before being cataloged, let alone studied.

The *subtropical* desert biome is the driest biome in South America and is considered the driest desert in the world, with an average annual precipitation of less than 0.25 inch (4 millimeters). The desert biome is restricted primarily to the west coast of South America from less than 10 degrees south of the equator to approximately 30 degrees south. Dry conditions prevail from the coast to relatively high elevations in the Andes. The Atacama Desert, in northern Chile, and the Patagonian desert, in central Chile, are the most notable South American deserts. Smaller desert regions also occur in the rain shadow portions of the Andes.

Next on the moisture scale are the *savanna* biomes. Savanna occurs in two distinctly different areas of South America. The largest savanna region includes three distinctive regions: the cerrado; the Pantanal; and farther south, in southern Brazil, Uruguay, and northern Argentina, the grassland called the Pampas. The other savanna region, the *llanos*, is found in lower-elevation areas of Venezuela and Colombia.

Although a few of the forests in South America are dry, most are rain forests, receiving annual precipitation from 79 inches to 118 inches (2,000-3,000 millimeters). The Amazon rain forest, the world's largest, accounts for more than three-fourths of the rain-forest area in South America. One of the most species-rich areas of the world, it is being rapidly destroyed by logging, ranching, and other human activities. Smaller rain forests are located along the southeastern coast of Brazil and in the northern part of Venezuela.

Covering much smaller areas are a small *mediterranean* region in central Chile characterized by cool, wet winters and warm, dry summers. In the far south of Chile and Argentina is a small area of temperate forest, becoming alpine tundra in the far south. Temperatures are relatively cool and mild year-round (except in the far south, where it can be extremely cold in the winter.

Plants of the Subtropical Desert

In the Atacama Desert, one of the world's driest, some moisture is available, but it is limited to certain zones. Coastal regions below 3,280 feet (1,000 meters) receive regular fog (called *camanchacas*). Rainfall is so low in the Atacama Desert that even cacti (which normally store water) can hardly acquire enough water from rainfall alone, so many plants, including *bromeliads*, receive a portion of their water from the fog. At midelevation areas there is no regular fog; thus, there is almost no plant cover. At higher elevations, the rising air has cooled sufficiently to produce moderate amounts of rainfall, although the vegetation is still desertlike. Shrubs typically grow near streambeds, where their roots can reach a permanent source of water.

The Atacama Desert often appears barren, but when a good dose of moisture becomes available *ephemerals* change its appearance, seemingly over-

PhotoDisc

An aerial view of the rain forest in Guyana. In addition to rain forests, South America is home to flora of such diverse biomes as tropical savannas, extremely dry deserts, temperate forests, and alpine tundra.

night. Ephemerals are typically annuals that remain dormant in the dry soil as seeds. When moisture increases, they quickly germinate, grow, flower, and set seed before dry conditions prevail again. In the days and weeks following a good rain, many grasses appear and provide a backdrop for endless varieties of showy flowers, many endemic to (found only in the region of) the Atacama Desert. Among the showier flowers are species of *Alstroemeria* (commonly called irises, although they are actually in the lily family) and *Nolana* (called pansies, although they are members of a family found only in Chile and Peru).

Conditions in the Patagonian desert are less harsh. The vegetation ranges from tussock grasslands near the Andes to more of a shrub-steppe community farther east. Needlegrass is especially abundant throughout Patagonia, and cacti are a common sight. In the shrub-steppe community in the eastern Patagonian desert, the shrubs *quilembai* and the cushionlike *colapiche* are common. Where the soil is salty, saltbush and other salt-tolerant shrubs grow.

Plants of the Tropical Savanna Biome

The cerrado region of east central Brazil and southward is not only the largest savanna biome of South America but also one of the most romanticized of the world's savannas. As in the Old West of North America, the grasslands of Brazil have cowboys who traditionally have used the cerrado for farming and cattle ranching. With ever-increasing pressure from agriculture, the cerrado is now under attack in various ways. Extensive fertilization, associated with modern agriculture, planting of trees for timber production, and the introduction of foreign species, especially African grasses, have all begun to change the cerrado. Frequent fires also have taken their toll. The cerrado contains more than ten thousand species of plants, with 44 percent of them endemic. As much as 75 percent of the cerrado has been lost since 1965, and what remains is fragmented. A number of conservation groups are trying to save as much as possible of what remains.

Two other savanna regions farther south are the Pantanal and the Pampas. Although the Pantanal is a savanna, during the rainy season it becomes a *wetland* and is a haven for aquatic plants. Later, the Pantanal dries out and grasslands appear in place of the water. This unique area is under attack by a variety of human activities, including navigation and artificial drainage projects, mining, agriculture, and urban waste.

The Pampas, like the great prairies that once covered central North America, is composed almost solely of grass. Trees and shrubs grow near bodies of water, but everywhere else grass predominates. Cattle ranching and wheat and corn farming are the primary occupations of the area and are thus the primary threat. Because the area is farther south than the Pantanal, it has a more temperate climate. Pampas grass from this area has been exported as an ornamental plant.

The last major savanna region is the *llanos*, located at lower elevations in the drainage area of the Orinoco River in Venezuela and Colombia. This area has pronounced wet and dry seasons. At the lowest elevations, treeless grasslands persist after the water from the rainy season subsides. On the higher plains is a scattering of smaller trees. The mauritia palm can also be found here in poorly drained areas.

Plants of the Tropical Forest Biome

The Amazon rain forest is the largest contiguous rain forest in the world. It is so large and so lush with tree growth that its *transpiration* is actually responsible, in part, for the wet climate of the region. Plant diversity is so great here that no comprehensive plant guide currently exists for many parts of the Amazon rain forest. Of tens of thousands of plant species, a large number have never been described.

This one-of-a-kind botanical treasure is being destroyed at a rapid pace of between 5,000 and 10,000 square miles (13,000-26,000 square kilometers) per year. The causes for this destruction are primarily logging, agriculture, and cattle ranching. A common practice for preparing an area for cattle ranching or farming has been to simply burn the forest, not even necessarily logging it first, and then to allow the grass and other vegetation or crops grow in its place. The soils of the rain forest are so poor, however, that this practice usually depletes the soil within a few years, and the land becomes a useless wasteland. Mining and oil drilling have also taken their toll.

The Amazon rain forest is an extremely complex biome. The main plant biomass is composed of trees, which form a closed canopy that prevents much of the sunlight from reaching the forest floor. Consequently, the forest floor has little herbaceous growth, and most smaller plants tend to grow as *epiphytes* on the branches and trunks of trees. Common epiphytes in the Amazon rain forest include orchids, bromeliads, and even some cacti. There is a large diversity of bromeliads, ranging from small, inconspicuous species to larger species, such as tank bromeliads that can collect significant amounts of water in their central whorl of leaves. The water in these plants can contain a whole miniature ecosystem, complete with mosquito larvae, aquatic insects, and frogs. Ferns are another significant member of the epiphyte community. Some larger species of ferns, often called tree ferns, also grow in the understory. Lianas, or vines, are a prominent component as well.

The trees that form the canopy are stratified into three fairly discrete levels. The lowest two levels are the most crowded, and the highest level comprises extremely tall trees, often referred to as *emergent trees* because they stand out randomly above the fairly continuous lower two layers. Many of the tallest trees are buttressed at the base, an adaptation that seems to give them greater stability. Beneath the canopy, there are some smaller palms, shrubs, and ferns, but they only occur densely where there is a break in the canopy that allows in greater light.

Some rain-forest species of trees are well known, primarily because of their economic value. A favorite tree for use in furniture is the mahogany. Because its wood is highly prized, many species of mahogany are becoming rare or extinct. The South American rain forests are also the original source of rubber. Brazil had a monopoly on rubber until seeds were smuggled out and planted in Malaysia. Synthetic rubber has now replaced natural rubber for many applications. Another popular tree is the Brazil nut tree, an abundant food source that has been exploited by suppliers of mixed nuts. The native cacao tree produces fruits from which cacao beans are extracted, then processed to make chocolate.

Every year during the rainy season, the lowest elevation areas of the Amazon rain forest are flooded with several feet of water, which recedes after a few months. The trees flourish in this flooding cycle. A few even have unique adaptations, such as producing fruits that are eaten by fish, thus assuring the spread of their seeds. The flooding can be so extensive in some areas that the water reaches the lower parts of the canopy.

Coastal tropical rain forests also occur in northwestern and southeastern South America. There are a high number of endemic species in each of these forests. Some tree species are so rare that they may be found in an area of only a few square miles and nowhere else. Where the tropical rain forest meets the ocean, mangrove trees have become adapted to the tidal environment. Mangroves have prop roots, which make the trees look like they are growing on stilts. They also frequently have special root structures that extend above the water at high tide and allow the roots to breathe. Mangrove trees are also extremely salt-tolerant.

Plants of the Mediterranean and Temperate Forest Biomes

One of the world's five mediterranean climate regions is found in central Chile. This climate is characterized by warm, dry summers and cool, wet winters. The vegetation, called *matorral*, is composed primarily of leathery-leaved, evergreen shrubs that are well adapted to the long summer drought. The matorral is the only mediterranean area that has bromeliads. At lower elevation areas, somewhat inland, many of the shrubs are drought-deciduous; that is, they drop their leaves in the summer. In more inland parts of this biome, the espino tree is common.

Because South America extends so far south, it actually has a small region containing temperate forests. These forests range from temperate rain forest to drier temperate forest, and in all cases are typically dominated by southern beeches. The undergrowth is dominated by small evergreen trees and shrubs. Fuchsias, which are valued the world over for their showy flowers, are common in the undergrowth. Although not rich in species, the temperate rain forests of southern South America can be lush. In the far south, before the extreme climate restricts the vegetation to alpine tundra, a region of elfin woodlands predominates. These woodlands can be nearly impenetrable, with the densest growth often associated with patches of tall bamboo.

Bryan Ness

See also: Deforestation; Deserts; Grasslands; Logging and clear-cutting; Mediterranean scrub; Rainforest biomes; Rain forests and the atmosphere; Rangeland; Rice; Savannas and deciduous tropical forests; South American agriculture; Sustainable forestry.

Sources for Further Study

Gentry, Alwyn H., ed. *Four Neotropical Rainforests*. New Haven, Conn.: Yale University Press, 1993. A comparative study of four neotropical rain forests. Six sections describe the rainforest sites and forest dynamics.

Gentry, Alwyn H., and Adrian G. Foryth. *A Field Guide to the Families and Genera of Woody Plants of Northwest South America: Colombia, Ecuador, Peru*. Washington, D.C.: Conservation International, 1996. Unlike many field guides, which rely for their identifications on flowers and fruits that are present only during certain seasons, this book focuses on characteristics such as bark, leaves, and odor that are present year-round. The guide is filled with clear illustrations, step-by-step keys to identification, and a wealth of previously unpublished data.

Kircher, John. *A Neotropical Companion*. 2d ed. Princeton, N.J.: Princeton University Press, 1997. An introduction to the American tropics, the lands of Central and South America, their rain forests and other ecosystems, and the creatures that live there. Revised and expanded to incorporate the abundance of new scientific information that has been produced since the book was first published in 1989.

Maxwell, Nicole. *Witch-Doctor's Apprentice: Hunting for Medicinal Plants in the Amazon*. New York: Citadel Press, 1990. A memoir of Maxwell's travels in the Amazon with the triple goal of conserving the jungle, improving the lot of humanity, and having a great deal of fun.

Plotkin, Mark J. *Tales of a Shaman's Apprentice: An Ethnobotanist Searches for New Medicines in the Amazon Rain Forest*. New York: Penguin Books, 1994. Plotkin's account of wandering through the Amazonian jungles focuses on local knowledge about plants, whose uses range from the mundane to the magical. The rain forests of the world, Plotkin notes, are a great natural resource and intercultural pharmacy of cures and therapies both known and yet unvisited.

Rice, Richard E., Raymond E. Gullison, and John W. Reid. "Can Sustainable Management

Save Tropical Forests?" *Scientific American* 276, no. 4 (April, 1997): 44-49. In a quest to reconcile conservation with the production of tropical timber, a study finds that sustainability of the forests is problematic.

Young, Kenneth R., and Blanca Leon. "Nature on the Rebound." *Americas* 51, no. 1 (January/February, 1999): 32-34. Botanical exploration in Rio Abiseo National Park is both difficult and rewarding; the beauties of nature in the park are discussed.

Zimmerer, Karl S. "The Ecogeography of Andean Potatoes." *Bioscience* 48, no. 6 (June, 1998): 445-454. Zimmerer discusses how ecogeography offers key insights about the viability of agricultural biodiversity by analyzing the occurrence and adaptive nature of diverse types of primary crops. The diversity and farming of Andean potatoes is studied as an example of food-plant ecogeography.

SPECIES AND SPECIATION

Categories: Classification and systematics; ecology; evolution

A species is any group of organisms recognized as distinct, with members able to interbreed with one another and produce fertile offspring. Speciation is the evolutionary process whereby a species comes into existence.

Species are distinct kinds of organisms, in that the organism can be recognized as different from other kinds of similar organisms by a combination of characteristic shapes, sizes, behaviors, physiology, or other attributes. For instance, white oaks can be recognized as different from other oak species by growth form and habitat, along with a combination of leaf, fruit, and bark characteristics. To be useful as a diagnostic feature, a characteristic must lend itself to measurement and remain relatively constant generation to generation. Such a hereditary pattern implies that members of a species share a common pool of genetic information.

Although most field guides and keys for species identification are based largely on measurable morphological traits, the *biological species concept*, attributed to Ernst Mayr in the 1940's, defines species as groups of interbreeding populations reproductively isolated from other such groups. Sometimes a greater range of variation can be observed in large, dispersed populations than is found between similar but reproductively isolated species. What, then, determines when a species is formed?

Unfortunately there is no clear-cut answer. Investigators may use different criteria or assign variable levels of importance to characteristics; therefore, a population of plants may be assigned species status by one expert and varietal status by another. While this is confusing to those searching for a name, the problem is indicative of the dynamic and changing nature of life. Because evolution is an ongoing process, some groups of plants are expected to be in transition.

The processes that contribute to variation in large, sexually reproducing populations also are responsible for the origin of species. Isolation and selection of genetically based variation are the only additional requirements. Speciation generally is conceived to involve the separation, isolation, and divergence of a genetic pool of information. Plants that share a common pool of genetic information are split into two pools that remain isolated until identifiable genetic differences accumulate. Classifying a segment of this pool as a species requires recognition of significant genetic differences and the relative isolation of the population.

Allopatric Speciation

The physical separation of a gene pool into populations that are geographically or spatially isolated is termed *allopatric speciation*. The physical separation could be the result of continental drift, uplift or subsidence of landmasses, glaciers, flooding, or radical dispersal of population members. Dispersal of a few fertile population members to an isolated island is a good example. The resulting gene pool, although small, is immediately isolated. For example, if the fruiting inflorescence of a common grass were transported on the struts of an airplane to a small, isolated island, any plants resulting from germination of those seeds would represent a new population, now isolated from the parent population on the distant landmass where the seeds originated. A drastic change in gene frequency could, and likely would, result: A gene present in one out of one thousand in the original population might, in the new, isolated location and within the smaller gene pool, increase in frequency from 0.05 percent to 25 percent if present in one of two plants forming the invading population. This sudden change in frequency is referred to as the *founder's effect*. When such a population is introduced into a new environment, it may rapidly change and give rise to a number of additional new species. The latter process is known as *adaptive radiation* and is credited for the assemblage of unique species often found in isolated areas.

Sympatric Speciation

It is possible for segments of a parent gene pool to diverge without spatial separation, ultimately becoming reproductively isolated. This process is termed *sympatric speciation*. Examples of situations that may lead to sympatric speciation include disruption of pollination that could result from different flowering times among members of a population. Such differences could be caused by variation in soil, moisture, or exposure. One example of this process is associated with the genus *Achillea*, or milfoil. In the late 1940's, investigators separated clumps of two species of milfoil to produce genetically identical clones that were subsequently transplanted to different elevations in the Sierra Nevada. Plants grown during this process exhibited a wide range of morphological variation. Not only did they look different; they flowered at different times. Thus, even genetically identical individuals of one species can be reproductively isolated if the timing of flowering precludes visitation by a common pollinator.

Polyploidy

Genetic variation can occur rapidly through an increase in chromosome number. Although recombination of genes brought about by sexual reproduction provides the greatest amount of the observable variation in a population, abrupt and large-scale change is also associated with a process called *polyploidy*. Through polyploidy, a plant can have its entire set of chromosomes multiplied. If the increase in chromosome number is brought about when chromosomes fail to separate after they duplicate during meiosis, an individual's number of chromosomes may double. That process is known as *autopolyploidy*. An increase in chromosome number associated with hybridization resulting from combination of two separate sets of chromosomes is termed *allopolyploidy*. Although typically sterile, an allopolyploid may duplicate its combined set of chromosomes, resulting in a fertile autoallopolyploid.

Although polyploidy occurs naturally in many plants, the results so frequently are associated with desirable changes in the bloom of ornamentals that polyploidy is often deliberately induced by chemical treatment. It is estimated that one-half or more of all flowering plant species may have arisen through some form of polyploidy.

Hybridization

Offspring produced from the interbreeding of related species, *hybridization*, may be either sterile or fertile. If the offspring are sterile, the genetic pools of parental species are not altered, because genes are not able to flow between the two. If the offspring are fertile, as are hybrids of crosses between North American and European species of sycamores, interbreeding among fertile hybrids and parent populations can provide a bridge for the merging of genetic information. With time, the flow of such genetic material can lead to an increase in variation, a decrease in interspecific differences, and an interesting taxonomic problem as to when the two parent species should be reclassified as one. Currently geographic isolation maintains the genetic integrity of these parental species.

Recombination Speciation

Although the original source of variation is permanent change in DNA (mutation), *recombination* of genetic material through sexual reproduction accounts for the vast majority of variation between generations. The result of this variation, acted upon by natural selection, leads to changes in the frequencies of genes within a population. The latter is often stated as a definition of evolution. Rapid selection of recombinants has been proposed as the method of speciation by which the anomalous sunflower, *Helianthus anomalus*, was produced from hybrids of two other species of sunflower, *Helianthus annulus* and *Helianthus petiolaris*. Studies, in fact, have duplicated the proposed process. After several generations of natural selective pressure, experimentally produced hybrids of *Helianthus annulus* and *Helianthus petiolaris* were demonstrated to be the genetic equivalents of naturally occurring *Helianthus anomalus*.

Sterile Hybrids

Plant hybrids often are sterile because newly combined hereditary material is so different that chromosomes will not pair during meiosis. Viable gametes, therefore, are not produced. This apparent genetic blind alley does not always translate into a lack of evolutionary success, however. Many sterile hybrids demonstrate an ability to reproduce *apomictically*, meaning they reproduce vegetatively rather than sexually. Species of blackberry, aspen, and many grasses reproduce by apomictic methods. Some, like dandelion, may have populations that are fertile and populations of infertile hybrids that produce seed with embryos that were produced asexually. Apomixis obviously does not promote variation, but it may permit expansion of populations where there is little chance for cross-pollination, as in areas of high disturbance or environmental stress.

Reproductive Isolation

Once a gene pool has separated and diverged into populations identifiable as separate species by genetic or morphological traits, the resulting segments of the pool must remain separated. As long as they are separate and distinct, populations can be labeled as separate species.

Barriers that prevent the flow of hereditary material may be classed as prezygotic or postzygotic. *Prezygotic mechanisms* prevent successful fertilization and include geographic isolation, temporal isolation (flowering at different times), mechanical isolation (flowers structurally different), and incompatible gametes. *Postzygotic mechanisms* prevent production of fertile adults and include hybrid inviability (offspring that do not live to sexual maturity), hybrid sterility, and offspring of hybrids that are weak.

John F. Logue

See also: Adaptations; Angiosperm evolution; Coevolution; Evolution: convergent and divergent; Evolution: gradualism vs. punctuated equilibrium; Evolution of plants; Gene flow; Genetic drift; Genetics: mutations; Hybrid zones; Hybridization; Nonrandom mating; Population genetics; Reproductive isolating mechanisms; Selection; Systematics and taxonomy; Systematics: overview.

Sources for Further Study

Moore, Randy, et al. *Botany*. 2d ed. New York: WCB/McGraw Hill, 1998. This basic textbook devotes an entire chapter to speciation. The preceding chapter on evolution provides good background. It is well organized and contains references to Internet sites, suggested readings, ample photographs, diagrams, and color drawings.

Raven, Peter H., Ray F. Evert, and Susan E. Eichhorn. *Biology of Plants*. 6th ed. New York: W. H. Freeman/Worth, 1999. The chapter on evolution and speciation contains examples, diagrams, photographs, and clear illustrations.

Stearns, Stephen C., and Rolf F. Hoekstra. *Evolution: An Introduction*. New York: Oxford University Press, 2000. Provides a separate chapter on speciation plus many cross-referenced sections in the index. Easy to understand for most introductory students, with examples and illustrations as well as an excellent bibliographical reference section.

Townsend, Colin R., et al. *Essentials of Ecology*. Malden, Mass.: Blackwell Science, 2000. A general ecology textbook that includes good sections on speciation and evolution within species. Well written text does not require a technical background. Plant and animal examples are used.

SPERMATOPHYTA

Categories: *Plantae*; taxonomic groups

The Spermatophyta *are those members of the* Tracheobionta *that produce seeds. At one time this category was considered to be natural, composed only of closely related groups with homologous seeds. It is now known that seeds evolved many different times during evolutionary history and that different groups of extant seed plants may or may not be closely related to other groups of extinct or extant spermatophytes.*

Scientists once thought that the seed plants were monophyletic, derived from a single common ancestor. It is now known they are polyphyletic, derived from more than one ancestral group. The distinctions can be seen in the variations among seed structures and functions.

Seed Structure

A seed is a complex reproductive structure composed of multiple generations of tissue that develop from an *ovule*. The outer layer of the seed is usually composed of thick-walled cells of the parental sporophyte that form a *seed coat*. Internal to the seed coat is typically a layer of stored food, which may consist of *megagametophyte* tissue or of a unique tissue of a new generation called *endosperm*. At the center of the seed is a new embryonic plant that emerges to become the seedling when the seed germinates. Scientists recognize two basic types of seed plants: gymnosperms and angiosperms. Gymnosperms are plants with "naked seeds." That is, the ovule is exposed to the environment and is not covered by a protective layer. In angiosperms, by contrast, the ovule is contained within a specialized structure, the *carpel*.

The outer layer of a seed, the seed coat, develops from the outer layers of the ovary, the *integument*. The integument is composed of sporophyte tissue of the parent plant. As the integument develops, it forms an enlarging sheath around the central ovule tissue, the *nucellus*. The integument does not completely enclose the nucellus but leaves a small pore, the *micropyle*, through which pollen grains or the pollen tube will move prior to fertilization. A *megaspore mother cell* within the nucellus undergoes meiosis to produce four megaspores. In gymnosperms, only one megaspore is functional, and it develops into a female gametophyte containing archegonia. In angiosperms, one, two, or all four of the megaspore nuclei may contribute to the female gametophyte.

In gymnosperms, a single egg forms within each archegonium. Pollen grains are drawn into the micropyle, and the pollen germinates to form a *pollen tube*. In most gymnosperms, the pollen tube grows through the nucellus until it reaches the archegonium, where the sperm are released to fertilize the egg. The fertilized egg, or *zygote*, grows into an *embryo*, a new sporophyte generation. Meanwhile, the integument enlarges to close the micropyle and hardens to form the seed coat. The remaining tissue of the female gametophyte functions as a store of food to nourish the embryo until the seed germinates. Because multiple archegonia are formed in the female gametophyte, more than one egg may be fertilized within a single ovule, but usually only a single embryo grows to full size.

In angiosperms the entire female gametophyte usually consists of a seven-celled *embryo sac*. Three cells, the egg and two *synergids*, lie near the micropyle. Three cells, the *antipodals*, form on the end opposite the micropyle. The remaining central cell contains two nuclei, the *polar nuclei*. Pollen germinates on the outside of the carpel in a specialized region called the *stigma*. The pollen tube grows down through the carpel tissue toward the ovules and eventually grows into the micropyle and through the nucellus of one ovule until it reaches a synergid. The sperm nuclei are discharged into the synergid cell, which subsequently degenerates. One sperm goes on to fertilize the egg, while the second fertilizes both polar nuclei. The fertilized egg develops into a diploid embryo in a manner similar to that of gymnosperms. The product of the other sperm and the two polar nuclei is a triploid cell, the primary endosperm. The endosperm cell also undergoes mitosis to form a multicellular triploid endosperm tissue, which functions directly as a stored food reserve in monocots. In dicots, the endosperm typically degenerates, and the nutrients are absorbed by the cotyledons of the embryo, which then serve as a food reserve.

Seed Function

Reproduction by means of seeds confers a number of advantages to seed plants for colonizing terrestrial habitats. First, the seed coat provides physical protection from predators and from dehydration. In many cases seeds, protected by the seed coat, pass unharmed through the digestive tracts of animals who have eaten them. In fact, the physical abrasion of chewing and stomach acid action associated with passage through an animal's digestive tract may be necessary to soften the coat and permit seed germination. The seed coat may also promote *dormancy* through chemical inhibition or by limiting water uptake. In this way the seed coat helps the seed survive periods of environmental stress, such as dry or cold seasons during which plant growth cannot occur. During the period of dormancy, seeds can be dispersed to new areas by a number of means, such as lofting through the air, floating on water, or being transported in an animal's digestive tract or on its body.

The seed will *germinate* only when conditions are suitable for growth. Stored food provides a nutrient supply to support early growth of the seedling. Most seeds germinate belowground, where the environment is moister and the temperature more uniform than on the soil surface. Thus early growth occurs before the seedling is exposed to light and is able to photosynthesize. Finally, the typical embryo in the seed has already formed root and shoot primordia, which are ready to function as soon as they absorb water and begin to grow.

Marshall D. Sundberg

See also: Angiosperm evolution; Angiosperm life cycle; Angiosperm plant formation; Angiosperms; Conifers; Cycads and palms; Dormancy; Flower structure; Flower types; Germination and seedling development; Ginkgos; Gnetophytes; Gymnosperms; Plant life spans; Pollination; Reproduction in plants; Roots; Seeds; Stems; *Tracheobionta*.

Sources for Further Study

Bell, Peter, and Alan Hemsley. *Green Plants: Their Origin and Diversity*. 2d ed. New York: Cambridge University Press, 2000. Includes a current interpretation of the phylogenetic relationships of all the green plants, including the vascular plants.

Foster, Adriance S., and Ernest M Gifford, Jr. *Morphology and Evolution of Vascular Plants*. New York: W. H. Freeman, 1996. This remains the classic text on the comparative structure of vascular plants.

SPICES

Categories: Economic botany and plant uses; food

Spices are a group of plant products used to impart flavor to foods. Unlike herbs, spices are generally derived not from the leafy and other green portions of the plant, but rather from the seeds, fruits, flower parts, bark, or rhizomes. Spices are also generally distinguished from herbs by the greater strength, intensity, or pungency of their flavors.

For thousands of years, humans have added spices and herbs to their foods in order to improve taste. Before modern refrigeration and canning, the first signs of spoiling could be masked by adding parts of aromatic plants—spices and herbs. In modern times, spices and herbs are added only to enhance the flavor of foods.

Spices are distinguished from herbs based upon what part of a plant is used and the way it is prepared. Generally spices are flavorings from dried seeds, fruits, or flower parts. A few spices are from dried bark and rhizomes. Spices can be used as intact dried fruits or seeds, but most often they are ground into a powder used in food preparation. Herbs, by contrast, are typically flavorings from leaves or other green parts of plants. Spices are used in a wide variety of dishes, from sweet to savory: desserts, breads, pickles, fruits, meats, and vegetables. Different spices are characteristic of different ethnic and regional cuisines.

In the late twentieth century, with the rise in global communications and travel as well as mass movement of migrant populations, the "typical" spices of North American cuisine increased and became more diverse, particularly in regions where many different cultures came together.

Allspice, from the myrtle family, is prepared from the dried fruit of a West Indian tree, *Pimenta dioica*. Whole allspice is used in preparing pickles and relishes and to flavor meats such as pot roast; the powder is used in cakes, cookies, pies, and mincemeat.

Anise seeds, from the celery family, come from the herbaceous plant anise, *Pimpinella anisum*, from the Mediterranean region. Whole anise seeds are used as flavorings in stews, pot roasts, and some Chinese meat dishes. Crushed or powdered anise seeds are used as flavorings in Italian cookies, cakes, and liqueurs, as well as German breads.

Cardamom, from the ginger family, is prepared from the seed pods of large herbaceous cardamom plants, *Elettaria cardamonum*, native to India. The dried pods are crushed and used in many Indian and African meat and vegetable dishes, in rice dishes, and in Scandinavian breads.

Cinnamon, from the laurel family, is made from the dried bark of several species of small tropical trees in the genus *Cinnamomum*. The bark is cut into 3- or 4-inch lengths and is permitted to curl. These dried "cinnamon sticks" often flavor hot cider and other beverages as well as meat stocks. Ground cinnamon is widely used to flavor cookies, pies, cakes, sweet rolls, and candies and is added to Mexican meat dishes.

Cloves, from the myrtle family, are the dried, unopened flower buds of a tropical tree, *Eugenia caryophyllata*, possibly native to the Moluccas Islands. The dried buds are ground for use in cookies, gingerbread, pies, and cakes. They are used whole in pickle brines or to flavor ham and other meats.

Coriander, from the celery family, comes from the dried seeds of the herbaceous plant *Coriandrum sativum*, a Mediterranean native. The seeds are used either whole in pickles and preserves or ground in cookies and puddings.

Cumin, from the celery family, comes from the dried seeds of the herbaceous plant *Cuminum cynimum*, also native to the Mediterranean region. Powdered cumin is widely used to flavor stews, fish and meat dishes, curries, and Mexican dishes, and it is one of the main ingredients of commercial chili powder.

Ginger, from the ginger family, is prepared from the thick, underground rhizomes (often called ginger root) of a large perennial, herbaceous plant, *Zingiber officinale*, native to islands of the Pacific. Powdered ginger, from the dried rhizome, is widely used in cookies, gingerbread, and other baked goods. Fresh ginger rhizome is used in many Asian dishes, and sweetened crystallized ginger is used in pickles or is eaten as candy.

Mace and nutmeg, from the *Myristicaceae* family, are two spices that come from the seeds of the nutmeg tree, *Myristica fragrans*, native to the Moluccas Islands. The hard seed has a covering of strips of reddish material; dried and ground, that material is called mace. The rest of the seed becomes nutmeg. Both mace and nutmeg are used in cakes, cookies, pies, and puddings.

Mustard, from the mustard family, comes from the dried seeds of herbaceous mustard plants. *Brassica sinapis* is the most widely used species. Mustard has been used in cooking for a very long time; its native origins are unknown. Ground to a powder, mustard is used in sauces and meat dishes, and whole mustard seeds are used in pickles.

Paprika, from the nightshade family, is a spice prepared from the dried fruits of a sweet pepper, *Capsicum annuum*, a native of Central America. Paprika is used to flavor a variety of meat, vegetable, egg, and cheese dishes.

Black and white pepper, from the pepper tree family, are both produced from the dried fruits (peppercorns) of *Piper nigrum* shrubs, native to India. Black pepper is made by picking and drying the unripe fruits. White pepper is prepared from riper fruits that are soaked in water. After the outer fruit wall is rubbed off, a smooth white seed remains. The dried peppercorns are either ground or shipped whole to be ground at the time of use. Pepper is used in all sorts of meat, cheese, vegetable, and salad dishes.

PhotoDisc

Red pepper, from the nightshade family, is made into a spice by grinding the dried fruits of any of several species of the genus Capsicum, *native to tropical regions of the Americas. Peppers are used to flavor meat and vegetable dishes characteristic of Mexican, Caribbean, Indian, Chinese, Thai, and many other cuisines.*

Red pepper, from the nightshade family, is a spice made by grinding the dried fruits of any of several species of hot peppers or chilies of the genus *Capsicum,* native to tropical regions of the Americas. *Capsicum frutescens* produces cayenne pepper. Ground or crushed, the peppers are used to flavor meat and vegetable dishes characteristic of Mexican, Caribbean, Indian, Chinese, Thai, and many other cuisines.

Saffron, from the iris family, is the most expensive spice, made from the stigmas of the flowers of a crocus plant, *Crocus saturis,* native to southern Europe and Asia. On top of the female part of the flower, the stigmas consist of three slender filaments; it takes about seventy-five thousand dried stigmas to make one pound of saffron, and each must be picked by hand. Used intact or ground, saffron adds a deep yellow color and a slightly bitter taste to breads, sauces, soups, meat, and rice dishes. Turmeric (Ginger family) is prepared from the dried rhizome of a perennial herbaceous plant, *Curcuma longa,* native to the East Indies. Turmeric gives color and a slightly sweet flavor to pickles, cakes, cookies, salad dressings, sauces and meat dishes.

John P. Shontz

See also: Herbs.

Sources for Further Study

Miloradovich, M. *The Art of Cooking with Herbs and Spices: A Handbook of Flavors and Savors.* Garden City, N.Y.: Doubleday, 1950. Includes a history of spices and herbs as well as recipes.

Morris, S., and L. Mackley. *The Spice Book*. Singapore: Southwater, an Imprint of Anness Publishing, 2000. Descriptions and uses of a wide variety of spices.
Tighe, E. *Woman's Day Encyclopedia of Cookery*. 12 vols. New York: Fawcett Publications, 1965. Each of the common spices is described, and several recipes using that spice are included with the description.
Walker, H., ed. *Spicing up the Palate: Studies of Flavourings, Ancient and Modern*. Proceedings of the Oxford Symposium of Food and Cookery 1992. Totnes, Devon, England: Prospect Books, 1993. A series of essays about the history of spices in cooking, treating where, when, how, and why spices are grown.

STEMS

Category: Anatomy

The stem is the part of a vascular plant that supports the leaves and reproductive structures, often at some distance above the ground.

Stems have two chief functions: to support various plant structures and to transport water and nutrients. Most stems support the leaves and reproductive organs at some distance above the ground. Such elevation is important to both leaves, whose function is to trap light energy for *photosynthesis*, and flowers, because being positioned above the ground helps attract *pollinators*. Fruits also benefit from being held aloft because height improves the chances of long-distance seed dispersal.

While being off the ground has advantages for shoot-borne organs, it also has one big disadvantage. These organs are far from their source of water, the soil, and therefore run the risk of drying out unless adequate water can be supplied. Therein lies the second function of stems: to conduct a supply of water from the roots to the organs. Stems also conduct nutrients in both directions between the roots and organs of the plant.

Structure

Unlike roots, which branch at irregular intervals and do not bear organs such as leaves or reproductive structures, stems have a definite pattern of organization that consists of alternating *nodes* and *internodes*. Each node is a point on a stem at which an organ is attached. Adjacent nodes are separated by an internode, a zone that lacks any attached organs. The part of the stem closest to the ground is called the *proximal end*, while the part farthest away is called the *distal end*. In many plants, internodes at the proximal end are the longest. Toward the distal end, the internodes tend to be progressively shorter.

An actively growing stem has an *apical meristem* at its extreme distal end. In this region, new stem tissue is produced. In the apical meristem, cells actively divide and elongate, causing the stem to become longer. That elongation process is called *primary growth*. Immature leaves are also produced at the apical meristem. Immediately after the new cells enlarge, they undergo differentiation to form primary tissue.

Newly formed stems tend to be soft, because the walls of the young cells are thin. Moreover, the stems are typically green because of the presence of photosynthesizing cells on the periphery. Such soft, green stems are termed *herbaceous*. These herbaceous stems often turn hard and woody through time.

At each node, the angle that is formed when a leaf attaches to the stem is called an *axil*. At that juncture is found an embryonic shoot, called an *axillary bud*. Normally, axillary buds remain dormant for some time. Eventually, however, many buds break dormancy and elongate to form a lateral shoot, called a *branch*.

Different species of plants have different patterns of *leaf arrangement*. Most plants display an alternate arrangement, in which only one leaf is inserted at a node. Plants with alternate leaves include oak trees, goldenrod, geraniums, and lilies. Plants with two leaves at a node, in what is called an opposite arrangement, include maples, lilacs, phlox, and mints. Finally, many plants, such as trillium, catalpa, and bedstraw, have three or more leaves at a node, in what is called a whorled arrangement.

Some plants exhibit very little internodal development. The result is that the plant consists of a cluster of leaves right at the soil surface. These are called rosette plants and are represented by dandelion, hawkweed, and plantain.

Woody Plants

Some plants produce an herbaceous stem that will develop a secondary growth of hard, woody tissue. These *woody plants* can be categorized as *trees*, which have only a single main stem emerging from the ground, or as *shrubs*, which have several stems coming from the ground. The development of woody tissue from herbaceous tissue results from a process called *lignification*, in which the walls of cells inside the stem become thick and hard because of the presence of a material called lignin.

The lignified inner cells of the stem are collectively called wood. Lignification causes the stem to become more rigid, allowing it to support more weight. In temperate climates with pronounced cold winters, an herbaceous stem becomes lignified during its first winter. In more moderate climates, stems become lignified gradually. In either case, the outside of the stem becomes covered with water-impermeable bark.

Woody plants that grow in temperate areas vary in their rates of growth from one season to another. Stems grow most rapidly during the spring and summer. The rate of growth declines markedly at the end of summer and during fall. Little, if any, growth occurs during winter.

Woody plants, especially *deciduous* ones, have a few features that are lacking in herbaceous plants, which have unprotected stems. First, when a leaf drops off a woody plant at the end of the growing season, it leaves a scar on the stem. The shapes of leaf scars vary from one species to another; some are circular, others are triangular, still others look like small lines. A second feature found on a woody plant is a *terminal bud*. That structure usually forms when the apical meristem stops actively growing at the end of the summer. The terminal bud is typically dormant during the winter and is often enclosed by one or more hard, modified leaves called *terminal bud scales*. Those scales protect the bud against injury by cold and predators.

When warm weather or abundant rainfall returns, the bud breaks its dormancy, and a new stem is produced. As the young shoots begin to elongate, the bud scales fall off, resulting in scars that encircle the stem. Thus, the bud scales mark the juncture between the growth that occurs in two successive years. Because a new set of bud scale scars is produced at the beginning of every growing season, one can tell the age of a stem by counting the number of sets of bud-scale scars from the base to the extreme distal end.

Finally, the bark that encircles woody stems does not allow air to pass between the inside of the stem and the outside. That poses a problem for the cells immediately under the bark, which need oxygen to survive. To help overcome this problem, many woody stems contain raised corky dots called *lenticels* scattered over the surface. The cells of the lenticels are spongy and allow air to diffuse to the living cells beneath.

Modified Stems

Many plants have stems that are horizontal instead of erect. *Stolons*, also called *runners*, are horizontal stems that lie above the soil surface. Stolons are found on strawberry plants, for example. *Rhizomes* are horizontal stems that lie below the soil surface. Ferns, irises, milkweeds, and goldenrods produce rhizomes. Both stolons and rhizomes often produce *adventitious roots*, which are roots that form along a stem or anywhere else roots typically do not grow. *Tubers* are thickened rhizomes that store starch. The potato is probably the best-known tuber. Interestingly, plants such as kohlrabi have thickened tuberlike stems that are borne above ground.

A *bulb* is a budlike modified stem that has sets of thick leaves closely overlapping one another (as an onion does). A *corm* is similar to a bulb except that the leaves are thin and papery, and they surround a stem that is short and fleshy. Gladiolus and crocus both produce corms.

Some plants, especially those that grow in desert regions, have thick, succulent stems that serve to store and conserve water. The stem is either padlike (such as that of a prickly-pear cactus) or barrel-shaped (such as that of a saguaro cactus) and is typically covered by a thick layer of green photosynthesizing cells and sharp spines. Asparagus and a few other plants produce stems called *cladophylls*, which are flattened, highly branched, and capable of photosynthesis.

Some branch stems, such as those of hawthorn, are modified to form sharp structures called *thorns*. In contrast, the sharp structures on the surfaces of rose and blackberry stems, commonly referred to as thorns, are actually *prickles*, while sharpened leaves such as those of holly are called *spines*. Finally, some woody plants, such as birches, produce stubby side branches called *spur shoots* that grow only a few millimeters each year.

Growth

Three types of primary tissue form the stem's interior: *dermal tissue*, *vascular tissue*, and *ground tissue*. Dermal tissue forms a thin layer surrounding the herbaceous stem and protects it from drying out. Vascular tissue is composed of *xylem* and *phloem* and serves to conduct water, minerals, and organic substances throughout the plant. In most plants, it forms a ring within the stem. Woody plants have a ring of cells called the *vascular cambium* located within the vascular tissue. The vascular cambium actively divides during the growing season, producing wood to the inside and bark to the outside, and causes the stem to increase in width. Finally, ground tissue has two main functions: photosynthesis and storage. It is found both at the extreme core of the stem and in a ring immediately inside the dermal tissue but outside the vascular tissue.

Many plants vary from these basic patterns. For example, in plants such as lilies and grasses, the cells of the vascular tissue occur in small clusters that are interspersed with the ground tissue. In woody plants, the vascular tissue makes up the bulk of the stem.

Many internal (hormonal) and external (environmental) factors affect stem growth. At least two classes of hormones appear to increase the growth of stems when applied to the plant: *auxins* and *gibberellins*. Auxins cause individual cells to become longer, whereas gibberellins stimulate both cell division and elongation. Gibberellins appear to be important in the life cycles of *biennials*, which are plants that typically spend their first year as a rosette and then flower in their second year. Before flowering, the second-year plant must produce an erect flowering stem, a process called *bolting*. When a biennial bolts, the new internodes are long rather than short, and gibberellins appear to cause this elongation.

Many environmental factors can influence stem growth, but light and physical damage to the apical meristem are perhaps the most dramatic. Plants that grow in light that comes from one side (instead of from the top) tend to bend toward the light, a process called *phototropism*. Plants that grow in complete darkness tend to have stems that are abnormally long and thin and are yellow in color. Such stems are said to be *etiolated*.

Finally, when the apical meristem is intact, lateral buds typically remain dormant. When, however, the meristem is removed by herbivory or clipping, at least one of the lateral buds is released from dormancy, producing a branch that then becomes the new meristem. The process by which an intact apical meristem inhibits the growth of the laterals is termed *apical dominance*. Research has shown that phototropism, etiolation, and apical dominance are mediated through the hormone auxin.

Kenneth M. Klemow

See also: Bulbs and rhizomes; Cacti and succulents; Dormancy; Flowering regulation; Garden plants: shrubs; Growth and growth control; Growth habits; Hormones; Leaf anatomy; Leaf arrangements; Plant tissues; Roots; Shoots; Wood.

Sources for Further Study

Browse, Philip McMillan. *Plant Propagation*. New York: Simon & Schuster, 1988. This spiral-bound book guides the nonspecialist through the various techniques of plant propagation with a detailed, step-by-step approach supported by clear illustrations. Numerous techniques for layerage, grafting, and making cuttings are clearly explained.

Gartner, Barbara L., ed. *Plant Stems: Physiology and Functional Morphology*. San Diego: Academic Press, 1995. Chapter topics include opportunities and constraints in the placement of flowers and fruits; stems and fires; and chemical defense against herbivores. Mostly black-and-white photographs, with two pages of color plates.

Kaufman, Peter B. *Plants: Their Biology and Importance*. New York: Harper & Row, 1989. This book was written to introduce nonbiology students to the basic biology and economic aspects of plants. Chapters 3-5 describe the basic structures of stems and their modifications. Chapter 26 discusses the environmental and hormonal factors that regulate the development of stems and other plant organs.

Raven, Peter, Ray F. Evert, and Susan E. Eichhorn. *Biology of Plants*. 6th ed. New York: W. H. Freeman/Worth, 1999. An excellent introduction to all phases of modern botanical science; includes sections on cell biology, energetics, genetics, diversity, structure, evolution, and ecology. The external and internal structure of the shoot is detailed. Also contains an authoritative review of the roles that various external factors and hormones play in stem growth.

Stern, Kingsley R. *Introductory Plant Biology*. 8th ed. Boston: McGraw-Hill, 2000. A fine general text covering all phases of plant science, including stems, with special attention paid to woody stems and their commercial use as lumber and fiber. The process of stem growth and how it is influenced by the environment is also covered.

STRIP FARMING

Categories: Agriculture; economic botany and plant uses

Strip farming is the growing of crops in narrow, systematic strips or bands to reduce soil erosion from wind and water and otherwise improve agricultural production.

The origins of strip farming can be traced to the *enclosure movement* of postmedieval Great Britain. Landlords consolidated the small, fragmented strips of land farmed by tenant peasants into large block fields in an effort to increase agricultural production. Peasant plots were typically 1 acre in size: 220 yards, or one furlong in length (the distance a team of oxen can plow before resting) and 22 yards in width (the amount one team of oxen can plow in one day). After enclosure, fields were 100 or more acres in size. Larger fields were more productive but were also more exposed to wind and water erosion and nutritional exhaustion.

As agricultural production gradually shifted to new lands in the Americas and colonial Africa, farmers continued to use large-field farming techniques and developed large-field plantations. By the early twentieth century, all readily tilled lands had been opened by the plow and were suffering the effects of water and wind erosion. Strip farming, also known as strip cropping, was developed as a *soil conservation* measure during the 1930's. During the 1960's strip farming became an important tool to prevent water and air pollution and improve wildlife habitat.

Agricultural Hazards

Wind erosion begins when wind velocity at 1 foot (0.3 meter) above soil level increases beyond 13 miles (21 kilometers) per hour. *Saltation* and *surface creep* also allow soil to move. In saltation, small particles are lifted off the surface. These articles travel ten to fifteen times the height to which they are lifted, then spin downward with sufficient force to dislodge other soil particles and break earth clods into smaller particles. Surface creep occurs when particles too small to be lifted move along the surface in a rolling motion. The wider the field, the greater the cumulative effect of saltation and surface creep. These factors can lead to an avalanche of soil particles across the widest fields even during moderate wind gusts.

Water erosion begins when raindrops or flowing water suspends soil particles above the surface and transports them downslope by splash or *runoff*. Ice crystals expand, then contract when melted, dislodging soil particles and making them available for both water and wind erosion. Water also *leaches* nutrients and chemicals from the soil, causing the soil to experience both nutrient loss and an increase in salts and acids.

The U.S. Department of Agriculture computes annual soil loss from agricultural and developed land using the formula A = RKLSCP. In this formula, *A* equals annual soil loss, *R* equals the amount of rainfall on the plot, *K* equals the erosion factor for the type of soil on the plot, *L* equals the length of the slope on which the plot is located, *S* equals the angle of the slope, *C* equals the type of crop or soil cover on the plot, and *P* equals the presence of management conservation practices such as buffers, terraces, and strip farming. Soil loss tolerances are calculated for each plot. The tolerance is the amount of soil that can be lost without reducing productivity. Loss tolerances range from 1 to 5 tons per acre per year. Farmers and developers reduce soil losses to tolerance levels by reducing soil exposure to wind and rain and by utilizing conservation practices, such as strip farming.

Alleviating Hazards

Strip farming reduces field width, thus reducing erosion. Large fields are subdivided into narrow, cultivated strips. Planting crops along the contour lines around hills is called *contour strip cropping*. Planting crops in strips across the top of predominant slopes is called *field stripping*. Crops are arranged so that a strip of hay or sod (such as grass, clover, or alfalfa) or a strip of close-growing small grain (such as wheat or oats) is alternated with a strip of cultivated row crop (such as tobacco, cotton, or corn). Rainwater runoff or blown dust from the row-crop strip is trapped in this way as it passes through the subsequent strip of hay or grain, thus reducing soil erosion and pollution of waterways. Contour or field strip cropping can reduce soil erosion by 65 to 75 percent on a 3 to 8 percent slope.

Cropping in each strip is usually *rotated* each year. In a typical four-strip field, each strip will be cultivated with a cover crop for one or two years, grain for one year, and row-crop planting for one year. Each strip benefits from one or two years of nitrogen replenishment from *nitrogen-fixing* cover crops, such as alfalfa, and each strip benefits from one year of absorbing nutrient and fertilizer runoff from the adjacent row-crop strip.

Strip widths are determined by the slope of the land: the greater the slope, the narrower the strips. In areas of high wind, the greater the average wind velocity, the narrower the strips. The number of grass or small-grain strips must be equal to or greater than the number of row-cropped strips.

Terraces are often constructed to reduce the slope of agricultural land. At least one-half of the land between each terrace wall is cultivated with grass or a close-growing crop. Diversion ditches are often used to redirect water from its downhill course across agricultural land. These ditches usually run through permanently grassed strips, through downhill grass waterways constructed across the width of the strips, and through grassed field borders surrounding each field.

Gordon Neal Diem

See also: Agriculture: modern problems; Erosion and erosion control; Slash-and-burn agriculture; Soil conservation; Soil management; Soil salinization; Sustainable agriculture.

Sources for Further Study

Francis, Charles, et al. "Strip Cropping Corn and Grain Legumes: A Review." *American Journal of Alternative Agriculture* 1, no. 4 (1986): 159-164. Results of experimental studies on strip cropping yields versus conventional techniques.

Hart, John Fraser. *The Rural Landscape*. Baltimore, Md.: Johns Hopkins University Press, 1998. A description of historic land division and cropping systems.

STROMATOLITES

Categories: Algae; bacteria; evolution; paleobotany

Stromatolites are laminated, sedimentary fossils formed from layers of blue-green algae (also known as blue-green bacteria or cyanobacteria). Located throughout the world, these ancient remnants of early life have revealed much about the "age of algae."

Stromatolites are the most common megascopic fossils, contained within ancient rocks dating to 3.5 billion years in age. In both the living and fossil form, they are created by the trapping and binding of sediment particles and the precipitation of calcium carbonate to the sticky surface of matlike filaments grown on a daily cycle by blue-green algae (also known as cyanobacteria). Modern stromatolites are found throughout the world; they are of particular use in the creation of hydrocarbon reservoirs, in geologic mapping, and as indicators of paleoenvironments.

Distribution of Stromatolites

In the geologic record, stromatolites are the most abundant of fossils found in rocks dating to the Precambrian era, that period of time from the origin of Earth (approximately 4.6 billion years ago) up to 544 million years ago. The oldest fossil stromatolites are contained in the 3.3-billion- to 3.5-billion-year-old Warrawoona group of rocks in Australia. Close in age are the 3.4-billion-year-old stromatolites of the Swaziland group of South Africa. Somewhat more removed are those associated with the 2.5-billion- to 2.8-billion-year-old Bulawayan Limestone, also in Africa.

All these examples originated in the Archean eon, the first recorded period of geologic history, extending from approximately 4 billion to 2.5 billion years ago. During the Proterozoic eon, immediately following the Archean eon and extending to 544 million years ago, stromatolites became prolific. This Proterozoic expansion is probably reflective of the initial development of continental landmasses and associated warm, photic continental shelf regions, as plate tectonics became a controlling process in the early development of the earth's crust.

Throughout the Archean and Proterozoic eons, blue-green algae underwent a steady and progressive state of biologic evolution, recognized today as the singular, common megascopic fossil of the Precambrian time period. For this reason, the Precambrian is often referred to as the "age of algae," or, more specifically, the "age of blue-green algae." Throughout the Phanerozoic eon, defined as 544 million years ago to the present, blue-green algae underwent minimal evolution, probably because their evolutionary state had become adapted to a variety of environments, reducing the need for further diversification.

Stromatolitic-building algae maintained their dominance of the aquatic world during the Early Phanerozoic eon (544 million to about 460 million years ago). With the rather abrupt appearance, however, of shelled, grazing, and cropping invertebrates in the early Phanerozoic eon, blue-green algae began to decline in significance. Today, as compared to their Precambrian domination, they have, on a relative scale, become endangered.

Geographically, fossil stromatolites are ubiquitous on every continent, especially within sedimentary carbonate rock sequences older than 460 million years. On southeastern Newfoundland, stromatolites built by blue-green algae of the genus *Girvanella* are found in conglomerate and limestone strata of the Bonavista Formation (approximately 550 million years in age). In the Transvaal region of South Africa, delicately banded stromatolitic structures compose one of the most widespread of early Proterozoic shallow-water carbonate deposits in the world, extending over an area exceeding 100,000 square kilometers.

In nearby Zambia, algal stromatolites are closely associated with rock sequences containing economic levels of copper and cobalt. Upper Permian

(250 million years ago) stromatolite horizons can be traced over an area of northern Poland exceeding 15,000 square kilometers. Miocene age (15 million years old) algae of the species *Halimeda* compose the limestone-forming rocks of the island of Saipan in the Mariana Islands of the Pacific Ocean. In North America, fossil algal-bearing rocks include the 2-billion-year-old Gunflint (Iron) Formation of Ontario, Canada, and the well-developed stromatolitic horizons of Early Paleozoic era (450 million years ago) composing the Ellenberger Formation of Oklahoma and Texas.

Classification and Identification

The classification and identification of stromatolites are often concluded on the basis of overall morphology, particularly the size, shape, and internal construction of the specimen. The relevant literature makes use of a variety of morphological terms, including the adjectives "frondose" (leaflike), "encrusting," "massive," "undulatory" (wavelike), "columnar," "laminar," "domed," "elliptical," and "digitigrade" (divided into fingerlike parts). Through the study of modern blue-green algae, it is suggested that three environmental criteria are of importance in stromatolite geometry development. These are direction and intensity of sunlight, direction and magnitude of water current, and direction of sediment transport. As an example, the extant elliptical stromatolites of Shark Bay, western Australia, are oriented at right angles to the shoreline as the result of strong current-driven wave and scour action. Under certain environmental conditions, cyanobacteria growth surfaces are not preserved, producing fossil algal structures characterized by a lack of laminae (leaf blades). These structures are termed *thrombolites,* in contrast to laminar-constructed stromatolites.

NOAA/National Undersea Research Program Collection

Stromatolites are formed by a slow buildup of microbial mats trapping sediment particles and the precipitation of calcium carbonate. Modern stromatolites are found throughout the world.

While stromatolites are generally described as megascopic in size, discussion of specific dimensions relates both to laminae thickness and to overall size. Stromatolite laminae of the Precambrian-aged Pethei Formation, an outcropping along the shores of Great Slave Lake in the Northwest Territories of Canada, are both fine and coarse in dimension. The coarse-grained layers, formed of lime-mud pellets and calcium and magnesium carbonate rhombs, are principally less than 5.0 millimeters in thickness. The fine-grained laminae, composed of calcium carbonate clay and silt-sized particles, are, on average, only 0.5 millimeter thick.

In size, individual stromatolites can range from centimeters up to several meters. Fossils of the common Precambrian genus *Conophyton* occur in a range of sizes, from pencil-sized shapes to columns up to 10 meters in diameter. Subspherical varieties of stromatolite-like structures, formed by the accretion of successive gelatinous mats of blue-green algae and generally less than 10 centimeters in diameter, are termed *oncolites*. Stromatolitic complexes in the Great Slave Lake district measure 80 meters long by 45 meters wide by 20 meters in thickness and can be continuously traced for distances exceeding 160 kilometers.

Albert B. Dickas

See also: Algae; Cyanobacteria; Evolution of cells; Fossil plants; Paleoecology.

Sources for Further Study

Dott, R. H., Jr., and D. R. Prothero. *Evolution of the Earth*. 6th ed. Dubuque, Iowa: McGraw-Hill, 2002. A well-illustrated textbook on the geologic history of Earth. Contains several sections on stromatolites, their significance to Precambrian history, and their role in the understanding of very early forms of life.

Gebelein, C. D. "Biologic Control of Stromatolite Microstructure: Implications for Precambrian Time Stratigraphy." *American Journal of Science* 274 (1974): 575-598. Lengthy paper discusses the various environmental factors, including water and wind currents, that affect the detailed structure of modern stromatolitic organisms. Such knowledge is useful in worldwide mapping of stromatolite horizons. Written for undergraduate college students.

Levin, H. L. *The Earth Through Time*. 6th ed. Fort Worth, Tex.: Saunders College Publishing, 1999. This college-level textbook discusses fossil stromatolites of the Proterozoic eon and their importance as heliotrophic rock structures. Places these fossils within the context of the evolution of prokaryotic organisms. An easy-to-read source complete with numerous photographs.

McMenamin, Mark. *The Garden of the Ediacara: Discovering the First Complex Life*. New York: Columbia University Press, 1998. Entertaining study of the earliest complex life-forms on the planet details the author's work on these organisms. Written for the student but understandable by the general reader.

Schopf, J. William, ed. *Major Events in the History of Life*. Boston: Jones and Bartlett, 1992. Excellent overview of the origin of life, the oldest fossils, and the early development of plants and animals. Written by specialists in each field but at a level suitable for high school students and undergraduates. Although technical language is used, most terms are defined in the glossary.

Stewart, W. N. *Paleobotany and the Evolution of Plants*. 2d ed. New York: Cambridge University Press, 1993. Discusses the preservation, preparation, and age determination of fossil plants, the geologic fossil record, and the Precambrian eon, when stromatolites were common throughout the aquatic world. Suitable for college-level readers.

Walter, M. R. "Understanding Stromatolites." *American Scientist*, September/October, 1977, 563-571. An in-depth paper on stromatolites, thrombolites, and oncolites. Discusses means of development of differing morphologies, modern analogues, and uses in stratigraphic mapping of stromotolitic-bearing rocks. Contains a complete reference section. Written for the high school science reader.

_______, ed. *Stromatolites*. New York: Elsevier, 1976. Contains papers on zonation, paleoenvironments, environmental diversity, morphologies, origin, geologic significance, and distribution of fossil stromatolites. Several papers discuss modern algal stromatolites and the organisms that build them. A valuable reference for the serious student.

SUCCESSION

Categories: Ecology; ecosystems; genetics; reproduction and life cycles

Succession is the progressive and orderly replacement of one biological community by another until a relatively stable, self-maintaining community is achieved.

Succession is an important ecological phenomenon because it allows the maximum variety and number of species to occupy a given area through time and leads to the establishment of an ecologically stable *climax community* that represents the most complex and diverse biological system possible, given existing environmental conditions and available energy input. As succession proceeds, significant changes occur in species composition, nutrient cycling, energy flow, productivity, and stratification. Changes also occur within the climax community; however, these changes act to maintain the climax, not alter it.

Immature communities tend to have high populations of a few species that are relatively small and simple. *Biomass* (weight of living material) is low, and nutrient conservation and retention are poor. *Food chains* are short, and available energy is shared by few species. Community structure is simple and easily disrupted by external forces. As communities mature, larger and more complex organisms appear, and there is a higher *species diversity* (number of different species). Biomass increases, and nutrients are retained and cycled within the community. The greater number of species results in more species interactions and the development of complex *food webs*. Community productivity (conversion of solar energy to chemical energy), initially high in immature communities, becomes balanced by community respiration as more energy is expended in maintenance activities.

Stages of Succession

The entire sequence of communities is called a sere, and each step or community in the sequence is a *seral stage*. The climax community is in balance, or *equilibrium*, with the environment and displays greater stability, more efficient nutrient and energy recycling, a greater number of species, and a more complex community structure than that of each preceding seral stage.

Each seral stage is characterized by its own distinctive forms of plant and animal life, which are adapted to a unique set of chemical, physical, and biological conditions. Excepting the climax community, change is the one constant shared by all seral stages. Changes can be induced by abiotic factors, such as erosion or deposition, and by biotic factors, modification of the environment caused by the activities of living organisms within the community.

These self-induced factors bring about environmental changes detrimental to the existing community but conducive to invasion and replacement by more suitably adapted species. For example, lichens are one of the first colonizers of barren rock outcrops. Their presence acts to trap and hold windblown and water-carried debris, thereby building up a thin soil. As soil depth increases, soil moisture and nutrient content become optimal for supporting mosses, herbs, and grasses, which replace the lichens. These species continue the process of soil-building and create an environment suitable for woody shrubs and trees.

In time, the trees overtop the shrubs and establish a young forest. These first trees are usually shade-intolerant species. Beneath them, the seeds of the shade-tolerant trees germinate and grow up, eventually replacing the shade-tolerant species. Finally, a climax forest community develops on what once was bare rock, and succession ends.

Primary and Secondary Succession

The sere just described—from barren rock to climax forest—is an example of *primary succession*. In primary succession, the initial seral stage, or pioneer community, begins on a substrate devoid of life or unaltered by living organisms. Succession that starts in areas where an established community has been disturbed or destroyed by natural forces or by human activities (such as floods, windstorms, fire, logging, and farming) is called *secondary succession*.

An example of secondary succession occurs on abandoned cropland. This is referred to as old-field succession and begins with the invasion of the abandoned field by annual herbs such as ragweed and crabgrass. These are replaced after one or two years by a mixture of biennial and perennial herbs, and by the third year the perennials dominate. Woody shrubs and trees normally replace the perennials within ten years. After another ten or twenty years have passed, a forest is established, and ultimately, after one or two additional seral stages in which one tree community replaces another, a climax forest emerges.

Both primary and secondary succession begin on sites typically low in nutrients and exposed to extremes in moisture, light intensity, temperature, and other environmental factors. Plants colonizing such sites are tolerant of harsh conditions, are characteristically low-growing and relatively small, and have short life cycles. By moderating the environmental conditions, these species make the area less favorable for themselves and more favorable for plants that are better adapted to the new environment. Such plants are normally long-lived and relatively large. Secondary succession usually proceeds at a faster rate than primary succession, because a well-developed soil and some life are already present.

AP/Wide World Photos

Each stage of a community's succession is characterized by its own distinctive life-forms. For example, this ancient bog in North Carolina was formed when emergent plants at the water's edge extended their stems and leaves above water, trapping sediment and adding organic matter. A soil rich in partially decomposed organic matter and saturated with water accumulated.

Aquatic Environments

Succession can also take place in aquatic environments, such as a newly formed pond. The pioneer community consists of microscopic organisms that live in the open water. Upon death, their remains settle on the bottom and join with sediment and organic matter washed into the pond. An accumulation of sediment provides anchorage and nutrients for rooted, submerged aquatic plants such as pondweeds and waterweeds. These add to the buildup of sediment, and as water depth decreases, rooted, floating-leaved species such as water lilies prevent light from reaching the submerged aquatics and eliminate them.

At the water's edge, emergent plants rooted in the bottom and extending their stems and leaves above water (cattails, rushes, and sedges) trap sediment, add organic matter, and continue the filling-in process. The shallow margins fill first, and eventually the open water disappears and a marsh or bog forms. A soil rich in partially decomposed organic matter and saturated with water accumulates. As drainage improves and the soil becomes raised above the water level, trees and shrubs tolerant of wet soils invade the marsh. These act to lower the water table and improve soil aeration. Trees suited to drier conditions move in, and once again a climax community characteristic of the surrounding area develops.

Influence of Climate

The American ecologist Frederic E. Clements (1874-1945) believed that the characteristics of a climax community were determined solely by regional climate. According to Clements, all communities within a given climatic region, despite initial differences, eventually develop into the same climax community. Some seral stages might be abbreviated or skipped entirely, while others could be lengthened or otherwise modified; however, the end result would always be a single climax community suited to the regional climate. This phenomenon is called *convergence*, and Clements's single-climax concept is known as the *monoclimax theory*.

Some ecologists have found the monoclimax theory to be simplistic and have offered other theories. One of these, the *polyclimax theory*, holds that, within a given climatic region, there could be many climaxes. It was noted that in any single climatic region, there were often many indefinitely maintained communities that could be considered separate and distinct climaxes. These developed as a result of differences caused by soil type, soil moisture, nutrients, slope, fire, animal activity (grazing and browsing), and other factors. Clements countered that these would eventually reach true climax status if given enough time and proposed terms such as *subclimax* (a long-lasting seral stage preceding the climax) and *disclimax* (a nonclimax maintained by continual disturbance) to describe such situations.

A third theory, the *climax pattern concept*, views the climax as a single large community composed of a mosaic or pattern of climax vegetation instead of many separate climaxes or subclimaxes. Numerous habitat and environmental differences account for the patterns of populations within the climax; no single factor such as climate is responsible.

While there is little doubt about the reality of succession, it is apparently not a universal phenomenon. For example, disturbed areas within tropical rain forests do not undergo a series of seral stages leading to reestablishment of the climax community. Instead, the climax is established directly by the existing species. Nevertheless, in most regions succession is the mechanism by which highly organized, self-maintained, and ecologically efficient communities are established.

Steven D. Carey

See also: Community-ecosystem interactions; Ecology: concept; Ecosystems: overview; Genetic drift; Reforestation; Reproductive isolating mechanisms; Selection; Species and speciation.

Sources for Further Study

Bazzaz, F. A. *Plants in Changing Environments: Linking Physiological, Population, and Community Ecology*. New York: Cambridge University Press, 1996. Discusses how disturbance changes the environment, how species function, coexist, and compete for resources, and how species replace one another over time.

Brewer, Richard. *The Science of Ecology*. Fort Worth, Tex.: Saunders College Publishers, 1994. A clearly written text suitable for high school reference and college use. Includes a chapter on community change and succession. Contains biographical sketches of ecology pioneers Frederic E. Clements and Henry Chandler Cowles. Discusses ecosystem development through the earth's history. Illustrated with black-and-white photographs and line drawings. Includes glossary, bibliography.

Perry, David A. *Forest Ecosystems*. Baltimore: Johns Hopkins University Press, 1994. Textbook for students, land managers, scientists, and policymakers. Topics range from the world's forest types and subdisciplines of ecology to more complex discussions of such things as productivity, succession, nutrient cycling, and stability.

Smith, Robert L. *Ecology and Field Biology*. 6th ed. San Francisco: Benjamin Cummings, 2001. The chapter "Succession" is a clear and informative treatment of the subject. Good discussion of land use and effects on succession as well as succession of human communities. The latter part of the chapter discusses ecosystem development through time with emphasis on continental drift. Glossary and extensive bibliography. Illustrated with black-and-white photographs and line drawings.

SUGARS

Categories: Economic botany and plant uses; food; nutrients and nutrition; photosynthesis and respiration

Refined from sugarcane, sugar beets, and corn, sugars are a major and vital part of nutrition and provide the basic molecular structure for most living matter.

Sugars, through the process called cellular respiration, are the primary power sources used to produce adenosine triphosphate (ATP), the energy exchange molecule that sustains all life. Energy is stored in organisms as starch or fat but is burned as sugar. Sugars include some of the simplest carbohydrates, and they are building blocks of more complicated molecules. Common sugars include glucose, fructose, sucrose, maltose, and lactose. Alcohol (ethanol) derived from sugar fermentation is an important product to the food and beverage, fuel, and drug manufacturing industries. The primary sugar derived from plants is sucrose, a disaccharide composed of two simpler sugars, glucose and fructose, joined together.

History of Use in Food

Demand for sugar and other food flavorings was particularly strong before canning and refrigeration. People often had to eat partially spoiled food, and they eagerly sought ways to improve its flavor. Items with high sugar levels, such as berries and grapes, were especially prized.

Sugarcane (*Saccharum officinarum y officinarum*) has a stem that can be squeezed to deliver a sucrose-rich syrup, and the leftover woody material (*bagasse*) can be dried and burned to boil off water. Molasses syrup or solid sugar can be made in this way. Arab traders brought sugarcane to Europe in the 1100's. By the 1500's European colonization in Brazil and the Caribbean provided rich growing fields. So important and valuable were sugar crops that in 1800 the new countries of Haiti and the United States had similar gross national products.

However, the cravings and rivalries that created sugar empires caused their decline. England and its allies fought a series of wars with France in the late 1700's and early 1800's, and British blockades kept molasses out of Europe. France's Napoleon responded by offering a prize for a process to produce sugar from a European-grown plant. A sugar beet process won, and cane sugar was never again as centrally important.

Yearly, sucrose production from sugarcane and sugar beets is more than 100 million metric tons. Both crops are excellent soil conditioners. Sugarcane is a tall, periodically harvested grass, so it limits soil erosion that would occur in bare ground. Meanwhile, cane roots steadily grow and increase humus in the soil. Sugar beets require plowing and

A worker harvests sugarcane near Mariel, Cuba. Sucrose production for food use comes from both sugarcane and sugar beets.

are dug from the ground during harvesting, but they, too, leave extensive roots.

When sucrose prices rose in the 1950's, high-fructose sweeteners were developed as alternatives. Enzymes are used to break starches into fructose. Corn syrup, one fructose sweetener, is cheaper than sucrose and has displaced much U.S. sucrose use. Elsewhere, fructose syrups are made from wheat, rice, tapioca, and cassava.

Sugar and Alcohol

The fermentation of alcohol has been a major aspect of sugar use since antiquity. Fungi called *yeasts* convert sugars into ethyl alcohol (ethanol); the yeasts can produce only mild alcohol levels. Alcohol, although medically classified as a depressant, can provide a short-term energy boost. It also acts as a mild poison that desensitizes the central nervous system, allowing drinkers to feel relaxed. Ancient Egyptians and Mesopotamians fermented grains into beer, while Greeks and Phoenicians traded wine from grapes. In the Middle Ages, alchemists experimented with distillation, a process in which a substance boiled out of one substance is cooled back into liquid elsewhere. Distillation transformed beers and wines into whiskeys and brandies, alcoholic drinks several times more potent.

Caribbean sugar and alcohol formed one leg of the "triangle trade" from the 1600's through the early 1800's. New Englanders sold fish for Caribbean molasses, fermented it, and distilled it into rum. They traded rum in Africa for slaves, sold largely in the Caribbean.

Fuel and Other Uses

Ethyl alcohol can burn more efficiently than gasoline in the internal combustion engines used in automobiles. However, because gasoline has historically been cheaper, research and development work on ethanol as a fuel was minimal until the 1970's. The energy crisis of 1973 involved oil shortages and soaring prices, and it stimulated many experimental ethanol programs. Most experiments were abandoned when oil prices dropped in the mid-1980's, but Brazil persevered in a national program of sugar-cane alcohol fuel.

Even Brazilian ethanol is only barely economically competitive with fossil fuels. The major problem is that energy expenditures in the manufacture of ethanol include fuel for tending the fields and gathering cane, energy lost to yeasts (about half, although the yeasts do yield high-protein by-product), and another half of the remainder expended for distilling the material to 95 percent alcohol. Suggested improvements include developing more efficient yeasts and performing the distillation process under partial vacuum, which would allow continuous processing rather than batch processing.

Theoretically, the most efficient way to use sugar energy would be the development of electrical fuel cells that would take energy from sugar, just as living organisms do. Losses from yeast digestion and distillation would be eliminated, and a fuel cell might achieve 50 percent efficiency, rather than the 25 percent efficiency of internal combustion engines, yielding eight times more energy. This approach could create an energy revolution if the technical problems could be overcome. (Energy-efficient processes must be developed for *saccharification*, or hydrolysis, of cellulose from wood and garbage.)

Sugars can also be nutrients for generating other products. Specialized groups of other cells, such as juice-producing cells from oranges or fiber-producing cells from cotton, can be cultured with sugar nutrients, for example, and genetic engineering has developed yeasts that produce specialty chemicals such as catalysts.

Roger V. Carlson

See also: ATP and other energetic molecules; Carbohydrates; Ethanol; Grasses and bamboos; Microbial nutrition and metabolism; Respiration; Yeasts.

Sources for Further Study

Daviss, Bennett. "Power Lunch." *Discover* 16, no. 3 (March, 1995): 58-63. Describes development of an electrochemical battery powered by bacteria; these sugar-eating "living fuel cells" could burn waste products to light cities.

Peach, W. N., and James A. Constantin. *Zimmermann's World Resources and Industries*. 3d ed. New York: Harper and Row, 1972. Summarizes the history and economic context of sugar production.

Ross, James R. *Fuel Alcohol: How to Make It, How to Use It*. New York: St. Martin's Press, 1981. An ethanol fuel how-to book.

SUSTAINABLE AGRICULTURE

Categories: Agriculture; disciplines; economic botany and plant uses; environmental issues

Sustainable agriculture is the practice of growing and harvesting crops in a manner that will allow the land to return to its precultivation state. It seeks to maintain the quality of air, soil, and water resources for future human use and for wildlife.

Most twentieth century agricultural practices were based upon continued economic growth. This practice demonstrated dramatic increases in production but had negative impacts on the environment through the losses of plant and animal *habitats*, depletion of soil nutrients, and increases in pollution of water supplies. The concept of sustainable development is based on using renewable resources and working in harmony with ecosystems. The World Commission on Environment and Development described the concept of sustainable development as being able "to meet the needs of the present without compromising the ability of future generations to meet their own needs." Sustainable agriculture strives to manage agricultural activities in such a way as to protect air, soil, and water quality as well as conserve wildlife habitats and *biodiversity*.

Problems Caused by Agriculture

Water pollution is one of the most damaging and widespread effects of modern agriculture. Runoff from farms accounts for more than 50 percent of sediment damage to natural waterways. Cleanup of the chemicals and nutrients associated with this runoff in the United States costs an estimated $2 billion to $16 billion per year. Heavy application of nitrogen fertilizers and pesticides has raised the potential for groundwater contamination. Feedlots that concentrate manure production lead to further groundwater contamination. Several of the most commonly used pesticides have been detected in the groundwater of at least one-half of the states in the United States. In addition, growing highly specialized *monoculture* crops, which requires a heavy reliance on agricultural chemicals, has depleted the natural organic nutrients that were formerly rich in North American *topsoils*.

Research has shown that many of the farm-based chemical agents, pesticides, fertilizers, plant-growth regulators, and antibiotics are now found in the food supply. These chemicals can be harmful to humans at moderate doses, and chronic effects can develop with prolonged exposure at lower doses. Further, widespread pesticide use has been shown to severely stress animals other than the target pest, including bee populations. Pesticides have often led to resurgences of pests after treatment, occurrences of secondary pest outbreaks, and resistance to pesticides in the target pest.

Because of these growing problems, many American farmers are turning to sustainable agriculture. The U.S. federal government has offered guidance for this transition through the Sustainable Agriculture Farm Bill, passed by Congress in 1990. The bill provides that sustainable agriculture, through an integrated system of plant and animal production practices, can, over the long run, meet human food and fiber needs, enhance environmental quality and natural resources, make the most efficient use of nonrenewable resources, maintain economic viability of farm operations, and enhance the quality of life for farmers and consumers.

Water Conservation

Water is one of the most important resources for agriculture and for society as a whole. In the western United States, it allows arid lands to produce crops through irrigation. In California, limited surface water supplies have caused overdraft of groundwater and the consequent intrusion of salt water, which causes a permanent collapse of aquifers. In order to counteract these negative effects, sustainable farmers in California are improving water conservation and storage methods, selecting drought-resistant crop species, using reduced-volume irrigation systems, and managing crops to reduce water loss. Drip and trickle irrigation can also be used to dramatically reduce water usage and water loss while helping to avoid such problems as soil salinization.

Salinization and contamination of groundwater by pesticides, nitrates, and selenium can be temporarily managed by using tile drainage to remove water and salt. However, this often has adverse affects on the environment. Long-term solutions include conversion of row crops to production of drought-tolerant forages and the restoration of wildlife habitats.

Contour Plowing and Terracing

One of the most important aspects of sustainable agriculture is soil conservation. Water runoff from a field having a 5 percent slope has three times the water volume and eight times the soil erosion rate as a field with a 1 percent slope. In order to prevent excessive erosion, farmers can leave grass strips in the waterways to capture soil that begins to erode. *Contour plowing*, which involves plowing across the hill rather than up and down the hill, helps capture overland flow and reduce water runoff. Contour plowing is often combined with strip farming, where different kinds of crops are planted in alternating strips along the contours of the land. As one crop is harvested, another is still growing and helps recapture the soil and prevent water from running straight down the hill. In areas of heavy rainfall, tiered ridges are constructed to trap water and prevent runoff. This involves a series of ridges constructed at right angles to one another. Such construction blocks direct runoff and allows water to soak into the soil.

Another method of soil conservation is *terracing*, in which the land is shaped into level shelves of earth to hold in the water and soil. To provide further stability to soil, soil-anchoring plants are grown on the edges of the terraces. Terracing, although costly, can make it possible to farm on steep hillsides. Some soils that are fairly unstable on sloping sites or waterways can require that perennial species of grasses be planted to protect the fragile soil from cultivation every year.

Green Manure

Farmers also use *green manure*—crops that are raised specifically to be plowed under—to introduce organic matter and nutrients into the soil. Green manure crops help protect against erosion, cycle nutrients from lower levels of the soil into the upper layers, suppress weeds, and keep nutrients in the soil rather than allowing them to leach out. Legumes such as sweet clover, ladino clover, and alfalfa are excellent green-manure crops. They are able to extract nitrogen from the air into the soil and leave a supply of nitrogen for the next crop that is grown. Some crops, such as beans and corn, can cause high soil erosion rates because they leave the ground bare most of the year. In sustainable farming, crop residues are left on the ground after harvest. Residues help reduce soil evaporation and even excessive soil temperatures in hot climates. Many farmers choose to use cover crops rather than residue crops. Which cover crop to use depends on which geographical area farmers live in and if they wish to control erosion, capture nitrogen in the fall, release nitrogen to the crop, or improve soil structure and suppress weeds.

Cover Crops

When planting crops with high nitrogen requirements, such as tomatoes or sweet corn, farmers can use cover crops such as hairy vetch or clover. Both these cover crops decompose and release nutrients into the soil within one month. To fight erosion, a farmer might choose a rapid-growing cover crop, such as rye. Rye provides abundant ground cover and an extensive root system below the soil to stop erosion and capture nutrients. Alfalfa, rye, or clover can be planted after harvest to protect the soil and add nutrients and can then be plowed under at planting time to provide a green manure for the crop. Cover crops can also be flattened with rollers, and seeds can be planted in their residue. This gives the new plants a protective cover and discourages weeds from overtaking the young plants. Use of natural nitrogen also reduces the risk of water contamination by agricultural chemicals.

Reduced Tillage

Sustainable agriculture emphasizes the use of reduced tillage systems. There are three reduced tilling systems that sustainable farmers use to disturb the soil as little as possible. Minimum till involves using the disc of a chisel plow to make a trench in the soil where seeds are planted. Plant debris is left on the surface of the ground between the rows, which helps further prevent erosion. Several sustainable planting techniques help prevent soil erosion. Conser-till farming uses a coulter to open a slot just wide enough to insert seeds without disturbing the soil. No-till planting involves drilling seeds into the ground directly through the ground cover or mulch. When mulch is still in place, a narrow slit can be cut through the cover or crop residues in order to plant the new crops.

AP/Wide World Photos

This field is part of a farm in Tolona, Illinois, which practices strip-tilling, thereby combining the environmental benefits of minimal disturbance to the soil (no-tilling) with the farming advantages of conventional tillage.

Crop Rotation vs. Monoculture

Planting the same crop every year on the same field can result in depleted soils. In order to keep the soil fertile, nitrogen-depleting crops (such as sweet corn, tomatoes, and cotton) should be rotated every year with legumes, which add nitrogen to the soil. Planting a winter cover crop, such as rye grass, protects the land from erosion. Such cover crops will, when plowed under, provide a nutrient-rich soil for the planting of a cash crop. *Crop rotations* improve the physical condition of the soil because of variations in root depth and cultivation differences.

In nature, plants grow in mixed meadows, which allow for them to avoid insect infestations. Agricultural practices that use monoculture place a great quantity of the food of choice in easy proximity of the insect predator. Insects can multiply out of proportion when the same crop is grown in the field year after year. Since most insects are instinctively drawn to the same home area every year, they will not be able to proliferate and thrive if crops are rotated and their crop of choice in not in the same field the second year.

Crop rotation not only helps farmers use fewer pesticides but also helps to control weeds naturally. Some crops and cultivation methods inadvertently allow certain weeds to thrive. Crop rotations can incorporate a successor crop that eradicates the weeds. Some crops, such as potatoes and winter squash, work as cleaning crops because of the different style of cultivation that is used on them. Pumpkins planted between rows of corn will help keep weeds at bay.

Integrated Pest Management

Most sustainable farmers use *integrated pest management* (IPM) to control insect pests. Using IPM techniques, each crop and its pest is evaluated as an ecological system. A plan is developed for using cultivation, biological methods, and chemical methods at different timed intervals. Although effective, profitable, and safe, the IPM techniques have been widely adopted only for a few crops, such as tomatoes, citrus, and apples.

The goal of IPM is to keep pest populations below the size where they can cause damage to crops. Fields are monitored to gauge the level of pest damage. If farmers begin to see crop damage, they put cultivation and biological methods into effect to control the pests. Techniques such as vacuuming bugs off crops are used in IPM. IPM encourages growth and diversity of beneficial organisms that enhance plant defenses and vigor. Small amounts of pesticides are used only if all other methods fail to control pests. It has been found that integrated pest management, when done properly, can reduce inputs of fertilizer, lower the use of irrigation water, and reduce preharvest crop losses by 50 percent. Reduced pesticide use can cut pest-control costs by 50 to 90 percent and increases crop yield without increasing production costs.

Toby R. Stewart and Dion Stewart

See also: Agriculture: modern problems; Biofertilizers; Biopesticides; Composting; Erosion and erosion control; Hydroponics; Integrated pest management; Monoculture; Organic gardening and farming; Soil conservation; Soil management; Strip farming; Sustainable forestry.

Sources for Further Study

Langer, Richard W. *Grow It: The Beginner's Complete, in-Harmony-with-Nature Small Farm Guide, from Vegetable and Grain Growing to Livestock Care*. New York: Saturday Review Press, 1972. An easy-to-read, hands-on book explaining how to use and apply sustainable farming methods to both farm and garden.

National Research Council Board on Agriculture. *Alternative Agriculture*. Washington, D.C.: National Academy Press, 1989. Provides pages of detailed information on sustainable agricultural methods.

President's Council on Sustainable Development. Sustainable Agriculture Task Force. *Sustainable Agriculture: Task Force Report*. Washington, D.C.: Author, 1994. Lists the goals and strategies for initiating sustainable agriculture in the United States.